T0305505

Scheduling in Supply Chains Using Mixed Integer Programming

Scheduling in Supply Chains Using Mixed Integer Programming

Tadeusz Sawik

AGH University of Science and Technology, Kraków, Poland

A JOHN WILEY & SONS, INC. PUBLICATION

Published by John Wiley & Sons, Inc., Hoboken, New Jersey
Published simultaneously in Canada

For general information on our other products and services or for technical support, please contact our Customer Care Department within the United States at (800) 762-2974, outside the United States at (317) 572-3993 or fax (317) 572-4002.

Wiley also publishes its books in a variety of electronic formats. Some content that appears in print may not be available in electronic formats. For more information about Wiley products, visit our web site at www.wiley.com.

Library of Congress Cataloging-in-Publication Data:

Sawik, Tadeusz.
Scheduling in supply chains using mixed integer programming / Tadeusz Sawik.
p. cm.
Includes bibliographical references and index.
ISBN 978-0-470-93573-6 (cloth)
1. Assembly-line methods—Data processing. 2. Business logistics—Data processing. 3. Production scheduling—Data processing. 4. Integer programming. I. Title.
TS178.4.S287 2011
658.701'51977--dc22

2010047228

Printed in Singapore

10 9 8 7 6 5 4 3 2 1

To Bartek,
Kappa,
Siasia, and
Toranoko with love,
and to the memory of
my parents

Contents

List of Figures

List of Tables

Preface

This book is a unified and systematic presentation of scheduling decision making in supply chains using mixed integer programming (MIP). The recent improvements in modeling, preprocessing, solution algorithms, MIP software, and computer hardware have made it possible to solve large-scale MIP models of scheduling problems, in particular, of scheduling in supply chains. The book demonstrates that MIP, widely used for long- and medium-term planning, can also be efficiently used for the short- and medium-term scheduling and integrated scheduling in customer-driven supply chains, that is, the supply chains in make-to-order discrete manufacturing and assembly environment.

The focus on state-of-the-art MIP modeling and solution approaches means focus on exact optimal or near optimal solutions that can hardly be found when commonly used heuristic algorithms are applied. Furthermore, the proposed MIP modeling and solution approaches allow the reader/user to determine optimal or near optimal schedules using commercially available software for MIP, which makes the decision-making independent of custom-made scheduling software.

It is not necessary to have detailed knowledge of integer programming and scheduling theory in order to go through this book. The knowledge required corresponds to the level of an introductory course in operations research (linear programming, production planning and scheduling, and basic probability for Chapters 9 and 10) for engineering, management, and economics students.

The book is addressed to practitioners and researchers on supply chain planning and scheduling, and to students in management, industrial engineering, operations research, applied mathematics, computer science, and the like at master's and PhD levels.

TADEUSZ SAWIK

Kraków, Poland
March 2011

Acknowledgments

The book is based on the results of my research on scheduling in supply chains by mixed integer programming over the last decade. I wish to acknowledge many anonymous reviewers for their comments and suggestions on my submissions to different international journals, including *Computers & Operations Research, European Journal of Operational Research, International Journal of Production Economics, International Journal of Production Research, International Transactions in Operational Research, Journal of Electronics Manufacturing, Mathematical and Computer Modelling*, and *Omega: The International Journal of Management Science*.

The material presented in this book is also partially based on the results of different research projects on scheduling in customer-driven supply chains, sponsored by Motorola over the period of 1999 to 2004. The book has benefited from numerous discussions at that time with Andreas Schaller and Tom Tirpak of the Motorola Physical Realization Research Center (formerly Motorola Advanced Technology Center), a corporate R&D lab in Motorola.

This project has been partially supported by the Polish Ministry of Science and Higher Education, research grant No. N N519 576338. Thanks are also due to the AGH University of Science & Technology for its support of research on scheduling and supply chain optimization over the last decade.

Finally, I wish to acknowledge Joanna Marszewska of Jagiellonian University for the cover design of the book.

Acknowledgments

Introduction

Supply chain management is a rapidly developing field of management science. The purpose of this book is to put forward and present, in a unified and systematic way, practical applications of mixed integer programming modeling and solution approaches to scheduling in customer-driven supply chains.

In a high-tech industry, a typical customer-driven supply chain may consist of a number of part manufacturers at several locations and one or more producers, where parts are supplied by the manufacturers and assembled into finished products, then distributed to customers to meet their demand. In such supply chains, productivity may vary from plant to plant and transportation time and cost are not negligible. Owing to a limited capacity of both the part manufacturers and the producers, the manufacturing and supply schedules for each supplier of parts and the assembly schedules for finished products should be coordinated in an efficient manner to achieve a high customer service level at low cost.

The purpose of supply chain scheduling is to optimize short- to medium-term decisions in supply chains, considering the trade-off between tangible economic objectives such as cost minimization or profit maximization and less tangible objectives such as customer satisfaction or customer service level. In addition to production scheduling, supply chain scheduling considers manufacturing and supply of materials and distribution of finished products, and includes additional decisions connected with functional, spatial, and intertemporal integration and coordination of schedules for these activities.

Short-term supply chain scheduling is typically concerned with the allocation of tasks and resources of a single facility and with detailed sequencing and timing decisions over a short time horizon (e.g., a shift or day) to complete a given number of jobs in such a way that one or more job completion time-related objectives are minimized. A typical single facility considered in a customer-driven supply chain is a single-stage set of parallel machines or a multistage flow shop or job shop with single or parallel machines.

In contrast, medium-term supply chain scheduling (also called planning) deals with the allocation of tasks and resources of one or more interconnected facilities over a longer time horizon (e.g., several shifts, a week, or month) to complete a number of customer orders for finished products in such a way that customer service level is maximized and one or more cost objectives are minimized.

The typical short-term, operational and medium-term, tactical decisions connected with scheduling in customer-driven supply chains are briefly described below.

1. **Short-Term, Operational Decisions**
   - *Loading and routing*—given a mix of products, the objective of the loading and routing problem is to allocate all tasks and component feeders and/or tools required for completion of each product among the machines with limited work space and/or limited tool magazines, and to select processing routes for a mix of products, so as to balance the machine workloads.
   - *Sequencing and scheduling*—detailed sequencing and timing of all tasks for each machine and each individual product so as to maximize the system's productivity, for example, to minimize the schedule length.
2. **Medium-Term, Tactical Decisions**
   - *Order acceptance/rejection*—decision whether or not to accept a customer request for quotation as a firm order. A request for quotation typically consists of the required quantities of ordered products and the requested delivery dates. A response to the customer request contains the quantity to be fulfilled, the date of delivery, and the price based on revenue management principles, which may involve penalties associated with deviations from the customer-requested quantities and dates.
   - *Due date setting*—quoting due date for each accepted order. The quoted due date is either identical with that requested by the customer or a later one if the requested due date cannot be met in view of the actual workload and available capacity. Typically, the due dates setting decisions aim at reaching a high service level (maximum number of orders to be fulfilled by the customer-requested due dates) and can also be based on revenue management principles. The quoted due dates are next considered as deadlines that cannot be violated when the orders are executed.
   - *Supplier selection and order allocation*—a single-period or multiperiod allocation of orders for materials among the selected suppliers, respectively with or without timing decisions. The objective is to determine what quantities of various materials to order, from which supplier, and in which periods (in a multiperiod setting) to minimize total ordering, purchasing, defect, and shortage costs.
   - *Customer order scheduling*—scheduling of accepted customer orders by the deadlines to achieve some additional objective, such as minimum cost of holding the inventory of finished products completed before the committed delivery dates.
   - *Material supply scheduling*—scheduling of material supplies coordinated with the schedules of customer orders to meet demand for materials at minimum cost. The material supply schedules should also be coordinated with the manufacturing schedules for material suppliers to minimize manufacturing, shipment, and inventory holding cost in the supply chain.

The following two main approaches are applied throughout this book for operational and tactical decision making: the ***integrated (simultaneous) approach***, in which all required decisions are made simultaneously using a complex, large monolithic MIP model; and the ***hierarchical (sequential) approach***, in which the required decisions are made successively, using hierarchies of simpler and smaller-sized MIP models.

For the above two approaches, the short- and medium-term decision-making problems can be formulated as below.

1. Short-Term Supply Chain Scheduling

- *Integrated (simultaneous) approach*—given a mix of products, the objective of integrated loading and scheduling is to simultaneously determine the allocation of tasks and component feeders among the machines and a detailed processing schedule for each individual product so as to complete all products and to maximize the system's productivity, for example, to minimize the schedule length.
- *Hierarchical (sequential) approach*—first, the allocation of tasks and component feeders and/or tools among the machines with limited work space and/or limited tool magazines is determined to balance the machine workloads, and then detailed sequencing and timing of all tasks for each product are found to maximize the system's productivity, for example, to minimize the schedule length, given the task assignments and fixed or alternative processing routes determined at the top level.

2. Medium-Term Supply Chain Scheduling

- *Integrated (simultaneous) approach*—the coordinated schedules for customer orders, material manufacturing, and material supplies, along with the supplier selection, are determined simultaneously, to achieve a high customer service level (i.e., a minimum number of tardy orders) or a high revenue (e.g., minimum lost revenue due to tardiness or full rejection of orders), at low cost, particularly at minimum cost of holding total supply chain inventory of materials and finished products. The quoted due dates are derived from the completion times of the scheduled orders, whereas the unscheduled orders are considered to be rejected.
- *Hierarchical (sequential) approach*—first, the order acceptance/due date setting decisions are made to select a maximal subset of orders that can be completed by requested due dates, and for the remaining orders delayed due dates are determined to minimize the lost revenue owing to tardiness or rejection of orders. The due dates determined must satisfy capacity constraints and are considered to be deadlines at the order scheduling level. Next, order deadline scheduling is performed to meet all committed due dates and to achieve some additional objective, for example, to minimize the cost of holding finished product inventory of orders completed before the deadlines. Finally, allocation of orders for materials among the suppliers and scheduling of material supplies (and material manufacturing), coordinated with the schedules for customer orders are accomplished to meet demand for the materials in such a way that the total cost of materials manufacturing, supply, and inventory holding is minimized.

Although the integrated approach contributes to the complexity of the underlying mathematical models and the decision-making procedures, it is capable of reaching a coordinated solution and a global optimum for the entire supply chain scheduling problem. The hierarchical approach does not guarantee that the overall optimal solution will be attained. In contrast, the complexity of the hierarchical approach can be significantly reduced; however, it may lead to infeasible solutions obtained at lower decision levels. For example, in the medium-term decision making, no feasible deadline schedule can be found at the customer order scheduling level, if available capacity is overestimated at the due date setting level. Therefore, additional coordinating constraints should be incorporated at different decision levels to avoid the infeasibility issue.

OUTLINE OF THE BOOK

The goal of this book is to enable a reader with a background in MIP approaches to modeling supply chain scheduling problems to apply the knowledge and tools provided to produce state-of-the-art models and solution approaches for tactical and operational problems of real-world supply chains in make-to-order discrete manufacturing environments. To model different scheduling problems, the following three types of decision variables are required:

- *Binary variables* such as assignment variables (e.g., machine assignment, task assignment, due date assignment), time-indexed assignment variables (e.g., order to time period assignment), selection variables (e.g., machine selection, supplier selection, customer order selection), sequencing variables (e.g., precedence variables), machine set-up or start-up variables, etc.
- *Integer variables* such as number of parts, number of machines, number of machine set-ups or start-ups, number of people, etc.
- *Continuous variables* such as timing variables (e.g., completion time or departure time of part, of batch, of customer order), lot sizing variables (e.g., production lot, transportation lot), fractional allocation variables (e.g., allocation of order among suppliers, allocation of order among time periods), cost variables (e.g., tail cost, value-at-risk), etc.

In this book, in addition to MIP models, IP (integer programming) models without continuous variables are presented and, in particular, binary programming models with 0-1 variables only.

The book is divided into three main parts. Part One (Chapters 1 to 4) addresses short-term scheduling and presents various MIP models and some heuristic algorithms for detailed scheduling in flexible assembly lines and general flexible assembly systems. Part One is comprised of these chapters:

- Chapter 1, Scheduling of Flexible Flow Shops. This chapter provides the reader with a variety of MIP formulations for scheduling flexible flow shops with single or parallel machines, with infinite, finite, or no in-process buffers and with continuous or limited machine availability. In addition, two fast constructive heuristics are presented for scheduling flexible flow shops with finite or with no in-process buffers.

- Chapter 2, Scheduling of Surface Mount Technology Lines. This chapter presents MIP formulations for general or batch scheduling in Surface Mount Technology (SMT) lines with continuous or with limited machine availability. The formulations can be applied for constructing optimal schedules for printed wiring board assembly by using commercially available MIP software, which is illustrated with a set of computational examples modeled after different real-world SMT lines. In addition, a fast improvement heuristic is presented, capable of scheduling different SMT line configurations.
- Chapter 3, Balancing and Scheduling of Flexible Assembly Lines. This chapter proposes MIP formulations for simultaneous or sequential balancing and scheduling of flexible assembly lines with infinite or finite in-process buffers. The balancing objective is to determine an allocation of assembly tasks for a mix of products among the assembly stations with limited work space for component feeders and tools so as to balance the station workloads, whereas the scheduling objective is to determine the detailed sequencing and timing of all assembly tasks for each individual product, so as to maximize the line's productivity. In particular, balancing and scheduling of SMT lines is considered and illustrated with computational examples modeled after real-world assembly lines in electronics manufacturing.
- Chapter 4, Loading and Scheduling of Flexible Assembly Systems. This chapter deals with MIP approaches to simultaneous or sequential loading and scheduling of a general flexible assembly system with single or parallel assembly stations, limited work space of assembly stations for component feeders assignment, infinite or finite in-process buffers, and a fixed or alternative assembly routing.

Part Two of the book (Chapters 5 to 10) is concerned with medium-term decision making in supply chains and with supply chain risk management and it presents MIP models and some MIP-based heuristics for supplier selection and order allocation, customer order acceptance and due date setting, material supply scheduling, and medium-term scheduling and rescheduling of customer orders in a make-to-order discrete manufacturing environment. In particular, Chapters 9 and 10 show that the MIP approach combined with scenario analysis and percentile measures of risk, value-at-risk (VaR), and conditional value-at-risk (CVaR), is capable of solving a hard discrete stochastic optimization problem of selection static or dynamic supply portfolio in the presence of supply chain disruption and delay risks. Part Two has six chapters:

- Chapter 5, Customer Order Acceptance and Due Date Setting in Make-to-Order Manufacturing. This chapter considers customer order acceptance/rejection and due date setting decisions combined with order scheduling over a rolling planning horizon in make-to-order discrete manufacturing.

A dual-objective time-indexed MIP formulation is proposed with the solution approach in which the decisions are directly linked with available capacity. The problem objective is to select a maximal subset of orders that can be completed by customer requested dates and to quote delayed due dates for the remaining acceptable orders to minimize the number of delayed orders, the total number of delayed products, or the total revenue as a primary optimality criterion and to minimize the total or maximum delay of orders, as a secondary criterion. A simple critical load index is introduced to quickly identify the system bottleneck and the overloaded periods.

- Chapter 6, Aggregate Production Scheduling in Make-to-Order Manufacturing. The purpose of this chapter is to present time-indexed MIP formulations and a lexicographic approach to a dual- or multi-objective aggregate production scheduling in a make-to-order discrete manufacturing environment. The primary scheduling objective is to allocate a set of customer orders with various due dates among planning periods to minimize the number of tardy orders and the secondary objectives are to level the total input and output inventory over a planning horizon, to level the aggregate production, to level the total capacity utilization, or to level the machine assignments, with limited earliness and tardiness for the early and tardy orders, respectively. A close relation between minimizing the total inventory level and the maximum earliness of customer orders is shown and used to simplify the inventory leveling problem. The basic MIP models are strengthened by the addition of some cutting constraints that are derived by relating the demand on required capacity to available capacity for each processing stage and each subset of orders with the same due date.
- Chapter 7, Reactive Aggregate Production Scheduling in Make-to-Order Manufacturing. This chapter introduces MIP-based heuristic algorithms for reactive scheduling in a dynamic, make-to-order discrete manufacturing environment. The problem objective is to update a medium-term aggregate production schedule subject to service level and inventory constraints, whenever the customer orders are modified or new orders arrive. Different policies are proposed, from a total rescheduling of all remaining orders through rescheduling only a subset of remaining orders awaiting material supplies to a non-rescheduling of all remaining orders.
- Chapter 8, Scheduling of Material Supplies in Make-to-Order Manufacturing. This chapter provides the reader with MIP formulations and enumeration schemes for a cyclic or flexible approach to material ordering and supply scheduling in make-to-order assembly. Computational examples modeled after a real-world make-to-order assembly in the electronics industry illustrate possible applications of the proposed MIP models for a flexible approach and the enumeration schemes for a cyclic approach.
- Chapter 9, Selection of Static Supply Portfolio in Supply Chains with Risks. This chapter presents a new portfolio approach for the problem of static (single-period) supplier selection and order allocation in a customer-driven

supply chain with operational and/or disruption risks. The selection problem with operational risks is formulated as a single- or multi-objective mixed integer program with the risk of defective or unreliable supplies controlled by the maximum number of delivery patterns (combinations of suppliers' delivery dates) for which the average defect rate or late delivery rate can be unacceptable. In order to mitigate the impact of disruption risks, the two popular percentile measures of risk, value-at-risk and conditional value-at-risk, are applied. The two different types of disruption scenarios are considered: scenarios with independent local disruptions of each supplier and scenarios with local and global disruptions that may result in disruption of all suppliers simultaneously. The resulting scenario-based optimization problem under uncertainty is formulated as a single- or bi-objective mixed integer program that can be solved using commercially available software for MIP.

- Chapter 10, Selection of Dynamic Supply Portfolio in Supply Chains with Risks. The portfolio approach presented in Chapter 9 is enhanced for a multi-period supplier selection and order allocation in supply chains with disruption and delay risks. The dynamic (multiperiod) selection of supply portfolio in the presence of combined low-probability, high-impact disruption risks and high-probability, low-impact delay risks is modeled using MIP formulations and scenario analysis. The proposed approach is capable of optimizing the dynamic supply portfolio and implicitly of scheduling material supplies, by calculating value-at-risk and minimizing conditional value-at-risk simultaneously. In addition to general scenarios of supplies with on-time, delayed, or disrupted deliveries, the two extreme cases of supplies are considered for which disruptions are combined either with on-time or with longest delay deliveries.

Part Three (Chapters 11 to 14) focuses on coordinated scheduling of manufacturing and supply of parts and assembly of products in supply chains with a single producer and single or multiple suppliers. The intertemporal coordination of medium- and short-term scheduling, as well as the spatial coordination of the producer's and suppliers' medium-term schedules, are considered. The two different approaches are applied and compared: the integrated approach with a monolithic large MIP model and a hierarchical approach with a sequence of MIP models. In order to better coordinate the various schedules or decision levels, different coordinating constraints are introduced in the MIP formulations.

- Chapter 11, Hierarchical Integration of Medium- and Short-Term Scheduling. This chapter considers an intertemporal coordination of medium-term, aggregate production scheduling and short-term, detailed machine assignment and scheduling in a make-to-order discrete manufacturing environment. A hierarchical decision-making framework with IP and MIP models is proposed and with the objective function that integrates maximization of the customer service level and best utilization of the production resources. Computational examples modeled on a real-world make-to-order flexible assembly system with multicapacity machines in the electronics industry illustrate the hierarchical integration of tactical and operational decision making in a customer-driven supply chain.

- Chapter 12, Coordinated Scheduling in Supply Chains with a Single Supplier. This chapter considers coordinated medium-term scheduling of manufacturing and supply of parts and assembly of products in a supply chain with a single supplier and single producer. The supplier manufactures and delivers product-specific parts to the producer where finished products are assembled according to customer orders, and then are delivered to the customers. The overall problem is how to coordinate manufacturing and supply of parts and assembly of products with respect to limited capacities and required customer service level such that the total supply chain inventory holding cost (or maximum total inventory level) and the production line start-up costs at the supplier stage and parts shipment costs are minimized. A monolithic large MIP model is proposed for the overall problem and compared with a hierarchy of interconnected smaller-sized MIP models.
- Chapter 13, Coordinated Scheduling in Supply Chains with Assignment of Orders to Suppliers. This chapter looks at coordinated multi-objective scheduling in supply chains with multiple suppliers in which a single supplier of parts is selected for each customer order. The objective functions integrate both the supply chain performance and the customer service level. For the multi-objective scheduling problem a complex, MIP monolithic model is presented and then its decomposition into a hierarchy of much simpler, single-objective MIP models is proposed. The MIP formulations are strengthened by introduction of various coordinating constraints to directly link assembly schedule of the producer and manufacturing and delivery schedules of suppliers. Computational examples modeled by real-world coordinated scheduling in a customer-driven supply chain of high-tech products are reported.
- Chapter 14, Coordinated Scheduling in Supply Chains without Assignment of Orders to Suppliers. This chapter looks at the similar coordinated scheduling in supply chains with multiple suppliers, but in a bi-objective setting and without selection of a single part supplier for each customer order. The total demand for parts is allocated among the suppliers so that each customer order can be partially provided with the required parts by different suppliers. In addition to maximum service level, maximum revenue is also considered as an alternative primary objective function, while minimization of the total inventory holding cost is the secondary objective function. An integrated approach with a large MIP model is compared with a hierarchical approach. In contrast to the hierarchical approach presented in Chapter 13, which begins with the combined due date setting and selection of a single parts supplier for each customer order, in this chapter the allocation of demand for parts among suppliers is postponed until the final scheduling of manufacturing and delivery of parts. Computational examples modeled by real-world coordinated scheduling in a customer-driven supply chain of high-tech products are reported.

The proposed models mainly address integrated scheduling decisions at the supply-production stage of a customer-driven supply chain, while decisions for the production-distribution stage are limited to customer order acceptance and due date setting.

The decision-making problems considered in this book are deterministic, with the exception of discrete stochastic optimization problems of single-period or multiperiod supplier selection and order allocation in supply chains with risks, presented in Chapters 9 and 10. In order to tackle the various dynamic disturbances that may occur in supply chains, a rolling planning horizon approach can be applied, to update the decisions made using the proposed deterministic models (e.g., Chapters 5, 6, and 7).

The proposed MIP approaches to operational, tactical, or integrated operational and tactical decision making in customer-driven supply chains are illustrated with many computational examples modeled on real-world supply chain scheduling problems in the high-tech industries. For example, MIP models are presented for balancing and/or scheduling of surface mount technology lines in electronics manufacturing, for scheduling of material supplies in electronics assembly, for the integrated short- and medium-term scheduling in a distribution center for high-tech products, for the coordinated scheduling of manufacturing and supply of custom-made components and assembly of electronic devices in a make-to-order discrete manufacturing environment. In the computational experiments reported throughout the book, an advanced algebraic modeling language AMPL (see Fourer et al., 2003) and the CPLEX solver have been used.

Notation used in this book is consistent within each part of the book, and some consistency of the notation is maintained among different parts. However, the diversity and the complexity of the models presented have made it difficult to develop a uniform notation throughout the book.

The parts and the chapters within each part are arranged in the order recommended for reading. However, the precedence relationship between Parts One and Two is much weaker than that between Parts Two and Three. On the other hand, strictly interconnected Chapters 9 and 10 of Part Two, in which discrete stochastic optimization problems of supply chain risk management are considered, can be read almost independently of the other chapters.

The reader interested in knowing more about scheduling theory is referred to the recent textbook by Baker and Trietsch (2009), which gives an excellent overview of the many aspects of machine scheduling, to an application-oriented book by Pinedo (2005), or to the more advanced books by Blażewicz et al. (1994, 2007) on computational and complexity aspects of machine scheduling and their applications to computer and manufacturing systems. For a general introduction to integer programming models and techniques, the reader is referred to Wolsey (1998), to the recent application-oriented book by D.-S. Chen et al. (2010), or to the seminal work in the field by Nemhauser and Wolsey (1999). Finally, a number of books cover supply chains in general, and some of these emphasize planning and scheduling, for example, Shapiro (2001), Miller (2002), Voss and Woodruff (2003), and Pochet and Wolsey (2006).

Part One

Short-Term Scheduling in Supply Chains

Chapter 1

Scheduling of Flexible Flow Shops

1.1 INTRODUCTION

This chapter deals with mixed integer programming (MIP) models for scheduling of flow shops, the design in which dedicated machines are arranged in series or in series and parallel, and in which a transportation system imposes a unidirectional flow of parts, with revisiting of machines not allowed. The serial configuration of machines is widely used in many industries, in particular, the serial/parallel configuration with several processing stages in series and one or more parallel machines in each stage, is common in a high-tech industry.

The proposed MIP models cover a wide range of flow shop configurations that can be encountered in modern supply chains. They include flow shops with single or with parallel machines, with infinite, finite, or no in-process buffers, with machines continuously available or machines with one or more intervals of unavailability. The following models are presented:

F1 for scheduling flow shops with single machines and infinite in-process buffers

FP for scheduling flow shops with parallel machines and infinite in-process buffers

F1B for scheduling flow shops with single machines and no in-process buffers

FPB for scheduling flow shops with parallel machines and finite in-process buffers

FPBD for scheduling flow shops with parallel machines, finite in-process buffers, and machine down times

All the above models consider makespan minimization as a main scheduling criterion that aims at reaching a high throughput of the flow shop. The models, however, can easily be enhanced for the other common criteria such as total completion time or maximum or total tardiness, if the due dates for some parts are given. The model enhancements are described in Section 1.2.3. The MIP models can also be easily

Scheduling in Supply Chains Using Mixed Integer Programming. By Tadeusz Sawik

enhanced for scheduling flow shops with nonnegligible transportation times between processing stages (Section 1.2.4) or for scheduling reentrant flow shops (Section 1.2.5), in which a part visits a set of stages more than once.

Finally, for a comparison with the proposed MIP models, two simple, constructive heuristics for scheduling flexible flow shops with finite or with no in-process buffers and with nonzero transportation times, are described in Section 1.3.

1.2 MIXED INTEGER PROGRAMS FOR SCHEDULING FLOW SHOPS

In this section basic MIP formulations are developed for scheduling flexible flow shops of different configuration.

1.2.1 Scheduling Flow Shops with Infinite In-Process Buffers

A regular flow shop consists of m machines in series with unlimited capacity buffers between the machines. In the line n parts of various types are processed (for notation used, see Table 1.1). Each part must be processed without preemption on each machine sequentially. That is, each part must be processed in stage 1 through stage m in that order. The order of processing the parts in every stage is identical and determined by an input sequence in which the parts enter the line, that is, a so-called permutation flow shop is considered (e.g., Baker and Trietsch, 2009).

Table 1.1 Notation: MIP Models for Scheduling Flexible Flow Shops

Indices

i	= processing stage, $i \in I = \{1, \ldots, m\}$
j	= processor in stage i , $j \in J_i = \{1, \ldots, m_i\}$
k	= part, $k \in K = \{1, \ldots, n\}$

Input parameters

m	= number of processing stages
m_i	= $\|J_i\|$—number of parallel processors in stage i
n	= number of parts
p_{ik}	= processing time for part k in stage i
Q	= a large positive constant not less than the schedule length

Decision variables

c_{ik}	= completion time of part k in stage i (timing variable)
d_{ik}	= departure time of part k from stage i (timing variable)
x_{ijk}	= 1, if part k is assigned to processor $j \in J_i$ in stage $i \in I$; otherwise $x_{ijk} = 0$ (assignment variable)
y_{kl}	= 1, if part k precedes part l in the processing sequence; otherwise $y_{kl} = 0$ (sequencing variable)

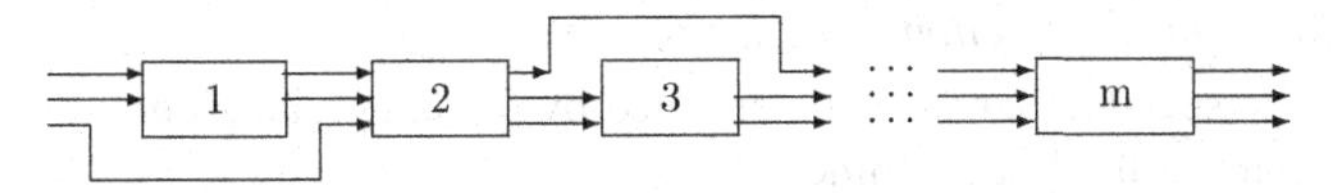

Figure 1.1 A general flow shop with single machines.

Let p_{ik} be the processing time on machine $i \in I$ of part $k \in K$. For every part k denote by c_{ik} its completion time in each stage i as well as its departure time from stage i.

Processing without preemption indicates that part k completed in stage i at time c_{ik} had started its processing in that stage at time $c_{ik} - p_{ik}$. Each part k completed in stage i at time c_{ik} immediately departs that stage for the next stage $i + 1$, if it is not occupied with another part; otherwise the part is transferred to the buffer with unlimited capacity between stages i and $i + 1$.

In contrast to the regular flow shop, in which each part requires m operations, each on a different machine, in a general flow shop (Fig. 1.1) parts may require fewer than m operations, not necessarily on adjacent machines. Then, the general case can be represented as a regular flow shop in which the processing times on some machines are zero.

The production schedule is specified by an input sequence in which the parts enter the line and are processed on each machine as well as by all completion times required for detailed scheduling of each individual part. The scheduling objective is to determine a permutation of parts such that all the parts are completed in a minimum time, that is, to minimize the makespan $C_{\max} = \max_{k \in K}(c_{mk})$, where c_{mk} denotes the completion time of part k in the last stage m.

The problem of scheduling a regular flow shop with single machines and infinite in-process buffers is formulated below as a mixed integer program *F1*.

Model F1: *Scheduling Flow Shops with Single Machines and Infinite In-Process Buffers*

Minimize

$$C_{\max} \tag{1.1}$$

subject to

1. *Part Completion Constraints*:
   - each part must be processed on the first machine and successively on all downstream machines,

$$c_{1k} \geq p_{1k}; \quad k \in K \tag{1.2}$$

$$c_{ik} - c_{i-1k} \geq p_{ik}; \quad i \in I, \ k \in K: i > 1 \tag{1.3}$$

2. *Part Noninterference Constraints*:
   - no two parts can be processed simultaneously on the same machine,

$$c_{ik} + Qy_{kl} \geq c_{il} + p_{ik}; \quad i \in I, k, l \in K: k < l \tag{1.4}$$

$$c_{il} + Q(1 - y_{kl}) \geq c_{ik} + p_{il}; \quad i \in I, k, l \in K: k < l \tag{1.5}$$

3. *Maximum Completion Time Constraints*:

- the schedule length is determined by the latest completion time of some part on the last machine,

$$c_{mk} \leq C_{\max}; \; k \in K \tag{1.6}$$

4. *Variable Nonnegativity and Integrality Conditions*:

$$C_{\max} \geq 0 \tag{1.7}$$

$$c_{ik} \geq 0; \; i \in I, k \in K \tag{1.8}$$

$$y_{kl} \in \{0, 1\}; \; k, l \in K: k < l. \tag{1.9}$$

The noninterference constraints (1.4) and (1.5) were constructed as follows. For any two parts k and l processed by the same machine i either part k precedes part l, and then processing of l cannot be started until processing of k is completed, or l precedes k, and then k cannot be started until l is completed. As a result, for all $i \in I$ and $k, l \in K: k \neq l$ a pair of the following disjunctive constraints must hold

$$c_{il} - p_{il} \geq c_{ik} \quad \text{or} \quad c_{ik} - p_{ik} \geq c_{il}.$$

These disjunctive constraints can be replaced with an equivalent pair of conjunctive constraints:

$$c_{il} + Q(1 - y_{kl}) \geq c_{ik} + p_{il}; \; i \in I, k, l \in K: k < l$$

$$c_{ik} + Qy_{kl} \geq c_{il} + p_{ik}; \; i \in I, k, l \in K: k < l$$

where y_{kl} is the binary sequencing variable defined below

$$y_{kl} = \begin{cases} 1, & \text{if part } k \text{ precedes part } l \text{ in the processing sequence} \\ 0, & \text{if part } l \text{ precedes part } k \text{ in the processing sequence,} \end{cases}$$

and Q is a large positive constant, not less than the schedule length $C_{\max}$.

Note that for all $k \neq l$, $y_{kl} + y_{lk} = 1$ or equivalently $y_{lk} = 1 - y_{kl}$, and hence in the above constraints it is sufficient to define y_{kl} for $k < l$.

In the next model, a flow shop with parallel machines is considered which consists of m processing stages in series with unlimited capacity buffers between the successive stages and each stage i, $(i = 1, \ldots, m)$ is made up of $m_i \geq 1$ parallel identical machines (Fig. 1.2).

The flow shop with parallel machines is also known as a hybrid flow shop, flow shop with multiple machines, flexible flow shop, flexible flow line, or multiprocessor flow shop. The problem of scheduling a flow shop with parallel machines may be seen

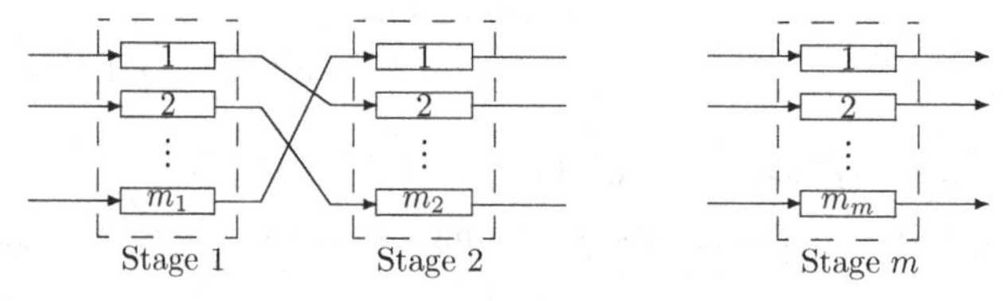

Figure 1.2 A flow shop with parallel processors.

as a combination of two particular types of scheduling problems: the parallel machine scheduling problem and the regular flow shop with single machines scheduling problem. The key decision of the problem of scheduling parallel machines is the assignment of parts to machines, whereas the key decision of scheduling a regular flow shop with single machines is the sequence of parts through the shop. Hence, the main decisions in the operation of the flow shop with parallel machines are to assign and schedule the parts to the machines in each stage, that is, to determine the order in which the parts are to be processed on the different machines of each stage.

In order to determine the assignment of parts to machines in each stage the following decision variables need to be added to the mixed integer programming formulations:

$$x_{ijk} = 1, \quad \text{if part } k \text{ is assigned to machine } j \in J_i \text{ in stage } i \in I;$$
$$\text{otherwise } x_{ijk} = 0.$$

The problem of scheduling a flow shop with parallel machines and infinite in-process buffers is formulated below as a mixed integer program *FP*.

Model FP: *Scheduling Flow Shops with Parallel Machines and Infinite In-Process Buffers*

Minimize (1.1) subject to

1. *Part Completion Constraints*: (1.2), (1.3)
2. *Maximum Completion Time Constraints*: (1.6)
3. *Machine Assignment Constraints*:
   - in every stage each part is assigned to exactly one machine,

$$\sum_{j \in J_i} x_{ijk} = 1; \quad i \in I, k \in K \tag{1.10}$$

4. *Part Noninterference Constraints*:
   - no two parts assigned to the same machine can be processed simultaneously,

$$c_{ik} + Q(2 + y_{kl} - x_{ijk} - x_{ijl}) \geq c_{il} + p_{ik}; \quad i \in I, j \in J_i, k, l \in K: k < l \tag{1.11}$$

$$c_{il} + Q(3 - y_{kl} - x_{ijk} - x_{ijl}) \geq c_{ik} + p_{il}; \quad i \in I, j \in J_i, k, l \in K: k < l \tag{1.12}$$

5. *Variable Nonnegativity and Integrality Conditions*: (1.7) to (1.9) and

$$x_{ijk} \in \{0, 1\}; \quad i \in I, j \in J_i, k \in K. \tag{1.13}$$

Part noninterference constraints (1.11) and (1.12) incorporate additional assignment variables x_{ijk} and x_{ijl}. For a given sequence of parts at most one constraint of (1.11) and (1.12) is active, and only if in stage i both parts k and l are assigned to the same machine j, i.e., only if $x_{ijk} = x_{ijl} = 1$ (which is always true for stages with a single machine). Otherwise, both (1.11) and (1.12) are inactive. In other words, non-interference constraints (1.11) and (1.12) prevent simultaneous processing of any two parts assigned in some stage to the same machine. However, two parts assigned to different parallel machines in the same stage can be processed in parallel, and then both constraints (1.11) and (1.12) are inactive.

In the above problem setting, all machines within each stage are considered to be identical. Therefore, the processing time of a part on each stage does not depend on the specific machine to which it is assigned. The proposed MIP models are also capable of scheduling flow shops with unrelated parallel machines, in which the processing times of a job in a stage depend on each specific machine within this stage, that is, with the input parameters p_{ik} replaced with p_{ijk} for each machine $j \in J_i$ and some or all stages $i \in I$. This may be due to the differences between the machines themselves, to the fact that a certain type of machine is better suited for processing a particular part, whereas others are not, or because the parts have some special characteristics and can only be assigned to machines that better physically meet them. Also, flexible flow shops with parallel uniform machines in some or all stages (e.g., Kyparisis and Koulamas, 2006) can be considered, where each machine $j \in J_i$ has an associated speed v_j, that is, when part k is processed in stage i by machine j, it requires p_{ik} / v_j time units to be completed.

1.2.2 Scheduling Flow Shops with Finite In-Process Buffers

In this section mixed integer programming models for scheduling are presented first for a special case of the flow shop with single machines and no buffers and then for a general case of the flow shop with parallel machines and finite in-process buffers.

A unified modeling approach is adopted with the buffers and machines jointly called processors. The buffers are viewed as special processors with zero processing times but with blocking. As a result the scheduling problem with finite in-process buffers can be converted into one with no buffers but with blocking, see Figure 1.2.

The blocking time of a processor with zero processing time represents part waiting time in the buffer represented by that processor. We assume that each part must be processed in all stages, including the buffer stages. However, zero blocking time in a buffer stage indicates that the corresponding part does not need to wait in the buffer. Let us note that for each buffer stage part completion time is equal to its departure time from the previous stage since the processing time is zero.

A flow shop with single machines and no in-process buffers represents a special type of regular flow shop with no store constraints in which there is only one machine in each stage, and every part visits every machine.

Assume that the flow shop consists of m machines in series with no intermediate buffers between the machines. The system produces n parts of various types and each part must be processed without preemption in each stage 1 through stage m in that order. Let p_{ik} be the processing time in stage i of part k. For every part k denote by c_{ik} its completion time in each stage i. A part completed on a machine blocks this

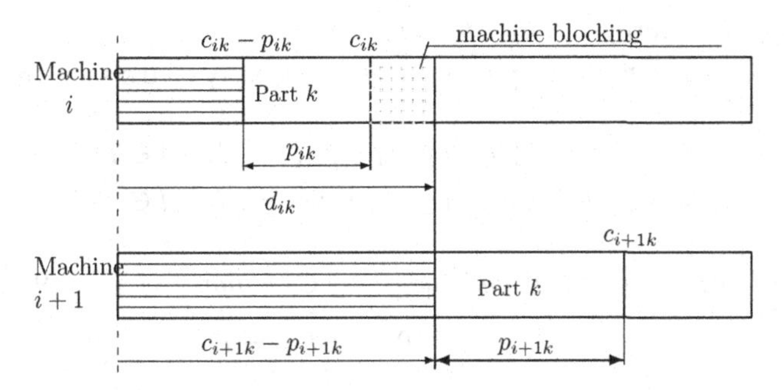

Figure 1.3 A partial schedule for part k in a flow shop with no in-process buffers.

machine until the next machine is available and denote by d_{ik} its departure time from stage i. This is a new variable that needs to be added to the previous models to account for the blocking scheduling that may occur when the intermediate buffers have limited capacity or there are no buffers at all.

A Gantt chart in Figure 1.3 shows a partial schedule for some part k in stages i and $i+1$. Processing without preemption indicates that part k completed in stage i at time c_{ik} had started its processing in that stage at time $c_{ik} - p_{ik}$. Part k completed in stage i at time c_{ik} departs at time $d_{ik} \geq c_{ik}$ to the next stage $i+1$. If at time c_{ik} machine $i+1$ is occupied, then machine i is blocked by part k until time $d_{ik} = c_{i+1k} - p_{i+1k}$ when part k can start processing on machine $i+1$.

The problem of scheduling the flow shops with no in-process buffers is formulated below as a mixed integer program *F1B*.

Model F1B: *Scheduling Flow Shops with Single Machines and Blocking (No In-Process Buffers)*

Minimize (1.1) subject to

1. *Part Completion Constraints*: (1.2), (1.3)
2. *Maximum Completion Time Constraints*: (1.6)
3. *Part Departure Constraints*:
   - each part cannot be departed from a machine until it is completed on this machine,
   - each part leaves the line as soon as it is completed on the last machine,

$$c_{ik} \leq d_{ik}; \ \ i \in I, k \in K : i < m \tag{1.14}$$

$$c_{mk} = d_{mk}; \ \ k \in K \tag{1.15}$$

4. *No Buffering Constraints*:
   - on every machine processing of each part starts immediately after its departure from the previous machine,

$$c_{ik} - p_{ik} = d_{i-1k}; \ \ i \in I, k \in K : i > 1 \tag{1.16}$$

5. *Part Noninterference Constraints*:

- no two parts can be processed simultaneously on the same machine,

$$c_{ik} + Qy_{kl} \geq d_{il} + p_{ik};\ i \in I,\ k, l \in K: k < l \qquad (1.17)$$

$$c_{il} + Q(1 - y_{kl}) \geq d_{ik} + p_{il};\ i \in I,\ k, l \in K: k < l \qquad (1.18)$$

6. *Variable Nonnegativity and Integrality Conditions*: (1.7)–(1.9), and

$$d_{ik} \geq 0;\ i \in I,\ k \in K. \qquad (1.19)$$

Inequalities (1.17) and (1.18) represent disjunctive constraints for scheduling with machine blocking, that is, they express interrelations between completion and departure times of any two different parts on the same machine i. No two parts can be performed on the same machine simultaneously. For any two different parts k and l either part k precedes part l, and then l cannot be started until k is departed from machine i (i.e., $c_{il} - p_{il} \geq d_{ik}$), or part l precedes part k, and then k cannot be started until l is departed (i.e., $c_{ik} - p_{ik} \geq d_{il}$). For a given sequence of parts only one constraint of (1.17) and (1.18) is active.

Finally, the most general problem of scheduling flow shops with parallel machines and finite in-process buffers is considered. Notation used to formulate the problem is shown in Table 1.1, where buffers and machines are jointly called processors. In the flow shop with parallel processors and no in-process buffers shown in Figure 1.2, processors represent either machines or buffers. If stage i is a buffer stage, then m_i is the number of buffers in that stage, otherwise m_i is the number of machines. Note that for the buffer stages all processing times are zero.

A Gantt chart in Figure 1.4 shows a partial schedule for some part k in stages i and $i+1$. Processing without preemption indicates that part k completed in stage i at time c_{ik} had started its processing in that stage at time $c_{ik} - p_{ik}$. Part k completed in stage i at time c_{ik} departs at time $d_{ik} \geq c_{ik}$ to an available processor in the next stage $i+1$. If at time c_{ik} all m_{i+1} processors in stage $i+1$ are occupied, then the processor in stage i is blocked by part k until time $d_{ik} = c_{i+1k} - p_{i+1k}$ when part k starts processing on an available processor in stage $i+1$. If $i+1$ is a buffer stage, then $p_{i+1k} = 0$, $d_{ik} = c_{i+1k}$, and $d_{ik+1} - d_{ik}$ is the waiting time of part k in the buffer stage $i+1$.

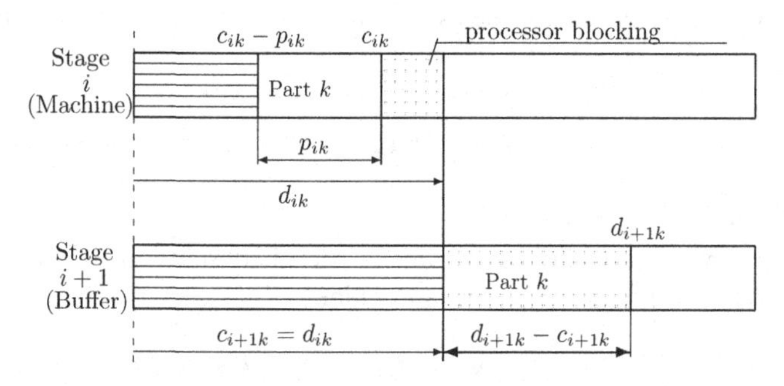

Figure 1.4 A partial schedule for part k in a flow shop with finite in-process buffers.

The problem of scheduling flow shops with parallel machines and finite in-process buffers is formulated below as a mixed integer program *FPB*.

Model FPB: *Scheduling Flow Shops with Parallel Machines and Blocking (Finite In-Process Buffers)*

Minimize (1.1) subject to

1. *Part Completion Constraints*: (1.2), (1.3)
2. *Maximum Completion Time Constraints*: (1.6)
3. *Processor Assignment Constraints*: (1.10)
4. *Part Departure Constraints*: (1.14), (1.15)
5. *No Buffering Constraints*: (1.16)
6. *Part Noninterference Constraints*:
   - no two parts assigned to the same processor can be processed simultaneously,

$$c_{ik} + Q(2 + y_{kl} - x_{ijk} - x_{ijl}) \geq d_{il} + p_{ik};\ \ i \in I, j \in J_i,\ k, l \in K\colon k < l \tag{1.20}$$

$$c_{il} + Q(3 - y_{kl} - x_{ijk} - x_{ijl}) \geq d_{ik} + p_{il};\ \ i \in I, j \in J_i,\ k, l \in K\colon k < l \tag{1.21}$$

7. *Variable Nonnegativity and Integrality Conditions*: (1.7)–(1.9), (1.13), (1.19).

Part noninterference constraints (1.20) and (1.21) incorporate assignment variables x_{ijk} and x_{ijl}. For a given sequence of parts at most one constraint of (1.20) and (1.21) is active, and only if in stage i both parts k and l are assigned to the same processor j (machine, if $p_{ijk} > 0$ or buffer, if $p_{ijk} = 0$), i.e., only if $x_{ijk} = x_{ijl} = 1$. Otherwise, both (1.20) and (1.21) are inactive. For any two parts assigned in some stage to the same processor (machine or buffer), the noninterference constraints (1.20) and (1.21) prevent simultaneous processing by the same machine or simultaneous occupying of the same buffer space. However, two parts assigned to different parallel processors (machines or buffers) in the same stage can be processed in parallel, and then both constraints (1.20) and (1.21) are inactive.

Model *FPB* for scheduling flexible flow shops with parallel processors and blocking (finite in-process buffers) is a general formulation and includes all the previous models as well as various special cases. For example, if $|J_i| = 1,\ \forall\, i \in I$ and if $p_{ik} > 0,\ \forall i \in I,\ k \in K$ model *FPB* reduces to model *F1B* for scheduling flow shops with single machines and no in-process buffers.

1.2.3 Alternative Objective Functions

In addition to the makespan, minimization of the total completion time of all parts, $C_{\text{sum}} = \sum_{k \in K} c_{mk}$ is sometimes considered in practice. The objectives $C_{\max}$ and C_{sum} aim at completing the production order as fast as possible and completing them fast in average, respectively. The total completion time represents average flow time and is also used as an implicit measure of work in process. Furthermore, where due dates D_k for completing of some parts $k \in K$ are given, then minimization of the maximum tardiness among all the parts, $T_{\max} = \max_{k \in K}\{0, c_{mk} - D_k\}$ or minimization of the total tardiness of all the parts, $T_{\text{sum}} = \sum_{k \in K} \max\{0, c_{mk} - D_k\}$ can be used as an optimality criterion of the processing schedule. The aim of $T_{\max}$ and T_{sum} is to minimize the maximum and the average delay, respectively, of part completion times with respect to given due dates.

If minimization of C_{sum} or T_{sum} is selected as an alternative objective function, then variable $C_{\max}$ and the *maximum completion time constraints* (1.6) should be removed from the MIP model. In addition, if T_{sum} is selected, new variables T_k that represent tardiness of each part k are introduced. The new objective function is to minimize $T_{\text{sum}} = \sum_{k \in K} T_k$ subject to the additional constraints that define the tardiness variables:

Part tardiness constraints

$$c_{mk} \le D_k + T_k; \ \ k \in K$$

$$T_k \ge 0; \ \ k \in K.$$

If in the objective function (1.1), $C_{\max}$ is replaced with $T_{\max}$, then the *maximum completion time constraints* (1.6) should be replaced with the *maximum tardiness constraints* to ensure that for each tardy part k, its tardiness $(c_{mk} - D_k)$ cannot exceed the maximum tardiness $T_{\max}$ to be minimized:

Maximum tardiness constraints

$$c_{mk} \le D_k + T_{\max}; \ \ k \in K$$

$$T_{\max} \ge 0.$$

1.2.4 Transportation Times

When in addition to processing times also transportation times between successive stages should be considered, then the *no buffering constraints* (1.16) should be replaced with the following constraints (1.22).

No buffering constraints with transportation times considered: in every stage processing of each part starts immediately after its arrival at this stage,

$$c_{ik} - p_{ik} = d_{i-1k} + q_{i-1}; \ \ i \in I, k \in K : i > 1 \qquad (1.22)$$

where q_i is the transportation time required to transfer a part from stage i to stage $i + 1$.

1.2.5 Reentrant Flow Shops

The proposed MIP models can also be applied for scheduling reentrant flow shops, where a part visits a set of stages more than once. For example, in a double-pass surface mount technology line (see Chapter 2) the double-sided printed wiring boards run twice through the same line, first to assemble the bottom side and then to assemble the top side. In order to extend an MIP for scheduling a double-pass reentrant flow shop, the number of parts is doubled to $2n$. A pair of parts $(k, k + n)$, $k = 1, \ldots, n$ represents the bottom and the top sides of board k. The release time for part $k+n$ cannot be less than the completion time of part k, that is, additional part completion constraints should be added for each part $k+n$, $k=1,\ldots,n$

Part completion constraints: each part $k+n$ can be started in the first stage only after completion of part k in the last stage,

$$c_{1,k+n} - p_{1,k+n} \geq c_{m,k}; \; k = 1,\ldots,n. \tag{1.23}$$

1.2.6 Scheduling Modes

The MIP models *F1*, *FP*, *F1B*, and *FPB* represent *general scheduling* problems, where any input sequence of parts is allowed. In order to reduce the complexity of the general scheduling problem, the following scheduling modes can also be considered.

- *Batch scheduling*, where parts of a given type are scheduled consecutively, and in addition:
  - the sequence of part types is fixed and equal to the optimal sequence determined for a Minimal Part Set (MPS) in the same proportion as the overall production target or
  - the sequence of part types is not determined *a priori*, but is obtained along with the optimal schedule for all parts.
- *Cyclic scheduling*, where a Minimal Part Set (MPS), in the same proportion as the overall production target, is repetitively scheduled and in addition:
  - the cycle of parts in an MPS is fixed and equal to the optimal sequence determined for the MPS or
  - the cycle of parts in an MPS is not determined *a priori*, but is obtained along with the optimal schedule for all parts.

The cyclic scheduling mode is often used when set up times are negligible and the demand for each part type remains constant over the scheduling horizon. As a result, inventory holding costs are reduced. Otherwise, the batch scheduling mode is applied, where it is more efficient to have long runs of identical parts to minimize sequence dependent set up times. While acyclic schedules (e.g., general or batch scheduling mode) often lead to the highest throughput or the smallest makespan, a cyclic schedule does not guarantee the maximum throughput or the minimum makespan.

MIP models for the general scheduling can also be used for the batch or cyclic scheduling mode after a simple addition of the constraints presented below.

Denote by

$$G,$$

$$K = \{1,\ldots,n\},$$

and

$$K_g = \left\{ \sum_{f \in G: f <= g-1} n_f + 1, \ldots, \sum_{f \in G: f <= g-1} n_f + n_g \right\}$$

the ordered sets of indices, respectively of all batches of parts, all individual parts, and all parts of type $g \in G$. (n_g and $n = \sum_{g \in G} n_g$ denote, respectively the number of parts type g and the total number of parts in the schedule.) Let $r_{ig} \geq 0$ be the processing time in stage i of part type $g \in G$.

Batch scheduling mode constraints:

$$y_{kl} = 1;\ g \in G,\ k \in K_g,\ l \in K_g{:}\,k < l \tag{1.24}$$

$$y_{kl} = y_{last(K_f),last(K_g)};\ f \in G,\ g \in G,\ k \in K_f,\ l \in K_g{:}\,f < g \tag{1.25}$$

$$c_{ik+1} \geq d_{ik} + r_{ig};\ i \in I,\ g \in G,\ k \in K_g{:}\,k < last(K_g),\ m_i = 1, \tag{1.26}$$

where $last(K_f)$ is last part in the ordered set K_f.

Equation (1.25) selects a sequence of processing different part types and constraints (1.24) and (1.26) ensure that boards of one type are processed consecutively.

In the cyclic scheduling mode, first the Minimal Part Set (MPS) should be determined, i.e., the smallest possible set of parts in the same proportion as the overall production target, which is to be repetitively processed in a cyclic mode. For example, if the production target is 100 units of part A, 300 units of part B and 500 units of part C, then MPS is one unit of A, three units of B, and five units of C, in total nine parts, which is to be repeated 100 times to meet the production target, using the same order of processing parts in each run of MPS. In the cyclic scheduling, all parts in an MPS can be ordered either arbitrarily, as in the general mode (e.g., C,C,A,B,C,B,C,B,C) or in batches of identical parts, as in the batch mode (e.g., C,C,C,C,C,A,B,B,B). The latter cyclic mode combined with the batch processing of identical parts in each run of MPS can be called a cyclic-batch scheduling mode.

Formally, the Minimal Part Set (MPS) is defined as

$$\{\underline{n}_g : g \in G\},$$

where

$$n_g = S\,\underline{n}_g\ \forall\, g \in G$$

and S is the greatest common divisor of integers $n_1, n_2, \ldots n_{|G|}$ ($|\cdot|$denotes the power of a set $\cdot$), i.e., a total of S runs of an MPS is required to meet the overall production target. Denote by $\underline{n} = \sum_{g \in G} \underline{n}_g$, the total number of parts in an MPS.

Cyclic scheduling mode constraints: (1.24) and

$$y_{kl} = y_{next(k,K_f,s\underline{n}_f),next(l,K_g,s\underline{n}_g)};\ f \in G,\ g \in G,\ k \in K_f,\ l \in K_g,$$
$$1 \leq s \leq S-1{:}\,f < g,\ 1 \leq ord(k, K_f) \leq \underline{n}_f,\ 1 \leq ord(l, K_g) \leq \underline{n}_g \tag{1.27}$$

$$c_{i,next(k,K_g,\underline{n}_g)} \geq d_{ik} + \sum_{f \in G} \underline{n}_f r_{if};\ i \in I,\ g \in G,\ k \in K_g{:}$$
$$k \leq last(K_g) - \underline{n}_g,\ m_i = 1 \tag{1.28}$$

$$c_{ik} \geq d_{il} + p_{ik};\ i \in I, f \in G, g \in G, k \in K_f, l \in K_g, 1 \leq s \leq S-1:$$
$$s\underline{n}_f < ord(k, K_f) \leq (s+1)\underline{n}_f, (s-1)\underline{n}_g < ord(l, K_g) \leq s\underline{n}_g, m_i = 1, \quad (1.29)$$

where $ord(k, K_f)$ denotes ordinal position of k in K_f, and $next(k, K_f, s\underline{n}_f)$ is the part type f, $s\underline{n}_f$ positions after part k in the ordered set K_f.

Equation (1.27) imposes the same sequence on processing parts in each run of an MPS, constraint (1.28) ensures periodic processing of every $\underline{n}_g$th part of each type g, and constraint (1.29) ensures that no part of a new run of MPS can be started on a machine until all parts from the previous run are departed.

Cyclic-batch scheduling mode constraints: (1.24), (1.28), (1.29) and

$$y_{kl} = y_{last(K_f),last(K_g)};\ f \in G, g \in G, k \in K_f, l \in K_g, 1 \leq s \leq S:$$
$$f < g, (s-1)\underline{n}_f < ord(k, K_f) \leq s\underline{n}_f, (s-1)\underline{n}_g < ord(l, K_g) \leq s\underline{n}_g. \quad (1.30)$$

Equation (1.30) selects the same cyclic sequence of processing minimal batches of different part types over all S runs of MPS.

If the production target consists of equal batch sizes for all part types, that is, $n_g = n/|G|\ \forall g \in G$, then $\underline{n}_g = 1\ \forall\ g \in G$. As a consequence, MPS is made of one unit of each part type, in total of $\underline{n} = |G|$ different parts, and the unit MPS set should be repeatedly processed $S = n/|G|$ times. In this case, the cyclic scheduling mode constraints reduce to the following set of constraints.

Cyclic scheduling mode constraints for a unit MPS: (1.24) and

$$y_{kl} = y_{last(K_f),last(K_g)};\ f \in G, g \in G, k \in K_f, l \in K_g:$$
$$f < g,\ ord(k, K_f) = ord(l, K_g) \quad (1.31)$$

$$y_{kl} = 1 - y_{next(k,K_f),l};\ f \in G, g \in G, k \in K_f, l \in K_g:$$
$$f < g,\ k < last(K_f),\ ord(k, K_f) = ord(l, K_g) \quad (1.32)$$

$$c_{i,next(k,K_g)} \geq d_{ik} + \sum_{f \in G} r_{if};\ i \in I, g \in G, k \in K_g:$$
$$k < last(K_g), m_i = 1, \quad (1.33)$$

where $next(k, K_f)$ is next part after k in the ordered set K_f.

Equations (1.31), (1.32) select the same sequence of processing different part types in successive runs of an MPS, and constraint (1.33) ensures that successive parts of one type are processed periodically.

In addition to the sequencing and timing constraints, the following assignment constraints should be added to the MIP formulations to model alternate assignment of parts to parallel machines, for example, the assignment by a shuttle device in a surface mount technology line (see, Chapter 2).

Assignment constraints for batch scheduling mode:

- in each stage with parallel machines, successive parts of one type are assigned to successive parallel machines,

$$x_{i,next(j,J_i),k+1} = x_{ijk};\ i \in I, j \in J_i,\ g \in G,\ k \in K_g:\\ k \leq last(K_g) - 1,\ m_i > 1, \tag{1.34}$$

where $next(j, J_i)$ is the next machine after j in the circular set J_i of parallel machines in stage i.

Assignment constraints for cyclic and cyclic-batch scheduling mode:

- in each stage i with parallel machines, corresponding parts in successive runs of MPS are assigned to every $\underline{n}$th machine in the circular set J_i of machines

$$x_{i,next(j,J_i,\underline{n}),next(k,K_g,\underline{n}_g)} = x_{ijk};\ i \in I, j \in J_i,\ g \in G,\ k \in K_g:\\ k \leq last(K_g) - \underline{n}_g\,,\ m_i > 1, \tag{1.35}$$

where $next(j, J_i, \underline{n})$ is the parallel machine in stage i, $\underline{n}$ $(= \sum_{g \in G} \underline{n}_g)$ positions after machine j in the circular set J_i of parallel machines, and $next(k, K_g, \underline{n}_g)$ is the part type g, $\underline{n}_g$ positions after part k in the ordered set K_g.

1.2.7 Computational Examples

In this subsection two numerical examples are presented to illustrate possible applications of the proposed MIP models.

The flexible flow shop configurations for the example problems are provided in Figures 1.5 and 1.6. The flow shop with no buffers in Figure 1.5 is made up of $m = 3$ processing stages, each representing a single machine or a set of parallel machines.

The following two variants of the three-stage flow shop with no buffers will be considered in the examples:

F3_1—the flow shop with single machines: $m_1 = m_2 = m_3 = 1$

F3_P—the flow shop with parallel machines: $m_1 = 2$, $m_2 = 3$, $m_3 = 2$

The flow shop with finite in-process buffers in Figure 1.6 consists of $m = 5$ stages, where stages $i = 1, 3, 5$ represent single or parallel machines and stages $i = 2, 4$ represent one or more buffers.

The following two variants of the five-stage flow shop with finite in-process buffers is considered in the examples:

F5_1—the flow shop with single processors: $m_1 = m_2 = m_3 = m_4 = m_5 = 1$

F5_P—the flow shop with parallel processors: $m_1 = 2$, $m_2 = 3$, $m_3 = 3$, $m_4 = 3$, $m_5 = 2$

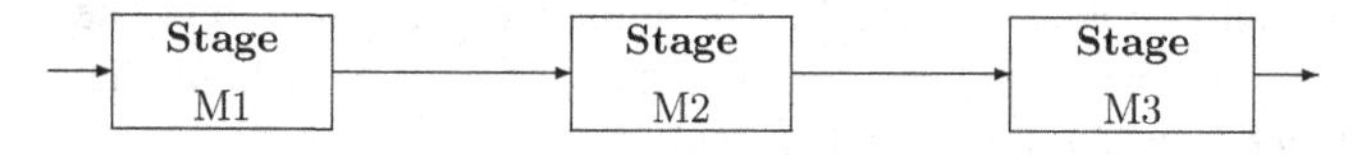

Figure 1.5 A three-stage flow shop with single machines and no in-process buffers.

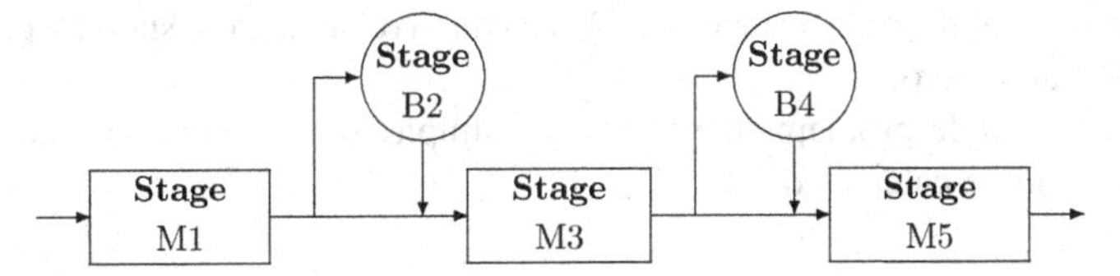

Figure 1.6 A five-stage flow shop with parallel machines and finite in-process buffers.

The production order consists of $n = 10$ parts, and the processing times p_{ik} are shown below for the three-stage and the five-stage flow shop, respectively:

$$[p_{ik}] = \begin{bmatrix} 4, 2, 1, 5, 4, 3, 5, 2, 1, 8 \\ 2, 5, 8, 6, 7, 4, 7, 3, 6, 2 \\ 2, 8, 6, 2, 2, 4, 2, 7, 1, 8 \end{bmatrix},$$

$$[p_{ik}] = \begin{bmatrix} 4, 2, 1, 5, 4, 3, 5, 2, 1, 8 \\ 0, 0, 0, 0, 0, 0, 0, 0, 0, 0 \\ 2, 5, 8, 6, 7, 4, 7, 3, 6, 2 \\ 0, 0, 0, 0, 0, 0, 0, 0, 0, 0 \\ 2, 8, 6, 2, 2, 4, 2, 7, 1, 8 \end{bmatrix}.$$

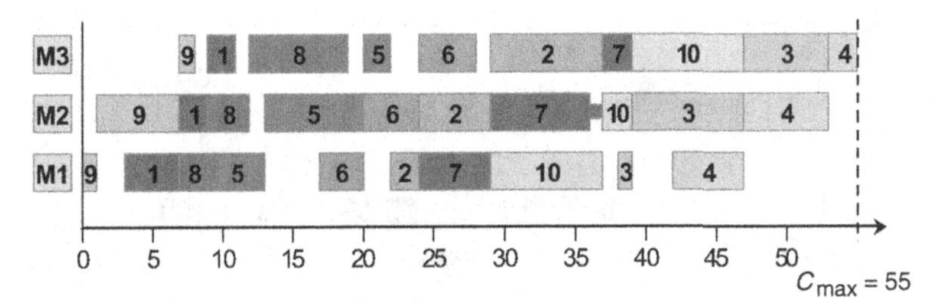

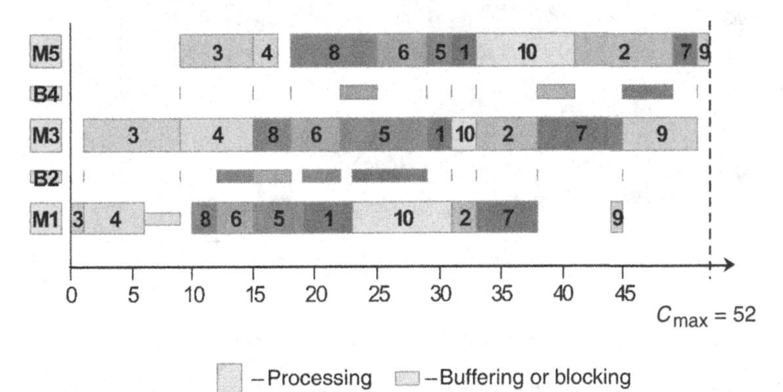

Figure 1.7 Optimal schedules for flow shops with single processors.

Let us note that for the buffer stages $i = 2, 4$ in the five-stage flow shop all processing times are equal to zero.

For the example problems the following simple lower bounds on the schedule length $C_{\max}$ can be calculated

$$LBC_{\max} = \max_{i \in I}\left\{\sum_{k \in K} p_{ik} + \min_{k \in K}\left(\sum_{h \in I : h < i} p_{hk}\right) + \min_{k \in K}\left(\sum_{h \in I : h > i} p_{hk}\right)\right\} = 52,$$

$$LBC_{\max} = \max_{i \in I}\left\{\left\lceil \sum_{k \in K} p_{ik}/m_i \right\rceil + \min_{k \in K}\left(\sum_{h \in I : h < i} p_{hk}\right) + \min_{k \in K}\left(\sum_{h \in I : h > i} p_{hk}\right)\right\} = 26,$$

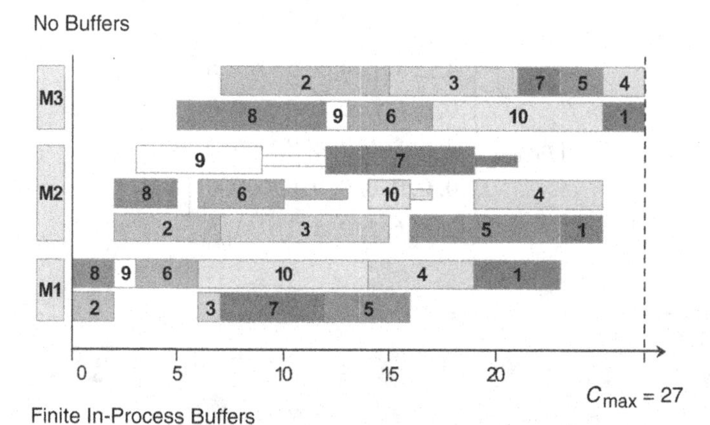

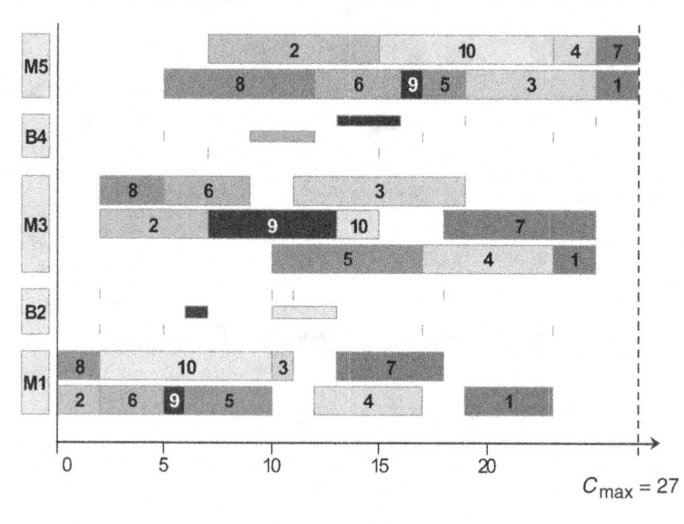

– Processing – Buffering or blocking

Figure 1.8 Optimal schedules for flow shops with parallel processors.

for the flow shop with single processors and with parallel processors, respectively ($\lceil \cdot \rceil$ is the smallest integer not less than $\cdot$). The above lower bounds are calculated as maximum over all stages of a stage workload and minimum flow time of a part through the remaining stages.

The optimal schedules for the examples were determined by solving the mixed integer programs *F1B* and *FPB* for the flow shop configurations *F3_1*, *F5_1*, and *F3_P*, *F5_P*, respectively. The solution values are $C_{\max} = 55$, $C_{\max} = 52$, $C_{\max} = 27$, and $C_{\max} = 27$, respectively, for configurations *F3_1*, *F5_1*, *F3_P*, and *F5_P*. The optimal schedules are shown on Gantt charts in Figures 1.7 and 1.8.

1.3 CONSTRUCTIVE HEURISTICS FOR SCHEDULING FLEXIBLE FLOW SHOPS

This section presents two fast constructive heuristics for scheduling flexible flow shops with finite in-process buffers or with no in-process buffers. The heuristics are not based on MIP; however, they use similar variables to determine the processing schedule and their performance can be compared to MIP approaches presented in the previous sections.

Unlike the exact MIP procedures capable of finding the proven optimal solution, heuristics are not able to find a proven optimal solution. In general, the heuristics can be divided into constructive procedures and improvement procedures. A constructive procedure is capable of constructing the solution from scratch step by step, which can be done either job-by-job or stage-by-stage, whereas an improvement procedure improves a given initial or constructed solution. The constructive heuristics proposed in this section are single-pass, part-by-part procedures, in which during every iteration a complete processing schedule is constructed for one part. The selection of the part and its complete schedule are based on the cumulative partial schedule obtained for all parts selected so far. The decisions in every iteration are made using a local optimization mechanism aimed at minimizing total idle time along the route of the selected part. For this reason, the heuristics will be called *Route Idle Time Minimization* (RITM) (Sawik, 1993) or *Route Idle Time Minimization–No Store* (RITM-NS) (Sawik, 1994, 1995b), respectively, for the flow shop with finite in-process buffers or the flow shop with no in-process buffers. Notation used to formulate the scheduling problems and the heuristic algorithms is introduced in Table 1.2.

1.3.1 A Fast Heuristic for Scheduling Flow Shops with Finite In-Process Buffers

The flexible flow shop under study consists of $m \geq 2$ processing stages in series with limited in-process buffers between the successive stages. Each stage i ($i = 1, \ldots, m$) is made up of $m_i \geq 1$ parallel identical machines. Ahead of each stage $i (i = 2, \ldots, m)$ there are b_i buffers, where each buffer can hold one part at a time (see Fig. 1.9).

Table 1.2 Notation: Heuristic Algorithms

Indices	
g	= part type, $g \in G$
i	= processing stage, $i \in I = \{1, \ldots, m\}$
j	= parallel machine in stage i, $j \in J_i = \{1, \ldots, m_i\}$
k	= part, $k \in K = \bigcup_{g \in G} K_g = \{1, \ldots, n\}$
l	= position in the loading sequence, $l = 1, \ldots, n$
Input parameters	
b_i	= number of buffers ahead of stage i $(i = 2, \ldots, m)$
K_g	= subset of parts type g
n_g	= demand for part type g, $(n_g = \|K_g\|)$
n	= total number of parts in the production order, $n = \sum_{g \in G} n_g$
P_{ig}	= processing time in stage i of part type g
p_{ik}	= P_{ig}, $k \in K_g$—processing time in stage i of part k
q_i	= transportation time from stage i to stage $i + 1$
Decision variables	
k_l	= part in position l $(l = 1, \ldots, n)$ in the loading sequence $[k_1, k_2, \ldots, k_n]$
f_{il}	= finish time of processing part k_l on machine in stage i
r_{il}	= release time of part k_l from stage i
s_{il}	= start time of part k_l on machine in stage i
Auxiliary variables	
t_{il}	= idle time on machine in stage i incurred by part k_l (waiting time for start of processing part k_l and blocking time by finished part k_l)
Ξ_{bil}	= total elapsed time, when buffer b $(b = 1, \ldots, b_i)$ ahead of stage i is available for assignment after the first l parts have been scheduled
ξ_{il}	= the earliest time at which a buffer ahead of stage i is available for assignment after the first l parts have been scheduled, $\xi_{il} = \min_{1 \le b \le b_i} \{\Xi_{bil}\}$
$\beta(i, l)$	= buffer ahead of stage i with the earliest available time after the first $l - 1$ parts have been scheduled, $\beta(i, l) = \arg\min_{1 \le b \le b_i} \{\Xi_{bi,l-1}\}$
Γ_{jil}	= total elapsed time, when machine j $(j = 1, \ldots, m_i)$ at stage i is available for assignment after the first l parts have been scheduled
γ_{il}	= the earliest time at which a machine in stage i is available for assignment after the first l parts have been scheduled, $\gamma_{il} = \min_{1 \le j \le m_i} \{\Gamma_{jil}\}$
$\mu(i, l)$	= machine in stage i with the earliest available time after the first $l - 1$ parts have been scheduled, $\mu(i, l) = \arg\min_{1 \le j \le m_i} \{\Gamma_{ji,l-1}\}$

The system produces different part types $g \in G$. Let $P_{ig} \ge 0$ be the processing time in stage i of part type g and let q_i be the transportation time required to transfer a part from stage i to stage $i + 1$.

A part completed in stage i is transferred either directly to an available machine in the next stage $i + 1$ (or another downstream stage depending on the part processing

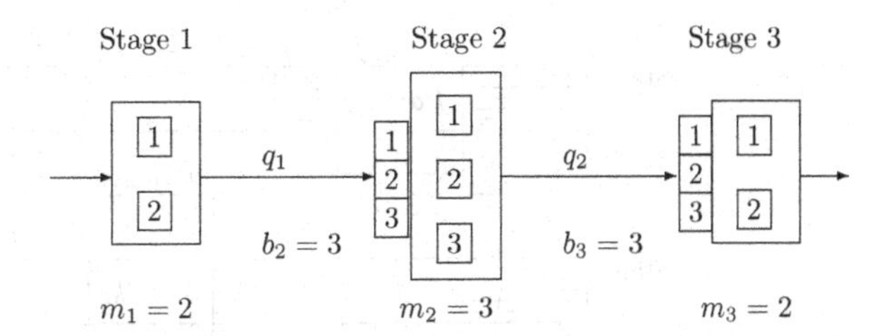

Figure 1.9 A three-stage flexible flow shop with finite in-process buffers.

route) or to a buffer ahead of that stage. If all b_{i+1} buffers ahead of stage $i + 1$ are occupied, then the machines in stage i are blocked by the completed parts until a buffer is available.

Given are demands n_g, $g \in G$ for all part types. The problem objective is to determine an assignment of all $n = \sum_{g \in G} n_g$ parts to machines in each stage over a scheduling horizon to meet the demand in minimum time, that is, to minimize the makespan $C_{\max}$.

The production schedule for all n parts is specified by the loading sequence (permutation of parts) $[k_1, k_2, \ldots, k_n]$ in which the parts enter the line, as well as the times required for detailed scheduling of each individual part. In particular, for each part k_l entering the line at position l ($l = 1, \ldots, n$) its start time s_{il}, finish time f_{il}, and release time r_{il} in each stage i should be determined. These variables satisfy the following formulas:

$$s_{1l} = \gamma_{1,l-1} \tag{1.36}$$

$$s_{il} = \max\{r_{i-1,l} + q_{i-1};\ \gamma_{i,l-1}\};\ i = 2, \ldots, m \tag{1.37}$$

$$f_{il} = s_{il} + p_{ik_l};\ i = 1, \ldots, m \tag{1.38}$$

$$r_{il} = \max\{f_{il},\ \xi_{i+1,l-1} - q_i\};\ i = 1, \ldots, m-1 \tag{1.39}$$

$$r_{ml} = f_{ml} \tag{1.40}$$

$$\xi_{il} = \min_{1 \le b \le b_i}\{\Xi_{bil}\};\ i = 2, \ldots, m \tag{1.41}$$

$$\gamma_{il} = \min_{1 \le j \le m_i}\{\Gamma_{jil}\};\ i = 1, \ldots, m \tag{1.42}$$

where the earliest available times Ξ_{bil} for buffers and Γ_{jil} for machines are calculated iteratively as follows

$$\Xi_{bil} = \begin{cases} \Xi_{bil-1} & \text{if } b \neq \beta(i, l) \\ s_{il} & \text{if } b = \beta(i, l) \end{cases} \tag{1.43}$$

$$\Gamma_{jik} = \begin{cases} \Gamma_{jil-1} & \text{if } j \neq \mu(i, l) \\ r_{il} & \text{if } j = \mu(i, l) \end{cases} \tag{1.44}$$

The formulae (1.43) and (1.44) were derived under the assumption that in every stage i part k_l is assigned to the buffer $\beta(i, l)$ and the machine $\mu(i, l)$ with the earliest available times (for additional interpretation of the variables introduced, see Fig. 1.10).

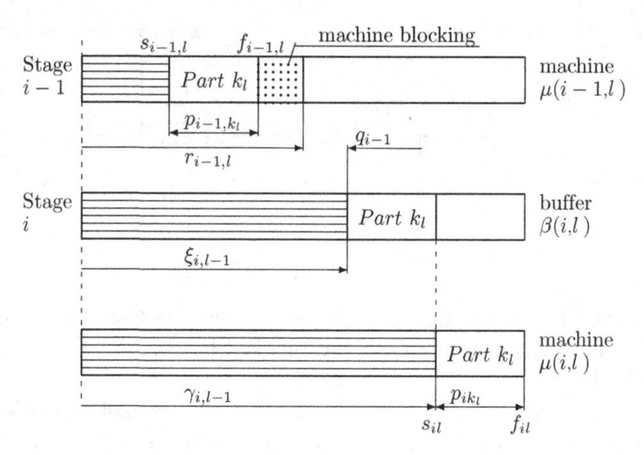

Figure 1.10 A partial schedule for part k_l.

Note that finish time f_{il} and release time r_{il} of the lth part in the loading sequence of the heuristic correspond to completion time c_{ik} and departure time d_{ik} of part k in the MIP models presented in Section 1.2.2.

Finally, notice that machine idle time t_{il} incurred in stage i by part k_l consists of two components:

- Time $(s_{il} - \gamma_{i,l-1})$ of machine waiting for start of processing part k_l
- Time $(r_{il} - f_{il})$ of machine blocking by finished part k_l

Hence, t_{il} can be expressed as follows

$$t_{il} = (s_{il} - \gamma_{i,l-1}) + (r_{il} - f_{il}) = r_{il} - \gamma_{i,l-1} - p_{ik_l};\ i = 1, \ldots, m \tag{1.45}$$

In order to minimize the makespan $C_{\max}$, the RITM heuristic presented in the next section assigns parts to the earliest available machines and aims at minimizing the total idle time $\sum_{i=1}^{m} \sum_{l=1}^{n} t_{il}$ along the routes of all n parts.

The RITM algorithm is a single-pass, part-by-part heuristic in which the loading sequence and the corresponding complete schedule are determined once. During every iteration a part for loading into the system is chosen as well as its complete processing schedule is determined. The decisions in every iteration are made based on the complete processing schedule determined for each part type waiting for loading. Given a cumulative partial schedule for the first $(l-1)$ parts selected so far, first the best route along the line is found as a sequence

$$[\mu(1, l), \beta(2, l), \mu(2, l), \ldots, \beta(m, l), \mu(m, l)] \tag{1.46}$$

of m machines and $m-1$ buffers (one in every stage) with the earliest available times. For each part type g waiting to enter the line the complete processing schedule is determined along the best route obtained. To evaluate the processing schedule for each part type considered for loading, the total duration of idle time t_g along the best route is determined:

$$t_g = \sum_{i=1}^{m} t_{ig} \tag{1.47}$$

Finally, the part type with the smallest total idle time t_g is selected for loading and its complete processing schedule is added to the cumulative partial schedule obtained so far.

Description of the algorithm RITM is given below ($\overline{G}$ denotes the set of part types loaded in the required number of parts).

Algorithm RITM

Step 0. Starting

1. Order all part types according to nonincreasing total processing times $p_g = \sum_{i=1}^{m} P_{ig}$, that is $p_1 \geq p_2 \geq \cdots \geq p_{|G|}$.
2. Set:
   $\Xi_{bi0} = 0,\ b = 1, \ldots, b_i,\ i = 2, \ldots, m.$
   $\Gamma_{ji0} = 0,\ j = 1, \ldots, m_i,\ i = 1, \ldots, m,$
   $\beta(i, 1) = 1,\ i = 2, \ldots, m,\ \mu(i, 1) = 1,\ i = 1, \ldots, m,$
   $\overline{G} = \emptyset,\ l = 1.$

Step 1. Selection of a part type for loading

1. For each part type $g \notin \overline{G}$ waiting for loading at position l determine start time s_{il}, finish time f_{il}, release time r_{il}, and duration t_{ig} of machine idle time, in every stage i on machine $\mu(i, l)$ with the earliest available time γ_{il-1}. Next, determine total idle time t_g.
2. Select for loading such a part type g_l that minimizes the total idle time, that is

$$g_l = \arg\min_{g \notin \overline{G}} \{t_g\}$$

To break ties, select the part type with the lowest number (i.e., with the largest total processing time).

Step 2. Determining complete processing schedule for the selected part type

For the selected part type g_l determine complete processing schedule (times s_{il}, f_{il}, r_{il}) by assigning it in every stage i to the buffer $\beta(i, l)$ and the machine $\mu(i, l)$ with the earliest available times, respectively ξ_{il-1} and γ_{il-1}, after the first $l - 1$ parts have been scheduled.
The processing schedule for a part $k_l \in K_{g_l}$ add to the cumulative partial schedule for the first $(l - 1)$ parts.

Step 3. Checking the state of completion of the production order

For $g = g_l$ set $n_g = n_g - 1$. If $n_g = 0$, then set $\overline{G} = \overline{G} \cup \{g\}$.
If $\overline{G} = G$, then terminate.

Otherwise determine for each buffer b ($b = 1, \ldots, b_i,\ i = 2, \ldots, m$) and for each machine j ($j = 1, \ldots, m_i,\ i = 1, \ldots, m$) the earliest available time, respectively Ξ_{bil} and Γ_{jil}, after the first l parts have been scheduled.

For each stage i find buffer $\beta(i, l+1) = \arg\min_{1 \le b \le b_i}(\Xi_{bil})$ and machine $\mu(i, l+1) = \arg\min_{1 \le j \le m_i}(\Gamma_{jil})$ with the earliest available time ξ_{il} and γ_{il}, respectively.

Set $l = l + 1$ and go to *STEP 1.*

To determine the computational complexity of the algorithm RITM notice that the algorithm requires n iterations and in every iteration a complete processing schedule is computed for one part according to the following procedure.

First, the best route is found as a sequence of at most m machines and $m-1$ buffers in the successive stages with the earliest available times. This step requires $O(M+B)$ computations since at most all M machines and B buffers are considered ($M = \sum_{i=1}^{m} m_i$ and $B = \sum_{i=2}^{m} b_i$ is the total number of machines and buffers, respectively). Then, for each of the remaining at most $|G|$ part types ($|\cdot|$ denotes the power of a set $\cdot$), the complete processing schedule is computed based on the best route found. This step requires $O(m|G|)$ computations to determine the times for the $|G|$ flow shop schedules on m machines. The best processing schedule (and by this a part for loading) with the smallest total idle time is next chosen and added to the cumulative partial schedule for all parts loaded so far.

Since the above procedure which requires $O(M+B+m|G|)$ computations is repeated in each of the n iterations, and the number of part types $|G| \le n$, the computational complexity of the algorithm RITM is $O(mn^2)$.

In order to evaluate the effectiveness of the proposed heuristic algorithm, the following lower bound on makespan can be used as a surrogate for the minimum makespan value

$$LBC_{\max} = \max_{1 \le i \le m} \left\{ \left\lceil \sum_{g \in G} n_g P_{ig}/m_i \right\rceil + \min_{g \in G}\left(\sum_{h \in I: h < i} p_{hg} \right) + \min_{g \in G}\left(\sum_{h \in I: h > i} p_{hg} \right) \right\} + \sum_{i \in I} q_i, \tag{1.48}$$

where $\lceil \cdot \rceil$ is the smallest integer not less than $\cdot$.

The above lower bound is the sum of total transportation time and maximum over all stages of a stage average workload plus minimum flow time of a part type in all upstream and downstream stages.

1.3.2 A Fast Heuristic for Scheduling Flow Shops with No In-Process Buffers

This subsection presents a single-pass heuristic algorithm for the scheduling of parts through a flexible flow shop with no in-process buffers (Sawik, 1995b). Since there are no buffers between the stages, intermediate queues of parts waiting in the system for their next operations are not allowed. The algorithm, called *Route Idle Time*

Minimization–No Store (RITM-NS), is a special variant of the RITM heuristic designed for scheduling flexible flow shops with finite in-process buffers and presented in the previous subsection. During every iteration a part for loading into the system is chosen as well as its complete processing schedule is determined. The decisions in every iteration are made based on the complete processing schedule determined for each part type waiting for loading. Given a cumulative partial schedule for the first $(l-1)$ parts selected so far, first the best route along the line is found as a sequence

$$[\mu(1, l), \mu(2, l), \ldots, \mu(m, l)]$$

of m machines (one in every stage) with the earliest available times. For each part type waiting for entering the line the complete processing schedule is determined along the best route. To evaluate the processing schedule for each part type k considered for loading, the total duration of machine idle time along the best route is determined. Finally, the part with the smallest total machine idle time is selected for loading and its complete processing schedule is added to the cumulative partial schedule obtained so far. The RITM-NS algorithm can be obtained from the RITM algorithm by removing from the latter all the buffer variables Ξ_{bil}, ξ_{il-1} and $\beta(i, l)$. A brief description of the RITM-NS algorithm is given below:

Algorithm RITM-NS

Step 0. Starting (as for RITM with buffer variables Ξ_{bil} and $\beta(i, l)$ deleted)

Step 1. Selection of a part type for loading (as for RITM with (1.39) replaced by $r_{il} = \max\{f_{il};\ \gamma_{i+1,l-1} - q_i\},\ i = 1, \ldots, m-1$)

Step 2. Determining complete processing schedule for the selected part (as for RITM with buffer variables ξ_{il-1} and $\beta(i, l)$ deleted)

Step 3. Checking the state of completion the production order (as for RITM with buffer variables Ξ_{bil}, ξ_{il-1} and $\beta(i, l)$ deleted).

The computational complexity of the RITM-NS algorithm is $O(mn^2)$, the same as that of the RITM heuristic.

1.3.3 Computational Examples

1.3.3.1 Flexible Flow Shop with Finite In-Process Buffers

First, an illustrative example for a three-stage flow shop with finite in-process buffers is presented. The flow shop (see Fig. 1.9) is made up of $m_1 = 2$ machines in stage $i = 1$, $m_2 = 3$ machines in stage $i = 2$, and $m_3 = 2$ machines in stage $i = 3$.

The number of buffers ahead of stage 2 and stage 3 are $b_2 = 3$ and $b_3 = 3$, respectively. Transportation times between stages are $q_1 = q_2 = 1$.

The production order consists of four part types and their corresponding production requirements (in number of parts) are $n_1 = 8$, $n_2 = 4$, $n_3 = 2$, and $n_4 = 3$. Therefore, the total number of parts to be scheduled is $n = 17$.

Processing times P_{ig} ($i = 1, 2, 3;\ g = 1, 2, 3, 4$) for each stage i and each part type g are shown below:

$$P_{11} = 5,\ P_{21} = 3,\ P_{31} = 7,$$
$$P_{12} = 2,\ P_{22} = 4,\ P_{32} = 6,$$
$$P_{13} = 3,\ P_{23} = 6,\ P_{33} = 1,$$
$$P_{14} = 1,\ P_{24} = 4,\ P_{34} = 2.$$

For the above data the lower bound on makespan is $LBC_{\max} = 51$ (1.48). The RITM heuristic constructs a schedule with the makespan $C_{\max} = 55$ and the loading sequence

$$[2, 2, 2, 2, 4, 3, 1, 3, 4, 4, 1, 1, 1, 1, 1, 1, 1].$$

Instead of using a Gantt chart, the constructed processing schedule is presented in the form of an assignment table (Table 1.3), where the assignment of parts to machines and buffers in each stage is indicated for every period of unit time duration. Numbers of the first and the last period of each assignment are given in the first column of the table.

The minimum values of total idle time t_{k_l} for every iteration $l = 1, \ldots, 17$ are the following:

$$8, 8, 4, 0, 0, 0, 0, 2, 1, 1, 0, 0, 1, 1, 3, 3, 4.$$

In order to better illustrate the solution procedure, some of the computations performed in iteration number 8 of the RITM algorithm are shown below as an example.

The partial loading sequence obtained during the first seven iterations is

$$[j_1, j_2, \ldots, j_7] = [2, 2, 2, 2, 4, 3, 1].$$

The partial schedule for $l = 7$ parts selected so far leads to the following values of times Ξ_{bil} ($b = 1, \ldots, b_i;\ i = 2, 3$) and Γ_{jil} ($j = 1, \ldots, m_i;\ i = 1, 2, 3$) at which, respectively, buffers and machines are available for assignment,

In stage 1: $\Gamma_{117} = 10,\ \Gamma_{217} = 7$

In stage 2: $\Xi_{127} = 11,\ \Xi_{227} = 7,\ \Xi_{327} = 9,\ \Gamma_{127} = 14,\ \Gamma_{227} = 11,\ \Gamma_{327} = 15$

In stage 3: $\Xi_{137} = 20,\ \Xi_{237} = 20,\ \Xi_{337} = 21,\ \Gamma_{137} = 22,\ \Gamma_{237} = 28$

The resulting earliest availability times: ξ_{ik} for buffer $\beta(i, 8)$ ahead of stage i ($i = 2, 3$), and γ_{ik} for machine $\mu(i, 8)$ in stage i ($i = 1, 2, 3$) are shown below:

$$\xi_{27} = 7,\ \xi_{37} = 20,\ \gamma_{17} = 7,\ \gamma_{27} = 11,\ \gamma_{37} = 22.$$

Therefore, the best route along the line is the following sequence of machines and buffers available at the earliest times:

$$[\mu(1, 8), \beta(2, 8), \mu(2, 8), \beta(3, 8), \mu(3, 8)] = [2, 2, 2, 1, 1].$$

In iteration 8, part k_8 for loading at position $l = 8$ has to be selected from among 10 remaining parts of three types $g = 1, 3, 4$. A complete processing schedule

Table 1.3 A Heuristic Schedule

	Assignment of parts to machines and buffers													
	Stage 1		Stage 2						Stage 3					
	Machines		Buffers			Machines			Buffers			Machines		
Periods from–to	1	2	1	2	3	1	2	3	1	2	3	1	2
1–3	2	2											
4	2	2				2	2						
5	4	3				2	2						
6	1	3	2			2	2	2					
7	1	3	2	4		2	2	2					
8	1	3				2	4	2					
9	1	3			3	2	4	2				2	2
10	1	3				2	4	3				2	2
11	4	4				2	4	3			2	2	2
12	1	1				1	3	3			2	2	2
13–14	1	1	4	4		1	3	3	2	4	2	2	2
15	1	1	4			4	3	3		4		2	2
16	1	1				4	3	4		4		2	2
17	1	1				4	3	4	3	4	1	2	2
18–19	1	1		1	1	4	3	4	3	4	1	2	2
20	1	1				1	1	4	3	4	1	2	2
21	1	1				1	1		3	4	1	4	3
22	1	1				1	1		3	4	4	4	1
23	1	1				1		1		4	4	3	1
24–25	1	1				1		1	1	1	4	4	1
26	1	1				1			1	1		4	1
27	1								1	1	1	4	1
28	1						1	1	1	1	1	1	1
29–30	1						1	1	1		1	1	1
31	1							1	1		1	1	1
32								1	1	1	1	1	1
33						1		1	1	1	1	1	1
34–35						1			1	1	1	1	1
36										1	1	1	1
37–41									1	1	1	1	1
42									1		1	1	1
43–48									1			1	1
49												1	1
50–55												1	

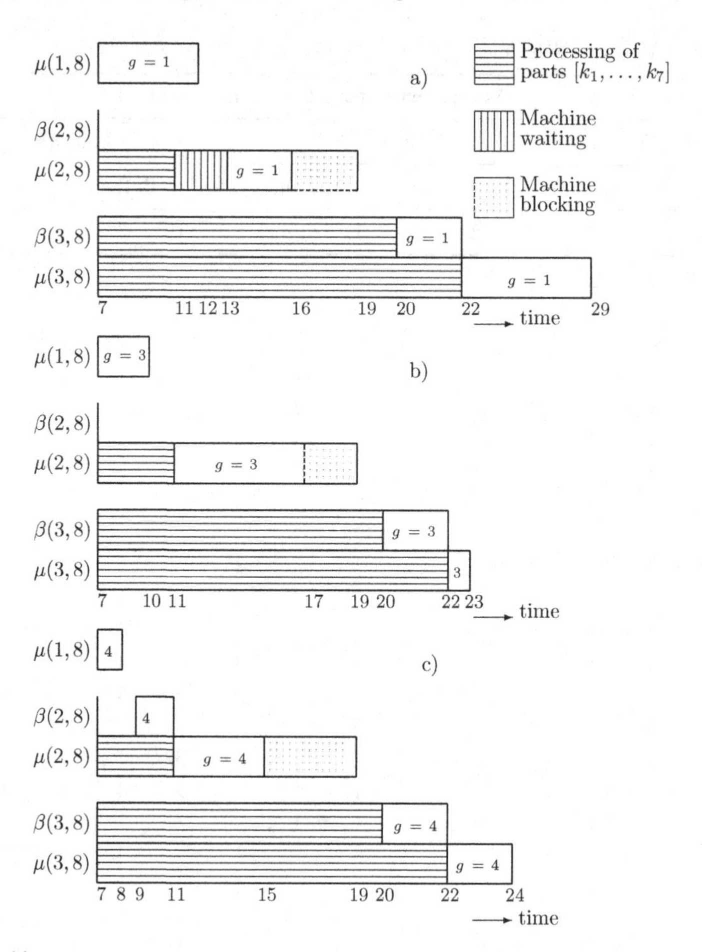

Figure 1.11 Processing schedules for part types $g = 1, 3, 4$ available for assignment to position 8 of the input sequence.

obtained along the best route for each of those three part types is shown in Figure 1.11, for part type 1 (Fig. 1.11 (a)), part type 3 (Fig. 1.11 (b)), and part type 4 (Fig. 1.11 (c)), respectively. The corresponding total idle times are $t_1 = 5$, $t_3 = 2$, $t_4 = 4$, and hence the minimum is $t_3 = 2$. Therefore, part type $g_8 = 3$ is selected for loading at position $l = 8$, and in iteration 8 its complete processing schedule shown in Figure 1.11 (b) is added to the cumulative partial schedule of the first seven parts.

1.3.3.2 Flexible Flow Shop with No In-Process Buffers

Below an illustrative example is presented for the same production order and the three-stage flexible flow shop with no in-process buffers, shown in Figure 1.12. The flow shop is made up of $m_1 = 2$ machines in stage $i = 1$, $m_2 = 3$ machines in stage $i = 2$, and $m_3 = 2$ machines in stage $i = 3$. Transportation times between stages are $q_1 = q_2 = 1$.

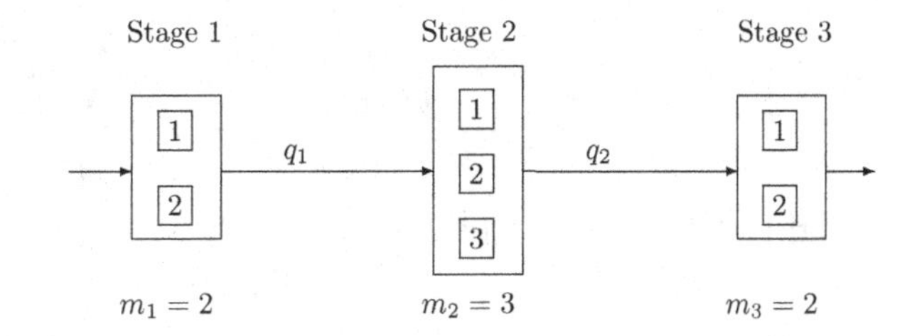

Figure 1.12 A three-stage flexible flow shop with no in-process buffers.

The lower bound on makespan is $LBC_{\max} = 51$ (1.48). The RITM-NS heuristic constructs a schedule with the makespan $C_{\max} = 52$ and the loading sequence

$$[4, 4, 2, 1, 3, 2, 3, 1, 1, 1, 1, 2, 1, 1, 1, 2, 4].$$

The optimal schedules for the two examples were obtained using the MIP model FPB with constraints (1.16) replaced by (1.22) to allow for inclusion of the transportation times between successive stages. The Gantt charts for the two optimal schedules are shown in Figures 1.13 and 1.14, where the three machine stages and the two buffers stages in Figure 1.13 are denoted by M1, M3, M5 and B2, B4, whereas the three machine stages in Figure 1.14 are marked with M1, M2, M3. The figures indicate that the optimal schedules have the same length, $C_{\max} = 52$.

It is interesting to compare the two heuristic schedules constructed for the two different flexible flow shops (with finite in-process buffers or with no in-process buffers) and the same production order. Unlike for the optimal schedules shown in

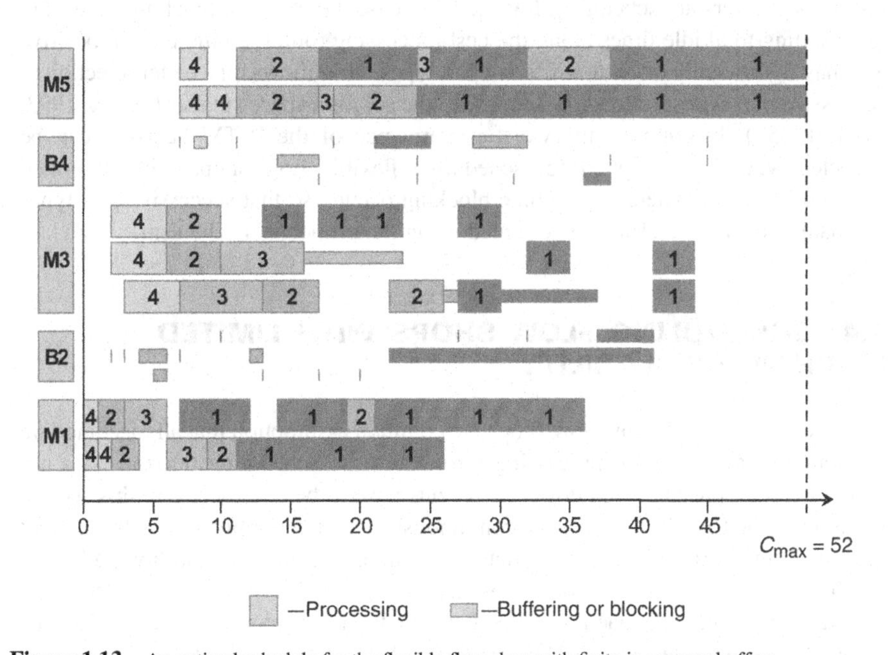

Figure 1.13 An optimal schedule for the flexible flow shop with finite in-process buffers.

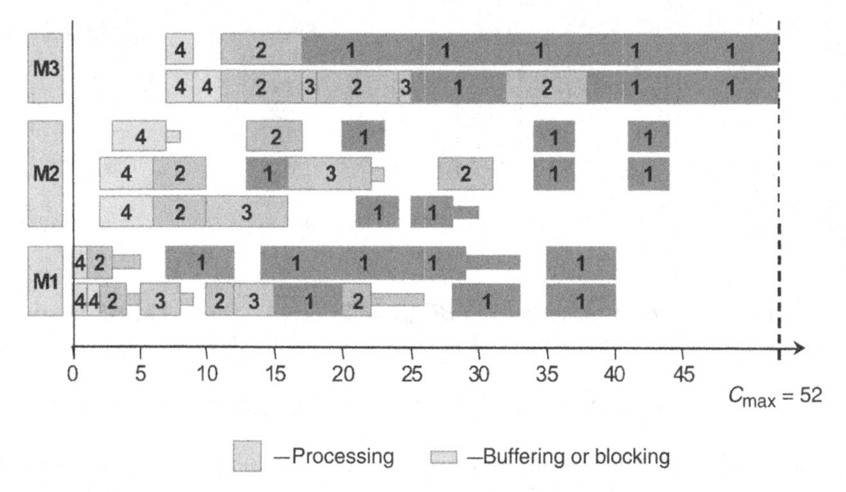

Figure 1.14 An optimal schedule for the flexible flow shop with no in-process buffers.

Figures 1.13 and 1.14, the heuristic schedule length for the more severe case, that is, with no in-process buffers, is shorter. The result of the RITM-NS heuristic outperforms that of the RITM heuristic in terms of the shorter makespans obtained. Such anomalies are sometimes encountered when heuristic algorithms are applied for scheduling, e.g., Blażewicz et al. (1994). As the machine blocking phenomenon may occur more frequently, the RITM heuristic mechanism for the selection of successive parts for loading may perform better when flexible flow shops with no in-process buffers are scheduled. For the latter case the machine blocking times and the resulting total idle times along the best processing routes are more often positive and may significantly differ for different part types. This allows for a better selection of successive part types for loading and makes the solution space smaller (Sawik, 1993, 1994, 1995b). In contrast, the worst performance of the RITM heuristic can be expected when it is applied for scheduling flexible flow shops with unlimited in-process buffers, where no machine blocking occurs, so that successive part types for loading cannot be clearly determined using the proposed mechanism.

1.4 SCHEDULING FLOW SHOPS WITH LIMITED MACHINE AVAILABILITY

In the scheduling of flexible flow shops the common assumption that all machines are continuously available for processing throughout the scheduling horizon does not apply if certain prescheduled downtime events have to be considered. In this section the assumption that all machines are continuously available for processing throughout the scheduling horizon has been restricted and limited machine availability is allowed.

The interval of machine nonavailability (scheduled downtime) is defined to be a period of time when the machine is not available to perform its intended function due to planned downtime events. The scheduled downtime state of the machine may include various planned events such as preventive maintenance, production tests, change of consumables/chemicals, setups, or prescheduled production runs to be completed during the upcoming periods.

Model *FPB* can be easily enhanced for scheduling with limited machine availability where machine downtimes are viewed as dummy parts preassigned to time intervals and machines. Each dummy part requires processing on one machine only. Then the real parts that must be completed in minimum time can be processed within the remaining free processing intervals.

Let $L \subset K$ be the set of dummy parts representing machine downtimes and let p_{il}, $a_{ijl} - p_{il}$, and a_{ijl} denote, respectively, the duration, the beginning, and the end of downtime $l \in L$ on machine $j \in J_i$ in stage i.

Model FPBD: *Scheduling Flow Shops with Parallel Machines, Finite In-Process Buffers, and Machine Downtimes*

Minimize (1.1) subject to

1. *Part Completion Constraints*: (1.2), (1.3)
2. *Part Departure Constraints*: (1.14), (1.15)
3. *No Buffering Constraints*: (1.16)
4. *Part Noninterference Constraints*: (1.20), (1.21)
5. *Processor Assignment Constraints*:
   - in every stage with parallel processors each real part is assigned to exactly one processor,
   - each dummy part is preassigned to an appropriate machine,

$$\sum_{j \in J_i} x_{ijk} = 1; \ \ i \in I, k \in K \setminus L \tag{1.49}$$

$$x_{ijk} = 1; \ \ i \in I, j \in J_i, k \in L\colon a_{ijk} > 0 \tag{1.50}$$

6. *Dummy Parts Completion and Departure Constraints*:
   - the completion and departure times of each dummy part are prefixed,

$$c_{ik} = a_{ijk}; \ \ i \in I, j \in J_i, k \in L\colon a_{ijk} > 0 \tag{1.51}$$

$$d_{ik} = a_{ijk}; \ \ i \in I, j \in J_i, k \in L\colon a_{ijk} > 0 \tag{1.52}$$

7. *Stages Bypassing Constraints by Dummy Parts*:
   - each dummy part bypasses all the stages that are not in its processing route,

$$c_{ik} = c_{i-1k}; \ \ i \in I, k \in L\colon i > 1, p_{ik} = 0 \tag{1.53}$$

$$d_{ik} = d_{i-1k}; \ \ i \in I, k \in L\colon i > 1, p_{ik} = 0 \tag{1.54}$$

$$x_{ijk} = 0; \ \ i \in I, j \in J_i, k \in L\colon a_{ijk} = 0 \tag{1.55}$$

8. *Maximum Completion Time Constraints*:
   - the schedule length is determined by the latest completion time of some real part in the last stage,
   - the maximum completion time is bounded from below by the average workload of each processing stage and the minimum processing time of

a part in all remaining stages,

$$c_{mk} \leq C_{\max}; k \in K \backslash L \tag{1.56}$$

$$C_{\max} \geq \sum_{k \in K \backslash L} p_{ik}/m_i + \min_{k \in K \backslash L} \left(\sum_{h<i} p_{hk} \right) + \min_{k \in K \backslash L} \left(\sum_{h>i} p_{hk} \right); i \in I. \tag{1.57}$$

9. *Variable Nonnegativity and Integrality Conditions*: (1.7) to (1.9), (1.13), (1.19).

The mixed integer program *FPBD* is a general formulation and allows for scheduling with an arbitrary pattern of machine availability.

1.5 COMPUTATIONAL EXAMPLES

In this section numerical examples and some computational results are presented to illustrate possible applications of the MIP models proposed for scheduling flexible flow shop with continuous or with limited machine availability.

The flexible flow shop configuration for the examples is provided in Figure 1.15 and it represents the front of a surface mount technology line for printed wiring board assembly in electronics manufacturing (for more details, see Chapter 2). The flow shop consists of $m = 5$ stages, where stage $i = 1$ is a single machine for screen printing, each stage $i = 3, 5$ represents two parallel machines for automatic placement of components, and each stage $i = 2, 4$ represents two intermediate buffers.

The production order consists of $n = 30$ parts of three types, 10 parts of each type. The processing times for each part type are shown below (for the buffer stages $i = 2, 4$ all processing times are equal to zero)

$$\begin{bmatrix} 10, & 10, & 10 \\ 0, & 0, & 0 \\ 56, & 59, & 74 \\ 0, & 0, & 0 \\ 53, & 54, & 55 \end{bmatrix}$$

The processing schedules were determined for the following two cases:

- *Batch scheduling*, where 10 parts of each type are assembled and the parts of a given type are scheduled consecutively. The optimal sequence of part types is obtained along with the optimal schedule for all parts.

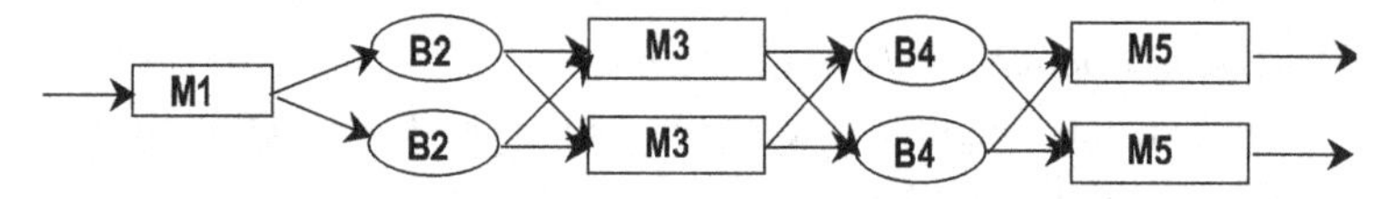

Figure 1.15 A flexible flow shop with parallel machines and finite in-process buffers.

- *Cyclic scheduling*, where 10 parts of each type are assembled and the parts of different types are scheduled alternately in a cyclic order. The optimal cycle of part types is obtained along with the optimal schedule for all parts.

1.5.1 Scheduling with Continuous Machine Availability

In this subsection the optimal schedules are presented for the example problem with continuous machine availability.

The processing schedules obtained are shown on Gantt charts in Figure 1.16, where parts of types 1, 2, and 3 are indicated with different shading. Processor blocking is indicated with a narrow bar. The optimal sequence of part types for batch scheduling is 3, 2, 1, and for cyclic scheduling 1, 2, 3. The solution values obtained are as

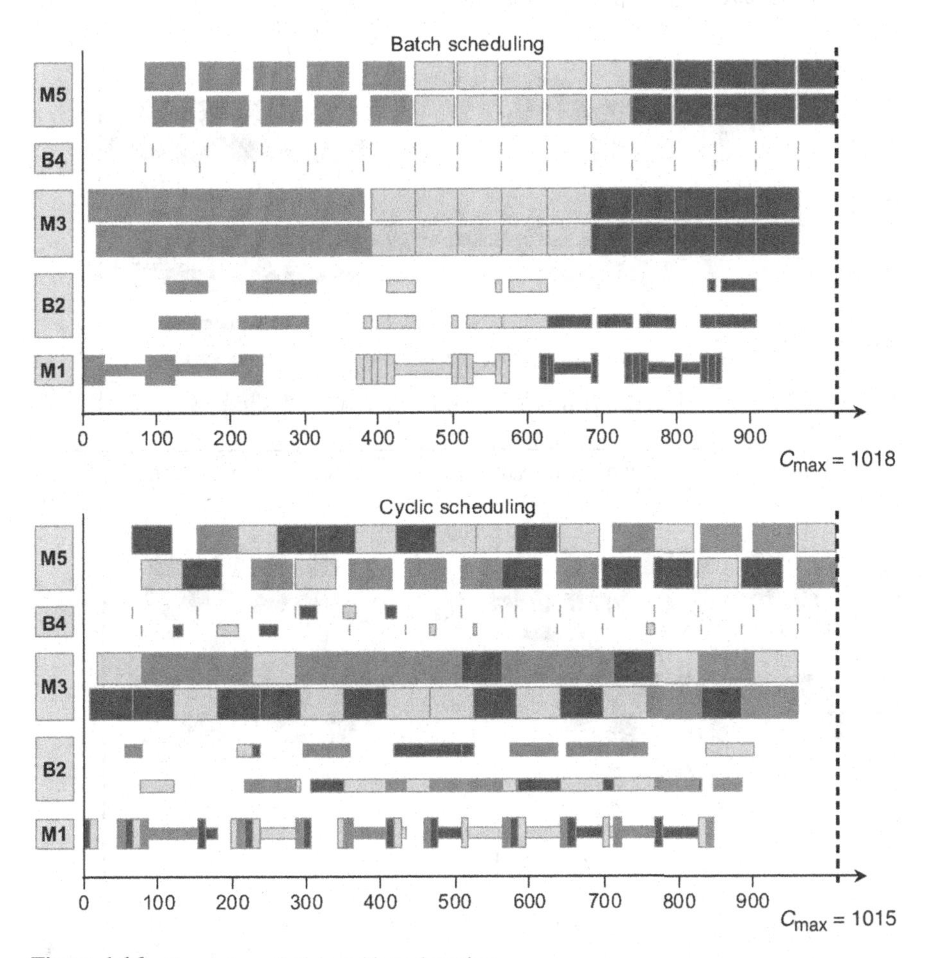

Figure 1.16 Processing schedules with no downtime.

follows: $C^*_{max} = 1018$ for batch scheduling, $C^*_{max} = 1015$ for cyclic scheduling. Model *FPB* was used to find the schedules.

1.5.2 Scheduling with Limited Machine Availability

In this subsection processing schedules with limited machine availability are presented for the example problem with a single interval of nonavailability on one machine only. In the example a nonavailable machine was either a single machine in stage 1 or parallel machine 1 in stage 3 or parallel machine 1 in stage 5, and the downtime state occurs in one of the following time intervals: [0,400), [400,800), [800,1200).

Three examples of processing schedules with downtime [400,800) obtained for three different locations in the line of nonavailable machines are shown on Gantt charts in Figures 1.17 to 1.19. Both batch and cyclic scheduling modes are illustrated. Model *FPBD* was used to find the schedules.

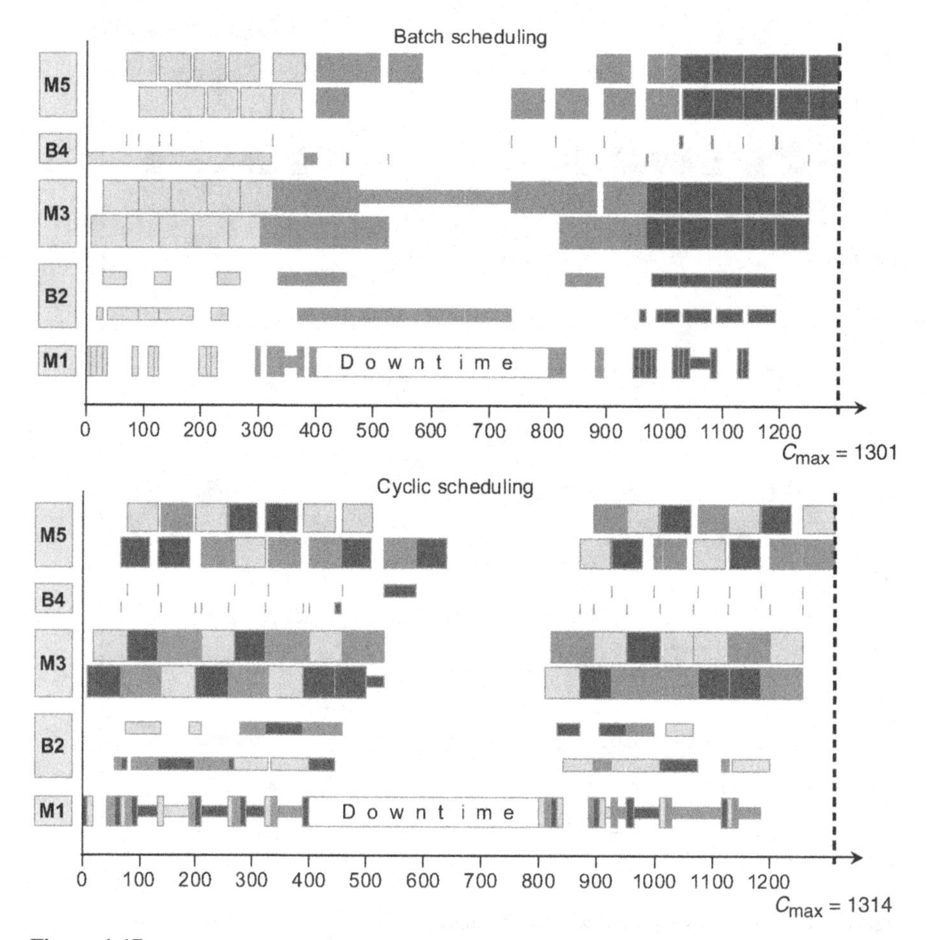

Figure 1.17 Processing schedules with downtime [400,800) in stage 1.

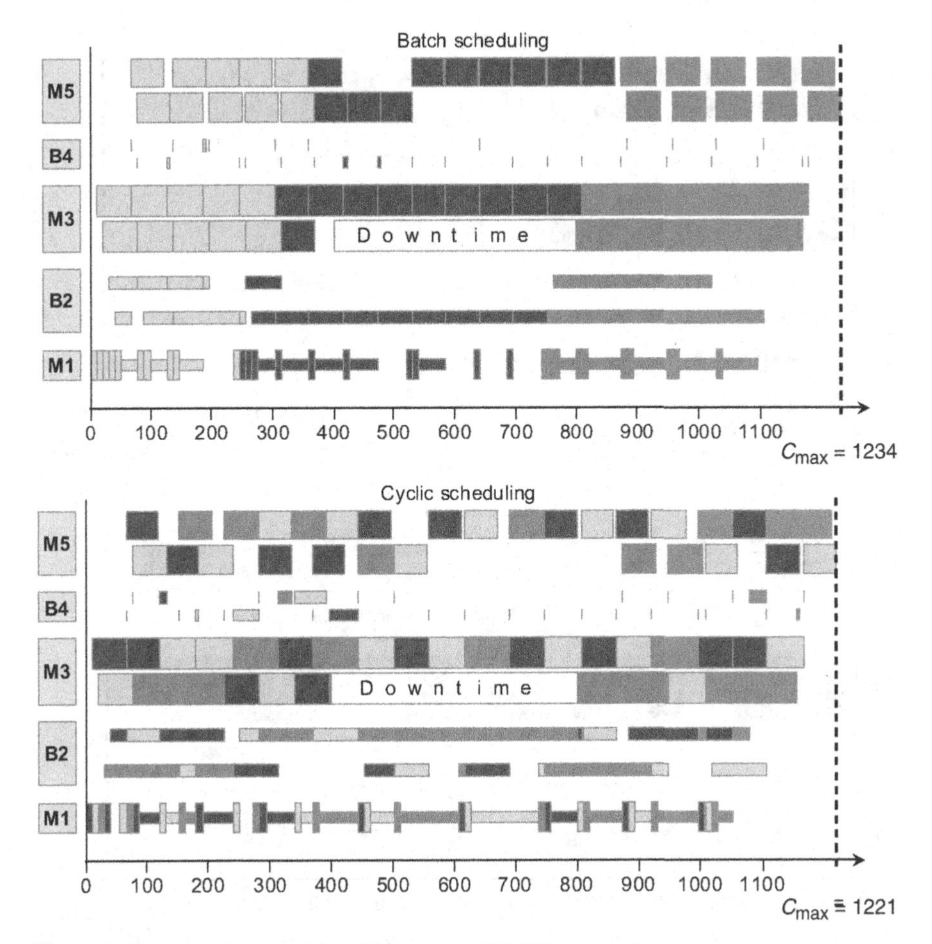

Figure 1.18 Processing schedules with downtime [400,800) in stage 3.

The characteristics of mixed integer programs *FPB* and *FPBD* for the example problems corresponding to Gantt charts in Figures 1.16 to 1.19, and the solution results are summarized in Table 1.4. The size of MIP models for the example problems is represented by the total number of variables, *Var.*, number of binary variables, *Bin.*, number of constraints, *Cons.*, and number of nonzero coefficients, *Nonz.*, in the constraint matrix. The last two columns of Table 1.4 give the lower bound *LB* and the makespan C^*_{max} or C^d_{max}, respectively, for continuous or limited machine availability.

Table 1.5 shows the relative increase of makespan $(C^d_{max} - C^*_{max})/C^*_{max}$ due to limited machine availability for various locations in the line of the downtime state. The computational results presented in Table 1.5 indicate that the relative increase of makespan depends on the position in the line of the nonavailable machine and position in the schedule of the corresponding interval of nonavailability. The more upstream the position of the nonavailable machine and the earlier the downtime state occurs, the greater is the relative increase of makespan. For a downstream machine, however, a later downtime state results in a greater relative increase of makespan.

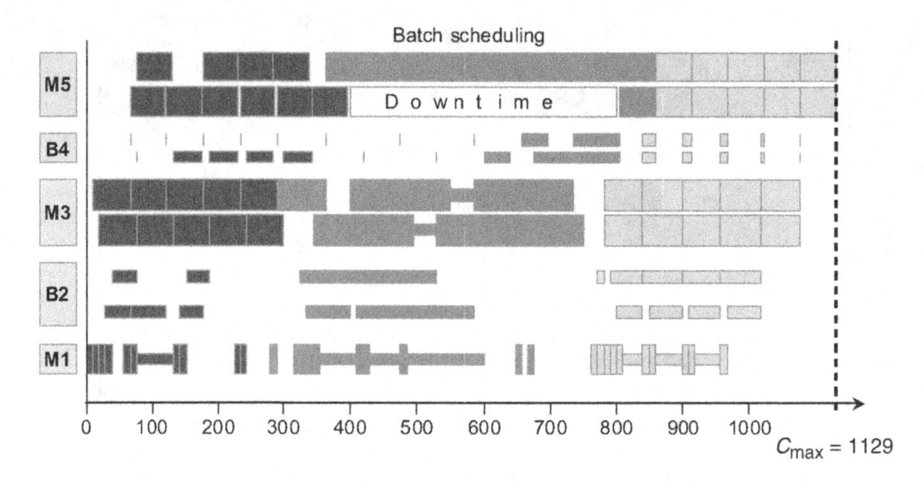

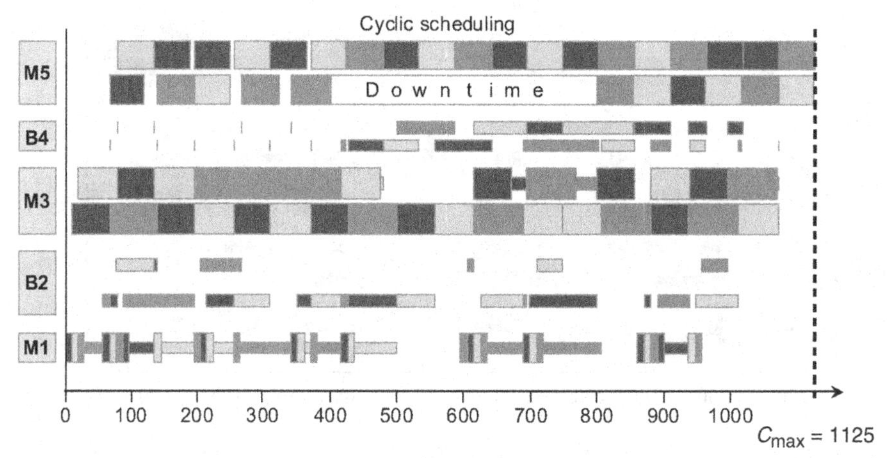

Figure 1.19 Processing schedules with downtime [400,800) in stage 5.

Table 1.4 Characteristics of *FPB*/*FPBD* Models and Solution Results

Problem	Var.	Bin.	Cons.	Nonz.	*LB*	C_{max}
Figure 1.16/Batch	841	540	8581	37,164	1008	1018
Figure 1.16/Cyclic	787	486	8311	35,544	1008	1015
Figure 1.17/Batch	871	570	8502	36,894	1232	1301
Figure 1.17/Cyclic	817	516	8178	35,116	1232	1314
Figure 1.18/Batch	871	570	8637	37,224	1208	1234
Figure 1.18/Cyclic	817	516	8367	35,604	1208	1221
Figure 1.19/Batch	871	570	8761	37,615	1076	1129
Figure 1.19/Cyclic	817	516	8491	35,965	1076	1125

Table 1.5 Relative Increase of Makespan

	$(C^d_{\max} - C^*_{\max})/C^*_{\max}[\%]$					
	Beginning: 0–400		Middle: 400–800		Final: 800–1200	
Downtime Scheduling Mode	Batch	Cyclic	Batch	Cyclic	Batch	Cyclic
Downtime in Stage 1	39.33	39.21	27.92	28.82	30.68	32.25
Downtime in Stage 3	20.06	19.02	21.93	20.49	17.60	15.29
Downtime in Stage 5	6.69	7.45	10.91	10.29	13.37	16.86

The computational experiments with various flow shop configurations have indicated that relative increase of makespan due to the downtime state depends on the location and capacity of the intermediate buffers as well as the relative position in the line of the stage with a nonavailable machine and the bottleneck stage with the largest workload. Sufficient capacity of in-process buffers immediately preceding and succeeding the downtime machine may significantly reduce the relative makespan increase. For instance, if in the example problem, capacity m_2 of the buffer stage 2 (see Fig. 1.15) increases from 2 to 12, then the downtime [400,800) of machine in stage 1 is fully compensated and the makespan decreases from 1301 to 1018 and from 1314 to 1015, respectively for batch and cyclic scheduling (see, Fig. 1.20). The additional buffer capacity results in the decrease of schedule length to the corresponding optimal value attained for continuous machine availability.

1.6 COMMENTS

MIP or IP models for scheduling, in particular flow shop scheduling, were originally devised around 1960. The models for the regular flow shop scheduling problem with single machines, unlimited buffers between successive machines, and permutation schedules can be divided into two broad classes depending on the type of variables and constraints used to assign jobs to various sequence positions. Wagner (1959) proposed IP formulation with sequence-position binary variables and machine idle times before and after processing jobs in each sequence position, whereas Manne (1960) introduced an alternative approach based on sequencing (precedence) variables and job noninterference disjunctive constraints, or their algebraic equivalents, to control the assignment of jobs to various sequence positions. The Manne model assures that only one of each pair of constraints can hold, so that job k either precedes job l somewhere in the processing sequence, or it does not, thus implying that job l precedes job k. For the assignment of jobs to sequence positions, Wilson (1989) uses the classic assignment problem and inequality constraints similar to those in the Manne model, to insure that the start of each job on each machine is no earlier than its finish on the previous machine and that the job in each position in the sequence does not start on a machine until the job in the preceding position in the sequence has completed its

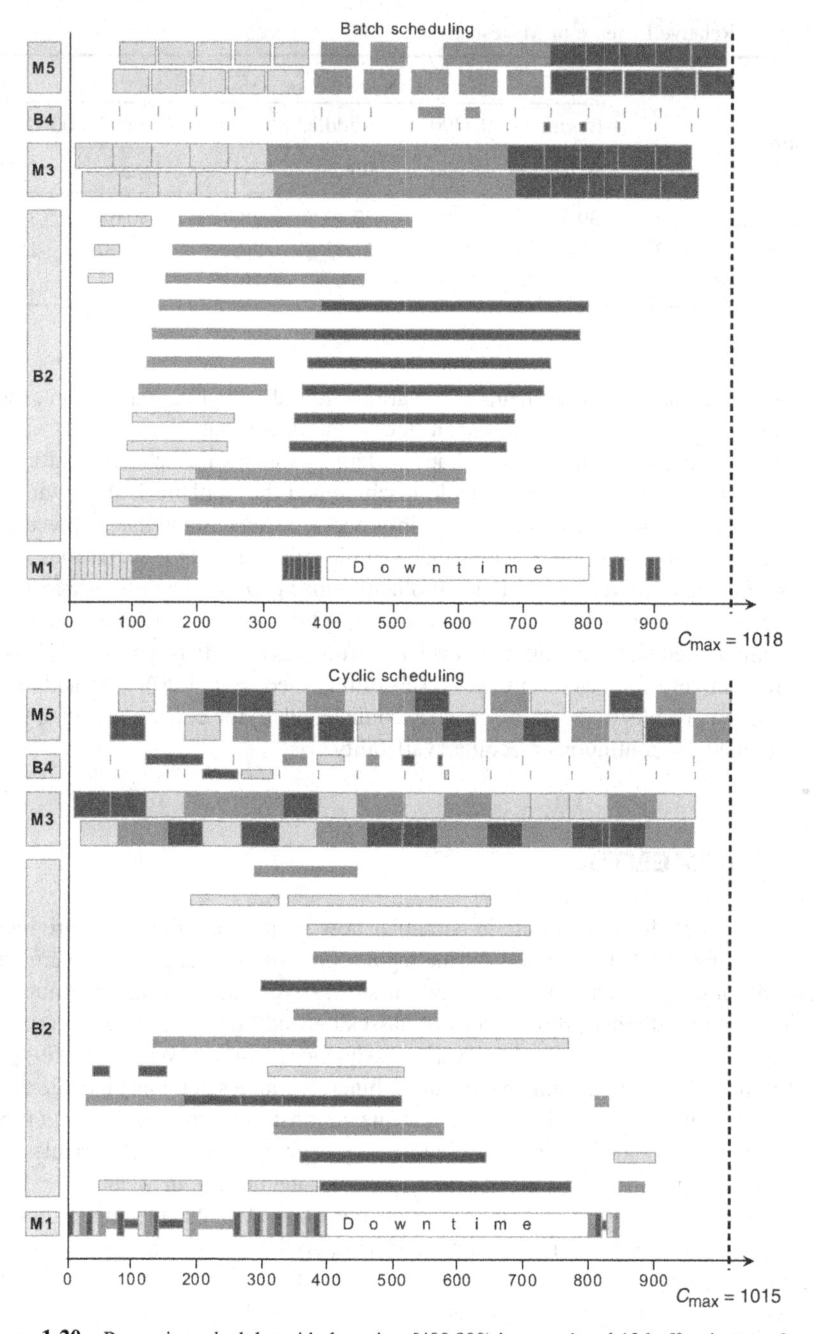

Figure 1.20 Processing schedules with downtime [400,800) in stage 1 and 12 buffers in stage 2.

processing on that machine. The sequence-position models are considered to be more efficient than the precedence models, for example, Tseng et al. (2004).

The first research papers about flexible flow shop scheduling appeared around 1970. Salvador (1973) published one of the pioneer papers on flexible flow shop scheduling by modeling the production system in the synthetic fibers industry as a no-wait scheduling of a flexible flow shop. Garey and Johnson (1979) showed that the flexible flow shop scheduling problem with makespan objective is NP-complete, and over the years the NP nature of most flexible flow shop scheduling problems has been shown. Therefore, a large number of heuristics and approximation algorithms have been proposed for scheduling different flexible flow shop configurations.

Blocking scheduling (e.g. Abadi et al., 2000) has received considerable attention from the study of McCormick et al. (1989), who consider a regular flow shop with finite capacity buffers between machines. A unified modeling approach has been adopted with the buffers viewed as machines with zero processing times to convert the flow shop scheduling problem with buffers into one with no buffers but with blocking, for example, McCormick et al. (1989) and Sawik (2000a).

The literature on cyclic (periodic) scheduling of Minimal Part Set (MPS) in flow shops is concentrated on the study of unpaced (with no exogenous limit on the cycle time) asynchronous (e.g., McCormick et al., 1989; Deane and Moon, 1992, Karabati and Kouvelis, 1996) or synchronous (e.g., Kouvelis and Karabati, 1999) configurations. MPS scheduling in paced assembly lines in which the cycle time is set exogenously, has been discussed in the book by Scholl (1998).

Double-pass reentrant flow shops, where a part visits a set of stages more than once were considered by various researcher, for example, Graves et al. (1983) and Tirpak (2000).

During the last decade, research in this area has focused on more realistic problems, including sequence-dependent set ups on machines (e.g., Jain et al., 1996; Liu and Chang, 2000; Kurz and Askin, 2004), limited availability of machines (e.g., Schmidt, 2000), etc. Reviews on flexible flow shop scheduling can be found in Vignier et al. (1999) or Linn and Zhang (1999). Kis and Pesch (2005) review exact methods for flexible flow shop scheduling problem with parallel identical machines to minimize makespan or total flow time, while H. Wang (2005) classifies the papers according to the solution procedure adopted (optimal, heuristics, and artificial intelligence). Finally, Quadt and Kuhn (2007) propose a taxonomy for flexible flow shop scheduling procedures focusing on heuristic procedures. The most recent comprehensive reviews can be found in Ribas et al. (2010) and Ruiz and Vazquez-Rodriguez (2010).

Introduction of limited machine availability constraints further complicates a flexible flow shop blocking scheduling problem. For example, the regular flow shop scheduling problem for two machines that can be solved in polynomial time by Johnson's rule (Johnson, 1954) becomes already NP-complete if there is a single interval of nonavailability on one machine only (see, Schmidt, 2000).

It can be observed that a lot of research work on scheduling flexible flow shops considers makespan minimization, while other common criteria, such as flow time (e.g., Rajendran and Chaudhuri, 1992; Guinet and Solomon 1996) or total tardiness (e.g., G. C. Lee and Kim, 2004; Sawik, 2005a) are less studied. Also it is observed that little research has been done concerning multicriteria problems, for example, Alfieri (2009).

With respect to solution approaches, branch and bound and MIP are the most frequently used exact procedures. At the time of Brah and Hunsucker (1991), instances of a very limited size with up to eight jobs and two stages with three parallel machines each could be solved within several hours of CPU time. More recently, the best branch and bound algorithms proposed are efficient to solve instances with 15 or 20 jobs and at most five stages. It can also be observed that the scheduling problems become more complex when the number of machines increases and when there is no single bottleneck stage. When a single bottleneck exists, then tight lower and upper bounds restrict the search effort required to find an optimal solution.

The constructive heuristics RITM and RITM-NS presented in Section 1.3 for scheduling flexible flow shops with machine blocking were developed by Sawik (1993, 1994, 1995b, 1999). The two algorithms belong to the best available constructive heuristics for scheduling flexible flow shops with machine blocking, for example, Nowicki and Smutnicki (1996) and Ribas et al. (2010).

EXERCISES

1.1 Denote by $m \leq \overline{m}$, the variable number of processing stages, not greater than $\overline{m}$, of a flow shop with single machines and infinite in-process buffers and by $\overline{C}$, an upper bound on the schedule length. Formulate a mixed integer program for finding the minimum-length flow shop such that a given set of part types is completed by the common due date $\overline{C}$.

1.2 Denote by $m_i \leq \overline{m}_i$, the variable number of parallel machines in stage $i = 1, \ldots, m$, not greater than $\overline{m}_i$, of an m-stage flow shop with parallel machines and infinite in-process buffers and by $\overline{C}$, an upper bound on the schedule length. Formulate a mixed integer program for finding the flow shop with minimum total number of parallel machines such that a given set of part types is completed by the common due date $\overline{C}$.

1.3 Consider batch scheduling in a flexible flow shop with parallel machines and finite in-process buffers, in which a different due date is given for each part type. Formulate the mixed integer program for scheduling the batches of part types to minimize the number of tardy part types.

1.4 In the computational examples in Section 1.3.3, Figures 1.13 and 1.14 indicate that the makespan of optimal schedule is identical for the flexible flow shops with finite in-process buffers and with no in-process buffers. Explain why the optimal makespans are identical.

1.5 Consider problem *FPBD* of scheduling a flow shop with parallel machines, finite in-process buffers, and machine downtimes. Suppose that for a particular problem instance no feasible solution exists with the schedule length not longer than a given bound $\overline{C}$, however, one may obtain a feasible solution by increasing the capacity of in-process buffering. Let $m_i \leq \overline{m}_i$ be the variable number of in-process buffers in stage $i \in I$ (such that $\sum_{k \in K} p_{ik} = 0$), not greater than $\overline{m}_i$. Formulate the mixed integer program for scheduling a flow shop with parallel machines, variable capacity of in-process buffers, and machine downtimes, such that the schedule length is not longer than $\overline{C}$ and the total capacity of in-process buffers is minimized.

Chapter 2

Scheduling of Surface Mount Technology Lines

2.1 INTRODUCTION

This chapter provides the reader with MIP formulations for scheduling surface mount technology (SMT) lines. The SMT has been widely used for the last decade in the manufacture of printed wiring boards (PWBs). SMT assembly involves the following basic processes: screen printing of solder paste on the bare board, automated placement of components, robotic or manual placement of large components, and solder reflow. A typical SMT line consists of several assembly stations in series and/or in parallel, separated by finite intermediate buffers. A conveyor system transfers the boards between the stations.

An SMT line is a practical example of a flexible flow shop. The line typically produces several different board types. Each board must be processed by at most one machine in each stage because of different routings for different board types. A board that has completed processing on a machine in some stage is transferred either directly to an available machine in the next stage or to a buffer ahead of that stage.

The two major short-term planning problems in electronics assembly are loading and scheduling. Given a mix of boards to be produced, the objective of the loading problem is to allocate assembly tasks and component feeders among the placement stations with limited working space, so as to balance the station workloads. In contrast, the objective of the scheduling problem is to determine the detailed sequencing and timing of all assembly tasks for each individual board, so as to maximize the line's productivity, which may be defined in terms of throughput or the assembly schedule length (makespan) for a mix of board types. The limited intermediate buffers between stations result in a *blocking scheduling* problem, where a completed board may remain on a machine and block it until a downstream machine becomes available. This prevents another board from being processed on the blocked machine.

In practice, scheduling of an SMT line is based on daily demands and a simple approach to executing a daily production plan is the use of batch scheduling, where

Scheduling in Supply Chains Using Mixed Integer Programming. By Tadeusz Sawik

boards of one type are processed consecutively. In a high-volume production, the production plan is often split into several identical sets of smaller batches of boards that are scheduled repeatedly. The smallest possible set of boards in the same proportion as the daily board mix requirements is called the Minimal Part Set (MPS), see, Section 1.2.6.

The following MIP models for scheduling SMT lines are presented in this chapter:

GESCH for general scheduling of an SMT line, where any input sequence of boards is allowed.

BASCH for batch scheduling of an SMT line, where boards of a given type are scheduled consecutively.

The proposed MIP models are capable of addressing the two basic questions of SMT line operations:

- What should be the sequence of boards entering the line?
- What should be the assignment of boards to parallel stations and buffers?

The above models are based on the MIP formulations proposed in Chapter 1. However, the formulation for general scheduling is an improvement over the previous models and incorporates new cutting constraints on decision variables. Furthermore, a new formulation is presented for batch scheduling with various specific cutting constraints. The formulations proposed can be applied for constructing optimal assembly schedules by using commercially available software for MIP. This is illustrated with numerical examples modeled after real-world SMT lines. In addition, Section 2.5 presents an improvement heuristic for scheduling a typical SMT line, subject to various additional requirements, which combines the tabu search technique and dispatching scheduling based on a pseudodynamic programming scheme.

2.2 SMT LINE CONFIGURATIONS

An automated SMT line includes three different processes in the following sequence: solder printing, component placement, and solder reflow. For the process of solder printing and reflow soldering, one machine per line is needed. The number of machines for the placement process can vary and depends on the number and type of components on the boards to be assembled. Basically, these electronic components can be divided into two major groups: small chip parts and large fine pitch parts. It is assumed that an SMT line contains at least one machine capable of placing each component group. The components can be assembled on one or both sides of a PWB. Various configurations of SMT lines can be encountered in electronics assembly. For example, there are single-pass lines, where one pass through the line is required to complete a board, or double-pass reentrant lines, where the double-sided boards run twice through the same line, first to assemble the bottom side and then to assemble the top side. In addition, each PWB can be transported and assembled as a set of boards in a panel.

High complexity of an SMT line scheduling problem is mainly caused by:

- Limited buffers that result in machine blocking and require separate board completion and board release time variables to be introduced for each board, machine, and buffer.
- Parallel processors that require additional binary assignment variables to be introduced for each board, machine, and buffer.
- Simultaneous assembly of different board types.
- Medium to high volume production that requires a significant number of boards to be included in the scheduling horizon.

The following are the basic SMT line configurations found in electronics assembly factories (e.g., Sawik et al., 2000, 2002):

1. SMT lines for single-sided boards
 1.1. SMT line with single stations
 1.2. SMT line with parallel stations
 1.3. SMT line with dual-conveyor
2. SMT lines for double-sided boards
 2.1. Single-pass SMT line
 2.2. Double-pass SMT line

2.2.1 SMT Lines for Single-Sided Boards

A simple SMT line with single stations is shown in Figure 2.1. In this basic configuration, all machines in the SMT line are connected in series. The line consists of a PWB loader, a solder printer, a reflow oven, and two placement machines (one for small and one for fine pitch components). The placement machines have to be adjusted to the product running by controlling the conveyor width, installing the proper feeders for components, as well as selecting the nozzle configuration to pick up the required components. Machines are separated by buffers and connected with conveyors. The assembly process is as follows: A tote of bare (preassembly) PWBs is brought to the beginning of the line, and a material loader loads each PWB separately on the conveyor. Each PWB is transported by the conveyor system through each machine in the line and then is stored again in a tote box. The loader and the tote box are used as the input and output buffers of the line. There are external buffers in front of and behind each placement machine, except the last one. In addition, every placement machine

Loader	Solder	External	BUF	Placement	BUF	External	BUF	Placement	BUF	Oven	Tote
	Printer	Buffer	FER	Machine	FER	Buffer	FER	Machine	FER		box

Figure 2.1 A simple SMT line with single stations.

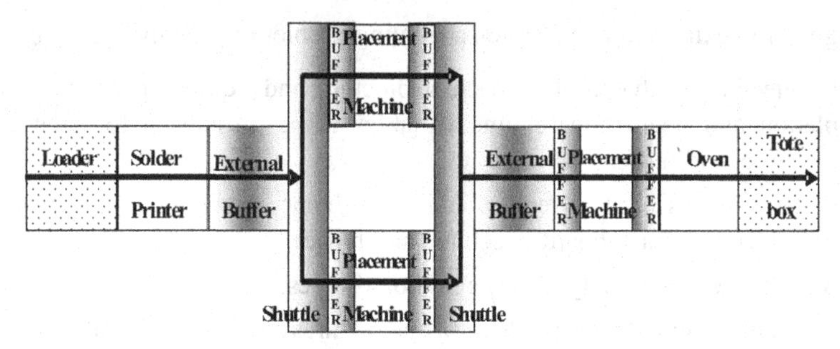

Figure 2.2 SMT line with parallel stations.

has its own internal input and output buffers of a fixed capacity. The internal and external buffers are shown in Figure 2.1 in gray.

A parallel station shown in Figure 2.2 would assemble twice as many parts as the single station in Figure 2.1. Because of that, the nonproductive operations of board loading and unloading would represent a smaller fraction of the total assembly time at the station, and the average time per placement would decrease. Therefore, this so called "placement density effect" tends to increase throughput in this line configuration. A line with parallel stations also provides redundancy in the system. If a machine breakdown occurs, processing can continue because an alternative routing exists.

To further reduce the effects of load and unload times and achieve higher throughput at each station, a dual-conveyor SMT line (Fig. 2.3) has been introduced. Each placement machine is equipped with a dual conveyor system that can operate in either synchronous or asynchronous mode. In synchronous mode, two panels are loaded at the same time. Thus, the loading time per panel is halved, and the number of placements in the assembly program is doubled. In asynchronous mode, a second panel can be loaded or unloaded while the first panel is being assembled.

2.2.2 SMT Lines for Double-Sided Boards

Generally there are two ways to produce double-sided boards, i.e., using a single-pass (continuous) line or a double-pass (reentrant) SMT line.

A single-pass SMT line (Fig. 2.4) consists of two lines linked together by a board flipping station. Each PWB is transported by the conveyor system through the complete line.

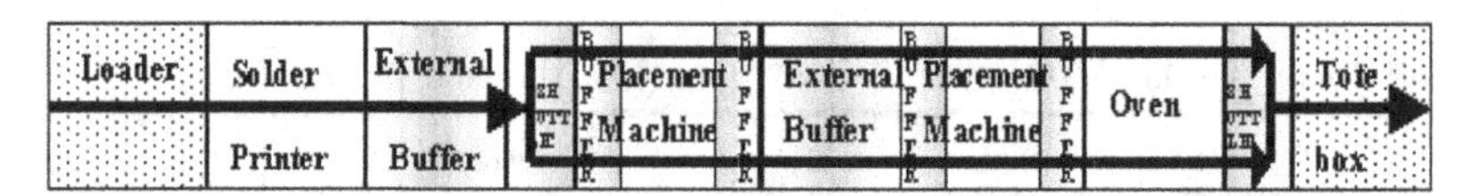

Figure 2.3 Dual-conveyor SMT line.

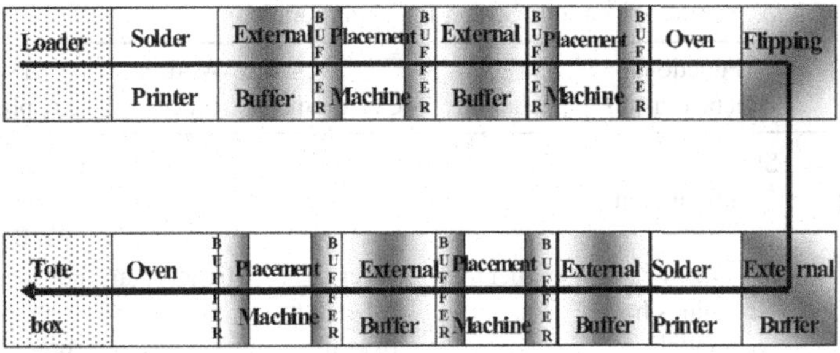

Figure 2.4 Single-pass SMT line.

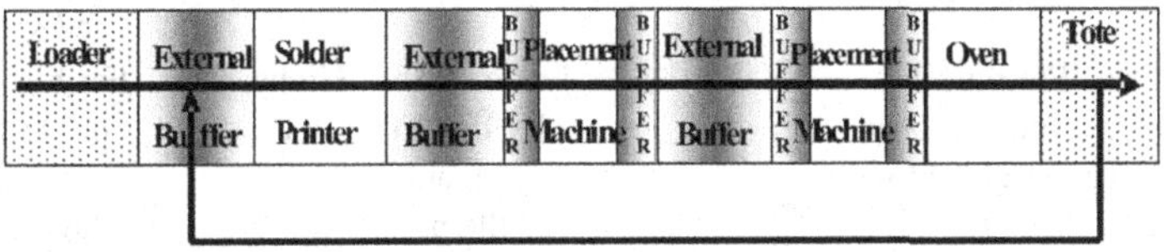

Figure 2.5 Double-pass SMT line.

After the first side of the PWB is completed in a double-pass (reentrant) SMT line (Fig. 2.5), the individual panels return to the front of the line, or the panels get collected in a cassette and then are returned as a batch to the front. At the beginning of the line, the PWB is flipped and inserted in the production flow again. During the second pass, the second side of the PWB is populated with components, and the finished products are collected in a tote and leave the line. Table 2.1 gives an overview and comparison of the different SMT line configurations.

2.3 GENERAL SCHEDULING OF SMT LINES

In this section an MIP model is presented for general scheduling of SMT lines. The mathematical formulation is based on the *FPB* model described in Chapter 1, where a unified modeling approach has been applied with the buffers viewed as machines with zero processing times. The notation used to formulate the problem is shown in Table 2.2, where buffers and machines are referred to as processors.

The SMT line under study consists of m processing stages in series. Each stage i, $(i = 1, \dots, m)$ is made up of $m_i \geq 1$ parallel identical processors (machines or buffers). Let J_i be the circular set of indices of parallel processors in stage i. The line produces various types of boards. Each board must be processed without preemption on exactly one processor in each of the stages sequentially. That is, each board must be processed in stage 1 through stage m in that order. The order of processing the boards in every stage is identical and determined by an input sequence in which the boards enter the line, that is, a so-called permutation flow shop is considered. The permutation flow

Table 2.1 Characteristics of Various SMT Line Configurations

Type of SMT line	Functional characteristics	Reliability	Cycle time	Work in process	Typical applications
1.1. SMT line with single stations	Standard configuration	–	–	–	For production of any type of PWB
1.2. SMT line with parallel stations	Placement machines have a parallel configuration	Higher than 1.1	Lower than 1.1 due to placement density effect	Higher than 1.1	For medium volume production (24 hours a day, 7 days a week), with high replenishment and setup times
1.3. SMT line with dual-conveyor	Placement machines are equipped with a dual-conveyor system	–	Lower than 1.1 due to elimination of the non-productive transport time	Higher than 1.1	For high volume production (24 hours a day, 7 days a week)
2.1. Single-pass SMT line	A manufacturing line for each side of the PWB	–	–	–	For high volume production (24 hours a day, 7 days a week)
2.1. Double-pass SMT line	PWB has to reenter the line	Higher than 2.1	Higher than 2.1	Lower than 2.1	For medium volume and high mix production

shop assumption eliminates overpassing among the boards and by this implicitly reduces board flow time, which cannot exceed the solder paste "pop" life time.

The generic scheduling model has the following structure:

Minimize maximum completion time subject to

1. *Assignment constraints for stages with parallel processors* to ensure that each board is assigned to exactly one processor and remains on the same conveyor until a shuttle stage.
2. *Board completion constraints* to ensure that each board is processed at all stages.
3. *Board departure constraints* to ensure that each board cannot be departed from a stage until it is completed in this stage and to ensure that each board leaves the line as soon as it is completed at the last stage.
4. *Board noninterference constraints* to ensure that no two boards are processed by the same processor simultaneously.
5. *No buffering constraints* to ensure that processing of each board at every stage starts immediately after its departure from the previous stage.

Table 2.2 Notation: MIP Models for Scheduling SMT Lines

Indices

g	= batch (board type), $g \in G$
i	= processing stage, $i \in I = \{1,\ldots,m\}$
j	= processor in stage i, $j \in J_i = \{1,\ldots,m_i\}$
k	= board, $k \in K = \{1,\ldots,n\}$

Input parameters

n_g	= size of batch g (number of boards of type g)
n	= total number of boards, $n = \sum_{g \in G} n_g$
p_{ik}	= processing time for board k in stage i
r_{ig}	= processing time for board type g in stage i
SH	= subset of shuttle stages, $SH \subset I$
$I(\sigma)$	= subset of stages with parallel processors, between two successive shuttles, immediately following shuttle stage $\sigma \in SH$
K_g	= subset of boards of type g

Decision variables

c_{ik}	= completion time of board k in stage i (timing variable)
d_{ik}	= departure time of board k from stage i (timing variable)
x_{ijk}	= 1, if board k is assigned to processor $j \in J_i$ in stage $i \in I$; otherwise $x_{ijk} = 0$ (assignment variable)
y_{kl}	= 1, if board k precedes board l; otherwise $y_{kl} = 0$ (sequencing variable for general scheduling)
Y_{fg}	= 1, if boards type f precedes boards type g; otherwise $Y_{fg} = 0$ (sequencing variable for batch scheduling)

6. *Maximum completion time constraints* to define the latest completion time of some board in the last stage, i.e., the makespan of a given production schedule.

The mixed integer program *GESCH* for general scheduling of an SMT line is similar to the *FPB* model and is presented below.

Model GESCH: *General Scheduling of an SMT Line*

Minimize

$$C_{\max} \tag{2.1}$$

subject to

1. *Board Assignment Constraints*:

$$\sum_{j \in J_i} x_{ijk} = 1; \quad i \in I, k \in K \tag{2.2}$$

$$x_{i_1,j_1,k} = x_{i_2,j_2,k}; \quad \sigma \in SH, i_1, i_2 \in I(\sigma), j_1 \in J_{i_1}, j_2 \in J_{i_2}, \quad k \in K: i_1 < i_2, j_1 = j_2 \tag{2.3}$$

2. *Board Completion Constraints*:

$$c_{1l} \geq p_{1l} + \sum_{k \in K: k<l} p_{1k} y_{kl} + \sum_{k \in K: k>l} p_{1k}(1 - y_{lk}); \ l \in K: m_1 = 1 \tag{2.4}$$

$$c_{1k} \geq p_{1k}; \ k \in K :: m_1 > 1 \tag{2.5}$$

$$c_{ik} - c_{i-1k} \geq p_{ik}; \ i \in I, \ k \in K: i > 1 \tag{2.6}$$

3. *Board Departure Constraints*:

$$c_{ik} \leq d_{ik}; \ i \in I, k \in K: i < m \tag{2.7}$$

$$c_{mk} = d_{mk}; \ k \in K \tag{2.8}$$

4. *Board Noninterference Constraints*:

$$c_{ik} + Q_{ikl}(2 + y_{kl} - x_{ijk} - x_{ijl}) \geq d_{il} + p_{ik}; \ i \in I, j \in J_i, k, l \in K: \ k < l \tag{2.9}$$

$$c_{il} + Q_{ilk}(3 - y_{kl} - x_{ijk} - x_{ijl}) \geq d_{ik} + p_{il}; \ i \in I, j \in J_i, k, l \in K: \ k < l \tag{2.10}$$

5. *No Buffering Constraints*:

$$c_{ik} - p_{ik} = d_{i-1k}; \ i \in I, k \in K: i > 1 \tag{2.11}$$

6. *Maximum Completion Time Constraints*:

$$c_{mk} \leq C_{\max}; \ k \in K \tag{2.12}$$

$$d_{ik} + \sum_{h \in I: h>i} p_{hk} \leq C_{\max}; \ i \in I, k \in K: i < m \tag{2.13}$$

$$\sum_{k \in K} p_{ik} x_{ijk} + \min_{k \in K} \left(\sum_{h \in I: h<i} p_{hk} \right) + \min_{k \in K} \left(\sum_{h \in I: h>i} p_{hk} \right) \leq C_{\max}; \ i \in I, j \in J_i: \sum_{k \in K} p_{ik} > 0 \tag{2.14}$$

$$c_{ml} - c_{1l} \leq C_{\max} - p_{1l} - \sum_{k \in K: k<l} p_{1k} y_{kl} - \sum_{k \in K: k>l} p_{1k}(1 - y_{lk}) - \sum_{k \in K: k<l} p_{mk}(1 - y_{kl}) - \sum_{k \in K: k>l} p_{mk} y_{lk}; \ l \in K: m_1 = 1, m_m = 1 \tag{2.15}$$

7. *Variable Nonnegativity and Integrality Conditions*:

$$C_{\max} \geq 0 \tag{2.16}$$

$$c_{ik} \geq 0; \ i \in I, k \in K \tag{2.17}$$

$$d_{ik} \geq 0; \ i \in I, k \in K \tag{2.18}$$

$$x_{ijk} \in \{0, 1\}; \ i \in I, j \in J_i, k \in K \tag{2.19}$$

$$y_{kl} \in \{0, 1\}; \ k, l \in K : k < l. \tag{2.20}$$

The objective function (2.1) represents the schedule length to be minimized. Assignment constraint (2.2) ensures that in every stage each board is assigned to exactly one processor. Constraint (2.3) ensures that each board remains on the same conveyor until the board reaches a shuttle stage. Constraint (2.4) or (2.5) ensures that each board is processed in the first stage, and (2.6) guarantees that it is also processed in all downstream stages. If the first stage is a single processor, then constraint (2.4) ensures that each board completion time in this stage is not less than the sum of processing times of all preceding boards; otherwise (2.5) is applied. However, (2.4) is more tight than a general constraint (2.5). Constraint (2.7) indicates that each board cannot be departed from every stage until it is completed in this stage, and equation (2.8) ensures that each board leaves the line as soon as it is completed in the last stage. Inequalities (2.9) and (2.10) represent disjunctive constraints for processor scheduling with blocking. No two boards can be performed on the same processor simultaneously. For any two different boards k and l assigned to the same processor in stage i either board k precedes board l, and then l cannot be started until k is departed from stage i (i.e., $c_{il} - p_{il} \geq d_{ik}$), or board l precedes board k, and then k cannot be started until l is departed (i.e., $c_{ik} - p_{ik} \geq d_{il}$). For a given sequence of boards at most one constraint of (2.9) and (2.10) is active, and only if both boards k and l are assigned to the same processor. Equation (2.11) indicates that processing of each board in every stage starts immediately after its departure from the previous stage. Constraint (2.12) defines the maximum completion time of all boards. Constraints (2.13) and (2.14) relate board departure times and processor workload, respectively, directly to makespan. Every board must be departed from a station sufficiently early in order to have all of its remaining tasks completed within the remaining processing time. Similarly, the total processing time of the boards assigned to a processor in each stage plus minimum processing times in all remaining stages must not exceed the makespan. Constraint (2.15) ensures that each board is processed within the time interval remaining after processing of all preceding boards and before processing of all succeeding boards. The flow time $c_{ml} - (c_{1l} - p_{1l})$ of each board $l \in K$ cannot be greater than the makespan $C_{\max}$ minus the sum of processing times of all preceding boards in the first stage

$$\sum_{k \in K: k < l} p_{1k} y_{kl} + \sum_{k \in K: k > l} p_{1k}(1 - y_{lk}),$$

and the sum of processing times of all succeeding boards in the last stage

$$\sum_{k \in K: k<l} p_{mk}(1 - y_{kl}) + \sum_{k \in K: k>l} p_{mk} y_{lk}.$$

Relating the assignment, sequencing, and timing variables in the cutting constraints (2.4) and (2.13) to (2.15) enables tighter time intervals available for processing of each board to be more precisely determined, which strengthens the MIP formulation.

The proposed modeling approach has enabled the sequencing variable y_{kl} $(k, l \in K : k \neq l)$ to be defined only for $k < l$ (2.20), and in the constraints (2.4) and (2.15) for single-machine stages, y_{lk} is replaced with $1 - y_{kl}$.

As a result y_{kl} for $k > l$ is eliminated from the model and the total number of variables y_{kl} has been reduced by half from $n(n-1)$ to $n(n-1)/2$ and also the number of constraints (2.9) and (2.10) was reduced by half.

The parameter Q_{ikl} in the board noninterference constraints is a large number not less than the schedule length, determined for stage i when board k precedes board l:

$$Q_{ikl} = \sum_{i \in I} \sum_{k \in K} p_{ik}/m_i - \sum_{h \in I: h>i} p_{hl} - \sum_{h \in I: h<i} p_{hk}; \quad i \in I, k, l \in K : k < l.$$

2.3.1 Special Cases

Model *GESCH* for scheduling SMT line is a general formulation and includes various special cases. Some of them are presented below.

1. SMT line with single processors. If $m_i = 1, \ \forall i \in I$ model *GESCH* reduces to scheduling a flow shop with single processors, including buffers.
2. SMT line with single machines and no in-process buffers. If $m_i = 1, \ \forall i \in I$ and $p_{ik} > 0, \ \forall i \in I, \ k \in K$ model *GESCH* reduces to scheduling a flow shop with single machines and no in-process buffers.
3. SMT line with parallel machines and no in-process buffers. If $p_{ik} > 0$, $\forall i \in I, \ k \in K$ model *GESCH* reduces to scheduling a flow shop with parallel machines and no in-process buffers.
4. Double-pass reentrant SMT line (see Section 1.2.5).
5. Dual-conveyor SMT line. The model *GESCH* can be enhanced for scheduling a dual-conveyor line as follows. Each placement machine with a dual conveyor is modeled as a series of the following four processing stages:
   - Two parallel input buffers, that is, one for each conveyor
   - One "dummy buffer" within the machine
   - One machine
   - Two parallel output buffers, that is, one for each conveyor

   In a machine with dual conveyors, the conveyor that does not contain the board being assembled functions as a "dummy buffer."

For subset $I2 \subset I$ of stages with a dual-conveyor and parallel processors, the following additional constraints are added to ensure that each board remains on the same conveyor until completion.

Assignment constraints for stages with parallel processors and a dual-conveyor:

$$x_{i_1 j_1 k} = x_{i_2 j_2 k};\ i_1, i_2 \in I2, j_1 \in J_{i_1}, j_2 \in J_{i_2}, k \in K: \\ i_1 \neq i_2, m_{i_1} = m_{i_2} = 2, j_1 = j_2. \qquad (2.21)$$

2.4 BATCH SCHEDULING OF SMT LINES

In order to reduce the complexity of the *general scheduling* problem, where any sequence of boards is allowed and to minimize machine setup to build a specific board, a *batch scheduling* mode is usually applied in practice, where all boards are scheduled in batches of boards of the same type and within the batch identical boards are processed consecutively. No setups are required between different boards or different batches of boards.

Let G, $K = \{1, \ldots, n\}$, and $K_g = \{\sum_{f \in G: f <= g-1} n_f + 1, \ldots, \sum_{f \in G: f <= g-1} n_f + n_g\}$ be the ordered sets of indices, respectively, of all batches of boards, all individual boards, and all boards of type $g \in G$. (n_g and $n = \sum_{g \in G} n_g$ denote, respectively, the number of boards of type g and the total number of boards in the schedule.) Let $r_{ig} \geq 0$ be the processing time in stage i of each board type $g \in G$ (see Section 1.2.6).

The objective is to determine an input sequence of batches and an assignment of boards to processors in each stage over a scheduling horizon to complete all the boards in minimum time.

The mixed integer program *BASCH* for batch scheduling of an SMT line is presented below. The formulation includes various cutting constraints that have been identified exploiting an SMT line configuration and some properties of batch processing.

Model BASCH: *Batch Scheduling of an SMT Line*

Minimize (2.1) subject to (2.2)–(2.8), (2.11)–(2.14), (2.16)–(2.19), and

1. *Board Assignment Constraints*:

$$x_{i,\, next(j,\, J_i),\, k+1} = x_{ijk};\ i \in I, j \in J_i, g \in G, k \in K_g: \\ k < last(K_g), m_i > 1 \qquad (2.22)$$

2. *Board Noninterference Constraints*:

$$c_{ik} + Q_{ifg}(2 + Y_{fg} - x_{ijk} - x_{ijl}) \geq d_{il} + r_{if}; \\ i \in I, j \in J_i, f, g \in G, k \in K_f, l \in K_g : f < g \qquad (2.23)$$

$$c_{il} + Q_{igf}(3 - Y_{fg} - x_{ijk} - x_{ijl}) \geq d_{ik} + r_{ig}; \\ i \in I, j \in J_i, f, g \in G, k \in K_f, l \in K_g : f < g \qquad (2.24)$$

3. *Maximum Completion Time Constraints*:

$$c_{ml} - c_{1l} \leq C_{\max} - \sum_{f \in G: f<g} n_f r_{1f} Y_{fg} - \sum_{f \in G: f>g} n_f r_{1f}(1 - Y_{gf})$$
$$- \left(l - \sum_{f=1}^{g-1} n_f\right) r_{1g} - \left(\sum_{f=1}^{g} n_f - l\right) r_{mg}$$
$$- \sum_{f \in G: f<g} n_f r_{mf}(1 - Y_{fg}) - \sum_{f \in G: f>g} n_f r_{mf} Y_{gf};$$
$$g \in G, l \in K_g : m_1 = 1, m_m = 1 \tag{2.25}$$

4. *Batch Processing Constraints*:

$$c_{ik+m_i} \geq d_{ik} + r_{ig};\ i \in I, g \in G, k \in K_g : k + m_i \leq last(K_g), m_i > 1 \tag{2.26}$$

$$c_{ik+1} \geq c_{ik};\ i \in I, g \in G, k \in K_g : k < last(K_g), m_i > 1 \tag{2.27}$$

$$c_{ik+1} \geq d_{ik} + r_{ig};\ i \in I, g \in G, k \in K_g : k < last(K_g), m_i = 1 \tag{2.28}$$

5. *Variable Nonnegativity and Integrality Conditions*:

$$Y_{fg} \in \{0, 1\};\ f, g \in G : f < g. \tag{2.29}$$

Assignment constraint (2.22) (cf. (1.34)) assigns successive boards of one type alternately to different parallel processors ($next(j, J_i)$ is the next processor after $j \in J_i$ in the circular set J_i of parallel processors in stage i). Constraints (2.23) and (2.24) are board noninterference constraints. No two boards can be performed on the same processor simultaneously. For a given sequence of batches only one constraint (2.23) or (2.24) is active, and only if both boards $k \in K_f$ and $l \in K_g$ are assigned to the same processor. Constraint (2.25) ensures that each board is processed within the time interval remaining after processing of all preceding boards and before processing of all succeeding boards. The flow time $c_{ml} - (c_{1l} - r_{1g})$ of each board $l \in K_g$ cannot be greater than the makespan $C_{\max}$ minus the sum of processing times of all preceding boards in the first stage

$$\sum_{f \in G: f<g} n_f r_{1f} Y_{fg} + \sum_{f \in G: f>g} n_f r_{1f}(1 - Y_{gf}) + \left(l - 1 - \sum_{f=1}^{g-1} n_f\right) r_{1g},$$

and the sum of processing times of all succeeding boards in the last stage

$$\left(\sum_{f=1}^{g} n_f - l\right) r_{mg} + \sum_{f \in G: f<g} n_f r_{mf}(1 - Y_{fg}) + \sum_{f \in G: f>g} n_f r_{mf} Y_{gf}.$$

The cutting constraint (2.22) fixes the assignment of boards to parallel processors and by this significantly reduces the computational complexity of the batch scheduling problem. Batch processing constraints (2.26) and (2.27) along with (2.22) ensure that boards of one type are processed consecutively in each stage with parallel processors, whereas consecutive processing of identical boards in each stage with a single processor is imposed by (2.28). Note that constraints (2.22) imitate function of a shuttle in an SMT line and tend to balance workloads of parallel stations.

The number of board noninterference constraints can be reduced for stages with at most two parallel processors, which is usually the case in SMT lines.

Board noninterference constraints for stages with at most two parallel processors:

$$c_{ik} + (Q_{ifg} + H_{ifk})(2 + Y_{fg} - x_{ijk} - x_{ij\,last(K_g)}) \geq d_{i\,last(K_g)} + r_{if} + H_{ifk} \; i \in I, j \in J_i, f, g \in G, k \in K_f : f < g \tag{2.30}$$

$$c_{il} + (Q_{igf} + H_{igl})(3 - Y_{fg} - x_{ij\,last(K_f)} - x_{ijl}) \geq d_{i\,last(K_f)} + r_{ig} + H_{igl}; \; i \in I, j \in J_i, f, g \in G, l \in K_g : f < g \tag{2.31}$$

$$c_{i\,first(K_f)} + (Q_{ifg} + T_{igl})(2 + Y_{fg} - x_{ij\,first(K_f)} - x_{ijl}) \geq d_{il} + r_{if} + T_{igl}; \; i \in I, j \in J_i, f, g \in G, l \in K_g : f < g \tag{2.32}$$

$$c_{i\,first(K_g)} + (Q_{igf} + T_{ifk})(3 - Y_{fg} - x_{ijk} - x_{ij\,first(K_g)}) \geq d_{ik} + r_{ig} + T_{ifk}; \; i \in I, j \in J_i, f, g \in G, k \in K_f : f < g. \tag{2.33}$$

For a given sequence of batches only one constraint (2.30) or (2.31) is active, and only if both boards $k \in K_f$ and $last(K_g)$ or $l \in K_g$ and $last(K_f)$ are assigned to the same processor. Likewise, either (2.32) or (2.33) is active, and only if both boards $l \in K_g$ and $first(K_f)$ or $k \in K_f$ and $first(K_g)$ are assigned to the same processor.

Parameters H_{ifk}, T_{ifk} and Q_{ifg} that tighten the board noninterference constraints are calculated as below.

$$H_{ifk} = \max\left\{0, \left\lfloor (k - \sum_{p=1}^{f-1} n_p - m_i)/m_i \right\rfloor\right\} r_{if}; \; i \in I, f \in G, k \in K_f \tag{2.34}$$

$$T_{ifk} = \max\left\{0, \left\lfloor (\sum_{p=1}^{f} n_p - k - m_i + 1)/m_i \right\rfloor\right\} r_{if}; \; i \in I, f \in G, k \in K_f \tag{2.35}$$

$$Q_{ifg} = \sum_{i \in I} \sum_{g \in G} n_g r_{ig}/m_i - \sum_{h \in I: h<i} r_{hf} - \sum_{h \in I: h>i} r_{hg}; \; i \in I, f, g \in G, \tag{2.36}$$

where H_{ifk} and T_{ifk} denote, respectively, the head and tail of board $k \in K_f$ in batch f in stage i, and Q_{ifg} is a large number not less than the schedule length determined for stage i when batch f precedes batch g.

The introduction of cutting constraints (2.13) to (2.15) in model *GESCH* and, in particular, constraints (2.22), (2.26) to (2.28), and (2.30) to (2.33) in model *BASCH*, strengthens the formulations and by this reduces the computational effort required to find the optimal solutions.

Note that the *GESCH* model for the general scheduling of SMT lines can also be used for the batch scheduling after a simple addition of batch scheduling mode constraints (1.24)–(1.26).

However, the *BASCH* model may outperform the enhanced *GESCH* model as the former exploits the special structure of the batch scheduling mode.

2.5 AN IMPROVEMENT HEURISTIC FOR SCHEDULING SMT LINES

In this section a fast improvement heuristic is presented for scheduling printed wiring board assembly in various SMT line configurations. One can distinguish between constructive and improvement heuristics. A constructive heuristic generates one complete solution by iteratively extending a partial schedule (see Section 1.3). In contrast, an improvement heuristic, also called a local search heuristic, starts from an initial feasible solution and searches in every iteration for some small changes in the solution that would lead to a better objective function value. In this section an improvement heuristic is proposed. The heuristic has a hierarchical structure based on the decomposition of the scheduling problem into two subproblems—sequencing and assignment/timing. The objective of the sequencing subproblem is to sequence all boards on processors in every stage, and the objective of the assignment/timing subproblem is to assign all boards to processors and to establish the times of all operations.

The two subproblems are solved sequentially. In every iteration first the sequencing procedure selects a new feasible input sequence to be evaluated and then the assignment and timing procedure determines the detailed schedule and the resulting makespan, as shown in Figure 2.6.

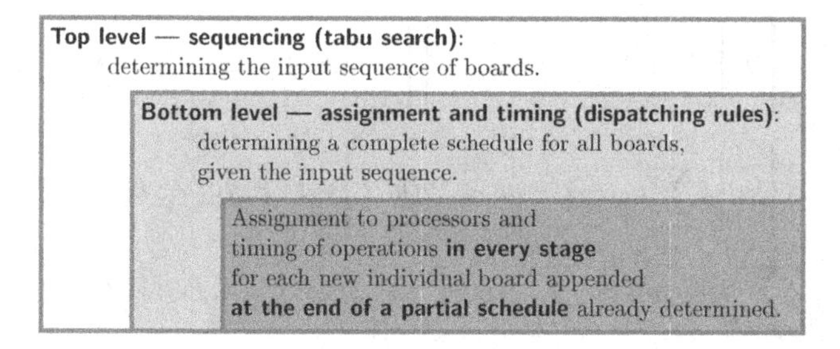

Figure 2.6 Hierarchical sequencing and assignment/timing.

The basic decision variables used in the scheduling heuristic are shown below.

α_{ij} = availability time of processor j in stage i, that is, the departure time of the last board assigned to this processor in the actual partial schedule,
S = input sequence $[k_1, \ldots, k_n]$ of boards entering the line, where
k_l = board at position l in the input sequence,
μ_{ik} = processor in stage i that processes board k.

2.5.1 Sequencing

The initial sequence is determined using a simple constructive heuristic described by Campbell et al. (1970). First, two stages with the highest workload $\sum_{k \in K} p_{ik}$ are found. Then, the input sequence is determined as the optimal solution of the classical two-stage flow shop with infinite buffers, where processing times are taken from the two bottlenecks.

The improvement heuristic searches over a huge set of sequences by small modifications of an existing solution called *moves*. The sequencing algorithm proceeds as follows. The procedure starts with an initial sequence and searches in its neighborhood that contains slightly different sequences for a sequence with the shortest or just shorter makespan. In order to determine makespan for each new sequence it is necessary to execute the assignment procedure. Then the search is applied in the new neighborhood, generated by the sequence found previously. The process is continued until some stopping conditions are satisfied.

The proposed heuristic modifies the sequence by exchanging the position of two boards. Let k_l be the board at position l in the sequence before a move, and (a, b) be an exchange of boards at positions a and b. If $a < b$, board a moves to the right, and the following new sequence is determined

$$[k_1, \ldots, k_{a-1}, k_b, k_{a+1}, \ldots, k_{b-1}, k_a, k_{b+1}, \ldots, k_n].$$

If $a > b$, board a moves to the left, and the new sequence is:

$$[k_1, \ldots, k_{b-1}, k_a, k_{b+1}, \ldots, k_{a-1}, k_b, k_{a+1}, \ldots, k_n].$$

There are $n(n-1)/2$ possible exchange moves.

An alternative type of move is insertion, where only one board changes its position. This method is preferable for the regular flow shop problems with single machines in every stage because a fast evaluation of the makespan is available (Nawaz et al.,1983). Unfortunately, this method cannot be applied to flexible flow shops such as SMT lines.

Successive improvement may lead the search to a local, rather than global, optimum. In addition there is the risk of cycling. Successively chosen solutions may build up a cycle so that after a while no new sequences are considered.

In order to avoid cycling or becoming trapped in a local optimum, various strategies have been developed. The most popular is the *tabu search* technique (Nowicki and Smutnicki, 1996, 1998). It introduces a memory of the search, called *tabu list*, to prohibit coming back to already considered solutions. The tabu search algorithm of Nowicki and Smutnicki, however, is incapable of scheduling flexible flow shops with limited in-process buffers, such as SMT lines.

In the proposed algorithm the tabu list is defined as a cyclic list of pairs $((v_1, w_1), (v_2, w_2), \ldots, (v_{l_{max}}, w_{l_{max}}))$ where v_l is a board at position l in the tabu list, w_l is its position in one of the previous sequences, and l_{max} is the length of the tabu list (in the computational experiments presented in Section 2.6, $l_{max} = 7$). After a new sequence has been chosen, two pairs have to be added at the end of the tabu list, and the two oldest pairs have to be removed from the beginning of the list. Any moves that include some pair from the tabu list are forbidden.

The search through the neighborhood is finished either after the enumeration of all neighbors or after some improved sequence has been found. The number of all neighbors is equal to $n(n-1)/2$. Although the enumeration makes it possible to find the best neighbor, the second alternative is better because of the relatively high computational time required for a single run of the assignment procedure. Moreover, in order to reduce the computational effort and accelerate improvement, the whole neighborhood search needs not to be carried out in a single iteration. Only n randomly chosen, and not forbidden, moves are tested.

Another important question is how to search over the defined neighborhood, that is, over the set of all two-element combinations (a, b) from the set $\{1, \ldots, n\}$. The best strategy is to increase the first position a from 1 to $n-1$ and decrease the second position b from n to $a+1$ for every value of a.

The stopping condition is based on the observation of changes in the objective function. If its value cannot be improved over 1000 new sequences, the search is stopped.

2.5.2 Assignment and Timing

Both the constructive and the improvement heuristic use the same assignment procedure. Its objective is to assign each board to some processor in every stage and then to determine the makespan by timing of all the operations. The proposed procedure is based on the assumption that the sequence of boards on all processors is identical to the input sequence, that is, a permutation flow shop problem is considered (e.g., Kochhar and Morris, 1987). The assumption of the permutation flow shop is particularly reasonable for scheduling SMT lines, where at each stage processing times of different boards are similar and hence changing the order of processing the boards at various stages would have no effect on the makespan. Furthermore, since the total flow time of each board is limited by the solder paste "pop" life time, the same sequence of boards should be maintained at each stage to eliminate overpassing among the boards and by this to minimize board waiting time in buffers.

In the case of single processors in every stage, there is nothing to be decided. Only the timing has to be done. One has to determine departure times d_{ik} for every board k

in every stage i. If there are parallel machines in some stages, assignment becomes an essential part of the optimization.

Timing is done successively for all boards according to the input sequence. The task of a single iteration is to append all operations of the next board in the sequence at the end of a partial schedule in order to minimize its completion time c_{mk}. When all operations of a new board k at position l in the sequence are to be appended, the partial schedule consists of all operations of preceding boards, that is, the boards at positions $(1, \ldots, l-1)$. New operations are added at the end of the partial schedule. Therefore the partial schedule can be represented by the processors' availability times α_{ij} after the first $l-1$ boards have been completed and departed from processors, that is, α_{ij} is equal to the departure time of the last board so far assigned to processor j in stage i. If all machines are identical, a good practical method is to choose the earliest available processors (buffers and machines) in every stage (see Section 1.3), that is,

$$\mu_{ik} = \arg\min_{j \in J_i} \alpha_{ij} \quad \forall i \in I.$$

Then, the timing can be successively done for all stages $i \in I$ using the formula:

$$d_{ik} = \max\{d_{i-1,k} + p_{ik}, \alpha_{i+1,\mu_{i+1,k}}\} \quad \forall i \in I.$$

The above equations hold also in the first and the last stage, where the boundary variables take on the following values: $d_{0k} = \alpha_{1,\mu_{1k}}$, $\mu_{m+1,k} = 1$, and $\alpha_{m+1,1} = 0$.

An important part of the heuristic implementation is how priority queues for buffers and machines are constructed so that the earliest available buffers and machines are found with minimum computational effort. These queues have been implemented as cyclic ordered lists. Because the number of elements in such a list for every stage is constant, it is convenient to use an array for its representation. Inserting a machine or buffer back into the queue starts at the end of the queue. Usually the sequence of elements in the list does not need to be changed. Only the variable holding the number of the first element has to be pushed by one position to the end of the list, that is, the first element becomes the last one. A sketch of the overall heuristic algorithm is shown in Figure 2.7.

2.5.3 Special Cases

The scheduling heuristic is a flexible approach that can be easily enhanced to satisfy various specific requirements of electronics assembly.

Batch scheduling. In order to schedule batches of boards instead of single boards the move that exchanges the positions of boards in the sequencing procedure has to be redefined. One has to exchange the positions of all boards of two batches in the sequence, not just two single boards. If there are many batches of the same type in a set of boards representing the daily product mix, the input sequence becomes a permutation with repetitions. This helps to reduce the computational effort required because there are many input sequences that differ only in terms of the positions of batches of the same type. All moves

HEURISTIC ALGORITHM – ENUMERATION

1 $C^*_{\max} \leftarrow$ very large number;
2 For all possible input sequences S'
3 ASSIGNMENT AND TIMING for schedule $H'(S')$;
4 If $C^*_{\max} > C_{\max}(H')$ then
5 $C^*_{\max} \leftarrow C_{\max}(H')$;
6 $S \leftarrow S'$; $H \leftarrow H'$;
7 Return S, H;

Variables:

H – complete schedule,

$S' \leftarrow \sigma(S)$ – *move* from sequence S to S',

Σ – set of all possible moves,

$\mathcal{T}$ – *tabu list* of moves.

HEURISTIC ALGORITHM – TABU SEARCH

1 Choose *initial input sequence* S;
2 ASSIGNMENT AND TIMING for schedule $H(S)$;
3 Repeat
4 $S' \leftarrow S$; $H' \leftarrow H$;
5 $k \leftarrow 0$
6 Repeat
7 $k \leftarrow k + 1$
8 Choose next move $\sigma \in \Sigma \setminus \mathcal{T}$;
9 Determine new sequence $S'' \leftarrow \sigma(S)$;
10 ASSIGNMENT AND TIMING for schedule $H''(S'')$;
11 If $C_{\max}(H') > C_{\max}(H'')$ then
12 $S' \leftarrow S''$; $H' \leftarrow H''$;
13 $\mathcal{T} \leftarrow \mathcal{T} \cup \{\sigma\}$;
14 Until $C_{\max}(H) > C_{\max}(H')$ or $k = n$;
15 $S \leftarrow S'$; $H \leftarrow H'$;
16 Until *Stop condition* satisfied;
17 Return S, H;

ASSIGNMENT AND TIMING

Input: Input sequence of boards S

1 Initialize empty partial schedule H;
2 For all boards $k \in K$, given the input sequence, do
3 For stage $i = 1$ to m do
4 Find earliest available processor $\mu_{ik}(H)$;
5 Determine departure time $d_{ik}(\mu_{ik})$;
6 Append board at the end of the partial schedule H;
7 Return H;

Figure 2.7 Scheduling heuristic.

that exchange positions of identical batches are forbidden since they do not effect the objective function. If the number of possible permutations is very small, e.g., when the number of batches is less than 10, the tabu search technique in the sequencing procedure can be replaced with a complete enumeration of batches.

Maximum flow time constraint. Total flow time of a board processed in an SMT line is limited by the solder paste "pop" life time. However, the maximum flow time constraint is not very tight in PWB assembly. The total processing time of a single board varies between 3 and 14 minutes, and the pop life time usually exceeds an hour. The proposed algorithm adds all operations of a new board at the end of the partial schedule. Therefore, after scheduling of a single board is finished, its total waiting time can be easily shortened or completely eliminated by shifting the operations forward.

Restricted transportation links. Restricted transportation links can be found in various SMT line configurations. For example, in SMT lines with parallel machines or in a dual-conveyor line, there are limited transportation links between processors in consecutive stages, as shown in Figure 2.8. Therefore, some combinations of buffers and machines do not generate feasible processing routes. In a general case, the assignment procedure would require a complete enumeration of all feasible processing routes. The departure time from a stage depends not only on the processing time in the stage and the departure time from the upstream stage, but also on the availability time of the earliest available processor in the downstream stage. Consequently, dynamic programming cannot produce an optimal solution to the assignment subproblem. Nevertheless, a dynamic programming scheme can be used to find a good heuristic solution. The assignment procedure for the line with restricted transportation links needs to be redesigned as follows. In the first step, a combination of machines is selected, that is, one for each processing stage. In the second step, all output and input buffers that belong to the selected machines are identified. Finally, the buffers in the remaining processing stages are selected. As a result the timing needs to be done for only one combination of machines and buffers.

Bypassing some stages. The assignment procedure for such types of boards that bypass some processing stages requires renumbering of processing stages so that consecutive stages are successively numbered.

Scheduled downtimes. Scheduled downtimes are viewed as preassigned dummy boards. Hence they do not conform with the basic assumption of the assignment procedure that all new boards are appended to the partial schedule at its end. The only way to handle scheduled downtimes in the proposed heuristic is to schedule all boards as if there were no downtimes, and then to reschedule when a board is assigned to a processor during its downtime. The rescheduling consists of two steps, first the availability time of the machine must be set to the end of the downtime, and next the assignment procedure has to go back to the

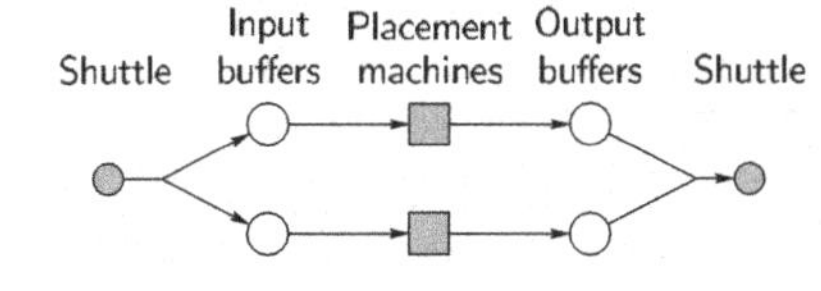

Figure 2.8 Parallel placement machines with input and output buffers.

upstream stage. For SMT assembly lines the procedure must go backward to the stage before the input buffer, and the availability time of the input buffer must also be set equal to the end of the downtime.

2.6 COMPUTATIONAL EXAMPLES

In this section five numerical examples are presented to illustrate application of the proposed optimization tools. In Examples 1 and 2 assembly schedules are determined by solving mixed integer programs representing some typical electronics assembly line configurations. The production mix consists of three board types and the assembly schedules are found for the following two cases:

- Batch scheduling, where 10 boards of each type are assembled, and all boards of a given type are scheduled consecutively. The optimal sequence of board types is obtained along with the optimal schedule for all boards.
- Cyclic scheduling, where 10 boards of each type are assembled, and the boards of different types are scheduled alternately in a cyclic order. The optimal cycle of board types is obtained along with the optimal schedule for all boards.

Examples 3 and 4 illustrate and compare applications of MIP and heuristic approaches. In the examples, the batch scheduling mode is considered with the input sequence of board types not fixed a priori and obtained along with the optimal schedule for all boards.

Finally, Example 5 illustrates a real world application of the proposed scheduling tool for improving configuration of an SMT line for a single type of printed wiring board.

EXAMPLE 2.1 *SMT Line with Parallel Stations*

The SMT line configuration for Example 2.1 is provided in Figure 2.9. The line consists of $m = 12$ stages, where stage $i = 1$ is a screen printer, each stage $i = 5, 9$ represents two parallel machines for automatic placement of components, stage $i = 12$ is a vision inspection station, stages $i = 3, 7, 11$ are shuttles, stage $i = 2$ is a single external buffer, and each stage $i = 4, 8$ and $i = 6, 10$ represents two internal input and output buffers, respectively.

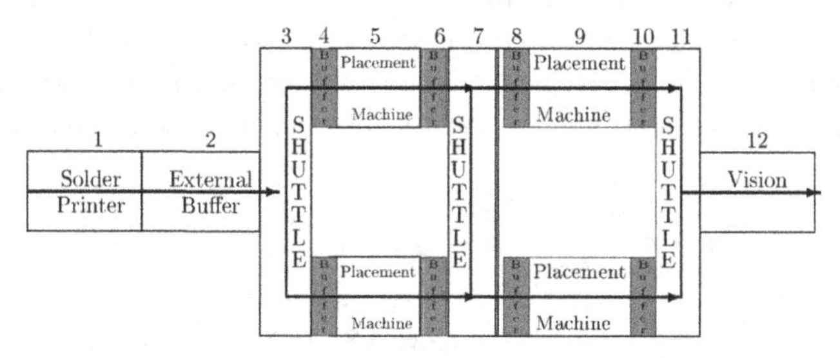

Figure 2.9 SMT line with parallel stations.

Table 2.3 Processing Times for an SMT Line with Parallel Stations

Board type	1	2	3
Screen printer P	10	10	10
Placement machine M1	56	59	74
Placement machine M2	53	54	55
Vision inspection machine V	20	20	20

The processing times p_{ik} for the boards are shown in Table 2.3 (for the buffer and shuttle stages $i = 2, 3, 4, 6, 7, 8, 10, 11$ all processing times are equal to zero).

The optimal schedules obtained for batch and cyclic scheduling are shown in Figure 2.10, where letters B, M, P, S, and V stand for Buffer, Machine for placement, Printer, Shuttle, and Vision inspection machine, respectively. Boards of types 1 and 2 are indicated, respectively, with gray and black shading, and boards type 3 with cross hatching. The optimal sequence of board types 1, 2, and 3 is identical for both batch and cyclic scheduling modes. The optimized makespans are as follows: $C_{\max} = 1051$ for batch scheduling, $C_{\max} = 1053$ for cyclic scheduling.

EXAMPLE 2.2 *SMT Line with a Dual-Conveyor*

In Example 2.2, optimal assembly schedules were determined for a dual conveyor line (see Fig. 2.3) with a screen printer, two placement machines, and an oven. The line consists of

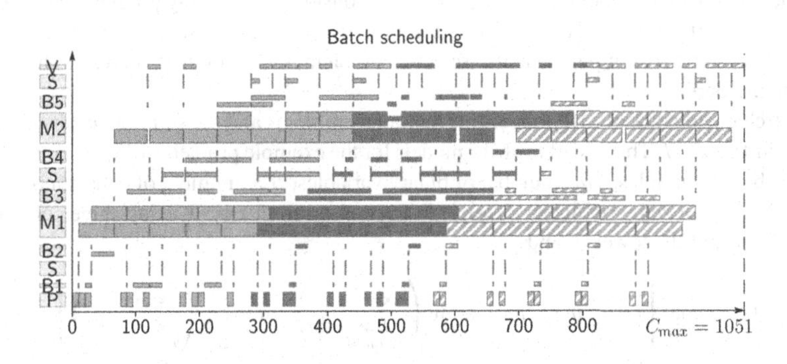

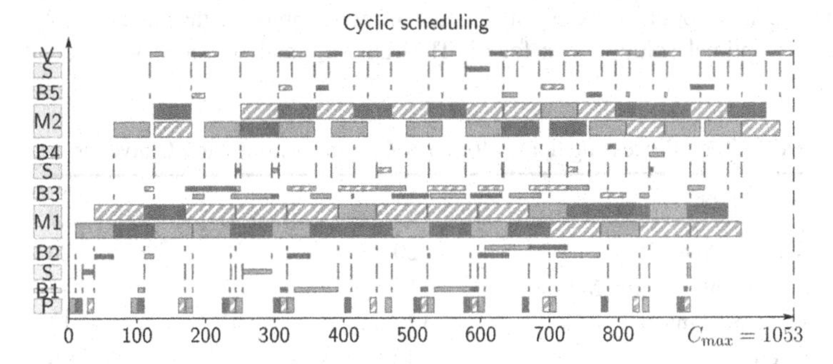

Figure 2.10 Assembly schedules for an SMT line with parallel stations.

$m = 14$ stages, where stage $i = 1$ is a screen printer, each stage $i = 6, 11$ represents a machine for automatic placement of components, stage $i = 13$ is a reflow oven, stages $i = 3, 14$ are shuttles, and stage $i = 2$ is a single external buffer. Each stage $i = 4, 7, 8, 9, 12$ represents two parallel buffers, one on each conveyor, and stages $i = 5, 10$ are single dummy buffers on the placement machines.

The processing times p_{ik} for the boards are shown in Table 2.4. (For the buffer and shuttle stages all processing times are equal to zero.)

The optimal schedules obtained for batch and cyclic scheduling are shown in Figure 2.11, where letters B, BM, M, O, P, and S stand for Buffer, Buffer on Machine, Machine for placement, Oven, Printer, and Shuttle, respectively. Boards of types 1 and 2 are indicated with light gray and dark gray, respectively, and boards of type 3 with cross hatching. The sequence of board of types was not fixed a priori. The optimal sequence of board types for batch scheduling is 1, 3, 2 and is the same as the optimal cycle of board types for cyclic scheduling. For Example 2.2, the same optimal makespan $C_{\max} = 3922$ was achieved for each scheduling mode.

EXAMPLE 2.3 *Factory with Single Stations*

The SMT line configuration for Example 2.3 is shown in Figure 2.12. The line consists of a loader, screen printer, four placement machines, and a vision inspection machine, in series separated by intermediate buffers.

The line represents a typical low-volume, medium-variety production system. For the industry scenario that was studied, 13 different board types are assembled in small batches. A daily production order consists of at most four different board types assembled in the line.

The input data for Example 2.3 were prepared considering the daily production of the line over a one-month horizon. Table 2.5 lists the processing times (in seconds) for boards, and Table 2.6 presents the input data for selected problem instances that represent five daily production orders.

The characteristics of MIP models for the same problems and the solution results are summarized in Table 2.7. The size of the MIP models for the example problems is represented by the total number of variables, *Var.*, number of binary variables, *Bin.*, number of constraints, *Cons.*, and number of nonzero coefficients, *Nonz.*, in the constraint matrix. The last three columns of Table 2.7 present the lower bound

$$LB = \max_{i\in I}\left\{\left\lfloor \sum_{g\in G} n_g r_{ig}/m_i \right\rfloor + \min_{g\in G}\left(\sum_{h\in I:h<i} r_{hg}\right) + \min_{g\in G}\left(\sum_{h\in I:h>i} r_{hg}\right)\right\} \tag{2.37}$$

on the makespan, the optimal makespan $C^*_{\max}$ and the node number in the branch-and-bound tree at which the optimal solution was found. The same optimal solutions were found by the heuristic.

Table 2.4 Processing Times for an SMT Line with a Dual Conveyor

Board type	1	2	3
Screen printer P	20	20	20
Placement machine M1	112	117	147
Placement machine M2	120	102	113
Oven 0	40	40	40

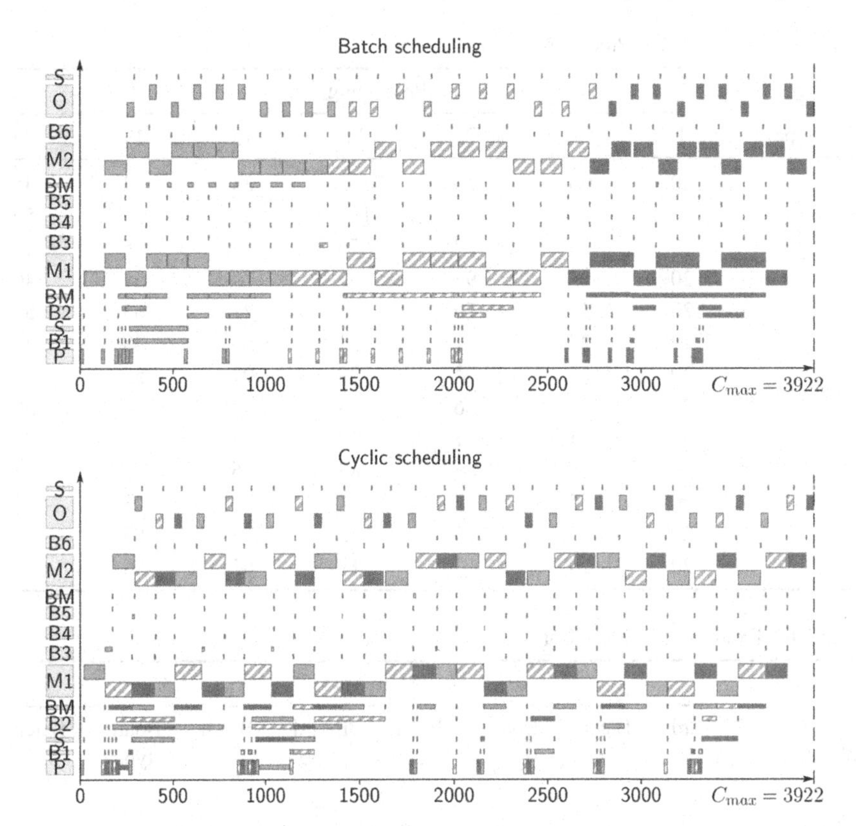

Figure 2.11 Assembly schedules for an SMT line with a dual conveyor.

Figure 2.13 shows a Gantt chart with the optimal batch schedule for Problem 5 of Example 2.3, where letter B stands for Buffer and M stands for Machine for board loading, screen printing, component placement, or vision inspection. Buffering or machine blocking is indicated with a narrow bar. The optimal input sequence of board types is 10, 7, 1, 4, and the optimal makespan $C^*_{\max} = 6925$.

The example clearly demonstrates that buffers are important to maximize throughput of an SMT line. The buffers ahead of a bottleneck stage (placement machine M5) are occupied most of the time and the first two machines in the line are frequently blocked.

– machine – buffer – many buffers

Figure 2.12 Factory with single stations.

Table 2.5 Example 2.3: Processing Times

Board type	Processing stage						
	1	3	7	11	15	19	23
1	20	25	123	45	38	62	45
2	20	25	155	156	28	58	50
3	20	25	67	56	36	35	45
4	20	25	93	95	–	51	40
5	20	25	76	111	41	63	50
6	20	25	87	93	52	48	45
7	20	25	34	78	92	55	45
8	20	25	66	28	34	–	30
9	20	25	141	90	49	–	40
10	20	25	86	83	56	22	45
11	20	25	98	84	36	43	45
12	20	25	176	175	76	65	50
13	20	25	–	17	67	28	45

Table 2.6 Example 2.3: Input Data

Problem no.	Daily mix							
	Board type	Batch size	Board type	Batch size	Board type	Batch size	Board type	Batch size
1	7	13	9	6	–	–	–	–
2	2	23	9	1	–	–	–	–
3	7	2	11	66	–	–	–	–
4	5	34	7	2	8	22	9	2
5	1	42	4	2	7	4	10	14

Table 2.7 Example 2.3: MIP Results for Daily Mix

Problem	Var.	Bin.	Cons.	Nonz.	*LB*	$C^*_{\max}$	Nodes
1	1085	590	2687	9649	1722	1722	13
2	1370	745	3442	12,284	3953	3967	0
3	3878	2109	9861	35,022	6789	6789	0
4	3427	1866	17,964	73,788	4869	5016	12
5	3541	1928	18,579	76,280	6923	6925	1

EXAMPLE 2.4 *Factory with Parallel Stations*

The SMT line configuration for Example 2.4 is shown in Figure 2.14. The line consists of a screen printer, three sets of two parallel placement machines and four shuttles routing the boards to the next available placement machine, a single placement machine, a vision inspection machine, and a final single placement machine, in series separated by intermediate buffers.

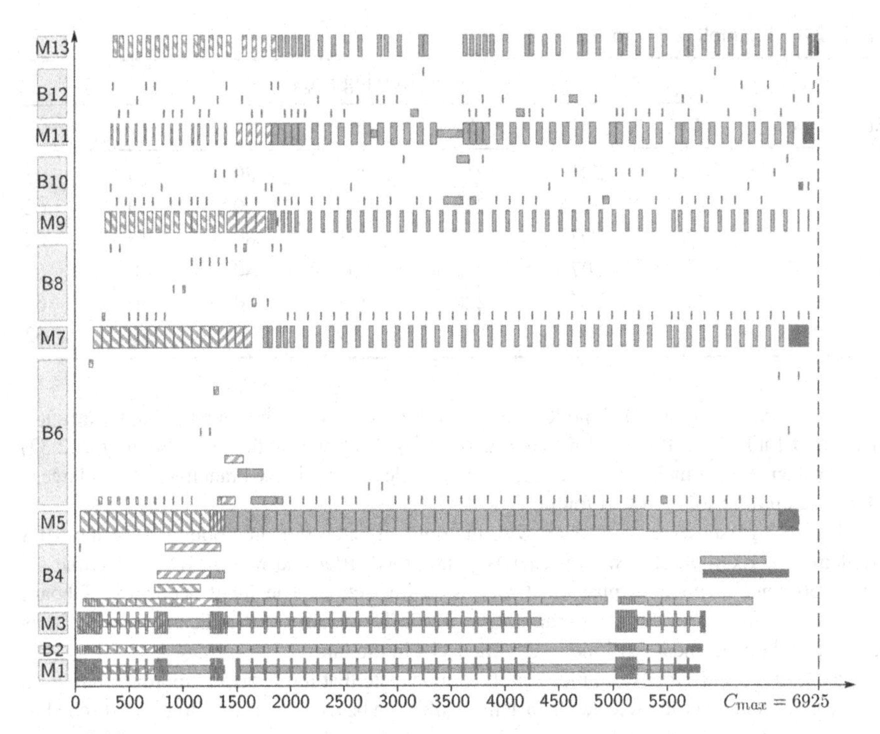

Figure 2.13 Batch schedule for an SMT line with single stations.

The line represents a typical high-volume, low-variety production system, in which six different board types are produced in medium to large batches. A daily production order consists of at most four different board types assembled in the line.

Table 2.8 lists the processing times (in seconds) for boards, and Table 2.9 presents the input data for selected problem instances that represent five daily production orders and the corresponding minimum part sets. The MPS production requirements represent $1/40$th, $1/40$th, $1/30$th, $1/100$th, and $1/40$th of the actual daily production order, respectively, for Problems 1, 2, 3, 4, and 5.

Table 2.10 shows solution results for daily and MPS requirements obtained using the heuristic.

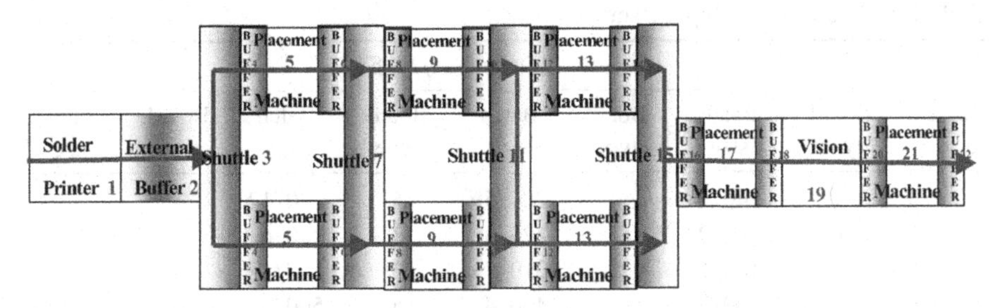

Figure 2.14 Factory with parallel stations.

Table 2.8 Example 2.4: Processing Times

	Processing stage						
Board type	1	5	9	13	17	19	21
1	22	207	213	204	80	40	62
2	22	208	220	204	80	40	62
3	22	207	224	191	80	40	62
4	22	207	213	204	80	40	62
5	22	207	220	204	80	40	62
6	22	184	196	199	80	40	62

The characteristics of MIP models for the MPS problems and the solution results are summarized in Table 2.11. The last three columns of Table 2.11 present the lower bound *LB* (2.37) on makespan, the optimal makespan C^*_{max}, and the node number in the branch-and-bound tree at which the optimal solution was found.

Figure 2.15 shows a Gantt chart with the optimal batch schedule obtained for the MPS Problem 2 of Example 2.4, where letter B stands for Buffer and M stands for Machine for screen printing, component placement, or vision inspection. The input sequence of board types is 5, 2, 1, 3, and the makespan $C^*_{max} = 3247$. The Gantt chart indicates that the three sets of parallel placement machines are bottlenecks in the line.

Both of the industrial examples clearly demonstrate that the heuristic approach based on tabu search and a set of the selected dispatching rules can be used in practice to support scheduling decision making in SMT lines.

EXAMPLE 2.5 *Configuring and Scheduling an SMT Line*

The optimization approach proposed in this chapter can be used to aid SMT line scheduling decisions such as the sequencing and timing of the release of boards into the line or the assignment of boards to parallel placement machines. The results of scheduling of an SMT line, and in particular, the throughput of an SMT line, depend on the line configuration. Example 2.5 shows that in addition to scheduling decision making, the scheduling tool can also be applied to find the best configuration of an SMT line.

Among the key structural characteristics of an SMT line are the location and size of the buffers between the placement machines and the arrangement of SMT machines of the same

Table 2.9 Example 2.4: Input Data

	Daily Mix/MPS							
Problem no.	Board type	Batch size	Board type	Batch size	Board type	Batch size	Board type	Batch size
1	3	240/6	4	200/5	5	480/12	–	–
2	1	80/2	2	120/3	3	240/6	5	480/12
3	1	180/6	2	210/7	3	510/17	–	–
4	3	300/3	4	400/4	5	500/5	–	–
5	3	1080/27	6	400/10	–	–	–	–

Table 2.10 Example 2.4: Heuristic Results for Daily Mix/MPS

Problem	LB	C^*_{max}
1	101,582/3127	101,662/3233
2	102,002/3137	102,082/3247
3	99,992/3915	100,072/3993
4	131,802/1914	131,882/1992
5	160,739/4583	161,217/4695

type. Specifically, there is the choice of "push" versus "pull" configurations of an SMT line. Factory observations have shown that when a set of identical SMT placement machines is arranged such that the machine with the longest cycle time is at the beginning of the line, work may wait in queues at the beginning of the line, but once it clears this portion of the line it is efficiently "pulled" through the remaining part of the line. The opposite configuration would require work to be "pushed" through queues that typically form at the end of the line. In order to maximize performance of an SMT line, the layout design decisions should be evaluated against the final assembly schedule.

Below, a real-world example is presented in which the scheduling tool is used for improving the configuration of an SMT line that assembles a new type of printed wiring board. Figure 2.16 shows the basic flow of the SMT line to be configured. The line consists of a screen printer, two chip-shooters for small components, and two placement machines for larger integrated circuits (ICs) and odd-form components, for example, metal shields and connectors.

Table 2.12 compares six different types of SMT line modifications, in terms of their estimated relative effect on the tact time, that is, the assembly time per board, and the throughput (TH), that is, the number of boards processed per shift. Results are shown for two different actual line configurations and products from real-world factories. The results are presented as a percent of the original tact times and throughputs, shown in the first row of Table 2.12 for the "push line".

As noted in Table 2.12, the different modifications are:

- Changing from a push to a pull line, by moving the slowest machine to the beginning of the line.
- Adding more buffer capacity before or after the bottleneck machine in the line.
- Balancing the cycle times of the chip-shooter and the IC-placer.
- Optimizing the cycle time of the bottleneck machine by applying engineering knowledge and additional machine-dependent optimization software.

Table 2.11 Example 2.4: MIP Results for MPS

Problem	Var.	Bin.	Cons.	Nonz.	*LB*	C^*_{max}	Nodes
1	1269	670	4778	18,250	3127	3233	20
2	1272	673	6265	25,433	3137	3247	46
3	1654	873	6276	23,892	3915	3993	8
4	664	351	2432	9400	1914	1992	21
5	2037	1074	5380	17,975	4583	4695	11

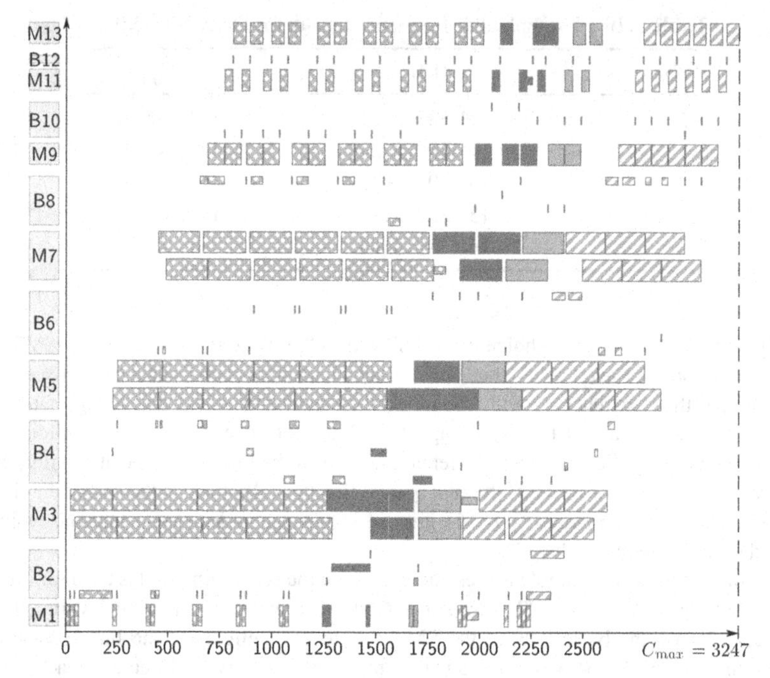

Figure 2.15 Batch schedule for an SMT line with parallel stations.

Theoretically, it would be possible to make all of the above modifications to a given line. However, one should realize that the net improvement would typically be less than the sum of the individual improvements listed in Table 2.12.

The scheduling tool was run to compare different scenarios, and in this way identify the best-in-class line configuration for a given product type. For example, the best configuration obtained for Line 1 was a "pull line" with additional buffer capacity between the placement

Table 2.12 Projected Impact of SMT Line Modifications

	Line 1		Line 2	
Modification	Tact time	TH	Tact time	TH
Push line	100%	100.0%	100%	100%
Pull line	98.7%	–	100%	–
Adding buffer before bottleneck machine	–	–	99.6%	100.5%
Line balancing	98.9%	–	93.8%	–
Optimizing bottleneck machine	98.2%	–	99.2%	–
Adding buffer after bottleneck machine	98.0%	103.7%	98.3%	102.0%

–, Data not available.

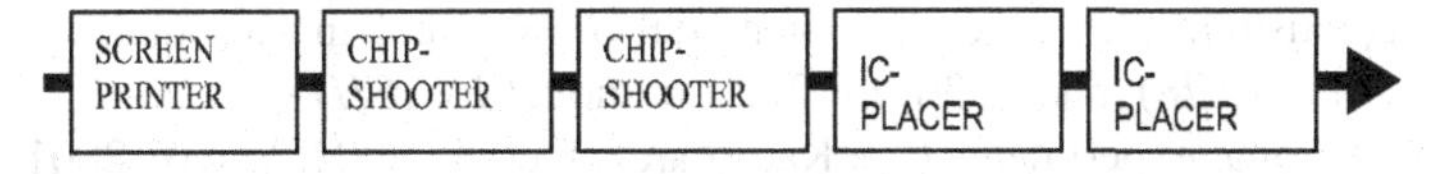

Figure 2.16 SMT line to be configured.

machines. The resulting tact time decreased to 94.2% of the original time and the throughput increased to 106% of the original number of boards per shift for the new product type. Though the throughput increased by only 6% in a high-volume production environment, even a few percent increase in throughput would typically correspond to large cost savings. Hence, the benchmark results presented in Table 2.12 should be considered to be very good.

2.7 COMMENTS

Scheduling of printed wiring board assembly has been considered by many researchers, including Ahmadi et al. (1995), Bard et al. (1994), Feo et al. (1995), Jin et al. (2002), Kim et al. (1996), Kumar and Li (1995), D.-Y. Liao et al. (1996), and Wittrock (1985, 1988).

The exact MIP approach presented in this chapter is capable of scheduling SMT lines by using commercially available software for integer programming. The approach benefits from some characteristics of the SMT assembly. The most important features that were exploited in the MIP formulations or that directly reduce the computational effort required are:

- A single station at the beginning and at the end of an SMT line
- At most two parallel placement machines in stages with parallel stations
- Batch scheduling mode
- Less than 10 different types of boards (batches of boards) to be simultaneously scheduled
- Similar processing times required for different types of boards

In practice, the similar processing times for different types of boards are typical for all processing stages, except placement stages. At placement stages, however, similar processing times are often obtained through the balancing decisions of an SMT line (see Chapter 3). As a result, total placement times for different types of boards at each placement stage tend to be equalized.

Since the batch sequencing problem in the two-machine flow shop with a finite intermediate buffer is NP-hard (see Agnetis et al., 1997, 1998), minimizing makespan in the batch scheduling of SMT lines is NP-hard as well.

The material presented in this chapter is mostly based on the results published by Sawik (2001, 2002d), Sawik et al. (2000, 2002), and Kaczmarczyk et al. (2004). The mixed integer programs *GESCH* for general scheduling, and in particular *BASCH* for batch scheduling, have been strengthened by the incorporation of various cutting constraints, for example, Sawik (2002d). Such constraints were identified exploiting the SMT line configuration by relating assignment, sequencing, and timing variables to more precisely calculate tighter time intervals available for processing of each board (constraints (2.13) to (2.15) in model *GESCH* and, in particular, constraints (2.22), (2.26) to (2.28), and (2.30) to (2.33) in model *BASCH*).

Computational experiments (see Kaczmarczyk et al., 2004; Sawik, 2001; Sawik et al., 2000, 2002) have indicated that the two MIP models can be applied to a variety of different real-world SMT line configurations and production scenarios with only small modifications to the constraint formulations or input data definitions.

The proposed improvement heuristic (see Kaczmarczyk et al., 2004), which combines a tabu search technique and dispatching scheduling based on a pseudo-dynamic programming scheme, is capable of scheduling a typical SMT line, subject to various additional requirements. In particular, the heuristic easily deals with a batch processing mode, limited machine availability, restricted transportation links, or bypassing of some processing stages. The most important part of the heuristic is the assignment and timing. The tabu search is used only for sequencing the boards. In most practical cases, however, the number of different types of boards to be scheduled was less than 10, and hence the tabu search can be simply replaced with a complete enumeration of all batches of boards.

EXERCISES

2.1 Modify the mixed integer program *GESCH* for general scheduling of an SMT line to account for constant transportation times between the stations and

(a) To minimize total completion time, $\sum_{k \in K} c_{mk}$, for a mix of boards.

(b) To minimize the maximum flow time, $\max_{k \in K}(c_{mk} - c_{1k})$, for a mix of boards, subject to an upper bound on makespan constraints.

(c) To minimize the makepan for a mix of boards, subject to the maximum flow time constraints.

2.2 Modify the mixed integer program *BASCH* for batch scheduling of an SMT line to account for constant transportation times between the stations and

(a) To minimize total completion time for a mix of boards.

(b) To minimize the maximum flow time for a mix of boards, subject to an upper bound on makespan constraints.

(c) To minimize the makespan for a mix of boards, subject to the maximum flow time constraints.

2.3 Formulate a mixed integer program for the problem of optimal allocation of in-process buffer capacities in an SMT line with single stations in such a way as to minimize the total completion time for a mix of boards.

2.4 Modify the mixed integer program of Exercise 2.3 to account for the prescheduled downtimes of some machines (see Section 1.4).

2.5 In the computational example in Section 2.6, Figure 2.11 indicates that the makespans of optimal batch and cyclic schedules are identical. Explain why the optimal makespans are identical.

Chapter 3

Balancing and Scheduling of Flexible Assembly Lines

3.1 INTRODUCTION

This chapter deals with balancing and scheduling of flexible assembly lines. A flexible assembly line (FAL) is a flexible flow shop made up of a set of assembly stations in series, each with limited work space for component feeders, with infinite or finite in-process buffers between the stations. The line is capable of simultaneously producing a mix of product types.

Balancing and scheduling are the two most important short-term planning issues. The balancing objective is to determine an allocation of assembly tasks for a mix of products among the assembly stations with limited work space so as to balance the station workloads. In contrast, the scheduling objective is to determine the detailed sequencing and timing of all assembly tasks for each individual product, so as to maximize the line's productivity, which may be defined in terms of throughput, makespan, average flow time for a given product, etc. The objective of the line balancing and scheduling considered in this chapter is to determine an assignment of assembly tasks to stations and an assembly schedule for all products so as to complete the products in minimum time. First, balancing and scheduling of a FAL with infinite in-process buffers is considered, and then balancing and scheduling of surface mount technology lines is modeled. In the FALs with infinite in-process buffers, the balancing and scheduling is investigated for fixed or alternative assembly routing, that is, for single or alternative task assignments. The following two approaches and the corresponding MIP models are proposed and compared: an integrated approach and a hierarchical approach. In the integrated approach the balancing and scheduling decisions are made simultaneously using a large, monolithic MIP model. In the hierarchical approach, however, a sequence of two MIP models is used, first to balance the

Scheduling in Supply Chains Using Mixed Integer Programming. By Tadeusz Sawik

station workloads and then to determine a detailed assembly schedule for prefixed task assignments and assembly routes.

The following MIP models for balancing and scheduling are presented in this chapter:

BAS1 for simultaneous balancing and scheduling of a flexible assembly line with infinite in-process buffers and fixed routing.

BAS for simultaneous balancing and scheduling of a flexible assembly line with infinite in-process buffers and alternative routing.

B1 for balancing of a flexible assembly line with infinite in-process buffers and fixed routing.

B for balancing of a flexible assembly line with infinite in-process buffers and alternative routing.

S|B for scheduling of a flexible assembly line with infinite in-process buffers and prefixed product assignments.

BASMT for simultaneous balancing and scheduling of an SMT line.

BSMT for balancing of an SMT line.

The above MIP models and the two solution approaches proposed are illustrated with a set of computational examples in Sections 3.2.3 and 3.3.3.

3.2 BALANCING AND SCHEDULING OF FLEXIBLE ASSEMBLY LINES WITH INFINITE IN-PROCESS BUFFERS

In this section the integrated and hierarchical approaches are applied for balancing and scheduling of a FAL with infinite in-process buffers. The problem objective is to determine an assignment of assembly tasks to stations for a mix of products and an assembly schedule for all products so as to complete the products in minimum time.

First, a monolithic MIP formulation is presented for the integrated approach to simultaneously balance and schedule the FAL, and then a sequence of two MIP models is proposed for the hierarchical approach to sequentially balance and schedule the line. The top level deals with line balancing, that is, allocation of assembly tasks among the stations for the upcoming scheduling horizon so as to balance the station workloads. The base level deals with machine scheduling, that is, detailed sequencing and timing of all tasks performed by the machines over the scheduling horizon, so as to minimize the makespan, given the task assignments and assembly routes determined at the top level. In the models proposed task duplications are considered and the case of multiple task assignments (alternative assembly routing) is compared with single task assignments (fixed assembly routing) for various line configurations.

3.2.1 Simultaneous Balancing and Scheduling

In this section two monolithic MIP models are presented for simultaneous balancing and scheduling of a flexible assembly line with single task assignments (model *BAS1*) and with alternative task assignments (model *BAS*).

Let us consider a FAL made up of m assembly stations $i \in I = \{1, \ldots, m\}$ and the L/U (loading/unloading) station. The stations are connected in series by a unidirectional transportation system (e.g., a conveyor) so that revisiting of stations is not possible. The line is assumed to have sufficient intermediate buffer capacities to temporarily store work in process so that blocking of stations by completed products never happens. This assumption will be restricted later in Section 3.3. In the line different types of assembly tasks $f \in F$ can be performed to simultaneously assemble various types of products $k \in K$. Let $I_f \subset I$ be the subset of stations capable of performing task f. Each station i has a finite work space b_i where a limited number of component feeders and gripper magazines can be placed. As a result only a limited number of assembly tasks can be assigned to one assembly station. Let a_{if} be the amount of station $i \in I_f$ work space required for assignment of task f. If station i is incapable of performing task f, that is, $i \notin I_f$, then $a_{if} > b_i$, by convention. Moreover, if $a_{if} = 0$ for $i \in I_f$, i.e., task f does not need any amount of station i work space, then task f is not associated with assembly of a new component.

Each product k requires a subset F_k of assembly tasks to be performed subject to in-tree precedence relations defined by the assembly plan for this product. The assembly plan is represented by the set R_k of immediate predecessor-successor pairs of assembly tasks (f, g) such that task $f \in F_k$ must be performed immediately before task $g \in F_k$. Finally, denote by q_{ifk} the assembly time required to perform task $f \in F_k$ of product k on station $i \in I_f$.

The objective of the problem is to determine an assignment of assembly tasks to stations for all products over the scheduling horizon so as to minimize the maximum completion time $C_{\max}$.

A feasible solution of the combined balancing and scheduling problem must satisfy the following basic types of constraints:

- Each assembly task type must be assigned to at least one station (alternative assignments) or to exactly one station (single assignments).
- The total space required for the tasks assigned to each station must not exceed the station finite work space available.
- Each task of every product must be performed on exactly one station.
- Each product must be successively routed to the stations where the required tasks have been assigned, subject to precedence relations defined by its assembly plan.
- Revisiting of stations is not allowed.
- Each station can perform at most one task at a time.

Table 3.1 Notation: Balancing and Scheduling of FAL with Infinite In-Process Buffers

Indices

i = assembly station, $i \in I = \{1, \dots, m\}$
f = assembly task, $f \in F$
k = product, $k \in K = \{1, \dots, n\}$

Input parameters

a_{if} = work space required for assignment of task f to station i
b_i = total work space of station i (number of tasks that may be assigned to station i, if all $a_{if} = 1$)
e_{ifk} = earliest completion time on station i for task f of product k
q_{ifk} = assembly time on station i for task f of product k
I_f = subset of stations capable of performing task f
F_k = subset of tasks required for product k
Q = a large positive constant not less than the schedule length
R_k = the set of immediate predecessor-successor pairs of tasks (f, g) for product k such that task f must be performed immediately before task g

Decision variables

c_{ifk} = completion time on station i of task f of product k (timing variable)
x_{ifk} = 1, if product k is assigned to station i to perform task f; otherwise $x_{ifk} = 0$ (assignment variable)
y_{kl} = 1, if product k precedes product l in the assembly sequence; otherwise $y_{kl} = 0$ (sequencing variable)
z_{if} = 1, if task f is assigned to station $i \in I_f$; otherwise $z_{if} = 0$ (assignment variable)

In order to model the balancing and scheduling problem, two types of binary assignment variables, binary sequencing variables and continuous timing variables are introduced (for notation used, see Table 3.1).

Model BAS1: *Balancing and Scheduling of a Flexible Assembly Line with Fixed Routing*

Minimize

$$C_{\max} \tag{3.1}$$

subject to

1. *Task Assignment Constraints:*

$$\sum_{i \in I_j} z_{if} = 1; \; f \in F \tag{3.2}$$

$$\sum_{f \in F} a_{if} z_{if} \le b_i; \; i \in I \tag{3.3}$$

2. *Precedence Constraints:*

$$\sum_{i \in I_g} c_{igk} \geq \sum_{i \in I_f} c_{ifk} + \sum_{i \in I_g} q_{igk} z_{ig}; \ k \in K, (f, g) \in R_k \tag{3.4}$$

$$\sum_{i \in I_f} i z_{if} \leq \sum_{i \in I_g} i z_{ig}; \ k \in K, (f, g) \in R_k \tag{3.5}$$

3. *Product Noninterference Constraints:*

$$c_{ifk} + Q(2 + y_{kl} - z_{if} - z_{ig}) \geq c_{igl} + q_{ifk};$$
$$k, l \in K, f \in F_k, g \in F_l, i \in I_f \cap I_g{:}k < l \tag{3.6}$$

$$c_{igl} + Q(3 - y_{kl} - z_{if} - z_{ig}) \geq c_{ifk} + q_{igl};$$
$$k, l \in K, f \in F_k, g \in F_l, i \in I_f \cap I_g{:}k < l \tag{3.7}$$

4. *Completion Time Constraints:*

$$c_{ifk} \leq C_{\max}; \ k \in K, f \in F_k, i \in I_f \tag{3.8}$$

$$e_{ifk} z_{if} \leq c_{ifk} \leq Q z_{if}; \ k \in K, f \in F_k, i \in I_f \tag{3.9}$$

5. *Variable Nonnegativity and Integrality Conditions:*

$$c_{ifk} \geq 0; \ k \in K, f \in F_k, i \in I_f \tag{3.10}$$

$$y_{kl} \in \{0, 1\}; \ k, l \in K{:}k < l \tag{3.11}$$

$$z_{if} \in \{0, 1\}; \ f \in F, i \in I_f \tag{3.12}$$

$$C_{\max} \geq 0, \tag{3.13}$$

where the earliest completion time on station i for task f of product k, e_{ifk}, is the sum of q_{ifk} and the minimum total assembly time of all tasks $f' \in F_k$ of product k that precede task f,

$$e_{ifk} = q_{ifk} + \sum_{f' \in F_k : f' \prec f} \min_{i \in I_{f'}} (q_{if'k}).$$

The objective function (3.1) represents the schedule length to be minimized. Constraint (3.2) ensures that each task is assigned to exactly one station. Constraint (3.3) is the station capacity constraint. Inequality (3.4) imposes precedence constraints on completion times of successive tasks of each product and (3.5) maintains for each product the precedence relations among its tasks in a unidirectional flow line with no revisiting of stations allowed. If task g is assigned to station i, then task f that is to be performed immediately before g must be assigned to a station h such that $h \leq i$. Product noninterference constraints (3.6) and (3.7) with binary variables y_{kl} and a $big - Q$ coefficient represent disjunctive constraints on the machines. No two products can be performed on the same station simultaneously. For a given sequence of products at most one constraint of (3.6) and (3.7) is active, and only if both task f

and task g are assigned to the same station i. Finally (3.8) defines the schedule length, and (3.9) imposes variable lower and upper bounds on completion times.

In model *BAS* presented below for balancing and scheduling of a FAL with alternative routing, additional product assignment variables x_{ifk} need to be introduced to select a unique assembly route for each product from a subset of alternative routes.

Model BAS: *Balancing and Scheduling of a Flexible Assembly Line with Alternative Routing*

Minimize (3.1) subject to

1. *Task Assignment Constraints:* (3.3) and

$$\sum_{i \in I_f} z_{if} \geq 1; \ f \in F \tag{3.14}$$

2. *Product Assignment Constraints:*

$$\sum_{i \in I_f} x_{ifk} = 1; \ \ k \in K, f \in F_k \tag{3.15}$$

$$x_{ifk} \leq z_{if}; \ \ k \in K, f \in F_k, i \in I_f \tag{3.16}$$

3. *Precedence Constraints:*

$$\sum_{i \in I_g} c_{igk} \geq \sum_{i \in I_f} c_{ifk} + \sum_{i \in I_g} q_{igk} x_{igk}; \ \ k \in K, (f, g) \in R_k \tag{3.17}$$

$$\sum_{i \in I_f} i x_{ifk} \leq \sum_{i \in I_g} i x_{igk}; \ \ k \in K, (f, g) \in R_k \tag{3.18}$$

4. *Product Noninterference Constraints:*

$$c_{ifk} + Q(2 + y_{kl} - x_{ifk} - x_{igl}) \geq c_{igl} + q_{ifk}; \quad k, l \in K, f \in F_k, g \in F_l, i \in I_f \cap I_g{:}k < l \tag{3.19}$$

$$c_{igl} + Q(3 - y_{kl} - x_{ifk} - x_{igl}) \geq c_{ifk} + q_{igl}; \quad k, l \in K, f \in F_k, g \in F_l, i \in I_f \cap I_g{:}k < l \tag{3.20}$$

5. *Completion Time Constraints:* (3.8) and

$$e_{ifk} x_{ifk} \leq c_{ifk} \leq Q x_{ifk}; \ \ k \in K, f \in F_k, i \in I_f \tag{3.21}$$

6. *Variable Nonnegativity and Integrality Conditions:* (3.10)–(3.13) and

$$x_{ifk} \in \{0, 1\}; \ \ k \in K, f \in F_k, i \in I_f. \qquad (3.22)$$

Constraint (3.14) ensures that each task is assigned to at least one station, and by this admits alternative assembly routes for products. Constraint (3.15) ensures that each task of every product is assigned to exactly one station. The routing constraint (3.16) ensures that products are assigned to the stations where the required tasks may be performed. Inequality (3.17) imposes precedence constraints on completion times of successive tasks of each product and (3.18) maintains for each product the precedence relations among its tasks in a unidirectional flow line with no revisiting of stations allowed. If product k is assigned to station i to perform task g, then to perform task f that is immediately preceding g, this product must be assigned to a station h such that $h \leq i$. Product noninterference constraints (3.19) and (3.20) with binary variables y_{kl} and a $big - Q$ coefficient represent disjunctive constraints on the machines. No two products can be performed on the same station simultaneously. For a given sequence of products at most one constraint of (3.19) and (3.20) is active, and only if both task f of product k and task g of product l are assigned to the same station i. Finally (3.21) imposes variable lower and upper bounds on completion times.

3.2.2 Sequential Balancing and Scheduling

In this section a two-level balancing and scheduling of a FAL is proposed and the corresponding MIP formulations are presented. In the two-level approach, first at the top level the balancing problem is solved and then at the base level a solution to the flow shop scheduling problem for prefixed task assignments is found (Fig. 3.1).

The levels are connected using a strict top-down hierarchy. The top-level objective of balancing the workloads is a commonly used surrogate objective that represents the base-level objective of minimizing the makespan. The top-level prefixes task assignments and assembly routes for each product for the entire scheduling horizon and by this transforms the balancing and scheduling problem into the permutation flow shop problem to be solved at the base level.

Below, the MIP models *B1* and *B* are presented for balancing of a FAL with fixed and alternative routing, respectively.

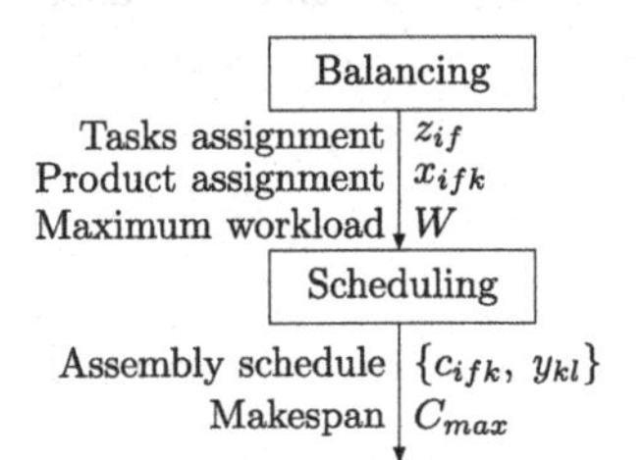

Figure 3.1 A two-level balancing and scheduling.

Model B1: *Balancing of a Flexible Assembly Line with Fixed Routing*

Minimize

$$W \tag{3.23}$$

subject to

1. *Task Assignment Constraints:* (3.2), (3.3)
2. *Precedence Constraints:* (3.5)
3. *Maximum Workload Constraints:*

$$\sum_{k \in K} \sum_{f \in F_k} q_{ifk} z_{if} \le W; \;\; i \in I \tag{3.24}$$

4. *Variable Nonnegativity and Integrality Conditions:* (3.12) and

$$W \ge 0. \tag{3.25}$$

The objective W, (3.23), is a measure of the line imbalance and represents workload of the bottleneck station defined by constraint (3.24).

Model B: *Balancing of a Flexible Assembly Line with Alternative Routing*

Minimize (3.23) subject to

1. *Task Assignment Constraints:* (3.3), (3.14)
2. *Product Assignment Constraints:* (3.15), (3.16)
3. *Precedence Constraints:* (3.18)
4. *Maximum Workload Constraints:*

$$\sum_{k \in K} \sum_{f \in F_k} q_{ifk} x_{ifk} \le W; \;\; i \in I. \tag{3.26}$$

5. *Variable Nonnegativity and Integrality Conditions:* (3.12), (3.22), (3.25).

Having solved the balancing problem $B1$ or B, the next step of the two-level approach is to determine a detailed assembly schedule by solving the flow shop scheduling problem $S|B$ for prefixed product assignments x^B_{ijk}. Note that in the balancing model $B1$, where each task type is assigned to exactly one station, the prefixed product assignments x^B_{ifk} are simply determined by z^B_{if}, that is,

$$x^B_{ifk} = z^B_{if} \; \forall k \in K, f \in F_k.$$

Model S|B: *Scheduling of a Flexible Assembly Line with Prefixed Product Assignments*

Minimize (3.1)

1. *Product Noninterference Constraints:*

$$c_{ifk} + Qy_{kl} \geq c_{igl} + q_{ifk};$$
$$k, l \in K, f \in F_k, g \in F_l, i \in I{:}k < l \text{ and } x^B_{ifk}x^B_{igl} = 1 \quad (3.27)$$
$$c_{igl} + Q(1 - y_{kl}) \geq c_{ifk} + q_{igl};$$
$$k, l \in K, f \in F_k, g \in F_l, i \in I{:}k < l \text{ and } x^B_{ifk}x^B_{igl} = 1 \quad (3.28)$$

2. *Precedence Constraints:*

$$c_{hgk} - q_{hgk} \geq c_{ifk};\ k \in K, (f, g) \in R_k, h, i \in I{:}h \neq i,\ x^B_{ifk}x^B_{hgk} = 1 \quad (3.29)$$
$$c_{igk} - q_{igk} = c_{ifk};\ k \in K, (f, g) \in R_k, i \in I{:}x^B_{ifk}x^B_{igk} = 1 \quad (3.30)$$

3. *Completion Time Constraints:*

$$c_{1gl} = e_{1gl} + \sum_{k \in K:k<l} \sum_{f \in F_k} p_{1fk}z^B_{1f}y_{kl} + \sum_{k \in K:k>l} \sum_{f \in F_k} q_{1fk}z^B_{1f}(1 - y_{lk});$$
$$l \in K, g \in F_l{:}x^B_{1gl} = 1 \quad (3.31)$$
$$c_{ifk} \leq C_{\max};\ k \in K, f \in F_k, i \in I_f{:}\ x^B_{ifk} = 1 \quad (3.32)$$
$$C_{\max} \geq W^B \quad (3.33)$$

4. *Variable Nonnegativity and Integrality Conditions:* (3.11) and

$$c_{ifk} \geq 0;\ k \in K, f \in F_k, i \in I_f{:}\ x^B_{ifk} = 1. \quad (3.34)$$

Product noninterference constraints (3.27) and (3.28) with binary variables y_{kl} and a *big* $- Q$ coefficient represent disjunctive constraints on the machines. No two products assigned to the same station can be performed simultaneously. For a given sequence of products only one constraint of (3.27) and (3.28) is active. Constraints (3.29) and (3.30) maintain for each product the precedence relations among its tasks. In particular, equation (3.30) ensures that successive tasks of each product assigned to the same station are performed contiguously with no breaks between them. Equation (3.31) denotes that on the first station of a FAL, products are assembled one by one with no idle time between successive tasks, i.e., start time of each task is equal to the sum of processing times of all preceding tasks assigned to this station. This constraint reflects the assumption of no-blocking scheduling as a result of infinite intermediate buffers between assembly stations. Finally, (3.32) defines the schedule length, and (3.33) imposes a lower bound on it, where W^B is the optimal value of the maximum workload W.

It is worth nothing that in order to minimize makespan in the flow shop scheduling problem, the station workloads should be balanced. Therefore, the proposed strict top-down hierarchy should produce optimal schedules in most cases and no iterative scheme with the exchange of information between the levels is required to improve the performance of the two-level approach.

3.2.3 Computational Examples

In this section numerical examples and some computational results are presented to illustrate possible application of the two approaches and the proposed MIP formulations.

The FAL configuration is provided in Figure 3.2. The line is made up of $m = 3$ assembly stations ($i = 1,2,3$), and one L/U station.

The production order consists of five products to be assembled of 10 types of components. The assembly plans for products are in the form of assembly task sequences (chains of tasks) to be performed (Fig. 3.3).

The assembly times ($q_{ifk} = q_{fk}, \forall i \in I_f$) are shown below:

$$[q_{fk}] = \begin{bmatrix} 4, 4, 0, 4, 4 \\ 2, 2, 2, 0, 0 \\ 2, 0, 2, 2, 2 \\ 2, 2, 2, 0, 0 \\ 0, 4, 4, 4, 4 \\ 2, 2, 0, 2, 2 \\ 0, 3, 3, 3, 3 \\ 5, 0, 5, 5, 5 \\ 0, 2, 2, 2, 2 \\ 0, 4, 4, 4, 4 \end{bmatrix}.$$

Work space required for task assignments:

$$[a_{if}] = \begin{bmatrix} 1, 2, 3, 1, 2, 3, 9, 9, 9, 9 \\ 11, 11, 11, 1, 2, 3, 1, 2, 3, 5 \\ 1, 2, 3, 10, 10, 10, 1, 2, 3, 5 \end{bmatrix}.$$

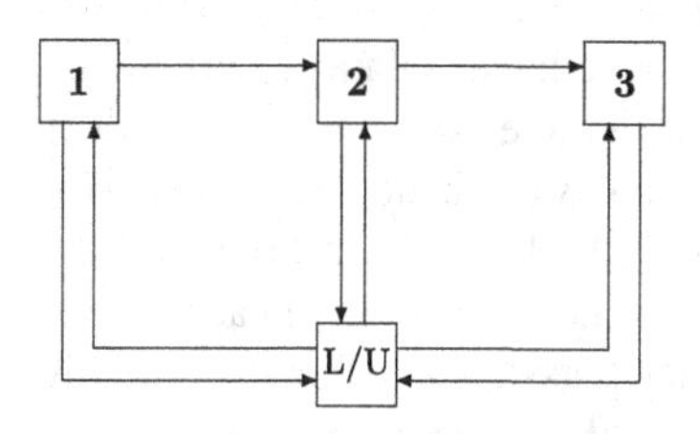

Figure 3.2 A flexible assembly line.

$k = 1:\quad L \longrightarrow 1 \longrightarrow 2 \longrightarrow 3 \longrightarrow 4 \longrightarrow 6 \longrightarrow 8 \longrightarrow U$

$k = 2:\quad L \longrightarrow 1 \longrightarrow 2 \longrightarrow 4 \longrightarrow 5 \longrightarrow 6 \longrightarrow 7 \longrightarrow 9 \longrightarrow 10 \longrightarrow U$

$k = 3:\quad L \longrightarrow 2 \longrightarrow 3 \longrightarrow 4 \longrightarrow 5 \longrightarrow 7 \longrightarrow 8 \longrightarrow 9 \longrightarrow 10 \longrightarrow U$

$k = 4:\quad L \longrightarrow 1 \longrightarrow 3 \longrightarrow 5 \longrightarrow 6 \longrightarrow 7 \longrightarrow 8 \longrightarrow 9 \longrightarrow 10 \longrightarrow U$

$k = 5:\quad L \longrightarrow 1 \longrightarrow 3 \longrightarrow 5 \longrightarrow 6 \longrightarrow 7 \longrightarrow 8 \longrightarrow 9 \longrightarrow 10 \longrightarrow U$

Figure 3.3 Graph of assembly sequences for products.

Work space available at stations:

$$b_1 = 8,\ b_2 = 10,\ b_3 = 9.$$

Note that, if station i is incapable of performing task f, then $a_{if} > b_i$.

First, the integrated approach was applied to find optimal solutions for simultaneous balancing and scheduling using the MIP models *BAS1* and *BAS*. Then, for comparison the assembly schedules were determined using the hierarchical approach presented in Section 3.2.2. For the example problem, both the approaches constructed the same optimal schedules. The schedules obtained for fixed and alternative routing are shown in Figure 3.4, where successive tasks of each product are indicated with the product number. The maximum workload W determined at the top level and the optimal makespan $C_{\max}$ achieved at the base level (as well as by using the integrated approach) are indicated in the figure.

In order to evaluate how the work space constraints influence the solution of the two approaches, the assembly schedules were determined for the example problem with varying work space available at each station. The results are summarized in Table 3.2, where the tightness of capacity constraint (3.3) is measured by the ratio $\sum_{i\in I} b_i / \sum_{f\in F} \min_{i\in I_f}(a_{if})$ of the total work space available to the total work space required, shown in column 2. The results indicate that both approaches give the same makespan in most cases. Some differences appeared only for fixed routing with tight capacity constraint and for alternative routing with large slack in the capacity constraint.

In order to further compare the performance of the integrated and hierarchical approaches with respect to number of stations, number of products, and number of tasks, additional numerical examples were generated with $m = 5$ or 10 stations, $n = 5$ or 10 products, $|F| = 10$ or 20 types of assembly tasks, and in total $\sum_{k\in K} |F_k| = 80$ tasks required to assemble all products.

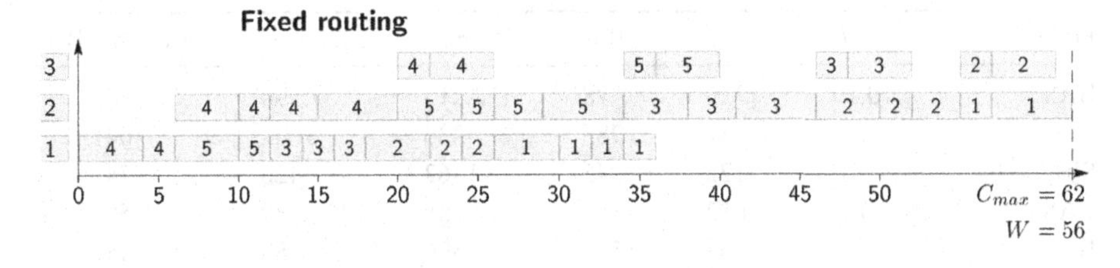

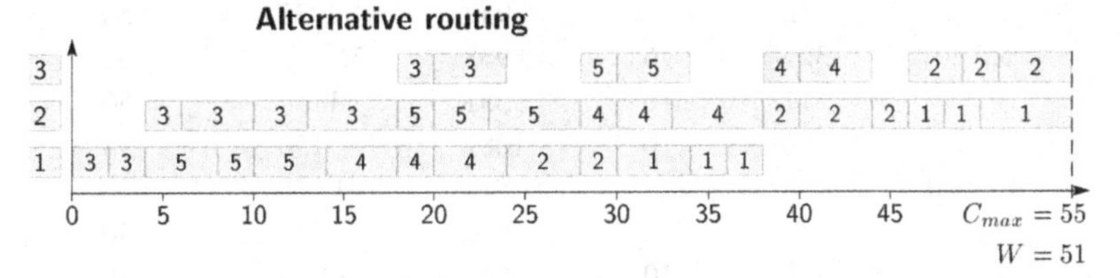

Figure 3.4 Assembly schedules for integrated/hierarchical balancing and scheduling.

Table 3.2 Influence of the Work Space Constraints

Work space	Work space tightness	Fixed routing			Alternative routing		
b_1, b_2, b_3	$\frac{\sum_{i\in I} b_i}{\sum_{f\in F} \min_{i\in I_f}(a_{if})}$	$C_{\max}^{(BAS1)}$	W	$C_{\max}^{(S\|B)}$	$C_{\max}^{(BAS)}$	W	$C_{\max}^{(S\|B)}$
16, 16, 16	2.09	57	44	57	45	39	49
15, 15, 15	1.96	57	44	57	45	39	52
12, 12, 12	1.56	57	44	57	45	39	52
8, 10, 9	1.17	62	56	62	55	51	55
8, 8, 8	1.04	62	56	62	62	56	62
7, 8, 8	1.00	62	56	62	–	–	–
6, 6, 11	1.00	62	56	66	–	–	–
6, 7, 10	1.00	57	44	57	–	–	–

In the examples with 10 task types additional stations were selected from the set of three stations of the first example. In the examples with 20 task types, capabilities of each station with respect to tasks $f = 11, \dots, 20$ were identical with capabilities of the corresponding station of the first example, with respect to tasks $f = 1, \dots, 10$, that is, $a_{if} = a_{i,f+10}, f = 1, \dots, 10$. Furthermore, if some product k requires task $f \in \{11, \dots, 20\}$ it also requires task $f - 10$, and the corresponding assembly times are assumed to be identical, that is, $q_{fk} = q_{f-10,k}$.

In the examples with 10 products, the work space of each station i is $b_i = 8$, whereas in the examples with 20 products total work space of all stations is $\sum_{i\in I} b_i = 60$. In the latter case the tightness of capacity constraint (3.3), $\sum_{i\in I} b_i / \sum_{f\in F} \min_{i\in I_f} (a_{if}) = 60/46$ is identical for all examples, while in the former case the tightness $8m/23$ decreases with the number of stations.

Table 3.3 Computational Results: Integrated vs. Hierarchical Approach

Problem	$m, n, \|F\|$	Var.	Bin.	Constr.	Nonz.	$C_{\max}$ or (W)
BAS	5,10,10	329	78	12,215	57,590	74
B		284	283	336	1044	(64)
S|B		122	45	1462	4420	74
BAS	10,10,10	612	111	24,288	115,180	56
B		567	566	596	2088	(44)
S|B		122	45	942	2732	56
BAS	5,5,20	327	76	10,938	52,250	90
B		317	316	351	1173	(56)
S|B		87	10	1081	3150	90
BAS	10,5,20	643	142	21,714	104,500	76
B		633	632	611	2346	(40)
S|B		87	10	779	2223	76

The computational results are summarized in Table 3.3, where characteristics m, n, $|F|$ of the examples are shown in column 2 of the table. Table 3.3 shows that for each example problem both approaches produce the same optimal solution.

The computational experiments described in Sawik (2002c) indicated that the proposed strict top-down hierarchy is capable of finding optimal assembly schedules in most cases. However, the CPU time required for the hierarchical approach is always much smaller than the corresponding CPU time for the integrated approach, in particular for larger problems. The CPU time for the monolithic model increases with the number of stations and with the slack in the capacity constraint (3.3).

3.3 BALANCING AND SCHEDULING OF SMT LINES

The two major short-term planning problems in electronics assembly are balancing and scheduling. Given a mix of boards to be produced, the objective of the balancing problem is to allocate assembly tasks and component feeders among the placement stations with limited number of feeder slots, so as to balance the station workloads. SMT line balancing can be accomplished if there are alternate placement machines at which components can be assembled. A typical SMT line has more than one placement machine for small components and more than one placement machine for large components.

The scheduling objective is to determine the detailed sequencing and timing of all assembly tasks for each individual board, so as to minimize the assembly schedule length (makespan) for a mix of board types. In order to best utilize an SMT line capability, balancing decisions should be evaluated against the final schedule for a mix of boards to be assembled. Therefore, it would be advantageous to simultaneously balance and schedule the SMT line. The resulting component feeder assignment and the corresponding placement times would better fit sequencing and timing constraints to minimize makespan, which is not possible when balancing and scheduling phases are separated.

The MIP models for scheduling of SMT lines are presented in Chapter 2. In this section simultaneous balancing and scheduling of an SMT line is compared with a typically used two-level, hierarchical approach, where first parts feeder assignments to placement machines are found to distribute the chip placement workload evenly over the line, and then the shortest assembly schedule is determined for a mix of boards, given component feeder allocation.

For both the integrated and the hierarchical approach, MIP formulations are proposed in Sections 3.3.1 and 3.3.2, and applied to solve a typical SMT line balancing and scheduling problem in Section 3.3.3.

The following eight basic categories of components can be met in electronics assembly (e.g., Tirpak, 2000).

- Basic components.
- Integrated circuits that typically require a special optical system and precision placement.

- Small integrated circuits that can be placed with a revolver head.
- Odd-shaped components that include connectors, which do not have a regular surface for easy pick-up with a vacuum nozzle.
- Shield components that include the metal shields used for electrically isolating parts of the PWB. Shields usually require precision placement and involve special precedence constraints so that shields are placed on a board after the components that are to be shielded beneath them.
- Direct chip attach components that require precision placement, special vision algorithms, and a flux dispensing step prior to placement.
- Manually placed components.
- CAD data, such as fiducial mark locations, a special category that does not involve placement of a component.

A placement machine typically can assemble more than one category of components. Various types of placement machines are met in electronics assembly. However, each machine consists of three basic structures: the feeder carrier, the placement head, and the board supporting system. A key element of the SMT line is the component placement machine, called the "chip shooter." Figure 3.5 shows a typical high-speed machine for placement of small components at speeds up to 48,000 components per hour. Components are picked from a movable feeder carrier that holds reels or cassettes with up to 10,000 of a given component type. The components are transported by the turret, past several stations for inspection and rotation, and then finally placed on the board at the preprogrammed X-Y location.

For placing larger components, fine-pitch components, and odd-shaped components requiring special nozzles, gantry-type placement machines are used, where the feeders are stationary and X-Y gantry carries components through the entire pick-and-place cycle.

Design specifications for each type of PWB include the number and sizes of particular components and their placement locations X-Y on the board. The number

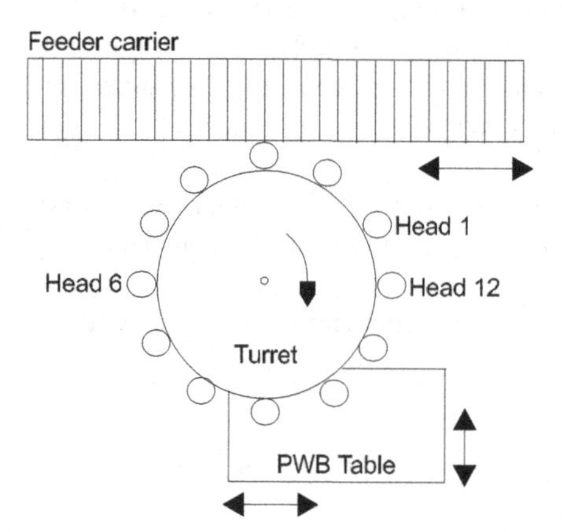

Figure 3.5 SMT placement machine for small components.

of placements may reach several hundred components, while the number of different component types on any PWB is typically much smaller. The component placement (pick-and-place) time consists of two parts: the time of picking the component from the feeder slot, carrying it to the vision camera and checking its orientation, and the time of moving the component from the vision camera to the placement position on the board. The placement time depends on placement location X-Y on the board. On the other hand, the difference in the traveling time from any two feeder slots to the vision system is relatively small and can be neglected. Therefore, for each placement station and each placement location on a given type of board it can be assumed that the placement time is known a priori.

3.3.1 Simultaneous Balancing and Scheduling of an SMT Line

In this section, MIP model *BASMT* is presented for simultaneous balancing and scheduling of an SMT line. The model addresses the three basic decisions:

- What should be the assignment of component feeders to placement stations with a limited number of feeder slots?
- What should be the assignment of boards to parallel stations and buffers?
- What should be the sequence and timing of boards entering the line?

The component feeder assignment determines processing times at each placement station for each type of PWB, and the scheduling decisions to optimize performance of an SMT line depend on these processing times. Notation used to formulate the problems is shown in Table 3.4, where both buffers and machines are referred to as processors.

Let us assume that the SMT line under study consists of m processing (placement and nonplacement) stages in series, where each stage consists of one or more parallel identical processors. Nonplacement stages include all internal input and output buffers, external buffers, shuttles as well as screen printer, reflow oven, vision inspection stations, etc. The line is capable of producing various types of boards using different types of components for PWB assembly. The feeder reels, which are tapes holding the components, are positioned in feeder slots of a feeder carrier at a placement station. The feeder carrier consists of a number of feeder slots. A reel containing wider components typically occupies two or more adjacent feeder slots in the feeder carrier. Let a_f be the number of feeder slots occupied by a reel with components of type f, and let b_i be the number of feeder slots of a feeder carrier at a placement machine in a placement stage $i \in IP$. Each component type is typically assigned to one placement stage. Duplication of components of the same type at different placement stages is not allowed. If in fact it is necessary, the duplicated component is regarded as a different type of component.

PWB design specifies the set of placement locations and for each location on the board the type of required component. Let q_{ifg} be the placement time in stage $i \in IP$ of

Table 3.4 Notation: Balancing and Scheduling of SMT Lines

Indices

f = component type, $f \in F$
g = board type, $g \in G$
i = processing stage, $i \in I = \{1, \ldots, m\}$
j = processor in stage i, $j \in J_i = \{1, \ldots, m_i\}$
k = board, $k \in K = \{1, \ldots, n\}$

Input parameters

a_f = number of feeder slots occupied by component type f
b_i = number of feeder slots at placement machine in stage $i \in IP$
q_{ifg} = total placement time in stage i of all required components of type f on board type g
r_{ig} = processing time for board type g in a nonplacement stage $i \notin IP$
F_g = subset of component types required for board type g
IP = subset of placement stages, $IP \subset I$
SH = subset of shuttle stages, $SH \subset I$
$I(\sigma)$ = subset of stages with parallel processors, between two successive shuttles, immediately following shuttle stage $\sigma \in SH$
K_g = subset of boards of type g
Q = a large positive constant not less than makespan

Decision variables

c_{ik} = completion time of board k in stage i (timing variable)
d_{ik} = departure time of board k from stage i (timing variable)
p_{ik} = processing time for board k in placement stage $i \in IP$
x_{ijk} = 1, if board k is assigned to processor $j \in J_i$ in stage $i \in I$; otherwise $x_{ijk} = 0$ (assignment variable)
y_{kl} = 1, if board k precedes board l; otherwise $y_{kl} = 0$ (sequencing variable)
z_{if} = 1, if component type f is assigned to stage $i \in IP$; otherwise $z_{if} = 0$ (assignment variable)

all required components of type $f \in F_g$ on board type $g \in G$, that is, q_{ifg} is a constant determined by the summation of placement times for all locations on the board type g that require component type f.

For each type of board, the total placement time in each stage depends on the assignment of required component feeders. Let p_{ik} be the total placement time in stage $i \in IP$ of each board $k \in K_g$ of type $g \in G$ and z_{if} component assignment binary variable (see Table 3.4). For each board type $g \in G$ and each placement stage $i \in IP$, the total placement time p_{ik}, $k \in K_g$ is a variable determined by the summation of the placement times for all required components (placements) $f \in F_g$ that have been assigned to this stage, i.e. $p_{ik} = \sum_{g \in G, f \in F_g : k \in K_g} q_{ifg} z_{if}$.

Finally, let $r_{ig} \geq 0$ be the fixed processing time of board type $g \in G$ in a nonplacement stage $i \notin IP$, ($r_{ig} = 0$ for each buffer stage i).

The balancing and scheduling objective is to determine an assignment of component feeders to placement stations and boards to processors in each stage over a scheduling horizon to complete all the boards in minimum time. The generic SMT line balancing and scheduling model has the following structure.

Minimize *maximum completion time* subject to

1. *Component feeder assignment constraints for placement stages* to ensure that each component type is assigned to exactly one placement stage and the number of occupied feeder slots at each placement station does not exceed number of available slots. The resulting processing times at placement stages are determined for each type of board.
2. *Board assignment constraints for stages with parallel processors* to ensure that each board is assigned to exactly one processor and remains on the same conveyor until a shuttle stage is reached.
3. *Board completion constraints* to ensure that each board is processed at all stages.
4. *Board departure constraints* to ensure that each board cannot be departed from a stage until it is completed in this stage and to ensure that each board leaves the line as soon as it is completed at the last stage.
5. *Board noninterference constraints* to ensure that no two boards are processed by the same processor simultaneously.
6. *No buffering constraints* to ensure that processing of each board at every stage starts immediately after its departure from the previous stage.
7. *Maximum completion time constraints* to calculate the maximum completion time, that is, the makespan of a given assembly schedule.

The monolithic MIP model for simultaneous balancing and scheduling of an SMT line is presented below.

Model BASMT: *Balancing and Scheduling of an SMT Line*

Minimize (3.1) subject to

1. *Component Assignment Constraints for Placement Stations*:

$$\sum_{i \in IP} z_{if} = 1; \; f \in F \tag{3.35}$$

$$\sum_{f \in F} a_f z_{if} \leq b_i; \; i \in IP \tag{3.36}$$

2. *Processing Time Constraints*:

$$p_{ik} = r_{ig}; \; g \in G, i \notin IP, k \in K_g \tag{3.37}$$

$$p_{ik} = \sum_{f \in F_g} q_{ifg} z_{if}; \; g \in G, i \in IP, k \in K_g \tag{3.38}$$

3. *Board Assignment Constraints*:

$$\sum_{j \in J_i} x_{ijk} = 1; \ i \in I, k \in K \tag{3.39}$$

$$x_{i_1, j_1, k} = x_{i_2, j_2, k}; \ \sigma \in SH, i_1, i_2 \in I(\sigma), j_1 \in J_{i_1}, j_2 \in J_{i_2}, k \in K: \ i_1 < i_2, j_1 = j_2 \tag{3.40}$$

4. *Board Completion Constraints*:

$$c_{1k} \geq p_{1k}; \ k \in K \tag{3.41}$$

$$c_{ik} - c_{i-1k} \geq p_{ik}; \ i \in I, \ k \in K{:}i > 1 \tag{3.42}$$

5. *Board Departure Constraints*:

$$c_{ik} \leq d_{ik}; \ i \in I, k \in K{:}i < m \tag{3.43}$$

$$c_{mk} = d_{mk}; \ k \in K \tag{3.44}$$

6. *Board Noninterference Constraints*:

$$c_{ik} + Q(2 + y_{kl} - x_{ijk} - x_{ijl}) \geq d_{il} + p_{ik}; i \in I, j \in J_i, k, l \in K: \ k < l \tag{3.45}$$

$$c_{il} + Q(3 - y_{kl} - x_{ijk} - x_{ijl}) \geq d_{ik} + p_{il}; i \in I, j \in J_i, k, l \in K: \ k < l \tag{3.46}$$

7. *No Buffering Constraints*:

$$c_{ik} - p_{ik} = d_{i-1k}; \ i \in I, k \in K{:}i > 1 \tag{3.47}$$

8. *Maximum Completion Time Constraints*:

$$c_{mk} \leq C_{\max}; \ k \in K \tag{3.48}$$

$$C_{\max} \geq LB \tag{3.49}$$

9. *Variable Nonnegativity and Integrality Conditions*:

$$c_{ik} \geq 0; \ i \in I, k \in K \tag{3.50}$$

$$d_{ik} \geq 0; \ i \in I, k \in K \tag{3.51}$$

$$p_{ik} \geq 0; \ i \in I, k \in K \tag{3.52}$$

$$x_{ijk} \in \{0, 1\}; \ i \in I, j \in J_i, k \in K \tag{3.53}$$

$$y_{kl} \in \{0, 1\}; \ k, l \in K{:}k < l \tag{3.54}$$

$$z_{if} \in \{0, 1\}; \ i \in IP, f \in F \tag{3.55}$$

The objective function (3.1) represents the schedule length to be minimized. Constraint (3.35) ensures that each component type is assigned to exactly one placement stage, and (3.36) that the total number of occupied slots does not exceed the total number of the feeder slots in each placement station. Equations (3.37) and (3.38) define fixed and variable processing times for each board, respectively, in a nonplacement stage and in a placement stage. Constraint (3.39) ensures that in every stage each board is assigned to exactly one processor. Constraint (3.40) ensures that each board remains on the same conveyor until a shuttle stage. Constraint (3.41) ensures that each board is processed in the first stage, and (3.42) guarantees that it is also processed in all downstream stages. Constraint (3.43) indicates that each board cannot be departed from a stage until it is completed in this stage, and equation (3.44) ensures that each board leaves the line as soon as it is completed in the last stage. Constraints (3.45) and (3.46) are board noninterference constraints. No two boards can be performed on the same processor simultaneously. For a given sequence of boards at most one constraint (3.45) or (3.46) is active, and only if both boards k and l are assigned to the same processor. Equation (3.47) indicates that processing of each board in every stage starts immediately after its departure from the previous stage. Finally (3.48) defines the maximum completion time, and (3.49) imposes a lower bound LB on it. The lower bound is calculated as below.

$$LB_1 = \max_{i \notin IP} \left\{ \sum_{g \in G} r_{ig}|K_g|/m_i + \min_{g \in G} \left(\sum_{h \notin IP: h \neq i} r_{hg} \right) \right\} + \min_{g \in G} \left(\sum_{f \in F_g} \min_{i \in IP} (q_{ifg}) \right) \tag{3.56}$$

$$LB_2 = \sum_{g \in G} |K_g| \left(\sum_{f \in F_g} \min_{h \in IP} (q_{hfg}) \right) \Big/ \sum_{h \in IP} m_h + \min_{g \in G} \left(\sum_{h \notin IP} r_{hg} \right) + \min_{g \in G, f \in F_g, i \in IP} (q_{ifg}) \tag{3.57}$$

$$LB = \max \{LB_1, LB_2\} \tag{3.58}$$

LB_1 and LB_2 denote lower bounds based on calculation of maximum completion time in nonplacement and placement stages, respectively. ($|K_g|$ is the number of boards of type $g \in G$.)

A large positive constant Q that is not less than optimal makespan can be calculated as follows ($|IP|$ is the number of placement stages).

$$Q = \max_{i \notin IP} \left[\sum_{g \in G} r_{ig}|K_g|/m_i \right] + \sum_{g \in G, f \in F_g} \max_{i \in IP} (q_{ifg})|K_g| \Big/ \sum_{i \in IP} m_i + (|IP| - 1) \max_{g \in G} \left(\sum_{f \in F_g} \max_{i \in IP} (q_{ifg}) \right), \tag{3.59}$$

where Q can also be used as an upper bound on optimal makespan.

The basic *BASMT* model can be strengthened by the incorporation of additional cut constraints. It is possible to generate such constraints by relating sequencing and timing variables to more precisely calculate tighter time intervals available for processing of each board. As a result the following constraints can be incorporated into the *BASMT* model (cf. (2.4), (2.13), (2.15) in Chapter 2):

$$c_{1l} \geq r_{1g} + \sum_{h \in G, k \in K_h : k < l} r_{1h} y_{kl} + \sum_{h \in G, k \in K_h : k > l} r_{1h}(1 - y_{lk});$$

$$g \in G, l \in K_g : 1 \notin IP, m_1 = 1 \tag{3.60}$$

$$d_{ik} + \sum_{h \in I : h > i} p_{hk} \leq C_{\max}; \; i \in I, k \in K : i < m \tag{3.61}$$

$$c_{ml} - c_{1l} \leq C_{\max} - r_{1g} - \sum_{h \in G, k \in K_h : k < l} r_{1h} y_{kl} - \sum_{h \in G, k \in K_h : k > l} r_{1h}(1 - y_{lk}) - \sum_{h \in G, k \in K_h : k < l} r_{mh}(1 - y_{kl}) - \sum_{h \in G, k \in K_h : k > l} r_{mh} y_{lk};$$

$$g \in G, l \in K_g : 1 \notin IP, m \notin IP, m_1 = 1, m_m = 1. \tag{3.62}$$

Constraint (3.60) (equivalent to (2.4)) arises from the observation that completion time of each board in the first stage cannot be less than the sum of its processing time and processing times of all preceding boards. Constraint (3.61) (equivalent to (2.13)) relates board departure times to makespan directly. Every board must be departed from a station sufficiently early in order to complete all of its remaining tasks within the remaining processing time. Finally, constraint (3.62) (equivalent to (2.15)) ensures that each board is processed within the time intervals remaining after processing of all preceding boards and before processing of all succeeding boards. Flow time $c_{ml} - (c_{1l} - p_{1l})$ of each board $l \in K$ cannot be greater than the makespan $C_{\max}$ minus the sum of processing times of all preceding boards in the first stage

$$\sum_{h \in G, k \in K_h : k < l} r_{1h} y_{kl} + \sum_{h \in G, k \in K_h : k > l} r_{1h}(1 - y_{lk})$$

and the sum of processing times of all succeeding boards in the last stage

$$\sum_{h \in G, k \in K_h : k < l} r_{mh}(1 - y_{kl}) + \sum_{h \in G, k \in K_h : k > l} r_{mh} y_{lk}.$$

It is worth remembering that r_{ig} is the fixed processing time of each board type g in a nonplacement stage $i \notin IP$ (see constraint (3.37)).

Constraints (3.60) and (3.62) require that the first and last stages are single nonplacement machines, which is usually met in practice (a typical SMT line begins with a screen printer and ends with an oven for solder reflow or a vision inspection station).

To reduce further the complexity of the *general scheduling* problem, where any sequence of boards is allowed and to minimize machine set up to build a specific board, a *batch scheduling* mode (see, Section 1.2.6) is usually applied, where boards of a given type are scheduled consecutively. In addition:

- The sequence of board types can be fixed and equal to the optimal sequence determined for a minimum set of boards in the same proportion as the production order; or
- The sequence of board types is not determined a priori, but is obtained with the optimal schedule for all boards.

3.3.2 Sequential Balancing and Scheduling of an SMT Line

In this section a hierarchical, two-level balancing and scheduling is proposed and the corresponding MIP formulations are presented. In the two-level approach, first at the top level the balancing problem is solved to best allocate component feeders among placement stages and then at the base level the shortest assembly schedule for a mix of board types is found for prefixed assignment of component feeders, see Figure 3.6.

The levels are connected using a pure top-down approach. The top-level objective of balancing the workloads is a commonly used surrogate objective that represents the base-level objective of minimizing the makespan. The top-level prefixes component feeders assignment and the corresponding processing times. The detailed schedule is next determined at the base level by solving the scheduling problem generated at the top level.

The MIP model for balancing of an SMT line is presented below.

Model BSMT: *Balancing of an SMT Line*

Minimize

$$W \tag{3.63}$$

subject to

1. *Component Assignment Constraints for Placement Stations:* (3.35), (3.36)
2. *Processing Time Constraints:* (3.37), (3.38)

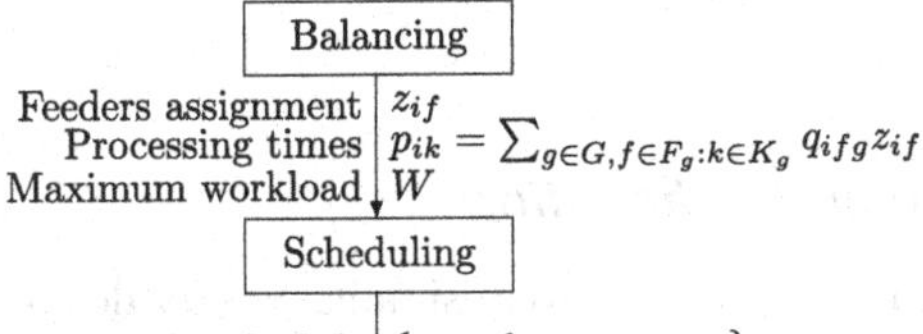

Figure 3.6 Two-level balancing and scheduling of an SMT line.

3. *Maximum Workload Constraints:*

$$\sum_{k\in K} p_{ik}/m_i + \min_{g\in G}\left(\sum_{h\notin IP:h\neq i} r_{hg}\right) + \min_{g\in G,f\in F_g}\left(\sum_{h\in IP:h\neq i} q_{hfg}\right) \leq W;\ i\in I \qquad (3.64)$$

4. *Variable Nonnegativity and Integrality Conditions:* (3.52), (3.55)

The objective W (3.63) defined by constraint (3.64) is a measure of the line imbalance and represents a lower bound on makespan.

Having solved the balancing problem, the next step of the two-level approach is to determine a detailed assembly schedule for prefixed feeder assignment z_{if} and the resulting processing times p_{ik}. The detailed assembly schedule can be found by using MIP model *GESCH* or *BASCH* (see Chapter 2), respectively, for general or batch scheduling of an SMT line.

3.3.3 Computational Examples

In this section simple numerical examples and some computational results are presented to illustrate application of the proposed MIP approach. In the computational experiments, the MIP model *BASMT* has been enhanced with the cut constraints, and the batch scheduling mode with a free sequence of board types has been applied. In the SMT line with parallel stations and the batch scheduling mode, successive boards of one type are assigned alternately to different parallel machines.

The SMT line configuration for the example is provided in Figure 2.9. The line consists of $m = 12$ stages, where stage $i = 1$ is a screen printer, each stage $i = 5,9$ represents two parallel machines for automatic placement of components, stage $i = 12$ is a vision inspection station, stages $i = 3,7,11$ are shuttles, stage $i = 2$ is a single external buffer, and each stage $i = 4,8$ and $i = 6,10$ represents two internal input and output buffers, respectively. The production order for the example consists of 30 boards of three types, 10 boards of each type. The boards are made up of 75 types of components. In the example, placement machines are identical, and hence placement times do not depend on machines, that is, $q_{ifg} = q_{fg},\ g \in G, f \in F_g, \forall i \in IP$. The component sizes $a_f, f = 1,\ldots,75$ and placement times $q_{fg},\ g = 1, 2, 3$ are shown in Table 3.5. Zero entries in the table indicates that particular types of components are not required to assemble some types of boards.

3.3.3.1 Simultaneous Balancing and Scheduling

The optimal feeder assignment and assembly schedules were simultaneously determined by solving MIP model *BASMT*. The number of feeder slots available at each placement machine M5 and M9 is $b_i = 100,\ i = 5, 9$.

Table 3.5 Component Sizes and Placement Times

Component type f	1	2	3	4	5	6	7	8	9	10
Component size a_f	1	2	3	1	2	3	1	2	3	5
Placement time q_{f1}	7	9	0	9	0	9	0	6	7	14
Placement time q_{f2}	7	0	10	9	0	0	8	0	7	14
Placement time q_{f3}	7	9	10	9	0	9	0	0	7	14
Component type f	11	12	13	14	15	16	17	18	19	20
Component size a_f	1	2	3	1	2	3	1	2	3	5
Placement time q_{f1}	7	9	0	9	0	9	0	6	7	14
Placement time q_{f2}	7	0	10	9	0	0	8	0	7	14
Placement time q_{f3}	7	9	10	9	0	9	0	0	7	14
Component type f	21	22	23	24	25	26	27	28	29	30
Component size a_f	1	2	3	1	2	3	1	2	3	5
Placement time q_{f1}	7	9	0	9	0	9	0	6	7	14
Placement time q_{f2}	7	0	10	9	0	0	8	0	7	14
Placement time q_{f3}	7	9	10	9	0	9	0	0	7	14
Component type f	31	32	33	34	35	36	37	38	39	40
Component size a_f	1	2	3	1	2	3	1	2	3	5
Placement time q_{f1}	7	9	0	9	0	9	0	6	7	14
Placement time q_{f2}	7	0	10	9	0	0	8	0	7	14
Placement time q_{f3}	7	9	10	9	0	9	0	0	7	14
Component type f	41	42	43	44	45	46	47	48	49	50
Component size a_f	1	2	3	1	2	3	1	2	3	5
Placement time q_{f1}	7	9	0	9	0	9	0	6	7	14
Placement time q_{f2}	7	0	10	9	0	0	8	0	7	14
Placement time q_{f3}	7	9	10	9	0	9	0	0	7	14
Component type f	51	52	53	54	55	56	57	58	59	60
Component size a_f	1	2	3	1	2	3	1	2	3	5
Placement time q_{f1}	7	9	0	9	0	9	0	6	7	14
Placement time q_{f2}	7	0	10	9	0	0	8	0	7	14
Placement time q_{f3}	7	9	10	9	0	9	0	0	7	14
Component type f	61	62	63	64	65	66	67	68	69	70
Component size a_f	1	2	3	1	2	3	1	2	3	5
Placement time q_{f1}	7	9	0	9	0	9	0	6	7	14
Placement time q_{f2}	7	0	10	9	0	0	8	0	7	14
Placement time q_{f3}	7	9	10	9	0	9	0	0	7	14
Component type f	71	72	73	74	75	–	–	–	–	–
Component size a_f	1	2	3	1	2	–	–	–	–	–
Placement time q_{f1}	7	9	0	9	0	–	–	–	–	–
Placement time q_{f2}	7	0	10	9	0	–	–	–	–	–
Placement time q_{f3}	7	9	10	9	0	–	–	–	–	–

Table 3.6 Processing Times for Simultaneous Balancing and Scheduling

Board type	1	2	3
Screen printer P	10	10	10
Placement machine M5	275	166	233
Placement machine M9	177	245	257
Vision inspection machine V	20	20	20

The following assignment of components was achieved: component feeders of type $f = 1, 2, 4, 6, 8, 9, 12, 18, 21, 22, 26, 27, 28, 31, 32, 34, 36, 37, 38, 39, 42, 46, 47, 48, 49, 50, 51, 54, 56, 57, 58, 59, 61, 66, 68, 69, 70, 74, 75$, were assigned to placement machine M5, and component feeders of type $f = 3, 5, 7, 10, 11, 13, 14, 15, 16, 17, 19, 20, 23, 24, 25, 29, 30, 33, 35, 40, 41, 43, 44, 45, 52, 53, 55, 60, 62, 63, 64, 65, 67, 71, 72, 73$, to placement machine M9. Table 3.6 shows the resulting total placement times at machines M5 and M9 for each type of board.

The optimal schedule obtained for batch scheduling is shown in Figure 3.7, where letters B, M, P, S, and V stand for Buffer, Machine for placement, Printer, Shuttle, and Vision inspection machine, respectively. Boards of types 1 and 2 are indicated, respectively with gray and black shading, and boards of type 3 with cross-hatching. Processor blocking is indicated with a narrow bar. The optimized makespan is $C_{\max} = 3611$.

3.3.3.2 Sequential Balancing and Scheduling

For a comparison, the two-level approach was applied to solve the same example problem. In the first stage, the optimal assignment of component feeders was determined

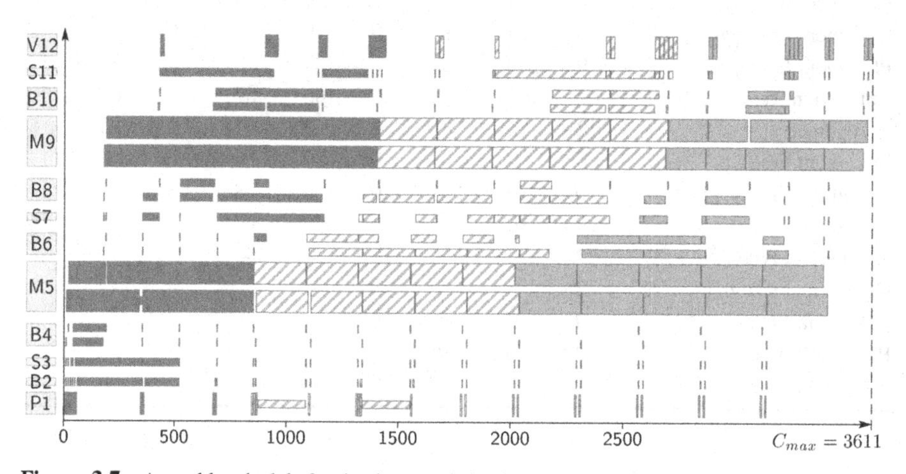

Figure 3.7 Assembly schedule for simultaneous balancing and scheduling.

Table 3.7 Processing Times for Two-Level Balancing and Scheduling

Board type	1	2	3
Screen printer P	10	10	10
Placement machine M5	198	218	260
Placement machine M9	254	193	230
Vision inspection machine V	20	20	20

by solving MIP model *BSMT*. The obtained component assignments are $f = 3, 4, 6, 9, 10, 13, 16, 17, 19, 21, 23, 26, 27, 28, 29, 33, 36, 37, 39, 42, 43, 46, 47, 48, 49, 53, 56, 57, 58, 59, 62, 63, 64, 66, 67, 68, 69, 72, 73, 74$, for machine M5, and $f = 1, 2, 5, 7, 8, 11, 12, 14, 15, 18, 20, 22, 24, 25, 28, 30, 31, 32, 34, 35, 38, 40, 41, 44, 45, 50, 51, 52, 54, 55, 60, 61, 65, 70, 71, 75$, for machine M9. The resulting total placement times are shown in Table 3.7 and the corresponding maximum workload is $W = 3421$.

In the second stage, given component feeder assignments and the resulting processing times p_{ik}, the optimal assembly schedule was determined by solving MIP model *GESCH* enhanced for the batch scheduling mode by the addition of constraints (1.24)–(1.26). Let us denote the enhanced *GESCH* model by *GESCH*+. Gantt chart with the schedule obtained is shown in Figure 3.8. The optimal makespan is $C_{\max} = 3633$.

To evaluate the effectiveness of the MIP approach for balancing and scheduling of SMT lines, the example problem was solved for different numbers of component feeders required to assemble three types of boards. The number of component types was equal to 20, 30, 50, 75, 100, 150, or 200, and the corresponding number of feeder slots b_i available at each placement station i was 30, 40, 70, 100, 140, 210, or 280, respectively.

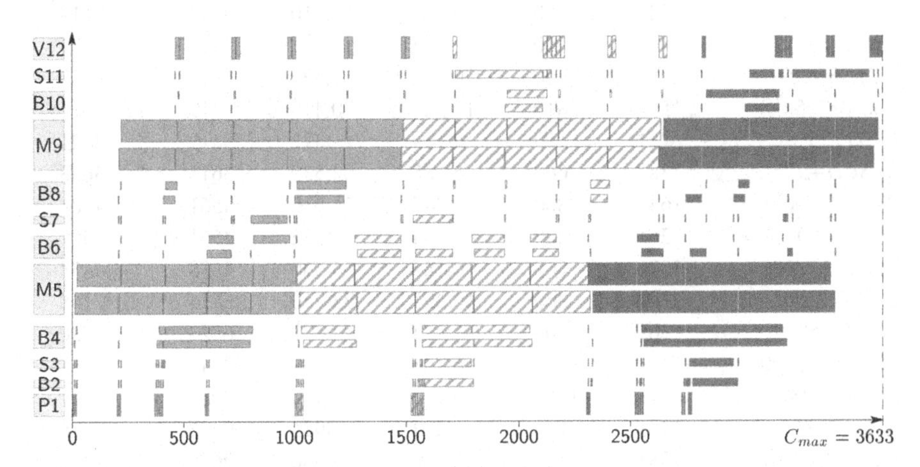

Figure 3.8 Assembly schedule for sequential balancing and scheduling.

The characteristics of MIP models *BASMT*, *BSMT*, and *GESCH+* for the example problems with various number of component feeders, and the solution results are summarized in Table 3.8. The size of the MIP models for the example problems is represented by the total number of variables, *Var.*, number of binary variables, *Bin.*, number of constraints, *Cons.*, and number of nonzero coefficients, *Nonz.*, in the constraint matrix. The last two columns of Table 3.8 present the lower bound *LB* on the makespan or on maximum workload and the makespan C_{max} or workload *W*.

The computational effort required to find assembly schedules was smaller for the two-level approach (Sawik, 2002b). The proven optimal solutions to the balancing *BSMT* and the scheduling *GESCH+* problems can be found within seconds even for a large number of component types. On the other hand, the integrated approach produces better solutions than the two-level approach and does not require much longer computation time to prove optimality. The size of the MIP model and computation time do not grow significantly with the number of different component types to be assigned. Therefore, the number of required component types does not restrict possible application of the integrated approach.

Table 3.8 Characteristics of MIP Models and Solution Results

Problem/feeders	Var.	Bin.	Cons.	Nonz.	*LB*	C_{max} or *W*
BASMT/20	1481	700	16,211	67,710	941	1000
BSMT/20	101	40	84	1002	941	941
GESCH+/20	1381	660	16,185	64,530	990	1020
BASMT/30	1501	720	16,171	68,276	1394	1476
BSMT/30	121	60	94	1442	1394	1396
GESCH+/30	1381	660	16,185	64,530	1437	1495
BASMT/50	1541	760	16,297	69,310	2299	2418
BSMT/50	161	100	114	2322	2299	2301
GESCH+/50	1381	660	16,185	64,530	2437	2457
BASMT/75	1591	810	16,322	70,410	3419	3611
BSMT/75	211	150	139	3422	3419	3421
GESCH+/75	1381	660	16,185	64,530	3613	3633
BASMT/100	1641	860	16,347	71,510	4561	4788
BSMT/100	261	200	164	4522	4561	4561
GESCH+/100	1381	660	16,185	64,530	4838	4858
BASMT/150	1741	960	16,397	73,710	6824	7148
BSMT/150	361	300	214	6722	6824	6826
GESCH+/150	1381	660	16,185	64,530	7241	7261
BASMT/200	1841	1060	16,417	75,910	9086	9527
BSMT/200	461	400	264	8922	9086	9086
GESCH+/200	1381	660	16,185	64,530	9648	9668

3.4 COMMENTS

This chapter shows that the MIP approach can be successfully used to solve hard combinatorial optimization problems of simultaneous balancing and scheduling flexible assembly lines, in particular balancing and scheduling of SMT lines in electronics assembly. The material presented in this chapter is an extension of the results discussed in Sawik (2000b, 2002b, 2002c).

The MIP approach has been widely used to solve assembly line balancing problems (e.g., Ghosh and Gagnon, 1989; Sawik, 1995a, 1999; Scholl, 1999; Dolgui et al., 2006; Guschinskaya and Dolgui, 2010), various setup optimization problems (e.g., Ahmadi et al., 1995; Balakrishnan and Vanderbeck, 1999; Jain et al., 1996; Kumar and Li, 1995), and also SMT line scheduling problems (e.g., Sawik, 2000a, 2001).

Of the two solution approaches proposed for balancing and scheduling in flexible assembly lines, the integrated approach formulates the problem as a large, monolithic MIP model to be solved to optimality, for example, Sawik (2000b, 2002a, 2002b). In contrast, the second approach, denoted as a hierarchical, partitions the problem into a hierarchy of two subproblems, that is, balancing and scheduling, and at best can optimize each of these subproblems independently, which results in suboptimization of a flexible assembly line operation, for example, Sawik (1998a, 1998b, 2000b, 2000c, 2002a, 2002b).

In the proposed model for balancing and scheduling of an SMT line the placement times are assumed to be known a priori. In more detailed studies, however, the component assignment and retrieval problem is considered with different retrieval times of components from different positions in the feeder carrier (e.g., Bard et al., 1994; Ellis et al., 2001; W. Wang et al., 2000).

The computational effort required to find feeder assignments and assembly schedules using the simultaneous approach is greater than that required for the hierarchical approach. However, the size of the mixed integer program *BASMT* and CPU time required to solve it do not grow significantly with the number of different component types assigned to placement machines in the SMT line.

The proposed models can also be applied for balancing and scheduling with additional precedence constraints imposed by placement sequences of electronic components for various board designs as well as for various types of alternative placement machines that can be met in different SMT line configurations.

EXERCISES

3.1 Assuming that each assembly station i is capable of performing only a subset $F^i \subset F$ of all assembly tasks, introduce appropriate changes in models *BAS* and *BAS1* for balancing and scheduling of a FAL.

3.2 Modify the mixed integer programs *BAS* and *BAS1* for balancing and scheduling of a FAL to account for constant transportation times between the stations and

(**a**) To minimize total completion time for a mix of products.

(**b**) To minimize the maximum flow time for a mix of products, subject to an upper bound on the makespan constraints.

(**c**) To minimize the makepan for a mix of products, subject to the maximum flow time constraints.

3.3 Consider models *BAS* and *BAS1* for balancing and scheduling of a FAL. Let $b_i \leq \overline{b}_i$ be the variable amount of work space at station $i \in I$, not greater than $\overline{b}_i$, and denote by $\overline{C}$ an upper bound on the makespan. Formulate the mixed integer program for balancing and scheduling of a FAL with variable work space for part feeders, such that the schedule length is not longer than $\overline{C}$ and the total work space required for component feeders is minimized.

3.4 Modify the mixed integer programs *B1*, *B* for balancing a FAL and *BSMT* for balancing an SMT line, for the different measures of line imbalance:

(**a**) Minimum $(W - \overline{W})$, where $\overline{W}$ is the average workload of a machine.

(**b**) Minimum of $(W - W_{\min})$, where $W_{\min}$ is the minimum workload of a machine.

(**c**) Maximum of $W_{\min}$.

3.5 Formulate a mixed integer program for the problem of optimal allocation of component feeders among assembly stations in a FAL in such a way as to minimize the length of the line (number of stations in series) for a mix of products, subject to the station maximum workload constraints.

Chapter 4

Loading and Scheduling of Flexible Assembly Systems

4.1 INTRODUCTION

This chapter deals with loading and scheduling of a general flexible assembly system. A flexible assembly system (FAS) is a network of assembly stages interconnected by transportation links, where each stage consists of one or more parallel identical stations. Each station has a finite work space for component feeders and finite capacity input and output buffers for temporary storage of products waiting for processing or for transfer between the stations. Limited capacity of in-process buffers helps to limit the average work-in-process, which is a typical performance measure of a FAS. However, when a station finishes processing a product and the output buffer is full the station is *blocked*, whereas if the input buffer is empty the station is *starved*. As a result the system throughput decreases. Blocking and starvation of assembly stations occur because of the difference in processing times of various products at the various assembly stages, for example, for a bottleneck stage the input buffer tends to become full while the output buffer tends to become empty. In order to avoid the station blocking, external buffers between the assembly stages are often introduced, and then the system can sometimes be considered as one with infinite in-process buffers.

A FAS is capable of simultaneously assembling a variety of product types in small to medium-sized batches and at a high rate comparable to that of conventional transfer lines designed for high-volume/low-variety manufacture. Unlike a flexible assembly line (FAL), in which a unidirectional flow of products is observed and revisiting of stations is not allowed, in a flexible assembly system multidirectional product flows occur and revisiting of stations is allowed. While a FAL is an example of a flexible flow shop, a FAS can be considered to be a flexible job shop type of production system.

Similar to FALs, loading and scheduling are the two major short-term planning issues in flexible assembly systems. Given a mix of products to be assembled, the objective of the loading problem is to allocate assembly tasks and component feeders among the assembly stations with limited work space and by this to select assembly routes for a mix of products, so as to balance the station workloads. In contrast, the

Scheduling in Supply Chains Using Mixed Integer Programming. By Tadeusz Sawik

objective of the scheduling problem is to determine the detailed sequencing and timing of all assembly tasks for each individual product, so as to maximize the system productivity, which may be defined in terms of the assembly schedule length (makespan) for a mix of products. The objective of the system loading and scheduling considered in this chapter is to determine an assignment of assembly tasks to stations and an assembly schedule for all products so as to complete the products in minimum time.

The following two approaches and the corresponding MIP models are proposed and compared: an integrated approach and a hierarchical approach. In the integrated approach the loading and scheduling decisions are made simultaneously using a large, monolithic MIP model. In the hierarchical approach, however, a sequence of two MIP models is used, first to balance the station workloads and then to determine the shortest assembly schedule for prefixed task assignments and assembly routes.

The following MIP models for loading and scheduling are presented in this chapter:

LAS1a for simultaneous loading and scheduling of a flexible assembly system with single stations, infinite in-process buffers and alternative routing

LAS1f for simultaneous loading and scheduling of a flexible assembly system with single stations, infinite in-process buffers and fixed routing

L1a for loading of a flexible assembly system with single stations and alternative routing

L1f for loading of a flexible assembly system with single stations and fixed routing

S|L1 for scheduling of a flexible assembly system with single stations, infinite in-process buffers and prefixed product assignments

LASPB for simultaneous loading and scheduling of a flexible assembly system with parallel stations, finite in-process buffers and no revisiting

LPB for loading of a flexible assembly system with parallel stations and no revisiting

S|LPB for scheduling of a flexible assembly system with parallel stations, finite in-process buffers, and prefixed task assignments with no revisiting

4.2 LOADING AND SCHEDULING OF FLEXIBLE ASSEMBLY SYSTEMS WITH SINGLE STATIONS AND INFINITE IN-PROCESS BUFFERS

This section presents an exact approach by MIP to simultaneous or sequential loading and scheduling of a flexible assembly system with single stations, infinite in-process buffers, and a fixed or alternative routing. The problem objective is to determine a single or alternative allocation of assembly tasks and component feeders among the stations and to find an assembly schedule for a mix of products so as to complete the products in minimum time. The simultaneous loading and scheduling of a FAS is compared with a typically used hierarchical, two-level approach, where first a single or alternative assignment of tasks and component feeders to assembly stations is found to balance the station workloads, and then the shortest assembly schedule is determined for a mix of products, given task allocation. For both the integrated and the hierarchical approach, the MIP formulations are provided and some computational examples are presented to illustrate applications of the two approaches.

4.2.1 Simultaneous Loading and Scheduling

In this subsection two monolithic MIP models *LAS1a* and *LAS1f* are presented for simultaneous loading and scheduling of a flexible assembly system with single stations, infinite in-process buffers, and alternative task assignments (alternative routing) and single task assignments (fixed routing), respectively.

Let us consider a FAS made up of m assembly stations $i \in I = \{1, \ldots, m\}$ and the L/U (loading/unloading) station. The stations can be connected by transportation paths that link any pair of assembly stations and revisiting of the stations is allowed. In the system different types of assembly tasks $f \in F$ can be performed to simultaneously assemble various types of products $k \in K$. Let $I_f \subset I$ be the subset of stations capable of performing task f. Each station i has a finite work space b_i where a limited number of component feeders and gripper magazines can be placed. As a result only a limited number of assembly tasks can be assigned to one assembly station. Let a_{if} be the amount of station $i \in I_f$ work space required for assignment of task f, for example, the amount of space required for different component feeders and/or gripper magazines. If station i is incapable of performing task f, that is, $i \notin I_f$, then $a_{if} > b_i$, by convention. Moreover, if $a_{if} = 0$ for $i \in I_f$, that is, task f does not need any amount of station i work space, then task f is not associated with assembly of a new component.

Each product k requires a subset F_k of assembly tasks to be performed subject to in-tree precedence relations defined by the assembly plan for this product. The assembly plan is represented by the set R_k of immediate predecessor-successor pairs of assembly tasks (f, g) such that task $f \in F_k$ must be performed immediately before task $g \in F_k$. Finally, denote by q_{ifk} the assembly time required to perform on station $i \in I_f$ task $f \in F_k$ of product k.

The system is assumed to have sufficient intermediate buffer capacities to temporarily store work in process so that blocking of stations by completed products never happens. This assumption will be restricted later in Section 4.3.

A feasible solution of the simultaneous loading and scheduling problem must satisfy the following basic types of constraints:

- Each assembly task type must be assigned to at least one station (alternative assignments) or to exactly one station (single assignments).
- The total space required for the tasks assigned to each station must not exceed the station finite work space available.
- Each task of every product must be performed on exactly one station.

- Each product must be successively routed to the stations where the required tasks have been assigned, subject to precedence relations defined by its assembly plan.
- Each station can perform at most one task at a time.

In order to model the simultaneous loading and scheduling, two types of binary assignment variables, binary sequencing variables and continuous timing variables are introduced (for notation used, see Table 4.1).

Model LAS1a: *Loading and Scheduling of a Flexible Assembly System with Single Stations and Alternative Routing*

Minimize maximum completion time

$$C_{\max} \tag{4.1}$$

Table 4.1 Notation: Loading and Scheduling of an FAS with Infinite In-Process Buffers

Indices

f	= assembly task, $f \in F$
i	= assembly station, $i \in I = \{1, \ldots, m\}$
k	= product, $k \in K = \{1, \ldots, n\}$

Input parameters

a_{if}	= work space required for assignment of task f to station i
b_i	= total work space of station i (number of tasks that may be assigned to station i, if all $a_{if} = 1$)
e_{ifk}	= earliest completion time on station i for task f of product k
q_{ifk}	= assembly time on station i for task f of product k
I_f	= subset of stations capable of performing task f
F_k	= subset of tasks required for product k
Q	= a large positive constant not less than the schedule length
R_k	= the set of immediate predecessor-successor pairs of tasks (f, g) for product k such that task f must be performed immediately before task g

Decision variables

c_{ifk}	= completion time on station i of task f of product k (timing variable)
x_{ifk}	= 1, if product k is assigned to station i to perform task f; otherwise $x_{ifk} = 0$ (product assignment variable)
y_{ifkgl}	= 1, if on station i task f of product k precedes task g of product l; otherwise $y_{ifkgl} = 0$ (sequencing variable for simultaneous loading and scheduling)
y_{fkgl}	= 1, if task f of product k precedes task g of product l when both tasks are assigned to the same station; otherwise $y_{fkgl} = 0$ (sequencing variable for scheduling, given task assignments)
z_{if}	= 1, if task f is assigned to station $i \in I_f$; otherwise $z_{if} = 0$ (task assignment variable)

subject to

1. *Task Assignment Constraints*:
   - each task is assigned to at least one assembly station (i.e., alternative routing is admitted),
   - the total space required for the tasks assigned to each assembly station cannot exceed the station finite work space available,

$$\sum_{i \in I_f} z_{if} \geq 1; \; f \in F \tag{4.2}$$

$$\sum_{j \in J} a_{if} z_{if} \leq b_i; \; i \in I \tag{4.3}$$

2. *Product Assignment Constraints*:
   - each product is assigned to exactly one station to perform each task,
   - each product can be routed to the stations where the required tasks can be performed,

$$\sum_{i \in I_f} x_{ifk} = 1; \; k \in K, f \in F_k \tag{4.4}$$

$$x_{ifk} \leq z_{if}; \; k \in K, f \in F_k, i \in I_f \tag{4.5}$$

3. *Product Completion Constraints*:
   - for each task of each product its completion time on a station cannot be less than the earliest completion time on this station, if the product is at all routed to this station to perform that task,
   - each task of each product cannot be started until its immediate predecessor task is completed,

$$e_{ifk} x_{ifk} \leq c_{ifk} \leq Q x_{ifk}; \; k \in K, f \in F_k, i \in I_f \tag{4.6}$$

$$\sum_{i \in I_g} c_{igk} - \sum_{i \in I_g} q_{igk} x_{igk} \geq \sum_{i \in I_f} c_{ifk}; \; k \in K, (f, g) \in R_k \tag{4.7}$$

4. *Product Noninterference Constraints*:
   - no two tasks can be performed on the same station simultaneously,

$$c_{ifk} + Q(2 + y_{ifkgl} - x_{ifk} - x_{igl}) \geq c_{igl} + q_{ifk};$$
$$k, l \in K, f \in F_k, g \in F_l, i \in I_f \cap I_g: k < l \tag{4.8}$$

$$c_{igl} + Q(3 - y_{ifkgl} - x_{ifk} - x_{igl}) \geq c_{ifk} + q_{igl};$$
$$k, l \in K, j \in F_k, g \in F_l, i \in I_f \cap I_g: k < l \quad (4.9)$$

5. *Maximum Completion Time Constraints*:
   - the schedule length is determined by the latest completion time of some product on some station,

$$c_{ifk} \leq C_{\max}; \; k \in K, f \in F_k, i \in I_f \quad (4.10)$$

6. *Variable Nonnegativity and Integrality Conditions:*

$$c_{ifk} \geq 0; \; k \in K, f \in F_k, i \in I_f \quad (4.11)$$
$$x_{ifk} \in \{0, 1\}; \; k \in K, f \in F_k, i \in I_f \quad (4.12)$$
$$y_{ifkgl} \in \{0, 1\}; \; k, l \in K, f \in F_k, g \in F_l, i \in I_f \cap I_g: k < l \quad (4.13)$$
$$z_{if} \in \{0, 1\}; \; f \in F, i \in I_f \quad (4.14)$$
$$C_{\max} \geq 0, \quad (4.15)$$

where the earliest completion time on station i for task f of product k, e_{ifk}, is the sum of q_{ifk} and the minimum total assembly time of all tasks $f' \in F_k$ of product k that precede task f,

$$e_{ifk} = q_{ifk} + \sum_{f' \in F_k: f' \prec f} \min_{i \in I_{f'}} (q_{if'k}).$$

The objective function (4.1) represents the schedule length to be minimized. Constraint (4.3) is the station capacity constraint. The routing constraint (4.5) ensures that products are assigned to the stations where the required tasks may be performed.

Constraints (4.6) impose variable lower and upper bounds on completion times. In particular, the left-hand-side constraint of (4.6) ensures that the first task of each product is completed. Constraint (4.7) maintains for each product the precedence relations among its tasks. Product noninterference constraints (4.8) and (4.9) with binary sequencing variables y_{ifkgl} and a *big*$-Q$ coefficient represent disjunctive constraints on the machines. For a given sequence of tasks at most one constraint of (4.8) and (4.9) is active, and only if both task f of product k and task g of product l are assigned to the same station i. Otherwise, both (4.8) and (4.9) are inactive. Finally (4.10) defines the schedule length.

In model *LAS1f* presented below for single task assignments where each task type is assigned to exactly one station (equation (4.16)), product assignment variables x_{ifk} are not required since a unique assembly route for each product is determined by task assignment variables z_{if}, i.e., $x_{ifk} = z_{if} \forall k \in K, f \in F_k$.

Model *LAS1f*: *Loading and Scheduling of a Flexible Assembly System with Single Stations and Fixed Routing*

Minimize (4.1) subject to

1. *Task Assignment Constraints*: (4.3) and

– each task is assigned to exactly one assembly station (i.e., alternative routing is not admitted),

$$\sum_{i \in I_f} z_{if} = 1; \ f \in F \tag{4.16}$$

2. *Product Completion Constraints*:
   – for each task of each product its completion time on a station cannot be less than the earliest completion time on this station, if the task is at all assigned to this station,
   – each task of each product cannot be started until its immediate predecessor task is completed,

$$e_{ifk} z_{if} \leq c_{ifk} \leq Q z_{if}; \ k \in K, f \in F_k, i \in I_f \tag{4.17}$$

$$\sum_{i \in I_g} c_{igk} - \sum_{i \in I_g} q_{igk} z_{ig} \geq \sum_{i \in I_f} c_{ifk}; \ k \in K, (f, g) \in R_k \tag{4.18}$$

3. *Product Noninterference Constraints*:

$$c_{ifk} + Q(2 + y_{ifkgl} - z_{if} - z_{ig}) \geq c_{igl} + q_{ifk}; \quad k, l \in K, f \in F_k, g \in F_l, i \in I_f \cap I_g\colon k < l \tag{4.19}$$

$$c_{igl} + Q(3 - y_{ifkgl} - z_{if} - z_{ig}) \geq c_{ifk} + q_{igl}; \quad k, l \in K, f \in F_k, g \in F_l, i \in I_f \cap I_g\colon k < l \tag{4.20}$$

4. *Maximum Completion Time Constraints*: (4.10)
5. *Variable Nonnegativity and Integrality Conditions*: (4.11), (4.13)–(4.15).

Product completion constraints (4.17) and (4.18) are equivalent to (4.6) and (4.7) in the *LAS1a* model, and product noninterference constraints (4.19) and (4.20) are equivalent to (4.8) and (4.9).

4.2.2 Sequential Loading and Scheduling

In this subsection, a hierarchical two-level loading and scheduling is proposed and the corresponding MIP formulations are presented. In the two-level approach, first at the top level the loading problem is solved and then at the base level solution to the standard job shop scheduling problem for prefixed product assignments is found, see Figure 4.1.

The levels are connected using a pure top-down approach. The top-level objective of balancing the workloads is a commonly used surrogate objective that represents the base-level objective of minimizing the makespan. The top-level prefixes task assignments and assembly routes for each product for the entire scheduling horizon and by this transforms the flexible job shop with single stations into the standard job shop problem. The detailed schedule is next determined at the base level by solving the standard job shop problem generated at the top level.

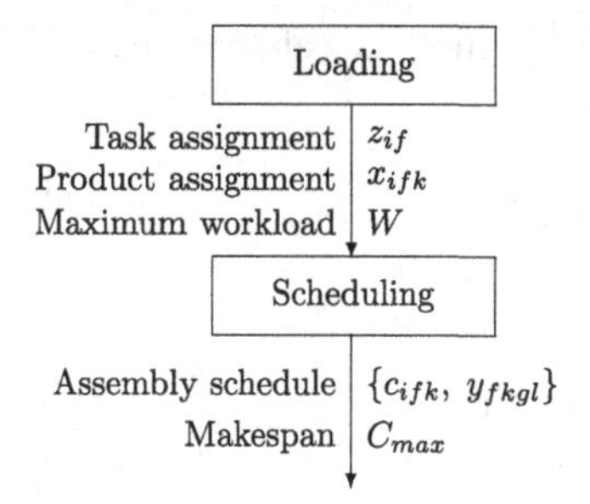

Figure 4.1 Sequential loading and scheduling with infinite in-process buffers.

Below, two MIP models $L1a$ and $L1f$ are presented for loading of a flexible assembly system single stations and with alternative task assignments (alternative routing) and with single task assignments (fixed routing), respectively.

Model *L1a*: *Loading of a Flexible Assembly System with Single Stations and Alternative Routing*

Minimize maximum workload

$$W \tag{4.21}$$

subject to

1. *Task Assignment Constraints:* (4.2), (4.3)
2. *Product Assignment Constraints:* (4.4), (4.5)
3. *Maximum Workload Constraints:*
   - for each station the total assembly time required to perform tasks for products assigned to this station cannot exceed the maximum workload to be minimized

$$\sum_{k \in K} \sum_{f \in F_k} q_{ifk} x_{ifk} \leq W; \; i \in I \tag{4.22}$$

4. *Variable Nonnegativity and Integrality Conditions:* (4.12), (4.14) and

$$W \geq 0. \tag{4.23}$$

The objective function W (4.21) is a measure of system imbalance and represents workload of the bottleneck station defined by constraint (4.22).

Model *L1f*: *Loading of a Flexible Assembly System with Single Stations and Fixed Routing*

Minimize (4.21) subject to (4.3), (4.14), (4.16) and

1. *Maximum Workload Constraints:*
   - for each station the total assembly time for all products and all tasks assigned to this station cannot exceed the maximum workload to be minimized

$$\sum_{k \in K} \sum_{f \in F_k} q_{ifk} z_{if} \leq W; \; i \in I. \tag{4.24}$$

Having solved the loading problem $L1a$ or $L1f$, the next step of the two-level approach is to determine a detailed assembly schedule by solving the corresponding job shop problem $S|L1$ for prefixed product assignments x^L_{ifk}. Note that for the fixed routing (model $L1$) $x^L_{ifk} = z^L_{if} \forall k \in K, f \in F_k$.

In the $S|L$ model presented below, sequencing variables y_{ifkgl} are replaced with variables y_{fkgl} (see Table 4.1), defined for each pair of tasks f of product k and g of product l that have been assigned to the same station i, i.e., $\sum_{i \in I} x^L_{ifk} x^L_{igl} = 1$.

Model $S|L1$: *Scheduling of a Flexible Assembly System with Single Stations and Prefixed Product Assignments*

Minimize (4.1) subject to

1. *Product Noninterference Constraints*:
   - no two products assigned to the same station can be performed simultaneously,

$$c_{ifk} + Qy_{fkgl} \geq c_{igl} + q_{ifk};$$
$$k, l \in K, f \in F_k, g \in F_l, i \in I: k < l, \; x^L_{ifk} x^L_{igl} = 1 \quad (4.25)$$

$$c_{igl} + Q(1 - y_{fkgl}) \geq c_{ifk} + q_{igl};$$
$$k, l \in K, f \in F_k, g \in F_l, i \in I: k < l, \; x^L_{ifk} x^L_{igl} = 1 \quad (4.26)$$

2. *Product Completion Constraints*:
   - completion time of each task of a product assigned to some station cannot be less than its earliest completion time on this station,
   - each task of each product cannot be started until its immediate predecessor task is completed,
   - successive tasks of each product assigned to the same station are performed contiguously,

$$c_{ifk} \geq e_{ifk}; k \in K, f \in F_k, i \in I_f: x^L_{ifk} = 1 \quad (4.27)$$

$$c_{hgk} - q_{hgk} \geq c_{ifk}; k \in K, (f, g) \in R_k, i, h \in I: i \neq h, x^L_{ifk} x^L_{hgk} = 1 \quad (4.28)$$

$$c_{igk} - q_{igk} = c_{ifk}; k \in K, (f, g) \in R_k, i \in I: x^L_{ifk} x^L_{igk} = 1 \quad (4.29)$$

3. *Maximum Completion Time Constraints*:
   - the schedule length is determined by the latest completion time among all products,
   - the schedule length cannot be less than the maximum workload,

$$c_{ifk} \leq C_{\max}; \; k \in K, f \in F_k, i \in I_f: x^L_{ifk} = 1 \quad (4.30)$$

$$C_{\max} \geq W^L \quad (4.31)$$

4. *Variable Nonnegativity and Integrality Conditions*:

$$c_{ifk} \geq 0;\ \ k \in K, f \in F_k, i \in I_f\colon x^L_{ifk} = 1 \tag{4.32}$$

$$y_{fkgl} \in \{0, 1\};\ \ k, l \in K, f \in F_k, g \in F_l\colon \sum_{i \in I} x^L_{ifk} x^L_{igl} = 1. \tag{4.33}$$

Product noninterference constraints (4.25) and (4.26) are defined only for such pairs of tasks f of product k and g of product l that are assigned to the same station i, that is, for $x^L_{ifk} x^L_{igl} = 1$. Constraints (4.28) and (4.29) maintain for each product the precedence relations among its tasks. W^L in (4.31) is the solution value of the corresponding loading problem $L1a$ or $L1f$.

4.2.3 Computational Examples

In this subsection numerical examples (Sawik, 2000b) are presented to illustrate possible application of the two approaches and the MIP formulations proposed.

The FAS configuration for the examples is provided in Figure 4.2. The system is made up of $m = 3$ assembly stations ($i = 1, 2, 3$), and one L/U station.

The production order consists of five products to be assembled of 10 types of components. The assembly plans for products are in the form of assembly sequences (chains of tasks) to be performed, see Figure 4.3.

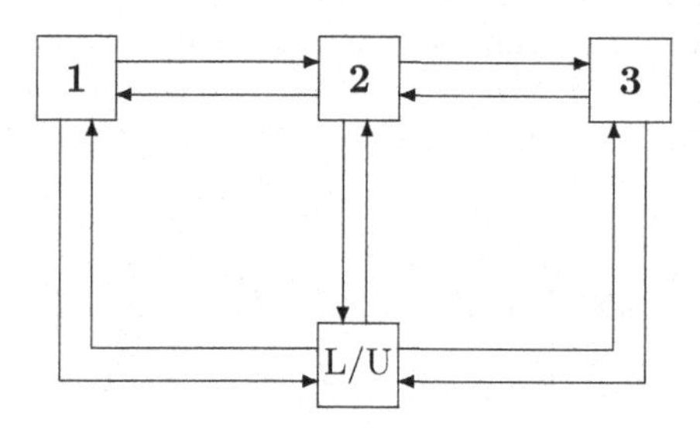

Figure 4.2 A flexible assembly system.

$k = 1:\ \ L \longrightarrow 1 \longrightarrow 2 \longrightarrow 3 \longrightarrow 4 \longrightarrow 6 \longrightarrow 8 \longrightarrow U$

$k = 2:\ \ L \longrightarrow 1 \longrightarrow 2 \longrightarrow 4 \longrightarrow 5 \longrightarrow 6 \longrightarrow 7 \longrightarrow 9 \longrightarrow 10 \longrightarrow U$

$k = 3:\ \ L \longrightarrow 2 \longrightarrow 3 \longrightarrow 4 \longrightarrow 5 \longrightarrow 7 \longrightarrow 8 \longrightarrow 9 \longrightarrow 10 \longrightarrow U$

$k = 4:\ \ L \longrightarrow 1 \longrightarrow 3 \longrightarrow 5 \longrightarrow 6 \longrightarrow 7 \longrightarrow 8 \longrightarrow 9 \longrightarrow 10 \longrightarrow U$

$k = 5:\ \ L \longrightarrow 1 \longrightarrow 3 \longrightarrow 5 \longrightarrow 6 \longrightarrow 7 \longrightarrow 8 \longrightarrow 9 \longrightarrow 10 \longrightarrow U$

Figure 4.3 Graph of assembly sequences for products.

The assembly times ($q_{ifk} = q_{fk}, \forall i \in I_f$) are shown below:

$$[q_{fk}] = \begin{bmatrix} 4, & 4, & 0, & 4, & 4 \\ 2, & 2, & 2, & 0, & 0 \\ 2, & 0, & 2, & 2, & 2 \\ 2, & 2, & 2, & 0, & 0 \\ 0, & 4, & 4, & 4, & 4 \\ 2, & 2, & 0, & 2, & 2 \\ 0, & 3, & 3, & 3, & 3 \\ 5, & 0, & 5, & 5, & 5 \\ 0, & 2, & 2, & 2, & 2 \\ 0, & 4, & 4, & 4, & 4 \end{bmatrix}$$

Work space required for task assignments:

$$[a_{if}] = \begin{bmatrix} 1, & 2, & 3, & 1, & 2, & 3, & 9, & 9, & 9, & 9 \\ 11, & 11, & 11, & 1, & 2, & 3, & 1, & 2, & 3, & 5 \\ 1, & 2, & 3, & 10, & 10, & 10, & 1, & 2, & 3, & 5 \end{bmatrix}$$

Work space available at stations:

$$b_1 = 8, \ b_2 = 10, \ b_3 = 9.$$

Note that, if station i is incapable of performing task f, then $a_{if} > b_i$.

First, the integrated approach was applied to find optimal solutions for simultaneous loading and scheduling using the MIP models *LAS1a* and *LAS1f*. The optimal assembly schedules are shown in Figure 4.4.

Then, for a comparison the assembly schedules for the example problems were determined using the hierarchical approach. The obtained schedules are shown in

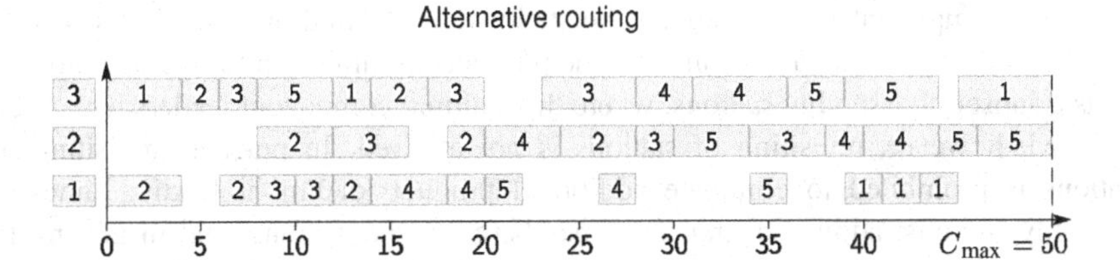

Figure 4.4 Assembly schedules for simultaneous loading and scheduling.

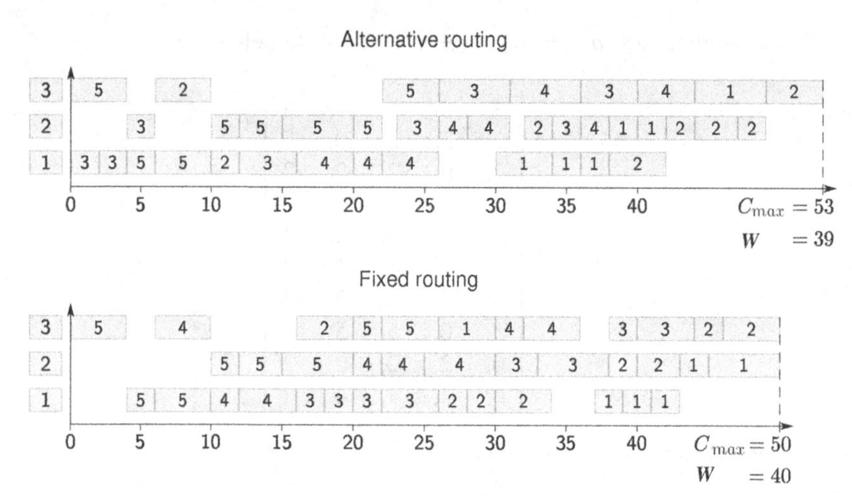

Figure 4.5 Assembly schedules for sequential loading and scheduling.

Figure 4.5, where the maximum workload W determined at the top level and makespan $C_{\max}$ achieved at the base level are indicated.

4.3 LOADING AND SCHEDULING OF FLEXIBLE ASSEMBLY SYSTEMS WITH PARALLEL STATIONS AND FINITE IN-PROCESS BUFFERS

This section presents an exact approach by MIP to simultaneous or sequential loading and scheduling of a flexible assembly system with parallel stations and finite in-process buffers. The problem objective is to determine an allocation of assembly tasks and component feeders among the stations and to find an assembly schedule for a mix of products so as to complete the products in minimum time. Each product visits a subset of assembly stations, where the required component feeders have been assigned; however, revisiting of stations is not allowed. In practice, revisiting of stations is prohibited to eliminate additional product loading/unloading times at assembly stations, additional transfer times between the stations, and in general to simplify product flow control, for example, Hopp and Spearman (1996).

The simultaneous loading and scheduling of an FAS is compared with a typically used hierarchical, two-level approach, where first assignments of tasks and component feeders to assembly stations are found to balance the station workloads and to eliminate revisiting of stations by products, and then the shortest assembly schedule is determined for a mix of products, given task allocation. For both the integrated and the hierarchical approach, the MIP formulations are proposed. Numerical examples and some computational results are presented in Section 4.3.3 to illustrate applications of the proposed approaches.

4.3.1 Simultaneous Loading and Scheduling

In this subsection, a MIP model *LASPB* is formulated for simultaneous loading and scheduling with no revisiting in an FAS with parallel stations and finite in-process buffers. The model addresses the three basic decisions:

- What should be the assignment of tasks and component feeders to assembly stations with limited work space?
- What should be the assignment of products to parallel stations and buffers?
- What should be the sequence and timing of products assigned to each station?

The task and component feeder assignment determines processing route as well as the resulting total processing times at each assembly station for each product, and the scheduling decisions depend on the selected routes and these processing times.

The FAS and the assembly process to be considered are assumed to have the following features.

- Each assembly stage consists of parallel identical assembly stations, where each station is assigned the same types of component feeders and is capable of performing the same types of assembly tasks.
- Each assembly station has internal input and output buffer space of a finite capacity, which may result in blocking of a station.
- Each assembly station has a finite work space where a limited number of component feeders can be placed.
- Transportation times between processing stages, loading and unloading times of products, and any changeover times of assembly stations are negligible.
- For each assembly task of each product, the processing times at each assembly station capable of performing it are known ahead of time.
- For each product a fixed assembly sequence (chain) of required tasks is known ahead of time.
- The processing route of each product consists of different assembly stages to be visited, that is, revisiting of stages is not allowed.
- Assembly tasks are nonpreemptive.
- All products are available for processing at time 0.

Let us consider an FAS made up of m processing stages $i \in I = I_A \cup I_B = \{1, \ldots, m\}$, for assembly ($I_A$) and for buffering ($I_B$). The processing stages are interconnected by transportation paths that link any pair of assembly stages. Transportation times between the stages, however, are assumed to be negligible.

Each assembly stage $i \in I_A$ consists of $m_i \geq 1$ parallel identical assembly stations and every station has its own internal input and output buffer of a fixed capacity. Denote by $i-1$ and $i+1$ the input and the output buffer stage of assembly stage $i \in I_A$ and by m_{i-1} and m_{i+1} the total number of input and output buffers, respectively.

Let J_i be the set of parallel assembly stations at stage $i \in I_A$, and for each station $j \in J_i$ let $B_{i-1}(j)$ and $B_{i+1}(j)$ be the subsets of input and output buffers, respectively. If, in addition to internal input and output buffers of each station, the system contains external buffers common for more than one station, then such buffers can be included in the corresponding subsets $B_{i-1}(j)$ and/or $B_{i+1}(j)$ of appropriate stations $j \in J_i$, $i \in I_A$.

In the example shown in Figure 4.9, a FAS system consists of $m = 9$ processing stages with $I_A = \{2, 5, 8\}$ and $I_B = \{1, 3, 4, 6, 7, 9\}$. Each assembly station has a single input and output buffer, that is, $m_{i-1} = m_i = m_{i+1}$, $i \in I_A$. For example, $m_4 = m_5 = m_6 = 3$, the set of parallel stations at stage $i = 5$ is $J_5 = \{1, 2, 3\}$ and the subsets of the corresponding input and output buffers are $B_4(j) = B_6(j) = \{j\}$, $j \in J_5$.

In the system different types of assembly tasks $f \in F$ can be performed to simultaneously assemble various types of products $k \in K$. Let F_k be the ordered set of assembly tasks required to complete product k, that is, F_k is a fixed assembly sequence of tasks for product k.

Each assembly station at stage $i \in I_A$ has a finite work space b_i where a limited number of component feeders and gripper magazines can be placed. As a result only a limited number of assembly tasks can be assigned to one assembly station. Let $I_f \subset I_A$ be the subset of assembly stages capable of performing task f, and let a_{if} be the amount of work space of assembly station at stage $i \in I_f$, required for assignment of task f. If assembly stations at stage i are incapable of performing task f, that is, $i \notin I_f$, then $a_{if} > b_i$, by convention. Moreover, if $a_{if} = 0$ for $i \in I_f$, that is, task f does not need any station work space at stage i, then task f is not associated with assembly of a new component. Finally, denote by $q_{ifk} > 0$ the assembly time required to perform in stage $i \in I_f$ assembly task $f \in F_k$ of product k.

The problem objective is to determine an allocation of assembly tasks and component feeders among the stations with limited work space and to find an assembly schedule for a mix of products so as to complete the products in minimum time with no revisiting of stages by products.

An assignment of assembly tasks to stations determines a processing route for each product, that is, a sequence of stations to be visited in order to complete the required sequence of tasks. The "no revisiting" requirement implies that for each product a subset of successive tasks is assigned to one assembly station, and hence each assembly station can be visited at most once by every product.

A unified modeling approach is adopted with the buffers viewed as machines with zero processing times (see Section 1.2.2). As a result the scheduling problem with buffers can be converted into one with no buffers but with blocking. The blocking time of a machine with zero processing time denotes product waiting time in the buffer represented by that machine. We assume that each product assigned to some assembly station must also visit the input and output buffers of that station. However, zero blocking time in a buffer stage indicates that the corresponding product does not need to wait in the buffer. Let us note that for each buffer stage, a product's completion time is equal to its departure time from the preceding stage, since the buffer processing time is zero.

Waiting of product in the input or output buffer connected with an assembly stage where the assembly task f is to be performed is referred to as buffering task f. Both assembly stations and buffers are referred to as processors, and both assembly and buffering tasks are referred to as processing tasks.

For each type of product the total processing time at each stage depends on the assignment of assembly tasks and the corresponding component feeders. Let p_{ik} be the total processing time at stage $i \in I_A$ of product k and z_{if} task assignment binary variable (see Table 4.2). The "no revisiting" requirement enables a subset of successive tasks of each product assigned to an assembly station to be performed contiguously, with no breaks between the tasks. Therefore, for each product k and each assembly stage $i \in I_A$, the assembly time p_{ik} is a variable determined by the

Table 4.2 Notation: Loading and Scheduling of an FAS with Finite In-Process Buffers

Indices

f	= processing task, $f \in F$
i	= processing stage, $i \in I = I_A \cup I_B = \{1, \ldots, m\}$
j	= processor in stage i, $j \in J_i = \{1, \ldots, m_i\}$
k	= product, $k \in K = \{1, \ldots, n\}$

Input parameters

a_{if}	= work space required for assignment of task f to assembly station at stage i
b_i	= total work space of each assembly station at stage $i \in I_A$ (number of tasks that may be assigned to each station at stage i, if all $a_{if} = 1$)
q_{ifk}	= processing time at stage i of task f of product k
J_i	= set of parallel processors at stage i
$B_{i-1}(j)$	= subsets of input buffers of assembly station $j \in J_i$ in stage $i \in I_A$
$B_{i+1}(j)$	= subsets of output buffers of assembly station $j \in J_i$ in stage $i \in I_A$
F_k	= ordered subset of tasks required for product k
I_A	= subset of assembly stages
I_B	= subset of buffer stages
I_f	= subset of assembly stages capable of performing task f
Q	= a large positive constant not less than the schedule length

Decision variables

c_{ik}	= completion time in stage i of product k (timing variable)
d_{ik}	= departure time from stage i of product k (timing variable)
p_{ik}	= total assembly time in stage i of product k
x_{ijk}	= 1, if product k is assigned to processor $j \in J_i$ in stage $i \in I$; otherwise $x_{ijk} = 0$ (product assignment variable)
y_{kl}	= 1, if product k precedes product l in the processing sequence; otherwise $y_{kl} = 0$ (product sequencing variable)
z_{if}	= 1, if task f is assigned to processing stage $i \in I_f$; otherwise $z_{if} = 0$ (task assignment variable)

summation of the assembly times for all tasks $f \in F_k$ that have been assigned to this stage, that is, $p_{ik} = \sum_{f \in F_k} q_{ifk} z_{if}$.

For every product k let c_{ik} denote its completion time at each stage i, and d_{ik} its departure time from stage i. Processing without preemption indicates that product k completed at stage i at time c_{ik} starts its processing at that stage at time $c_{ik} - p_{ik}$. Product k completed at stage i at time c_{ik} departs at time $d_{ik} \geq c_{ik}$ to an available processor at the next stage of its processing route. If at time c_{ik} all processors at the next stage are occupied, then the processor at stage i is blocked by product k until a downstream processor becomes available.

If a product k is successively processed at assembly stages h and i (see Fig. 4.6), then its corresponding completion and departure times at these stages satisfy the following relations

$$\begin{aligned} \ldots c_{h-1k} \leq d_{h-1k} < c_{hk} = d_{h-1k} + p_{hk} \leq d_{hk} = c_{h+1k} \leq d_{h+1k} = \\ = c_{i-1k} \leq d_{i-1k} < c_{ik} = d_{i-1k} + p_{ik} \leq d_{ik} = c_{i+1k} \leq d_{i+1k} \ldots \end{aligned}$$

A feasible solution of the FAS loading and scheduling problem must satisfy the following basic constraints.

- Each assembly task must be assigned to exactly one assembly stage.
- The total space required for the component feeders assigned to each assembly station must not exceed the station finite work space available.
- The tasks must be allocated among assembly stages in such a way that for each product precedence relations among the tasks are maintained with no revisiting of stages required.
- Each product must be assigned to exactly one station at each assembly stage where the required tasks have been assigned as well as to exactly one input and one output buffer of this station.
- For each product all of its tasks must be completed at the assembly stages where the tasks have been assigned, subject to precedence relations among the tasks.
- No two products assigned to the same processor can be processed simultaneously.
- At each stage processing of every product starts immediately after departure from the preceding stage of its processing route.
- A product does not visit the assembly stage where none of its required tasks have been assigned.
- The latest completion time of a final task of the last product determines the schedule length.

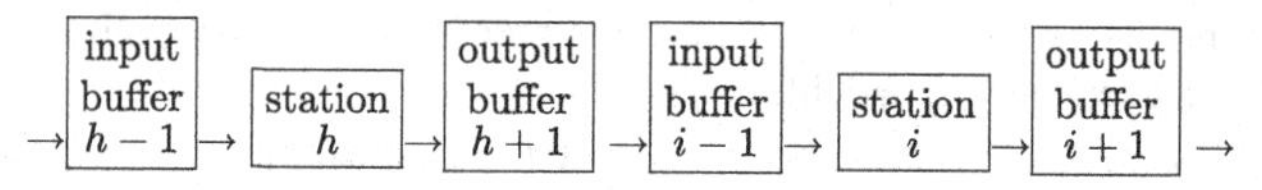

Figure 4.6 A partial processing route.

The mathematical formulation of the MIP model for simultaneous loading and scheduling of a flexible assembly system with parallel stations, finite in-process buffers, and revisiting of stages not allowed is presented below.

Model *LASPB*: *Loading and Scheduling of a Flexible Assembly System with Parallel Stations, Finite In-Process Buffers, and No Revisiting*

Minimize (4.1) subject to

1. *Task Assignment with No Revisiting Constraints*:
   - each task is assigned to exactly one assembly station,
   - the total space required for the tasks assigned to each assembly station cannot exceed the station finite work space available,
   - the buffering tasks are assigned to the input and output buffer of the assembly stage where the corresponding assembly task is assigned,
   - consecutive tasks of each product are assigned to the same assembly stage, so that revisiting of stages is not required,

$$\sum_{i \in I_f} z_{if} = 1; \ f \in F \tag{4.34}$$

$$\sum_{f \in F} a_{if} z_{if} \le b_i; \ i \in I_A \tag{4.35}$$

$$z_{if} = z_{hf}; \ f \in F, i \in I_f, h \in \{i-1, i+1\} \tag{4.36}$$

$$z_{if} \ge z_{ie} + z_{ig} - 1; \quad k \in K, e, f, g \in F_k, i \in I_e \cap I_f \cap I_g\colon \ e \prec f \prec g \tag{4.37}$$

2. *Product Assignment Constraints*:
   - in every assembly stage each product is assigned to exactly one processor, if at least one of its required tasks is assigned to this stage,
   - each product is assigned to one input and one output buffer of the assembly stage selected for assignment of required tasks,

$$z_{if} \le \sum_{j \in J_i} x_{ijk} \le \sum_{g \in F_k} z_{ig}; \ i \in I, k \in K, f \in F_k \tag{4.38}$$

$$\sum_{j \in J_i} x_{ijk} \le 1; \ i \in I, k \in K \tag{4.39}$$

$$x_{ijk} = \sum_{j' \in B_h(j)} x_{hj'k}; \ i \in I_A, h \in \{i-1, i+1\}, j \in J_i, k \in K \tag{4.40}$$

3. *Product Completion and Departure Constraints*:
   - each product must be completed in the first assembly stage of its processing route as well as in all successive stages, subject to the precedence relations among its tasks,
   - each product cannot be departed from a stage until it is completed in this stage,

- each product does not visit stages where none of its required tasks is assigned,

$$c_{ik} \geq \sum_{f \in F_k} q_{ifk} z_{if}; \; i \in I_{\text{first}(F_k)}, \; k \in K \tag{4.41}$$

$$c_{hk} + Q(2 - z_{if} - z_{hg}) \geq c_{ik} + \sum_{r \in F_k} q_{hrk} z_{hr};$$

$$k \in K, \; f,g \in F_k, \; i \in I_f, \; h \in I_g\colon \; i \neq h, f \prec \text{last}(F_k), \; g = \text{next}(f, F_k) \tag{4.42}$$

$$c_{ik} \leq d_{ik}; \; i \in I, \; k \in K \tag{4.43}$$

$$d_{ik} \leq Q \sum_{f \in F_k} z_{if}; \; i \in I, \; k \in K \tag{4.44}$$

4. *Product Noninterference Constraints*:

- no two products can be performed on the same processor simultaneously,

$$c_{ik} + Q(2 + y_{kl} - x_{ijk} - x_{ijl}) \geq d_{il} + \sum_{f \in F_k} q_{ifk} z_{if};$$

$$i \in I, j \in J_i, \; k, l \in K\colon \; k < l \tag{4.45}$$

$$c_{il} + Q(3 - y_{kl} - x_{ijk} - x_{ijl}) \geq d_{ik} + \sum_{f \in F_l} q_{ifl} z_{if};$$

$$i \in I, j \in J_i, \; k, l \in K\colon \; k < l \tag{4.46}$$

5. *Buffering Constraints*:

- each product arrives in an input buffer of an assembly stage immediately after its departure from the output buffer of the preceding assembly stage of its processing route,
- in every stage, assembly of each product starts immediately after its departure from the input buffer of this stage,
- each product arrives in the output buffer of an assembly stage immediately after its departure from this assembly stage,

$$c_{h-1k} + Q(2 - z_{if} - z_{hg}) \geq d_{i+1k}; \; k \in K, f, g \in F_k, i \in I_f, h \in I_g\colon \; i \neq h, f \prec \text{last}(F_k), \; g = \text{next}(f, F_k) \tag{4.47}$$

$$c_{h-1k} - Q(2 - z_{if} - z_{hg}) \leq d_{i+1k}; \; k \in K, f, g \in F_k, i \in I_f, h \in I_g\colon \; i \neq h, f \prec \text{last}(F_k), \; g = \text{next}(f, F_k) \tag{4.48}$$

$$c_{ik} - \sum_{f \in F_k} q_{ifk} z_{if} = d_{i-1k}; \; i \in I_A, \; k \in K \tag{4.49}$$

$$c_{i+1k} = d_{ik}; \; i \in I_A, \; k \in K \tag{4.50}$$

6. *Maximum Completion Time Constraints*:
   - the schedule length is determined by the latest departure time of some product from some stage,
   - the schedule length cannot be less than the average workload of assembly station at each assembly stage,
   - the schedule length cannot be less than the total assembly time of each product,

$$d_{ik} \leq C_{\max}; \; i \in I, k \in K \tag{4.51}$$

$$\sum_{k \in K} \sum_{\in F_k} q_{ifk} z_{if} / m_i \leq C_{\max}; \; i \in I_A \tag{4.52}$$

$$\sum_{i \in I_A} \sum_{f \in F_k} q_{ifk} z_{if} \leq C_{\max}; \; k \in K \tag{4.53}$$

7. *Variable Nonnegativity and Integrality Conditions*: (4.15) and

$$c_{ik} \geq 0; \; i \in I, k \in K \tag{4.54}$$

$$d_{ik} \geq 0; \; i \in I, k \in K \tag{4.55}$$

$$x_{ijk} \in \{0, 1\}; \; i \in I, j \in J_i, k \in K \tag{4.56}$$

$$y_{kl} \in \{0, 1\}; \; k, l \in K\colon k < l \tag{4.57}$$

$$z_{if} \in \{0, 1\}; \; i \in I, f \in F\colon i \in I_f \text{ or } i \in \{h-1, h+1\}, h \in I_f \tag{4.58}$$

Equation (4.40) ensures that the product is assigned to one input and one output buffer of the assembly station selected by (4.38) and (4.39). Assignment of a product to exactly one buffer in a buffer stage enables the product to wait in the buffer, if it is necessary, that is, when all processors in the next processing stage are not immediately available. Otherwise, product waiting time in the buffer is zero.

Constraint (4.42) maintains for each product the precedence relations among its tasks. If two successive tasks $f \in F_k$ and $g = \text{next}(f, F_k)$ of product k are assigned to different assembly stages i and h, then its completion time at the latter stage cannot be less than the total processing time at this stage and completion time at the previous stage. Otherwise (4.42) is inactive.

Constraint (4.44) ensures that a product does not visit stages where none of its required tasks is assigned. If none of the tasks of product k is assigned to stage i, then $z_{if} = 0$ for all $f \in F_k$ and (4.44) implies $d_{ik} = 0$. Then (4.43) implies $c_{ik} = 0$. In addition, (4.38) ensures that if for some assembly stage $i \in I$ and product $k \in K$, $z_{if} = 0$ for all $f \in F_k$, then $x_{ijk} = 0$ for all $j \in J_i$, that is, product k does not visit stage i. The above implies that additional cutting constraint $x_{ijk} \leq c_{ik}$; $i \in I, j \in J_i, k \in K$ might be added to model *LASBNB*.

For a given sequence of products at most one constraint of product noninterference constraints (4.45) or (4.46) is active, and only if both products k and l are assigned to the same processor. Otherwise, both (4.45) and (4.46) are inactive.

A pair of constraints (4.47) and (4.48) indicate that each product arrives in an input buffer $h-1$ of an assembly stage $h \in I_A$ immediately after its departure from the output buffer $i+1$ of the preceding assembly stage $i \in I_A$ of its processing route. If two successive tasks $f \in F_k$ and $g = \text{next}(f, F_k)$ of product k are assigned to different assembly stages i and h, i.e., $z_{if} = z_{hg} = 1$, then (4.47) and (4.48) generate equality $c_{h-1k} = d_{i+1k}$, otherwise (4.47) and (4.48) are inactive. Equation (4.49) ensures that in every stage $i \in I_A$ assembly of each product starts immediately after its departure from the input buffer $i-1$, and (4.50) ensures that each product arrives in the output buffer $i+1$ immediately after its departure from the assembly stage $i \in I_A$.

A large positive constant Q that is not less than the schedule length can be calculated as follows:

$$Q = \sum_{k \in K} \sum_{f \in F_k} \max_{i \in I_f} \{q_{ifk}/m_i\}.$$

Q can also be used as an upper bound on the optimal makespan.

When in addition to assembly times *transportation times* between the assembly stages should also be considered, then the buffering constraints (4.47) and (4.48) should be replaced with the following constraints:

Buffering Constraints with Transportation Times Considered:

– each product arrives in an input buffer of an assembly stage immediately after its transfer from the output buffer of the preceding assembly stage of its processing route,

$$c_{h-1k} + Q(2 - z_{if} - z_{hg}) \geq d_{i+1k} + q_{ih};\ k \in K, f, g \in F_k, i \in I_f, h \in I_g:$$
$$i \neq h, f \prec \text{last}(F_k), g = \text{next}(f, F_k)$$
$$c_{h-1k} - Q(2 - z_{if} - z_{hg}) \leq d_{i+1k} + q_{ih};\ k \in K, f, g \in F_k, i \in I_f, h \in I_g:$$
$$i \neq h, f \prec \text{last}(F_k), g = \text{next}(f, F_k),$$

where q_{ih} is the transportation time required to transfer a product from assembly stage i to assembly stage h.

Similar to (4.47) and (4.48), if two successive tasks $f \in F_k$ and $g = \text{next}(f, F_k)$ of product k are assigned to different assembly stages i and h, then the above constraints generate equality $c_{h-1k} = d_{i+1k} + q_{ih}$ (cf. (1.22) in Section 1.2.4), otherwise they are inactive. Note that for the buffer stage $i = h-1$, processing time $p_{h-1k} = 0\ \forall k$ and hence each product completion time is equal to its arrival time in the buffer.

4.3.2 Sequential Loading and Scheduling

In this subsection a hierarchical, two-level loading and scheduling is proposed and the corresponding MIP formulations are presented. In the two-level approach, first at the top level the loading problem is solved to best allocate assembly tasks among assembly stages so that no revisiting of stages by any product is required, and then at the base

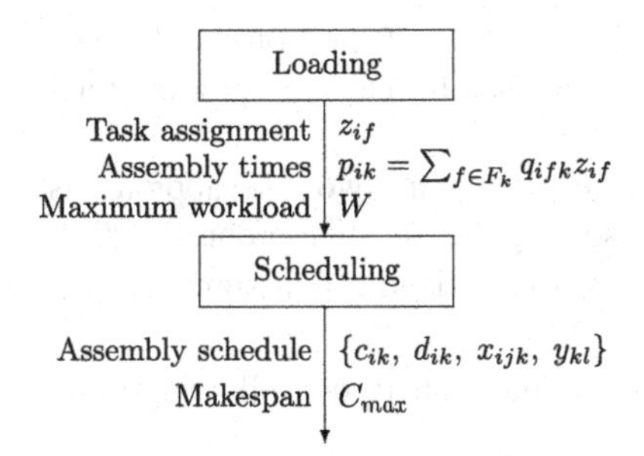

Figure 4.7 Sequential loading and scheduling with finite in-process buffers.

level the shortest assembly schedule for a mix of products is found for prefixed product assignment, see Figure 4.7. The solution of the loading problem at the top level creates a job shop problem with finite capacity buffers to be solved at the base level, where both problems are simpler than the integrated problem *LASPB* for simultaneous loading and scheduling.

The levels are connected using a pure top-down approach. The top-level objective of balancing the workloads is a commonly used surrogate objective that represents the base-level objective of minimizing the makespan. The top-level prefixes component feeder assignment, the corresponding processing routes with no revisiting, and the processing times for all products at assembly stages. The shortest schedule is next determined at the base level by solving the resulting job shop problem with finite in-process buffers.

Below, the MIP model *LPB* is presented for loading of a flexible assembly system with parallel stations and no revisiting.

Model *LPB*: *Loading a Flexible Assembly System with Parallel Stations and No Revisiting*

Minimize

$$W \tag{4.59}$$

subject to

1. *Task Assignment with No Revisiting Constraints*: (4.34) to (4.37)
2. *Processing Time Constraints*:

$$p_{ik} = \sum_{f \in F_k} q_{ifk} z_{if}; \;\; i \in I_A, k \in K \tag{4.60}$$

3. *Maximum Workload Constraints*:

$$\sum_{k \in K} p_{ik}/m_i \leq W; \;\; i \in I_A \tag{4.61}$$

4. *Variable Nonnegativity and Integrality Constraints*: (4.23), (4.58) and

$$p_{ik} \geq 0; \;\; i \in I, k \in K. \tag{4.62}$$

The objective function W (4.59) represents workload of the bottleneck stage defined by constraint (4.61), and equation (4.60) defines for each product the total processing time at every assembly stage.

Having solved the loading problem *LPB*, the next step of the two-level approach is to determine the shortest assembly schedule by solving the job shop problem $S|LPB$ with parallel stations and finite in-process buffers, for prefixed task assignments z_{if}^{L} and processing times p_{ik}.

A feasible solution of the scheduling problem must satisfy the following basic constraints.

- Each product must be assigned to exactly one station at each assembly stage where the required tasks have been assigned as well as to exactly one input and one output buffer of this station.
- For each product all of its tasks must be completed at the assembly stages where the tasks have been assigned, subject to precedence relations among the tasks.
- No two products assigned to the same processor can be processed simultaneously.
- At each stage processing of every product starts immediately after departure from the preceding stage of its processing route.
- A product does not visit the assembly stage where none of its required tasks have been assigned.
- The optimal makespan cannot be less than the maximum workload of assembly stations.

The mathematical formulation of the MIP model for scheduling of a flexible assembly system with parallel stations, finite in-process buffers, and prefixed task assignments is presented below.

Model $S|LPB$: *Scheduling a Flexible Assembly System with Parallel Stations, Finite In-Process Buffers and Prefixed Task Assignments with No Revisiting*

Minimize (4.1) subject to

1. *Product Assignment Constraints*:
   - in every stage to be visited by a product, the product is assigned to exactly one processor,
   - each product assigned to an assembly station is also assigned to one input and one output buffer connected to this station,

$$\sum_{j \in J_i} x_{ijk} = 1;\ i \in I, k \in K\colon \sum_{f \in F_k} z_{if}^{L} > 0 \tag{4.63}$$

$$x_{ijk} = \sum_{j' \in B_h(j)} x_{hj'k};\ i \in I_A, h \in \{i-1, i+1\}, j \in J_i, k \in K\colon$$

$$\sum_{f \in F_k} z_{if}^{L} > 0 \tag{4.64}$$

2. *Product Completion and Departure Constraints*: (4.43) and
   - each product must be completed at the first stage of its processing route as well as at all successive stages,

$$c_{ik} \geq p_{ik};\ i \in I_A, k \in K\colon z^L_{i,\text{first}(F_k)} = 1 \tag{4.65}$$

$$c_{hk} \geq c_{ik} + p_{hk};\ i, h \in I_A, k \in K, f \in F_k, g \in F_k\colon$$
$$i \neq h, f \prec \text{last}(F_k), g = \text{next}(f, F_k), z^L_{if} z^L_{hg} = 1 \tag{4.66}$$

$$c_{il} \geq p_{il} + \sum_{k \in K: k<l} p_{ik} y_{kl} + \sum_{k \in K: k>l} p_{ik}(1 - y_{lk});$$
$$i \in I_A, l \in K\colon m_i = 1, p_{il} > 0 \tag{4.67}$$

3. *Product Noninterference Constraints*:
   - no two products assigned to the same processor can be processed simultaneously,

$$c_{ik} + Q(2 + y_{kl} - x_{ijk} - x_{ihl}) \geq d_{il} + p_{ik};\ i \in I, j \in J_i, k, l \in K\colon$$
$$k < l, \left(\sum_{f \in F_k} z^L_{if}\right)\left(\sum_{f \in F_l} z^L_{if}\right) > 0 \tag{4.68}$$

$$c_{il} + Q(3 - y_{kl} - x_{ijk} - x_{ihl}) \geq d_{ik} + p_{il};\ i \in I, j \in J_i, k, l \in K\colon$$
$$k < l, \left(\sum_{f \in F_k} z^L_{if}\right)\left(\sum_{f \in F_l} z^L_{if}\right) > 0 \tag{4.69}$$

4. *Buffering Constraints*:
   - each product arrives in an input buffer of an assembly stage immediately after its departure from the output buffer of the preceding assembly stage of its processing route,
   - in every assembly stage, processing of each product starts immediately after its departure from the input buffer,
   - each product arrives in an output buffer immediately after its departure from the corresponding assembly stage,

$$c_{h-1k} = d_{i+1k};\ k \in K,\ f, g \in F_k, i \in I_f, h \in I_g\colon$$
$$i \neq h, f \prec \text{last}(F_k), g = \text{next}(f, F_k), z^L_{if} z^L_{hg} = 1 \tag{4.70}$$

$$c_{ik} - p_{ik} = d_{i-1k};\ i \in I_A, k \in K\colon p_{ik} > 0 \tag{4.71}$$

$$c_{i+1k} = d_{ik};\ i \in I_A, k \in K\colon p_{ik} > 0 \tag{4.72}$$

5. *Maximum Completion Time Constraints*: (4.51) and

$$C_{\max} \geq W^L \tag{4.73}$$

6. *Variable Nonnegativity and Integrality Conditions*: (4.57) and

$$c_{ik} \geq 0; \; i \in I, k \in K: p_{ik} > 0 \tag{4.74}$$

$$d_{ik} \geq 0; \; i \in I, k \in K: p_{ik} > 0 \tag{4.75}$$

$$x_{ijk} \in \{0, 1\}; \; i \in I, j \in J_i, k \in K: \sum_{f \in F_k} z_{if}^L > 0. \tag{4.76}$$

Constraint (4.66) maintains for each product the precedence relations among its tasks. (4.66) has been derived from (4.42), given task assignments. Inequality (4.67) for stages with a single processor indicates that completion time of each product l is not less than the sum of processing times at this stage of this product, p_{il}, and of all preceding products, $\sum_{k \in K: k<l} p_{ik}y_{kl} + \sum_{k \in K: k>l} p_{ik}(1 - y_{lk})$. The noninterference constraints (4.68) and (4.69) are defined only for stages to be visited by the products considered. Equation (4.70) has been derived from (4.47) and (4.48), given task assignments. Equations (4.71) and (4.72) are equivalent of (4.49) and (4.50), respectively. Finally, inequality (4.73) imposes a lower bound on the maximum completion time, where W^L is the solution value of the loading problem *LPB*.

4.3.3 Computational Examples

In this subsection numerical examples (Sawik, 2004) and some computational results are presented to illustrate possible application of the two approaches and the proposed MIP formulations.

The FAS configurations for the examples are shown in Figures 4.8, 4.9, and 4.10, respectively, for systems with single stations and single buffers, parallel stations and single buffers, and parallel stations and multiple buffers. The system is made up

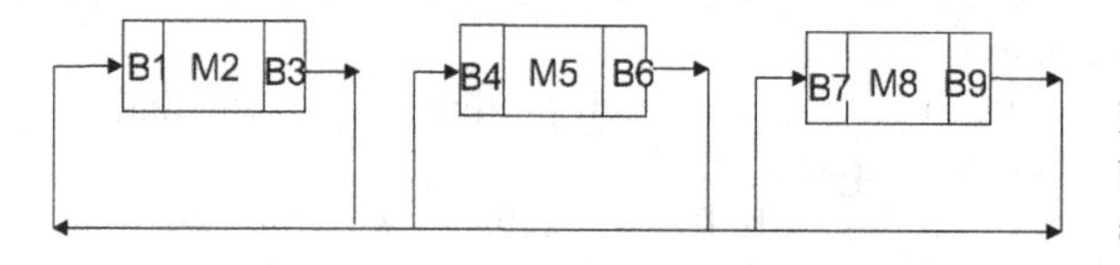

Figure 4.8 A flexible assembly system with single stations and single buffers.

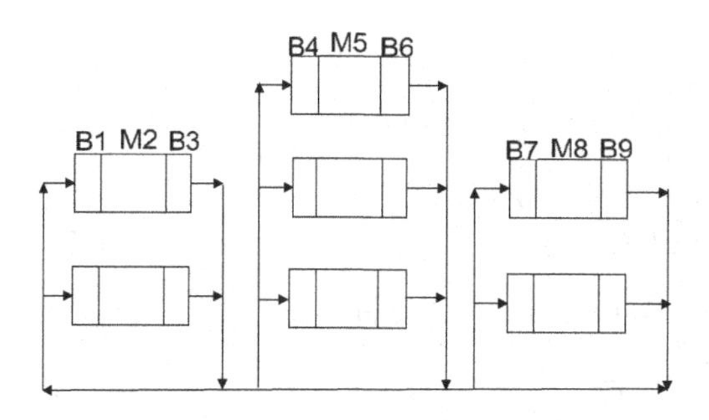

Figure 4.9 A flexible assembly system with parallel stations and single buffers.

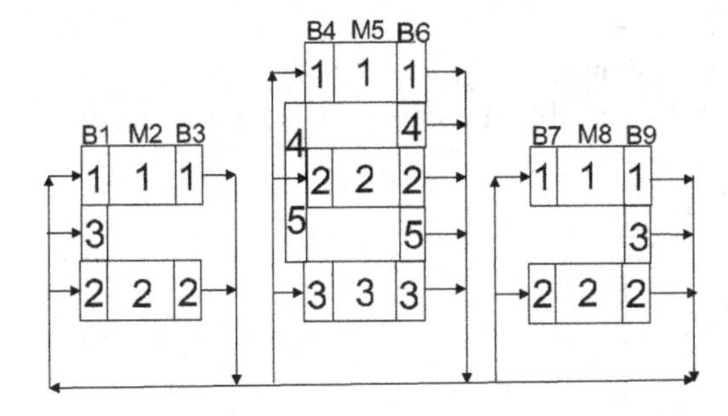

Figure 4.10 A flexible assembly system with parallel stations and multiple buffers.

$m = 9$ processing stages. The set of assembly stages is $I_A = \{2, 5, 8\}$ and the set of buffer stages is $I_B = \{1, 3, 4, 6, 7, 9\}$. Each assembly station has its internal input and output buffer of a unit capacity, and the system with multiple buffers has additional external buffers.

The system with parallel stations and single buffers (Fig. 4.9) consists of $m_i = 2$ parallel processors at stages $i = 1, 2, 3$, $m_i = 3$ parallel processors at stages $i = 4, 5, 6$ and $m_i = 2$ parallel processors at stages $i = 7, 8, 9$.

The system with parallel stations and multiple buffers (Fig. 4.10) consists of $m_1 = 3, m_2 = 2, m_3 = 2, m_4 = 5, m_5 = 3, m_6 = 5, m_7 = 2, m_8 = 2, m_9 = 3$ parallel processors in stages $i = 1, 2, 3, 4, 5, 6, 7, 8, 9$, respectively. The subsets of input and output buffers of each assembly station are $B_1(1) = \{1, 3\}$, $B_1(2) = \{2, 3\}$, $B_3(1) = \{1\}$, $B_3(2) = \{2\}$, $B_4(1) = \{1, 4, 5\}$, $B_4(2) = \{2, 4, 5\}$, $B_4(3) = \{3, 4, 5\}$, $B_6(1) = \{1, 4\}$, $B_6(2) = \{2, 4, 5\}$, $B_6(3) = \{3, 5\}$, $B_7(1) = \{1\}$, $B_7(2) = \{2\}$, $B_9(1) = \{1, 3\}$, $B_9(2) = \{2, 3\}$.

The production order consists of seven products to be assembled of 20 types of components. The ordered sets F_k, $k \in K$ of tasks (assembly sequences) required for each product k are shown below.

$$J_1 = (1, 2, 3, 4, 6, 8, 11, 12, 13, 14, 16, 18),$$
$$J_2 = (1, 2, 4, 5, 6, 7, 9, 10, 11, 12, 14, 15, 16, 17, 19, 20),$$
$$J_3 = (2, 3, 4, 5, 7, 8, 9, 10, 12, 13, 14, 15, 17, 18, 19, 20),$$
$$J_4 = (1, 3, 5, 6, 7, 8, 9, 10, 11, 13, 15, 16, 17, 18, 19, 20),$$
$$J_5 = (1, 3, 5, 6, 7, 8, 9, 10, 11, 13, 15, 16, 17, 18, 19, 20),$$
$$J_6 = (1, 2, 3, 4, 6, 8, 11, 12, 13, 14, 16, 18),$$
$$J_7 = (1, 2, 3, 4, 6, 8, 11, 12, 13, 14, 16, 18).$$

The subsets I_f of assembly stages capable of performing each task $f \in F$ are

$$I_1 = I_2 = I_3 = I_{11} = I_{12} = I_{13} = \{2, 8\},$$
$$I_4 = I_5 = I_6 = I_{14} = I_{15} = I_{16} = \{2, 5\},$$
$$I_7 = I_8 = I_9 = I_{10} = I_{17} = I_{18} = I_{19} = I_{20} = \{5, 8\}.$$

The assembly times ($q_{ifk} = q_{fk}, \forall i \in I_f,\ f = 1, \ldots, 20,\ k = 1, \ldots, 7$), work space required for component feeders assignment ($a_{if},\ i = 2, 5, 8,\ f = 1, \ldots, 20$), and the total work space ($b_i,\ i = 2, 5, 8$) available at each station are given below

$$[q_{fk}] = \begin{bmatrix} 4, 4, 0, 4, 4, 4, 4 \\ 2, 2, 2, 0, 0, 2, 2 \\ 2, 0, 2, 2, 2, 2, 2 \\ 2, 2, 2, 0, 0, 2, 2 \\ 0, 4, 4, 4, 4, 0, 0 \\ 2, 2, 0, 2, 2, 2, 2 \\ 0, 3, 3, 3, 3, 0, 0 \\ 5, 0, 5, 5, 5, 5, 5 \\ 0, 2, 2, 2, 2, 0, 0 \\ 0, 4, 4, 4, 4, 0, 0 \\ 4, 4, 0, 4, 4, 4, 4 \\ 2, 2, 2, 0, 0, 2, 2 \\ 2, 0, 2, 2, 2, 2, 2 \\ 2, 2, 2, 0, 0, 2, 2 \\ 0, 4, 4, 4, 4, 0, 0 \\ 2, 2, 0, 2, 2, 2, 2 \\ 0, 3, 3, 3, 3, 0, 0 \\ 5, 0, 5, 5, 5, 5, 5 \\ 0, 2, 2, 2, 2, 0, 0 \\ 0, 4, 4, 4, 4, 0, 0 \end{bmatrix}$$

$$[a_{if}, i = 2, 5, 8, f = 1, \ldots, 20] =$$

$$\begin{bmatrix} 1, 2, 3, 1, 2, 3, 17, 17, 17, 17, 1, 2, 3, 1, 2, 3, 17, 17, 17, 17 \\ 21, 21, 21, 1, 2, 3, 1, 2, 3, 5, 21, 21, 21, 1, 2, 3, 1, 2, 3, 5 \\ 1, 2, 3, 19, 19, 19, 1, 2, 3, 5, 1, 2, 3, 19, 19, 19, 1, 2, 3, 5 \end{bmatrix},$$

$$b_2 = 16,\ b_5 = 20,\ b_8 = 18.$$

Note that, if station i is incapable of performing task f, then $a_{if} > b_i$.

First, the monolithic model *LASPB* was applied to find the optimal solution for simultaneous loading and scheduling. Then, for comparison the hierarchical approach was applied to find assembly schedule using models *LPB* and *S|LPB*. For each of the example problems both approaches have constructed assembly schedules with minimum makespan C_{max}. The schedules obtained are shown in Gantt charts presented in Figures 4.11, 4.12, and 4.13, respectively, for FAS with single stations and with parallel stations with single or multiple buffers. In the figures, M indicates an assembly

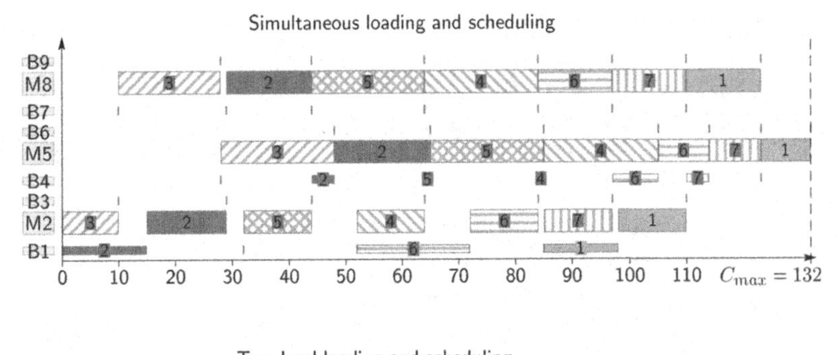

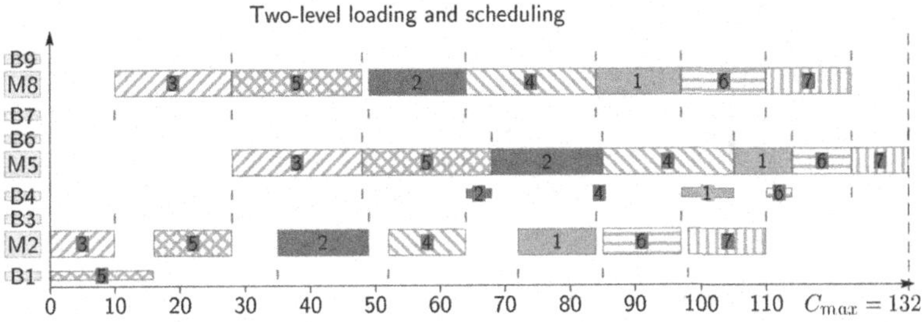

Figure 4.11 Assembly schedules for an FAS with single stations.

stage and B stands for a buffer stage, where machine blocking or buffering is indicated with a narrow bar. Products are numbered and in Figures 4.11 and 4.12 additionally indicated with different patterns. In the figures, the optimal makespans $C_{\max}$ are presented.

For the system with parallel stations and multiple buffers (Fig. 4.10), the production order consists of eight units of each product type, in total of 56 products. For this example, the optimal batch schedules were found, with products of the same type scheduled consecutively. In the Gantt charts shown in Figure 4.13, products of one type are indicated with product type number. The obtained sequence of product types is 3,1,4,2,5,6,7 for the integrated approach and 3,1,4,2,5,7,6 for the hierarchical approach. The optimal makespan is $C_{\max} = 467$ for both approaches.

The computational results for the examples are summarized in Table 4.3. The size of the various MIP models is represented by the total number of variables, *Var.*, number of binary variables, *Bin.*, number of constraints, *Constr.*, and number of nonzero coefficients, *Nonz.*, in the constraint matrix. The last two columns of the table give the optimal solution value of makespan $C_{\max}$ or maximum workload W and total number of nodes in the branch-and-bound tree until a proven optimal solution was reached.

4.4 COMMENTS

A flexible assembly system is an example of a flexible job shop-type production system with parallel machines and various additional limited resources. The literature

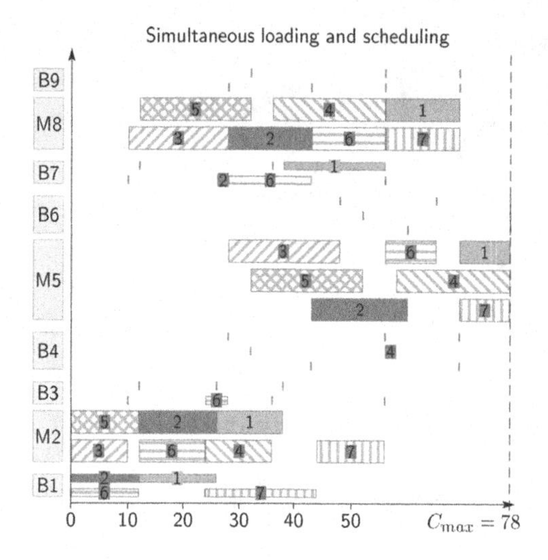

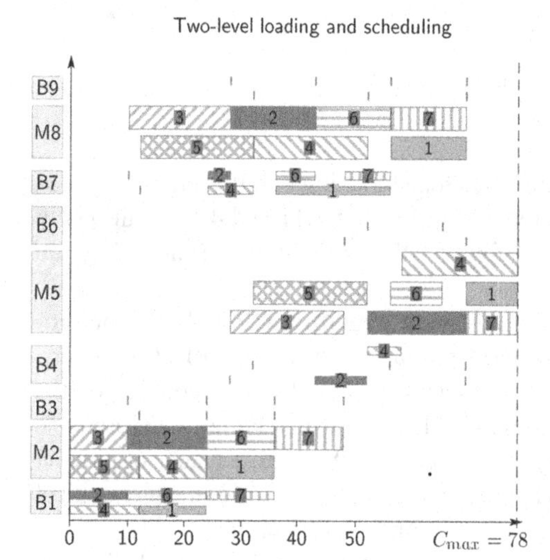

Figure 4.12 Assembly schedules for an FAS with parallel stations.

on MIP models for scheduling production orders in flexible job shops, that consider capacity constraints and precedence relations is very limited, for example, Greene and Sadowski (1986), Sawik (1990), Jiang and Hsiao (1994), Liu and MacCarthy (1997), K. Chen and Ji (2007), Örnek et al. (2010).

This chapter shows that the MIP approach is capable of solving to optimality a hard combinatorial optimization problem of loading and scheduling of a general flexible assembly system with limited work space of assembly stations for component feeder assignment, inifinite or finite capacity in-process buffers, and revisiting or no revisiting of stations allowed. The material presented in this chapter is an extension of the results discussed in Sawik (2000b, 2004). MIP models for simultaneous or

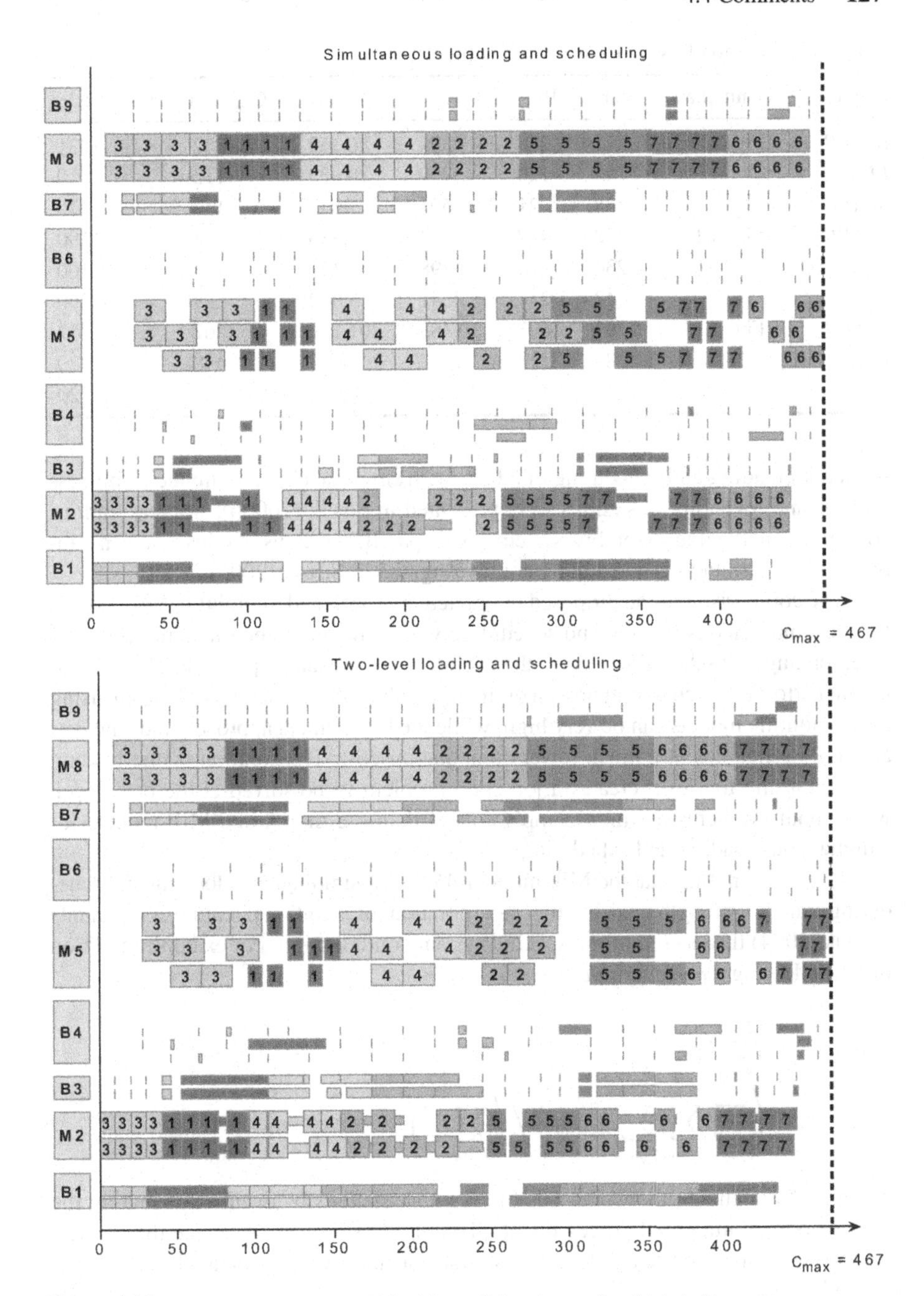

Figure 4.13 Batch schedules for an FAS with parallel stations and multiple buffers.

Table 4.3 Solution Results for the example problems

Problem	Gantt chart	Var.	Bin.	Constr.	Nonz.	$C^*_{\max}$ or (W^*)	Nodes
LASPB	Fig. 4.11	371	244	5394	20,152	132	1031
LPB		142	120	3929	10,625	(112)	0
S|LPB		148	21	595	1673	132	375
LASPB	Fig. 4.12	708	497	10,774	47,335	78	1390
LPB		234	198	6298	17,127	(65)	100
S|LPB		442	231	4641	13,062	78	260
LASPB	Fig. 4.13	2702	1693	30,483	201,199	467	57
LPB		289	120	1732	6043	(448)	0
S|LPB		2542	1533	21,903	98,938	467	2720

sequential loading and scheduling of a FAS with single stations, unlimited in-process buffers, and revisiting of stations were presented in Sawik (2000b), whereas various configuration flexible assembly systems with parallel stations, limited capacity in-process buffers, and no revisiting were investigated in Sawik (2004).

The comparison of the proposed integrated approach and monolithic MIP models for the simultaneous loading and scheduling with a more common hierarchical, two-level loading and scheduling proves the usefulness of the latter approach. The computational effort required to find proven optimal schedules for realistic problems by using the monolithic models can be very high, while the hierarchical approach finds the best assembly schedules at a much lower computation cost. The solution of the loading problem at the top level creates a job shop problem to be solved at the base level, where both problems are more simply solved than the monolithic MIP models for simultaneous loading and scheduling.

It is worth noting that the MIP model *LASPB* for simultaneous loading and scheduling can be strengthened by the incorporation of additional cutting constraints (Sawik, 2004) that account for a specific system configuration. For example, addition of the following *workload balancing constraints*

$$\sum_{k\in K}\sum_{f\in F_k} q_{ifk}z_{if}/m_i \le$$

$$\left(\sum_{h\in I_A}\sum_{k\in K}\sum_{f\in F_k} q_{hfk}z_{hf}\right)\Big/\left(\sum_{l\in I_A} m_l\right) + \max_{k\in K, f\in F_k} q_{ifk};\ i\in I_A$$

that aim at equalizing the workloads of assembly stations by evenly distributing the total workload among all processors at all assembly stages, may reduce the computation time required to reach the best solution for the FAS with parallel stations and single buffers.

Let us note that no revisiting constraints (4.37) are defined for at most $|F|(|F|-1)(|F|-2)/6$ triplets of tasks for each product and each assembly stage, and hence the number of "no revisiting" constraints in models *LASPB* and *LPB* is $O(|I_A||F|^3 n)$, where $|I_A|$, $|F|$, and n denote, respectively, the number of assembly stages, number of task types, and number of products. In some of the test instances the number of constraints (4.37) is greater than half of the total number of all the constraints, which indicates that no revisiting requirement significantly contributes to the computation time of the FAS loading and scheduling problem. The number of no revisiting constraints can be reduced if identical triplets of successive tasks for different products are considered only once. Then (4.37) should be redefined as follows:

$$z_{if} \geq z_{ie} + z_{ig} - 1; \ (e, f, g) \in \bigcup_{k \in K} \{e \in F_k, f \in F_k, g \in F_k: e \prec f \prec g\},$$

$$i \in I_e \cap I_f \cap I_g,$$

where $\bigcup_{k \in K} \{e \in F_k, f \in F_k, g \in F_k: e \prec f \prec g\}$ is the the union of triplets of successive assembly tasks for all products.

It is instinctively believed that computation time increases and computational feasibility decreases as the number of constraints increases. In practice, however, a formulation with a smaller number of constraints more often requires a longer computation time to find a proven optimal solution (e.g., Nemhauser and Wolsey, 1999; D.-S. Chen et al., 2010).

It should be pointed out that the performance of the MIP models may depend on the system configuration (e.g., single vs. parallel assembly stations, single vs. multiple in-process buffers, etc.), the cutting constraints that are included in the corresponding MIP models and the solver setting. It is possible to observe a slightly better or worse performance for some particular system configurations and some particular settings of the branch-and-bound algorithm. The experiments with various features of the CPLEX solver to speed up the solution process have indicated that the best results are obtained for various nondefault settings of the branch-and-bound algorithm. In most cases, the best results were obtained for a nearly depth-first branch-and-bound strategy for node selection and for the strong branching strategy with a limited number of different branches considered for different choices of branching variable. For such settings good feasible solutions were found more quickly and fewer nodes were required to reach the best solutions.

The flexible assembly system considered in this chapter is assumed to have sufficient material handling capacity so that products are delivered instantaneously from one assembly stage to another without transportation time involved or with a fixed transportation time. In a more general setting, a flexible assembly system can be considered to be made up of two interrelated subsystems: an assembly subsystem and a finite capacity materials handling subsystem, typically an automated guided vehicle system. The two subsystems are closely integrated so that the performance of one affects the other, and both transportation capacity and transportation times should be explicitly taken into account. While completion of each assembly task at an assembly station generates an arrival of some transportation task to the materials handling subsystem, completion of a transportation task by a vehicle determines a new assembly task for some assembly station. The objective of the combined machine and vehicle scheduling is to find an assignment of assembly tasks to machines for each product over a scheduling horizon as well as the associated time table for vehicle movements so as to complete production order and minimize some optimality criterion, for example, the schedule length with transportation times included.

Similar to loading and scheduling, there are two main approaches used for machine and vehicle scheduling in a general flexible assembly system, that is, Ulusoy and Bilge (1993), Bilge and Ulusoy (1995), Sawik (1996, 1999), C.-Y. Lee and Chen (2001):

- The integrated approach, in which a detailed assembly schedule for all products and the corresponding timetable for vehicle movements are determined simultaneously with no initial loading and routing decisions required.
- The hierarchical approach, in which first the machine loading and assembly routing problem is solved and then, given task assignments and assembly routes selected, detailed machine and vehicle schedules are found.

EXERCISES

4.1 Modify the mixed integer programs *L1a* and *L1f* for loading an FAS with single stations, for different measures of system imbalance:

(a) Minimum $(W - \overline{W})$, where $\overline{W}$ is the average workload of an assembly station.

(b) Minimum of $(W - W_{\min})$, where $W_{\min}$ is the minimum workload of an assembly station.

(c) Maximum of $W_{\min}$.

4.2 Modify the mixed integer program *LPB* for loading an FAS with parallel stations, for different measures of system imbalance:

(a) Minimum of $(W - W_{\min})$, where $W_{\min}$ is the minimum workload of an assembly station.

(b) Maximum of $W_{\min}$.

4.3 Modify the mixed integer program *LASPB* for loading and scheduling of an FAS, to allow for revisiting of stations.

4.4 In the computational examples in Section 4.2.3, Figures 4.4 and 4.5 indicate that for simultaneous loading and scheduling, makespans are identical for both alternative and fixed routing, whereas for sequential loading and scheduling, the makespan for alternative routing is greater than for the fixed routing. Explain the reasons for such results.

4.5 In the computational examples in Section 4.3.3, Figures 4.12 and 4.13 indicate that makespans for both simultaneous and sequential loading and scheduling are identical. Explain why the makespans are identical.

Part Two

Medium-Term Scheduling in Supply Chains

Chapter 5

Customer Order Acceptance and Due Date Setting in Make-to-Order Manufacturing

5.1 INTRODUCTION

In make-to-order manufacturing accepting or rejecting customer orders is often combined with due date setting. Accepting too many orders with customer-requested dates may increase demand on capacity above available capacity and as a result may increase lead time and decrease customer service level, that is, more orders are delivered after the requested dates. To reduce the number of rejected or delayed orders, a manufacturer should quote due dates for some orders later than the due dates requested by customers. Setting a due date later than the requested date, however, may result in a reduction of revenue, whereas fulfilling the order later than the quoted date may also result in loss of goodwill and sometimes even in contractual penalty costs. On the other hand fulfilling the order earlier then the quoted date may incur finished products inventory holding costs. Thus, the due date quoting problem should account for the three costs (e.g., Hegedus and Hopp, 2001): cost for quoting a due date later than the requested date, cost for fulfilling the order later than the quoted date, and finished products inventory costs for fulfilling the order too early.

The order acceptance and due date setting decisions can be made in either a real-time mode or a batch mode. For the real-time mode, a commitment due date is determined at the time of arrival of the customer order. For the batch mode, customer orders are collected into a "batch" and subsequently considered together to determine the committed due dates for all orders in the batch. While sometimes an initial due date quote is done in real time, the batch mode is commonly used in practice, for example, in the e-business order fulfillment systems, as actual resource allocation and hard order commitment are carried out, see C.-Y. Chen et al. (2001).

Scheduling in Supply Chains Using Mixed Integer Programming. By Tadeusz Sawik

This chapter presents a dual objective problem of order acceptance and due date setting over a rolling planning horizon in make-to-order manufacturing, and proposes a bi-criterion Integer Programming (IP) formulation for its solution. In the proposed model the order acceptance and due date setting decisions are directly linked with available capacity. The problem objective is to select the maximal subset of orders that can be completed by the customer-requested dates and to quote delayed due dates for the remaining acceptable orders to minimize the number of delayed orders or the total number of delayed products as a primary optimality criterion and to minimize the total or maximum delay of orders, as a secondary criterion. The delay of an order is defined as the positive difference between the due date committed by the manufacturer and the due date requested by the customer. Possible enhancements of the basic models are discussed, in particular a revenue management approach is proposed to maximize total revenue subject to service level constraints. In addition, an MIP model is provided for scheduling customer orders over a rolling planning horizon to minimize maximum inventory level.

The two approaches are proposed. An integrated approach, based on the weighted-sum monolithic model where the order acceptance and the due dates setting are determined simultaneously, and a hierarchical approach based on the lexicographic model, where first the maximal subset of acceptable customer orders is selected and then delayed due dates are determined for unrejected, acceptable orders to minimize their total or maximum delay.

The following time-indexed IP or MIP models for customer order acceptance/due date setting and for customer orders scheduling are presented in this chapter:

DDS for order acceptance and due date setting to minimize weighted sum of delayed orders or delayed products and total or maximum delay

AR for order acceptance/rejection to minimize number of delayed/rejected orders or delayed/rejected products

DD for due date setting for delayed orders to minimize total or maximum delay

SCO for scheduling customer orders to minimize maximum earliness

SCO1 for scheduling single-period customer orders to minimize maximum earliness

In Section 5.6 computational examples modeled after a real-world make-to-order flexible flow shop environment in the electronics industry are provided and, for comparison, the single-objective solutions that maximize total revenue subject to service level constraints are reported.

5.2 PROBLEM DESCRIPTION

The production system under study is a flexible flow shop that consists of m processing stages in series, where each stage $i \in I = \{1, \dots, m\}$ is made up of $m_i \geq 1$ parallel, identical machines. In the system various types of products are manufactured according to customer orders, where each product type requires processing in various stages; however, some products may bypass some stages. The customer orders are single product type orders.

Order acceptance and due date setting decisions over a rolling planning horizon are assumed to be made periodically upon the arrival of a number of orders in a specific time interval (the batching interval), given the set of already accepted orders remaining for processing and the remaining available capacity. The batching interval consists of a fixed number of σ most recent time periods (e.g., days) immediately preceding period t_1, when the optimization model is about to be executed, that is, the model is executed every σ time periods at $t_1 = 1, 1+\sigma, 1+2\sigma, 1+3\sigma, \ldots$. The problem objective is to plan activities over a planning horizon, which consists of the ensuing h $(h > \sigma)$ time periods (e.g., working days) of equal length (e.g., hours or minutes). Denote by $T = \{t_1, \ldots, t_1 + h - 1\}$ the set of planning periods covered in each model run.

Let J be the set of newly arrived customer orders collected over a batching interval, and $\tilde{J}$ the subset of previously accepted orders remaining for processing, to be completed by $t_1 + h - 1$. (Notice that all previously rejected orders are not considered any more.) Each order $j \in J$ (or $j \in \tilde{J}$) is described by a triple $[a_j, d_j$ (or $\tilde{d}_j$), n_j (or $\tilde{n}_j$)], where a_j is the order ready date (e.g., the earliest release period or the earliest period of material availability), d_j is the customer-requested due date (e.g., customer-required shipping date), $\tilde{d}_j \le t_1 + h - 1$ is the due date of order $j \in \tilde{J}$ committed by the manufacturer, n_j is the size of the order (required quantity of ordered product type), and $\tilde{n}_j$ is the remaining order size.

Let $p_{ij} \ge 0$ be the processing time in stage i of each product in order j, and let $q_{ij} = n_j p_{ij}$ (or $\tilde{q}_{ij} = \tilde{n}_j p_{ij}$) be the total processing time required to complete order $j \in J$ (or $j \in \tilde{J}$) in stage i. Denote by c_{it} the total processing time available in period t on each machine in stage i. The amount c_{it} takes into account the flow shop configuration of the production system and the production/transfer lot sizes. For each machine in stage i, c_{it} must take into account the time required for processing a single production lot at all upstream $1, \ldots, i-1$ and downstream $i+1, \ldots, m$ stages during the same planning period. As a result the available capacity c_{it} is smaller than simply the available machine hours in period t; c_{it} can be bounded as follows:

$$\underline{c}_i \le c_{it} \le \overline{c}_i, \tag{5.1}$$

where

$$\underline{c}_i = H - \max_{j \in J}\left(\sum_{l \in I: l < i} b_j p_{lj}\right) - \max_{j \in J}\left(\sum_{l \in I: l > i} b_j p_{lj}\right),$$

$$\overline{c}_i = H - \min_{j \in J}\left(\sum_{l \in I: l < i} b_j p_{lj}\right) - \min_{j \in J}\left(\sum_{l \in I: l > i} b_j p_{lj}\right).$$

H is the length of each planning period (e.g., working hours per day) and b_j is the production/transfer lot size for order j (i.e., order quantity n_j is split across multiple lots of size b_j).

The order acceptance and due date setting decisions are made for a set J of newly arrived customer orders collected over a batching interval, given the remaining available capacity. The problem objective is to select a maximal subset of orders $j \in J$ that can be completed by the customer-requested due dates and to quote delayed due dates for the remaining acceptable orders to minimize the number of delayed orders or delayed products as a primary optimality criterion and to minimize their total or maximum delay as a secondary criterion.

5.2.1 Critical Load Index

In this subsection a simple critical load index is introduced and some necessary conditions are derived for all customer orders to be accepted and for all requested due dates to be met.

When executing the model over time, after each batch execution, the remaining available capacity is converted into a fixed input for the next model run. Let $C_i(t, d)$ be the remaining cumulative capacity available in stage i in periods t through d, after deducting the capacity reserved for orders $j \in \tilde{J}$ that were previously committed in earlier model runs but whose production has not yet been completed, that is,

$$C_i(t, d) = m_i \sum_{\tau \in T: t \leq \tau \leq d} c_{i\tau} - \sum_{j \in \tilde{J}: t \leq \tilde{d}_j \leq d} \tilde{q}_{ij}; \; d, t \in T: t \leq d. \tag{5.2}$$

A necessary condition for meeting all customer-requested due dates is that for each processing stage i, each due date $d \leq t_1 + h - 1$ and each interval $[t, d]$, $t \in T: t \leq d$ ending with d, the demand on capacity does not exceed the available capacity, that is,

$$\Psi_i(d) = \max_{t \in T: t \leq d} \left(\frac{\sum_{j \in J: t \leq a_j \leq d_j \leq d} q_{ij}}{C_i(t, d)} \right) \leq 1; \; d \in T, i \in I \tag{5.3}$$

where $\Psi_i(d)$ is the cumulative capacity ratio for due date d with respect to processing stage i.

If $\Psi_i(d) \leq 1$, then for any period $t \leq d$ the cumulative demand on capacity in stage i of all the orders with due dates not greater than d and ready dates not less than t (the numerator in Equation (5.3)) does not exceed the cumulative capacity available in this stage in periods t through d (the denominator in Equation (5.3)).

When $\Psi_i(d) > 1$, then at least one order to be processed at stage i, with requested due date not later than d, must be delayed or rejected (if $d = t_1 + h - 1$) to meet available capacity constraints.

If all customer orders were continuously allocated among the consecutive time periods so that all periods could be filled exactly to their capacities, the necessary condition (5.3) could become sufficient for all orders to be completed by their due dates.

Denote by $\Psi(d)$ the cumulative capacity ratio for due date d

$$\Psi(d) = \max_{i \in I} \Psi_i(d); \;\; d \in T. \tag{5.4}$$

The ratio $\Psi(d), \; d \in T$ can be used as a simple critical load index to identify the bottleneck stages and the overloaded periods.

Note that, if all customer orders are ready at the beginning of the planning horizon, $(a_j = t_1 \; \forall j \in J)$, then $\Psi(t_1 + h - 1)$ is the cumulative capacity ratio for the entire horizon, i.e., the total capacity ratio. A necessary condition to have a feasible production schedule with all customer orders completed during the planning horizon is that the total capacity ratio is not greater than 1:

$$\max_{i \in I} \frac{\sum_{j \in J} q_{ij}}{C_i(t_1, t_1 + h - 1)} \leq 1. \tag{5.5}$$

The basic order acceptance and due date setting problem presented in the next section is applied for orders with requested due dates such that condition (5.3) is not satisfied. In this case the proposed model determines new, delayed due dates to satisfy (5.3) and, in addition, to attain certain optimality criteria. If, however, condition (5.3) holds for all customer-requested due dates, then the due dates can be met and the due date setting problem becomes trivial and need not be considered.

5.3 BI-OBJECTIVE ORDER ACCEPTANCE AND DUE DATE SETTING

In this section time-indexed IP and MIP models are proposed for the bi-objecive order acceptance and due date setting over a rolling planning horizon (Fig. 5.1). The two sets of models are proposed: a weighted-sum, monolithic model *DDS*, based on

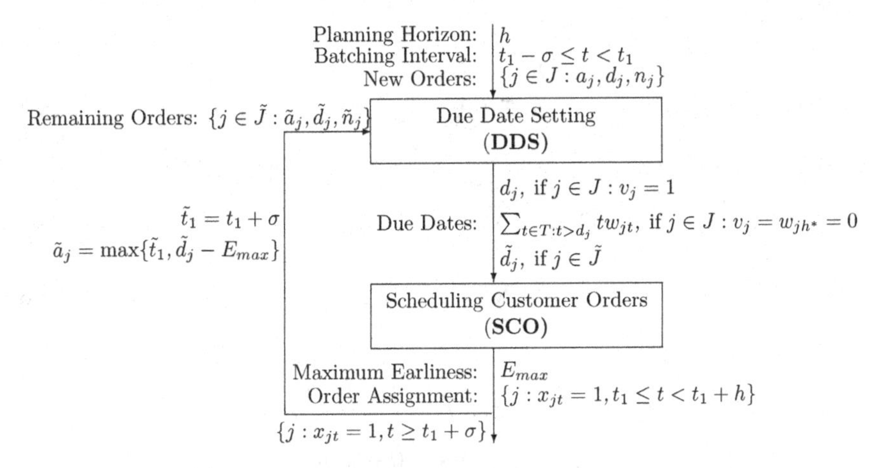

Figure 5.1 Order acceptance, due date setting, and scheduling of customer orders over a rolling planning horizon.

the scalarization approach, and a hierarchy of two models *AR* and *DD*, based on the lexicographic approach.

The primary objective of the order acceptance and due date setting problem is to maximize the customer service level, that is to minimize the number of delayed orders O_{sum}, that is, the orders for which the committed due dates are later than the customer-requested dates. Minimization of the number of delayed orders may often lead to a large number of delayed products since a high customer service level can be achieved by setting later due dates for a small number of large customer orders. Therefore, an alternative primary objective is to minimize the number of delayed products P_{sum}.

Similarly, two alternative secondary objective functions are considered: minimization of the total delay Q_{sum} of all orders or minimization of the maximum delay $Q_{\max}$ among all orders, where order delay is defined as the positive difference between the committed and the requested due date. While minimization of Q_{sum} aims at reducing the total delay of all postponed customer orders, the minimization of $Q_{\max}$ gives preference to a reduction of the maximum delay with respect to the requested due date of each individual order.

The orders that cannot be accepted in periods t_1 through $t_1 + h - 1$ due to insufficient capacity and hence should be rejected are assigned at a significant penalty to a dummy planning period $h^* = t_1 + h$ with infinite capacity. Let $T^* = T \cup \{t_1 + h\} = \{t_1, \ldots, t_1 + h - 1, t_1 + h\}$ be the enlarged set of planning periods with a dummy period $h^* = t_1 + h$ included.

The following two basic decision variables are introduced in the proposed MIP models (for notation used, see Table 5.1).

- *Order Acceptance Variable*: $v_j = 1$, if order j is accepted with its requested due date or $v_j = 0$ if order j needs to be delayed or rejected.
- *Due Date Assignment Variable*: $w_{jt} = 1$, if order j is assigned delayed due date t ($d_j < t < h^*$), or is rejected ($t = h^*$); otherwise $w_{jt} = 0$.

Now, the primary and the secondary objective functions f_1 and f_2 can be expressed as below.

$$f_1 \in \{O_{\text{sum}}, P_{\text{sum}}\} \tag{5.6}$$

$$f_2 \in \{Q_{\text{sum}}, Q_{\max}\}, \tag{5.7}$$

where

$$O_{\text{sum}} = \sum_{j \in J} (1 - v_j - w_{jh^*}) \tag{5.8}$$

$$P_{\text{sum}} = \sum_{j \in J} n_j(1 - v_j - w_{jh^*}) \tag{5.9}$$

$$Q_{\text{sum}} = \sum_{j \in J, t \in T: t > d_j} (t - d_j) w_{jt} \tag{5.10}$$

$$Q_{\max} = \max_{j \in J, t \in T: t > d_j} (t - d_j) w_{jt}. \tag{5.11}$$

Table 5.1 Notation: Order Acceptance and Due Date Setting

Indices		
i	=	processing stage, $i \in I$
j	=	customer order, $j \in J \cup \tilde{J}$
t	=	planning period, $t \in T$
Input parameters		
a_j, n_j	=	ready date, size of order j
c_{it}	=	processing time available in period t on each machine in stage i
$C_i(t, d)$	=	cumulative capacity available in stage i in periods t through d
d_j, $(\tilde{d}_j)$	=	customer requested (manufacturer quoted) due date for order $j \in J$, $(j \in \tilde{J})$
h	=	planning horizon
m_i	=	number of parallel, identical machines in stage i
p_{ij}	=	processing time in stage i of each product in order j
q_{ij}	=	$n_j p_{ij}$—total processing time required to complete order j in stage i
σ	=	length of batching interval
J	=	set of newly arrived customer orders collected over a batching interval
$\tilde{J}$	=	set of previously accepted customer orders remaining for processing
Decision variables		
v_j	=	1, if order j is accepted with customer requested due date; $v_j = 0$ if order j needs to be delayed or rejected (order acceptance variable)
w_{jt}	=	1, if order j is assigned delayed due date t, $(d_j < t < h^*)$ or is rejected $(t = h^*)$; otherwise $w_{jt} = 0$ (due date assignment variable)
x_{jt}	=	1, if customer order j is assigned to planning period t; otherwise $x_{jt} = 0$ (order-to-period assignment variable)
$0 \leq y_{jt} \leq 1$	=	fraction of multiperiod order j assigned to period t (multiperiod order allocation variable)
Objective functions		
$E_{\max}$	=	maximum earliness
O_{sum}, P_{sum}	=	total number of delayed orders, delayed products, respectively
$Q_{\max}$, Q_{sum}	=	maximum delay, total delay, respectively
R_{sum}	=	total revenue

Model DDS: *Order Acceptance and Due Date Setting to Minimize the Weighted Sum of Delayed Orders or Delayed Products and Total or Maximum Delay*

Minimize

$$\lambda_0 \sum_{j \in J} w_{jh^*} + \lambda_1 f_1 + \lambda_2 f_2 \tag{5.12}$$

where $\lambda_0 \gg \lambda_1 \geq \lambda_2$

subject to

1. *Order Acceptance and Due Date Setting Constraints*:
   - each customer order is either accepted with its requested due date, is assigned a delayed due date, or is rejected,

$$v_j + \sum_{t \in T^*: t > d_j} w_{jt} = 1; \quad j \in J \tag{5.13}$$

2. *Capacity Constraints*:
   - for any period $t \leq d$, the cumulative demand on capacity in stage i of all orders accepted with requested (or delayed) due dates not greater than d and ready dates (or requested due dates) not less than t must not exceed the cumulative capacity available in this stage in periods t through d

$$\sum_{j \in J: t \leq a_j \leq d_j \leq d} q_{ij} v_j + \sum_{j \in J} \sum_{\tau \in T: t \leq d_j < \tau \leq d} q_{ij} w_{j\tau} \leq C_i(t, d);$$
$$d, t \in T, i \in I: t \leq d \tag{5.14}$$

3. *Maximum Delay Constraints (if $f_2 = Q_{\max}$)*:
   - for each delayed order j with adjusted due date $t > d_j$, its delay $(t - d_j)$ cannot exceed the maximum delay $Q_{\max}$,

$$(t - d_j) w_{jt} \leq Q_{\max}; \quad j \in J, t \in T: t > d_j \tag{5.15}$$
$$Q_{\max} \geq 0 \tag{5.16}$$

4. *Integrality Conditions*:

$$v_j \in \{0, 1\}; \quad j \in J \tag{5.17}$$
$$w_{jt} \in \{0, 1\}; \quad j \in J, t \in T^*: t > d_j. \tag{5.18}$$

In the objective function (5.12), $\lambda_1 \geq \lambda_2$ as the primary objective of *DDS* is to minimize the number of delayed orders $f_1 = O_{\text{sum}}$ (5.8) or alternatively to minimize total number of delayed products $f_1 = P_{\text{sum}}$ (5.9), delivered after the customer-requested dates. The objective function is additionally penalized with $\lambda_0 \gg \lambda_1$ for each rejected order.

The time-indexed binary (or mixed binary, if $f_2 = Q_{\max}$) program *DDS* for order acceptance and due date setting determines feasible due dates using the capacity constraint (5.14), which is based on condition (5.3) for the feasibility of customer requested due dates. Constraints (5.13) and (5.14) ensure that each accepted order $j \in J$ (such that $w_{jh^*} = 0$) is completed on or before its requested due date d_j (if $v_j = 1$) or on its delayed due date $t > d_j$ (if $v_j = 0$ and $w_{jt} = 1$). If condition

(5.3) holds for all customer-requested due dates, then the due date setting problem *DDS* becomes trivial and the objective function (5.12) takes on zero value, since $v_j = 1 \; \forall j \in J$ and $w_{jt} = 0 \; \forall j \in J, \, t \in T^*$. Otherwise, delayed due dates are determined for some customer orders. Note that in (5.16), $Q_{\max}$ does not need to be constrained of being integer.

The solution to the integer program *DDS* determines the maximal subset $\{j \in J\colon v_j = 1\}$ of customer orders accepted with the customer-requested due dates d_j and the subsets of remaining orders: $\{j \in J\colon v_j = 0, \; w_{jh^*} = 0\}$—delayed orders and $\{j \in J\colon v_j = 0, \; w_{jh^*} = 1\}$—rejected orders.

Denote by D_j the requested or delayed due date for each newly arrived and accepted order $j \in J$, or committed due date for each previously accepted order $j \in \tilde{J}$ remaining for processing, i.e.,

$$D_j = \begin{cases} d_j & \text{if } j \in J\colon v_j = 1 \\ \sum_{t \in T: t > d_j} t w_{jt} & \text{if } j \in J\colon v_j = 0, \; w_{jh^*} = 0 \\ \tilde{d}_j & \text{if } j \in \tilde{J}. \end{cases} \tag{5.19}$$

5.4 LEXICOGRAPHIC APPROACH

Since $\lambda_1 \geq \lambda_2$ in the objective function (5.12), a lexicographic approach can also be applied to solve the bi-objective binary program *DDS*. Then *DDS* can be replaced by the time-indexed binary program AR and the time-indexed binary or mixed binary program DD to be solved sequentially (see Fig. 5.2).

Model AR: *Order Acceptance/Rejection to Minimize the Number of Delayed/Rejected Orders or Delayed/Rejected Products*

Minimize

$$f_1 \tag{5.20}$$

subject to

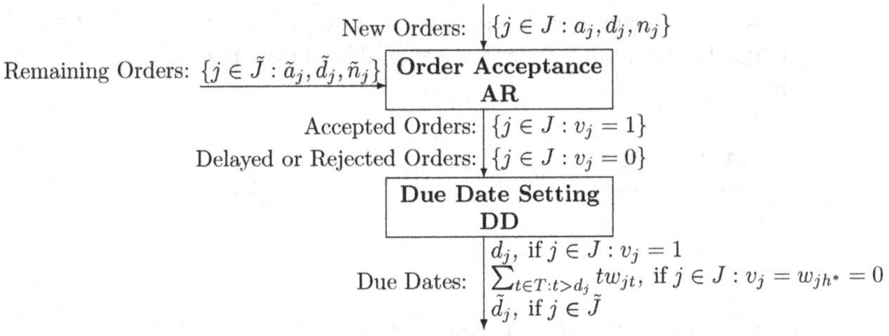

New Orders: $\{j \in J : a_j, d_j, n_j\}$

Remaining Orders: $\{j \in \tilde{J} : \tilde{a}_j, \tilde{d}_j, \tilde{n}_j\}$

Order Acceptance AR

Accepted Orders: $\{j \in J : v_j = 1\}$

Delayed or Rejected Orders: $\{j \in J : v_j = 0\}$

Due Date Setting DD

Due Dates: d_j, if $j \in J : v_j = 1$; $\sum_{t \in T: t > d_j} t w_{jt}$, if $j \in J : v_j = w_{jh^*} = 0$; $\tilde{d}_j$, if $j \in \tilde{J}$

Figure 5.2 Two-level order acceptance and due date setting.

1. *Capacity Constraints*:
   - for any period $t \leq d$, the cumulative demand on capacity in stage i of all accepted orders with due dates not greater than d and ready dates not less than t must not exceed the cumulative capacity available in this stage in periods t through d

$$\sum_{j \in J: t \leq a_j \leq d_j \leq d} q_{ij} v_j \leq C_i(t, d); \; d, t \in T, i \in I : t \leq d \tag{5.21}$$

2. *Integrality Conditions*: (5.17).

The solution to time-indexed binary program *AR* determines the minimal subset $J0 = \{j \in J: v_j = 0\}$ of delayed or rejected orders. New, delayed due dates for acceptable orders are determined using the time-indexed binary (or mixed binary, if $f_2 = Q_{\max}$) program presented below.

Model DD: *Due Date Setting for Delayed Orders to Minimize Total or Maximum Delay*

Minimize

$$f_2 + h \sum_{j \in J0} w_{j,h^*} \tag{5.22}$$

subject to

1. *Due Date Assignment Constraints*:
   - each order is either assigned a due date later than its requested due date or is rejected,

$$\sum_{t \in T^*: t > d_j} w_{jt} = 1; \quad j \in J0 \tag{5.23}$$

2. *Capacity Constraints*:
   - for any period $t \leq d$, the cumulative demand on capacity in stage i of all accepted orders with requested due dates not greater than d and ready dates not less than t, and of all delayed orders with adjusted due dates not greater than d and requested due dates not less than t must not exceed the cumulative capacity available in this stage in periods t through d

$$\sum_{j \in J0} \sum_{\tau \in T: t \leq d_j < \tau \leq d} q_{ij} w_{j\tau} \leq C_i(t, d) - \sum_{j \in J \setminus J0: t \leq a_j \leq d_j \leq d} q_{ij}; \quad d, t \in T, i \in I: t \leq d \tag{5.24}$$

3. *Maximum Delay Constraints (if $f_2 = Q_{\max}$)*: (5.15), (5.16)
4. *Integrality Conditions*: (5.18).

Objective function (5.22) is penalized with h periods of delay for each rejected order. Note that the base level problem is a valid formulation for the lexicographic optimization, only if a single optimal solution exists for the top level problem. If multiple optima (alternative minimal sets $J0$ of delayed and rejected orders) exist for the top-level problem *AR*, then the base-level problem *DD* (where a single set $J0$ is applied only) may produce weakly nondominated solutions with f_2 greater than those obtained by parameterizing on λ the weighted-sum program *DDS*. On the other hand, it is well known that the nondominated solution set of a multi-objective integer program such as *DDS* cannot be fully determined even if the complete parameterization on λ is attempted.

In order to eliminate the weakly nondominated solutions, the secondary objective function f_2 should be minimized over the solutions that minimize the primary objective function f_1. Then, the constraint set of the base-level problem *DD* should be replaced by the constraints of *DDS* with additional upper bound $f_1 \leq f_1^*$ on the corresponding primary objective function (5.8) or (5.9), where f_1^* is the optimal solution value to the top-level problem *AR*.

5.4.1 Model Enhancements

The models presented in this section can be modified or enhanced to consider additional features of the order acceptance and due date setting problem that can be encountered in practice. A few possible extensions of the models are proposed below.

1. *Modified Objective Functions.*
   - Maximization of total revenue. Sales departments often apply revenue management principles for order selection and due date setting. The objective is to maximize a revenue function, e.g., to maximize

$$R_{\text{sum}} = \sum_{j \in J} n_j r_j v_j + \sum_{j \in J, t \in T: t > d_j} n_j r_{jt} w_{jt} - \sum_{j \in J} n_j r_j^* w_{jh^*}, \tag{5.25}$$

where $r_j = r_{j,d_j}$ and r_{jt} is per unit revenue for order j accepted with customer-requested due date d_j and for order j with delayed due date $t > d_j$, respectively. For rejected orders, r_j^* is per unit loss of revenue. Most often customers value short lead times (due dates) over long lead times. Setting delayed due dates results in reduction of revenue. The revenue declines with an increase in the delay of committed due dates with respect to requested due dates, that is,

$$r_{jt} > r_{j,t+1}, \quad t \in T, t \geq d_j.$$

We assume that setting delayed due date results in reduction of revenue proportional to the delay (e.g., Bertrand and van Ooijen, 2000). Per unit revenue r_{jt} decreases by some percent for each day of delay $(t - d_j)$ of delivery with respect to customer-requested date d_j, for example,

$$r_{jt} = r_j(1 - \alpha_j(t - d_j)); \quad t \geq d_j,$$

where $0 < \alpha_j < 1$ is the rate of daily loss of revenue for order j. In addition, a fixed loss $\beta_j r_j$ $(0 < \beta_j < 1)$ of revenue may be applied for each delayed product in order j, i.e.,

$$r_{jt} = r_j(1 - \beta_j - \alpha_j(t - d_j)); \; t > d_j.$$

2. *Service Level Constraints.* If minimization of the number of delayed orders is replaced by another objective function, for example, maximization of total revenue (5.25), then the following constraint should be added to the modified model to maintain the required service level γ, $0 < \gamma \leq 1$, where γ is the fraction of nondelayed customer orders.

$$\sum_{j \in J} v_j \geq \gamma n \tag{5.26}$$

3. *Nonnegotiable Customer Due Dates.* Some customers specify requested due dates that cannot be delayed. Let $JN \subset J$ be the subset of customer orders with nonnegotiable due dates. A feasible solution must satisfy the following constraints:

$$v_j = 1; \; j \in JN \tag{5.27}$$

4. *Customer Due Date Windows.* The customer specifies a delivery time window, for example, acceptable latest delay of shipping date $\delta_{j\max}$, $j \in J$. Then the integer programs must include the following constraints:

$$tw_{jt} \leq d_j + \delta_{j\max}; j \in J, d_j < t \leq d_{j\max} \tag{5.28}$$

5. *Rush Orders.* For urgent orders a high priority $\pi_j > 1$ can be introduced in the objective function, for example,

$$\lambda_0 \sum_{j \in J} \pi_j w_{jh^*} + \lambda_1 \sum_{j \in J} \pi_j(1 - v_j - w_{jh^*}) + \lambda_2 \sum_{j \in J, t \in T: t > d_j} \pi_j(t - d_j) w_{jt} \tag{5.29}$$

where $\lambda_0 \gg \lambda_1 \geq \lambda_2$, and $\pi_j = 1$ for regular orders.

6. *Real-Time Mode.* The proposed integer programs can be applied in real-time mode upon arrival of each new order, given the set of already accepted orders waiting for processing and the remaining available capacity. In particular, the lexicographic approach that does not require as much computation time as the weighted-sum approach (see Section 5.6) is capable of quoting due date in real-time mode for each new order.

5.5 SCHEDULING OF CUSTOMER ORDERS

Model *DDS* (or a hierarchy of models *AR* and *DD*) is executed over a rolling planning horizon every σ time periods (the length of batching interval) to quote due dates for all newly arrived orders $j \in J$ collected over the most recent batching interval, given the previously accepted orders $j \in \tilde{J}$ remaining for processing. When simulating the execution of model *DDS* over time, the set $j \in \tilde{J}$ of previously accepted orders remaining for processing must be determined for each model run, which requires detailed scheduling of customer orders to be performed over a rolling planning horizon.

In this section the MIP model *SCO* is presented for a nondelayed scheduling of customer orders over a rolling planning horizon. The scheduling objective is to find an assignment of orders to periods over the horizon such that each order is assigned not later than its committed due date and the maximum earliness with respect to the due date among all orders is minimized.

The following two types of customer orders are considered:

1. Small, indivisible orders, where each order can be fully processed in a single time period. Small orders are referred to as single-period orders.
2. Large, divisible orders, where each order cannot be completed in one period and must be split into single-period portions to be processed in a subset of consecutive time periods. Large orders are referred to as multiperiod orders.

In practice, two types of customer orders are scheduled simultaneously. Denote by $J1 \subseteq J$, and $J2 \subseteq J$, respectively, the subset of newly arrived indivisible and divisible orders, where $J1 \cup J2 = J$, and $J1 \cap J2 = \emptyset$.

The basic decision variable for scheduling customer orders is order-to-period assignment variable x_{jt}, where $x_{jt} = 1$, if customer order j is assigned to planning period t; otherwise $x_{jt} = 0$. In addition, order allocation variable y_{jt} is required to schedule multiperiod orders, where $y_{jt} \in [0, 1]$ denotes the fraction of multiperiod order j assigned to period t.

Let $\tilde{J} = \tilde{J}1 \cup \tilde{J}2' \cup \tilde{J}2''$ be the set of previously accepted orders, where $\tilde{J}1$, $\tilde{J}2'$, and $\tilde{J}2''$ are the subset of previously accepted single-period orders waiting for processing, the subset of previously accepted multiperiod orders waiting for processing and the subset of previously accepted and uncompleted multiperiod orders remaining for completion, respectively.

It is assumed that the allocation over time of uncompleted multiperiod orders $j \in \tilde{J}2''$ (i.e., such that $0 < \sum_{t<t1} y_{jt} < 1$) remains unchanged, that is,

$$x_{jt} = \tilde{x}_{jt};\ y_{jt} = \tilde{y}_{jt};\ j \in \tilde{J}2'',\ t < t_1 + h,$$

where $\tilde{x}_{jt}$ and $\tilde{y}_{jt}$ are the assignments and the allocation of uncompleted multiperiod orders determined at the previous run of the scheduling model.

Model SCO: *Scheduling Customer Orders to Minimize Maximum Earliness*

Minimize

$$E_{\max} \tag{5.30}$$

subject to

1. *Order-to-Period Nondelayed Assignment Constraints*:

– each single-period order is assigned to exactly one planning period not later than its due date,

$$\sum_{t \in T: a_j \le t \le D_j} x_{jt} = 1; \; j \in J1 \cup \tilde{J}1 \tag{5.31}$$

– each multiperiod order waiting for processing is assigned to a subset of consecutive planning periods not later than its due date,

$$x_{j\lfloor(\tau_1+\tau_2)/2\rfloor} \ge x_{j\tau_1} + x_{j\tau_2} - 1; \; j \in J2 \cup \tilde{J}2', \; \tau_1, \tau_2 \in T: \; a_j \le \tau_1 < \tau_2 \le D_j \tag{5.32}$$

2. *Order Allocation Constraints*:

– each order waiting for processing must be completed not later than its due date,

$$\sum_{t \in T: a_j \le t \le D_j} y_{jt} = 1; \; j \in J \cup \tilde{J}1 \cup \tilde{J}2' \tag{5.33}$$

– each single-period order is completed in a single period,

$$x_{jt} = y_{jt}; \; j \in J1 \cup \tilde{J}1, \; t \in T: \; a_j \le t \le D_j \tag{5.34}$$

– each multiperiod order waiting for processing is allocated among all the periods that are selected for its assignment,

$$y_{jt} \le x_{jt}; \; j \in J2 \cup \tilde{J}2', \; t \in T: \; a_j \le t \le D_j \tag{5.35}$$

3. *Capacity Constraints*:

– in every period the demand on capacity at each assembly stage cannot be greater than the capacity available in this period,

$$\sum_{j \in J \cup \tilde{J}} n_j p_{ij} y_{jt} \le m_i c_{it}; \; i \in I, t \in T \tag{5.36}$$

4. *Maximum Earliness Constraints*:

– for each early order j assigned to period $t < D_j$, its earliness $(D_j - t)$ cannot exceed the maximum earliness $E_{\max}$ to be minimized,

$$(D_j - t)x_{jt} \le E_{\max}; \; j \in J \cup \tilde{J}1 \cup \tilde{J}2', t \in T: \; a_j \le t \le D_j \tag{5.37}$$

5. *Fixed Allocation Constraints*:

the allocation of each uncompleted multiperiod order remains unchanged,

$$x_{jt} = \tilde{x}_{jt}; j \in \tilde{J}2'', t \in T \tag{5.38}$$

$$y_{jt} = \tilde{y}_{jt}; j \in \tilde{J}2'', t \in T \tag{5.39}$$

6. *Nonnegativity and Integrality Conditions*:

$$x_{jt} \in \{0, 1\}; j \in J \cup \tilde{J}, t \in T\colon a_j \leq t \leq D_j \tag{5.40}$$

$$y_{jt} \in [0, 1]; j \in J \cup \tilde{J}, t \in T\colon a_j \leq t \leq D_j \tag{5.41}$$

$$E_{\max} \geq 0. \tag{5.42}$$

If single- and two-period orders are considered only, then (5.31) can be replaced by the following constraints to guarantee that each order is assigned to at most two consecutive periods:

$$x_{jt} + x_{jt+1} \leq 2; j \in J2 \cup \tilde{J}2', t \in T : a_j \leq t < D_j, \tag{5.43}$$

$$x_{jt} + x_{jt'} \leq 1; j \in J2 \cup \tilde{J}2', t \in T, t' \in T : a_j \leq t < D_j - 1,\ t' \geq t + 2. \tag{5.44}$$

The objective (5.30) minimizes the maximum earliness $E_{\max}$ (5.37) among all customer orders or, equivalently, the maximum difference between the order due date and its assignment period such that no tardiness of the customer orders with respect to committed due dates is ensured. The resulting assignment period can be considered to be the latest period of delivery of the required parts such that no tardiness of orders is ensured. If for some customer orders the required parts are delivered later than $E_{\max}$ periods ahead of the due date, that is, later than in period $\max\{t_1, D_j - E_{\max}\}$, the limited order earliness due to the later part availability could restrict a reallocation of the orders to earlier periods with surplus of capacity. As a consequence, tardy orders or even infeasible schedules could occur, with some customer orders unscheduled during the planning horizon.

An implicit objective of *SCO* is to minimize the maximum level of total input inventory of parts waiting for assembly and output inventory of finished products waiting for delivery to the customers (see Section 5.6). To minimize the maximum level of total input and output inventory, the ready date a_j of each customer order $j \in J \cup \tilde{J}$ for each run of model *DDS* can be replaced by the latest delivery date of the required parts, that is,

$$a_j = \max\{t_1, D_j - E_{\max}\} \tag{5.45}$$

Model *SCO* is executed over a rolling planning horizon every σ time periods. The solution to the mixed integer program *SCO* determines the assignment of customer orders to planning periods $t \in [t_1, t_1 + h)$ over the current planning horizon and by this the production schedule for customer orders assigned to periods in the next batching interval $[t_1, t_1 + \sigma - 1]$. As a result, the solution to *SCO* determines the set $\tilde{J} = \{j : x_{jt} = 1, t_1 + \sigma \leq t < t_1 + h\}$ of customer orders assigned to periods $[t_1 + \sigma, t_1 + h)$, that is, the set of orders remaining for processing over the next planning horizon and hence required for the next run of model *DDS* (see Fig. 5.1).

5.5.1 Scheduling Single-Period Orders

The MIP model *SCO* for assignment of single and multiperiod orders can be simplified when only single-period orders are considered. Then the order allocation variables y_{jt} are no longer required and model *SCO* for single-period orders can be rewritten as below.

Model SCO1: *Scheduling Single-Period Customer Orders to Minimize Maximum Earliness*

Minimize (5.30) subject to

1. *Order-to-Period Nondelayed Assignment Constraints*: (5.31)
2. *Capacity Constraints*:

$$\sum_{j \in J \cup \tilde{J}} n_j p_{ij} x_{jt} \le m_i c_{it}; \quad i \in I, t \in T \tag{5.46}$$

3. *Maximum Earliness Constraints*: (5.37)
4. *Nonnegativity and Integrality Conditions*: (5.40), (5.42).

5.6 COMPUTATIONAL EXAMPLES

In this section computational examples are presented to illustrate possible applications of the proposed approach. The examples are modeled after a real-world distribution center for high-tech products, where finished products are assembled for shipping to customers. The distribution center is a flexible flow shop made up of six processing stages with parallel machines, see Figure 5.3. The customer orders require processing in at most four stages: 1, 2, 3 or 4 or 5, and 6. All customer orders are single-period and single-product type orders.

A brief description of the production system, production process, products and the customer orders is given below.

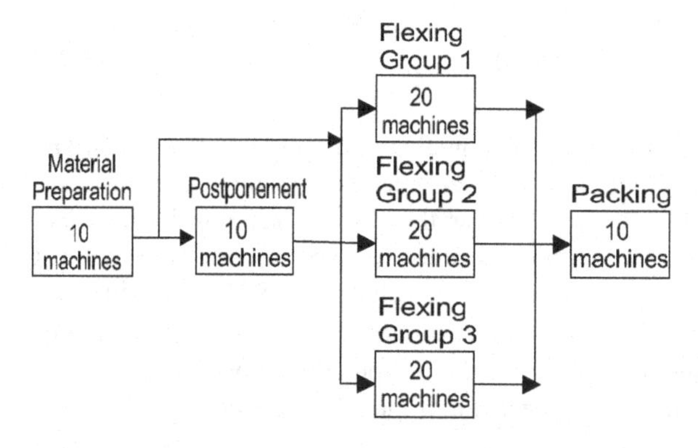

Figure 5.3 Distribution center.

1. *Production System*
   - six processing stages: 10 parallel machines in each stage $i = 1, 2$; 20 parallel machines in each stage $i = 3, 4, 5$; and 10 parallel machines in stage $i = 6$.
2. *Products*
   - 10 product types of three product groups, each to be processed on a separate group of machines (in stage 3 or 4 or 5),
3. *Processing Times (in Seconds) for Product Types*:

Product type/stage	1	2	3	4	5	6
1	20	0	120	0	0	15
2	20	0	140	0	0	15
3	10	0	160	0	0	10
4	15	5	0	120	0	15
5	15	10	0	140	0	15
6	10	5	0	160	0	10
7	15	10	0	180	0	15
8	20	5	0	0	120	15
9	15	0	0	0	140	10
10	15	0	0	0	160	10

4. *Length of the Planning Period (One Production Day)*: $2 \times 8 = 16$ hours.
5. *Length of the Batching Interval*: $\sigma = 5$ days.
6. *Planning Horizon*: $h = 20$ days.

In the computational experiments the MIP models *DDS* and *SCO1* were executed three times over a rolling planning horizon to quote due dates for orders collected over the three batching intervals:

- In period $t_1 = 1$, the due dates ranging from period 1 to period 20 are quoted for 641 customer orders collected over the first batching interval, before period 1.
- In period $t_1 = 6$, the due dates ranging from period 6 to period 25 are quoted for 75 customer orders collected over the second batching interval [1, 5].
- In period $t_1 = 11$, the due dates ranging from period 11 to period 30 are quoted for 92 customer orders collected over the third batching interval [6, 10].

The total of 808 orders are considered over the entire planning horizon [1, 30], each ranging from 5 to 9700 products of a single type. The total demand for all products is 551965. For the input data the necessary condition (5.5) to obtain a feasible schedule with all orders completed during the planning horizon is satisfied, and hence no order needs to be rejected. Furthermore, the input data indicates that stage 2 has significant over capacity and stages 3, 4, and 5 are bottlenecks.

Each run of the models *DDS* and *SCO1* assigns orders to planning periods over a 20 time-period horizon, which corresponds to the assumption that resource availability is fixed for 20 planning periods in advance. These resources can be reassigned in subsequent runs, that is, for a $\sigma = 5$ days long batching interval, the first and the second runs overlap in 15 time periods and some order assignments set in the first run can be changed in the second run (subject to the constraint that committed order due dates remain unchanged.)

In the computational experiments a single solution to *DDS* is sought for the weights $\lambda_1 \geq \lambda_2$, selected as nonnegative integers.

Table 5.2 Computational Results: Model *DDS*

Batching interval/ planning interval	Var.	Bin.	Cons.	Solution values
Objective function: $\min(10\,O_{sum} + Q_{sum})$				
$t < 1/[1, 20]$	7565	7565	1541	$O_{sum} = 8,\ Q_{sum} = 71$
$[1, 5]/[6, 25]$	250	250	109	$O_{sum} = 0,\ Q_{sum} = 0$
$[6, 10]/[11, 30]$	285	285	158	$O_{sum} = 2,\ Q_{sum} = 2$
Objective function: $\min(10\,O_{sum} + Q_{max})$				
$t < 1/[1, 20]$	7566	7565	8480	$O_{sum} = 8,\ Q_{max} = 14$
$[1, 5]/[6, 25]$	251	250	304	$O_{sum} = 0,\ Q_{max} = 0$
$[6, 10]/[11, 30]$	286	285	369	$O_{sum} = 2,\ Q_{max} = 1$
Objective function: $\min(P_{sum} + Q_{sum})$				
$t < 1/[1, 20]$	7565	7565	1541	$P_{sum} = 39795,\ Q_{sum} = 78$
$[1, 5]/[6, 25]$	250	250	112	$P_{sum} = 0,\ Q_{sum} = 0$
$[6, 10]/[11, 30]$	285	285	158	$P_{sum} = 2000,\ Q_{sum} = 4$
Objective function: $\min(P_{sum} + Q_{max})$				
$t < 1/[1, 20]$	7566	7565	8480	$P_{sum} = 39795,\ Q_{max} = 11$
$[1, 5]/[6, 25]$	251	250	301	$P_{sum} = 0,\ Q_{max} = 0$
$[6, 10]/[11, 30]$	286	285	369	$P_{sum} = 2000,\ Q_{max} = 1$
Objective function: $\max R_{sum}$				
$t < 1/[1, 20]$	7565	7565	1541	$R_{sum} = 409426$
$[1, 5]/[6, 25]$	250	250	111	$R_{sum} = 370400$
$[6, 10]/[11, 30]$	285	285	158	$R_{sum} = 315850$

$O_{sum} = \sum_{j\in J}(1 - v_j)$ = total number of delayed orders.
$P_{sum} = \sum_{j\in J} n_j(1 - v_j)$ = total number of delayed products.
$Q_{sum} = \sum_{j\in J, t\in T: t>d_j}(t - d_j)w_{jt}$ = total delay.
$Q_{max} = \max_{j\in J, t\in T: t>d_j}(t - d_j)w_{jt}$ = maximum delay.
$R_{sum} = \sum_{j\in J} n_j r_j v_j + \sum_{j\in J, t\in T: t>d_j} n_j r_{jt} w_{jt}$ = total revenue.
($r_j = 1$ and $r_{jt} = 0.80 - 0.02(t - d_j)$; $t > d_j,\ j \in J$).

The characteristics of MIP model *DDS* for various objective functions and for the subsequent batching and planning intervals are summarized in Table 5.2. The size of the integer program is represented by the total number of variables, Var., number of binary variables, Bin., and number of constraints, Cons. Table 5.2 presents solution values O_{sum} or P_{sum} of the primary objective function f_1, respectively, with $\lambda_1 = 10$ or $\lambda_1 = 1$ in (5.12), and Q_{sum} or Q_{max} of the secondary objective function f_2 with $\lambda_2 = 1$ in (5.12). In addition, the last part of Table 5.2 presents solution results with maximum revenue R_{sum} (5.25).

Table 5.2 indicates that optimal values for the primary objective functions O_{sum} or P_{sum} are identical for different secondary objective functions Q_{sum} and Q_{max} of the corresponding solutions. In order to reach feasibility, the surplus of demand exceeding available capacity in the beginning periods has been reallocated to later periods with excess of capacity in a similar way for both the secondary criteria. Table 5.2 demonstrates that for the customer orders collected in the second batching interval $[1, 5]$ all requested due dates are acceptable, that is, condition (5.3) holds over the planning horizon [6, 25], and hence the execution of model *DDS* was not necessary.

Table 5.3 Computational Results: Model *SCO1*

Planning interval	Var.	Bin.	Cons.	Solution value
Objective function of *DDS*: $\min(10\,O_{sum} + Q_{sum})$				
[1, 20]	5905	5904	6007	$E_{max} = 2$
[6, 25]	3113	3112	3211	$E_{max} = 0$
[11, 30]	3120	3119	3211	$E_{max} = 0$
Objective function of *DDS*: $\min(10\,O_{sum} + Q_{max})$				
[1, 20]	5957	5958	6059	$E_{max} = 2$
[6, 25]	3094	3093	3180	$E_{max} = 0$
[11, 30]	3174	3173	3265	$E_{max} = 0$
Objective function of *DDS*: $\min(P_{sum} + Q_{sum})$				
[1, 20]	5888	5887	5990	$E_{max} = 2$
[6, 25]	2742	2741	2832	$E_{max} = 0$
[11, 30]	3023	3022	3115	$E_{max} = 0$
Objective function of *DDS*: $\min(P_{sum} + Q_{max})$				
[1, 20]	6099	6098	6201	$E_{max} = 2$
[6, 25]	3802	3801	3893	$E_{max} = 0$
[11, 30]	2976	2975	3068	$E_{max} = 0$
Objective function of *DDS*: $\max R_{sum}$				
[1, 20]	6095	6094	6196	$E_{max} = 2$
[6, 25]	3096	3095	3176	$E_{max} = 0$
[11, 30]	2924	2923	3015	$E_{max} = 0$

E_{max} = maximum earliness.

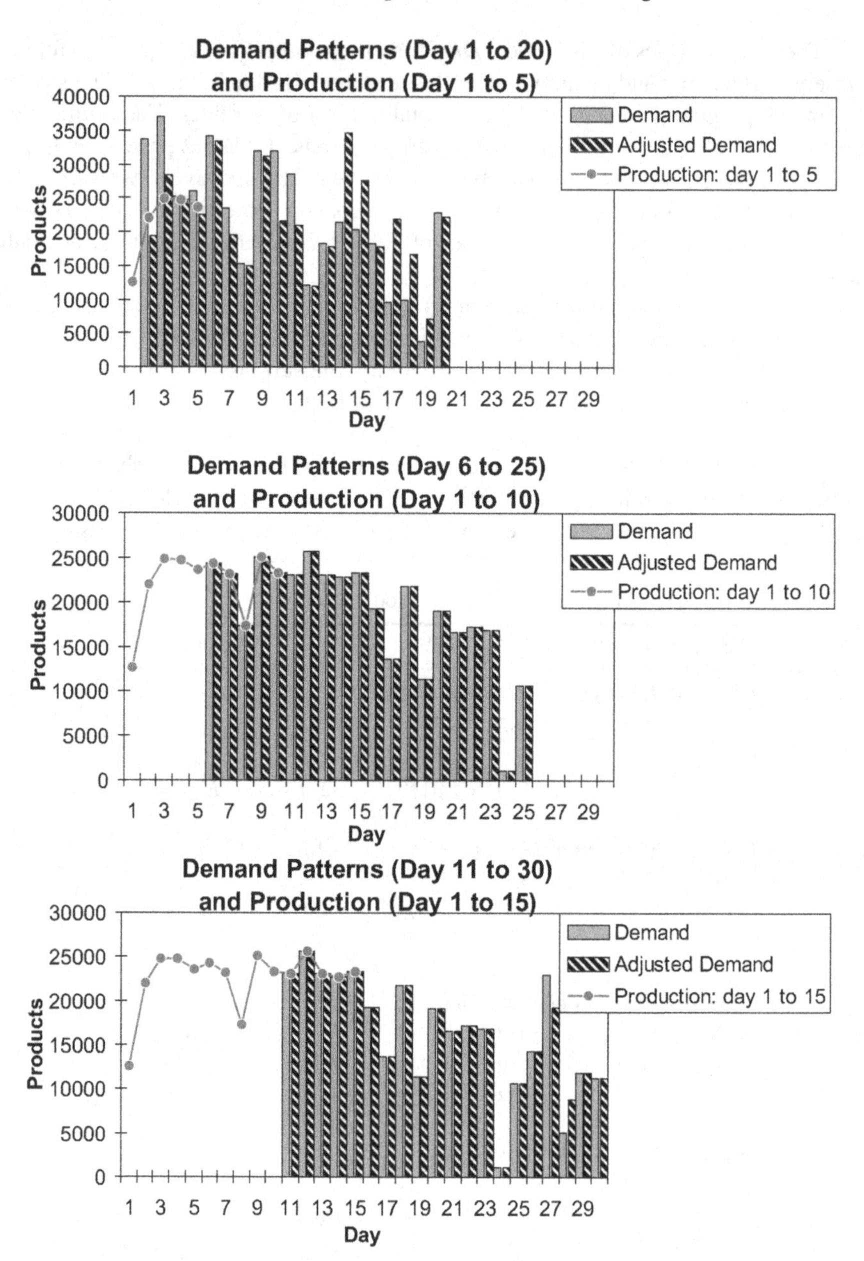

Figure 5.4 Demand patterns and production over a rolling planning horizon for minimum: $10\times$ delayed orders + total delay.

The characteristics of the MIP model *SCO1* for scheduling customer orders and the solution results are summarized in Table 5.3. For all objective functions of *DDS*, model *SCO1* yields identical maximum earliness $E_{\max} = 2$, $E_{\max} = 0$, and $E_{\max} = 0$

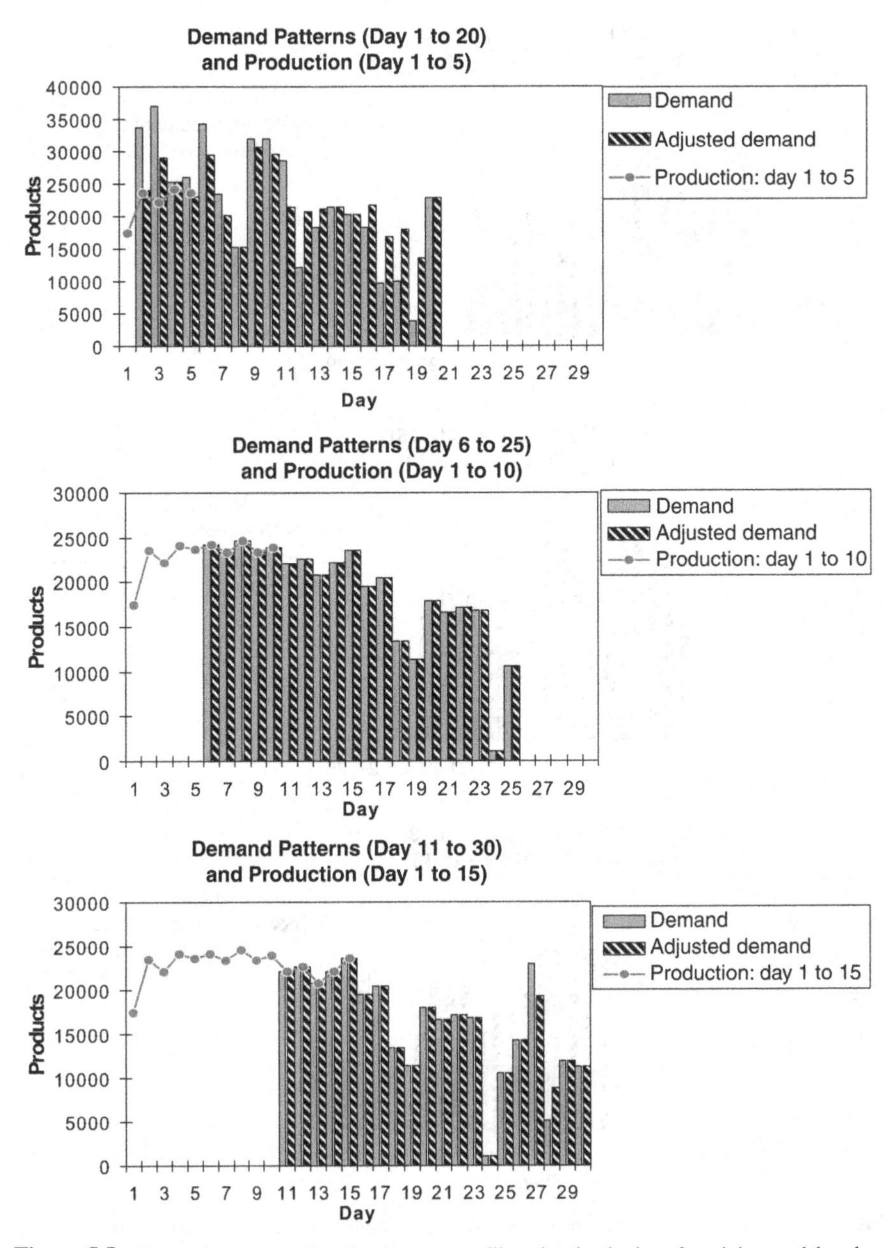

Figure 5.5 Demand patterns and production over a rolling planning horizon for minimum: delayed products + total delay.

for subsequent planning horizons, [1, 20], [6, 25], and [11, 30], respectively. The results indicate that for the example considered, all customer orders in the second and third rolling planning horizons can be completed on due dates (requested or committed).

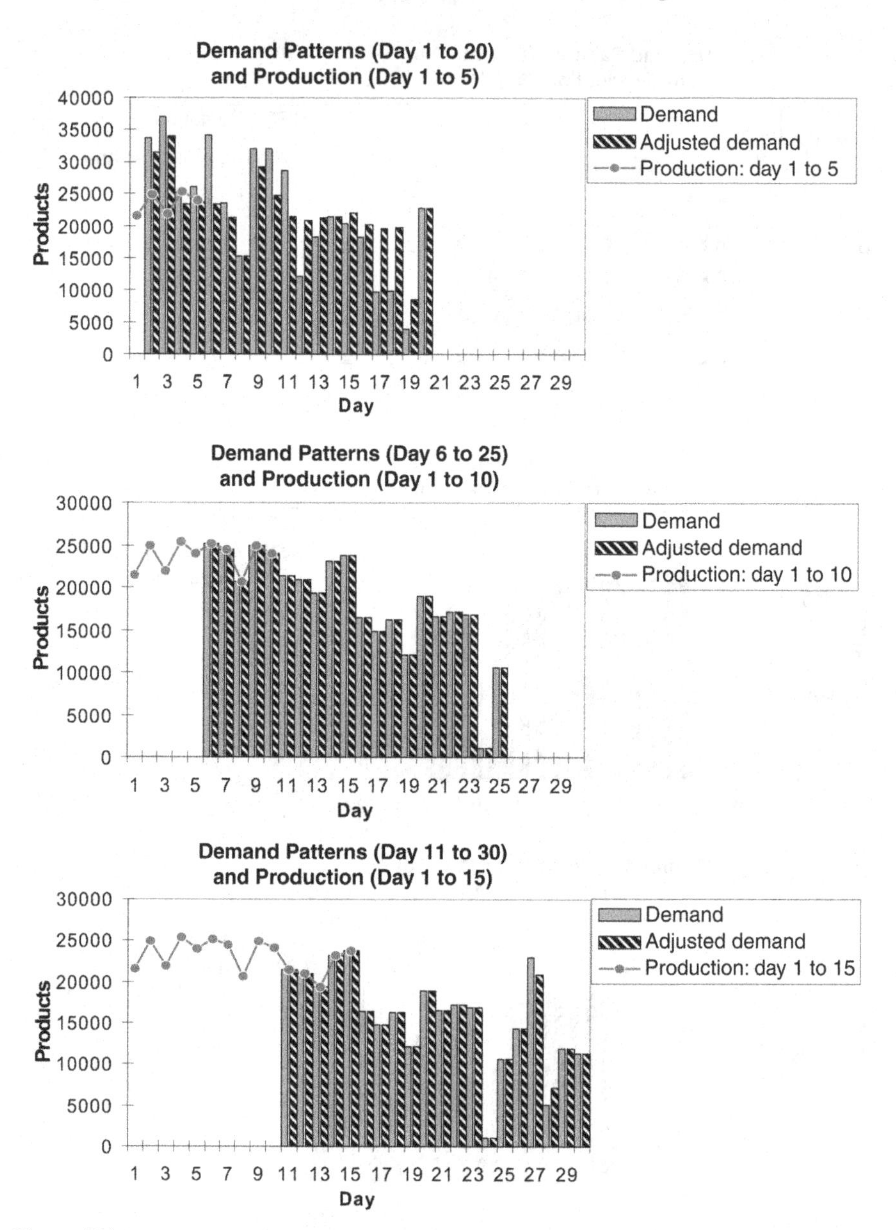

Figure 5.6 Demand patterns and production over a rolling planning horizon for maximum revenue.

Note that, in the computational experiments the number of customer orders collected before period $t = 1$ is much greater than those collected over the subsequent batching intervals. As a result, the MIP models *DDS* and *SCO1* for the the first interval [1, 20] of the rolling planning horizon have the greatest size.

Demand patterns and the aggregate production schedule over a rolling planning horizon for various objective functions are shown in Figures 5.4 to 5.6. Note the similar adjusted demand patterns for $\min(P_{\text{sum}} + Q_{\text{sum}})$ (Fig. 5.5) and for $\max R_{\text{sum}}$ (Fig. 5.6).

For comparison, Table 5.4 presents ex post solution results for various objective functions, obtained when the demand is known ahead of time for the entire monthly horizon. In particular, Table 5.4 presents ex post solutions for the objective of maximizing total revenue (5.25) subject to service level constraints (5.26). The resulting demand patterns are shown in Figure 5.7. Comparison of the ex post solutions (Table 5.4) with the corresponding results on a rolling horizon basis (Table 5.2)

Table 5.4 Computational Results: Ex Post Solutions

Objective function	Var.	Bin.	Cons.	Solution values
Weighted-sum approach: model *DDS*				
$100O_{\text{sum}} + Q_{\text{sum}}$	14,913	14,913	3019	$O_{\text{sum}} = 8$, $Q_{\text{sum}} = 71$
$100O_{\text{sum}} + Q_{\text{max}}$	14,914	14,913	17,141	$O_{\text{sum}} = 8$, $Q_{\text{max}} = 14$
Lexicographic approach: models *AR* and *DD*				
O_{sum}	808	808	48	$O_{\text{sum}} = 8$
Q_{sum}	112	112	294	$Q_{\text{sum}} = 73$
Q_{max}	113	112	406	$Q_{\text{max}} = 15$
Weighted-sum approach: model *DDS*				
$P_{\text{sum}} + Q_{\text{sum}}$	14,913	14,913	3019	$P_{\text{sum}} = 39795$, $Q_{\text{sum}} = 78$
$P_{\text{sum}} + Q_{\text{max}}$	14,914	14,913	17,141	$P_{\text{sum}} = 39795$, $Q_{\text{max}} = 12$
Lexicographic approach: models *AR* and *DD*				
P_{sum}	808	808	85	$P_{\text{sum}} = 39795$
Q_{sum}	965	965	374	$Q_{\text{sum}} = 88$
Q_{max}	966	965	1339	$Q_{\text{max}} = 12$
Maximization of total revenue: model *DDS*				
R_{sum}	14,913	14,913	3019	$R_{\text{sum}} = 537{,}132$
$R_{\text{sum}} \mid \gamma = 90\%$	14,913	14,913	3020	$R_{\text{sum}} = 537{,}108$
$R_{\text{sum}} \mid \gamma = 95\%$	14,913	14,913	3020	$R_{\text{sum}} = 537{,}106$
$R_{\text{sum}} \mid \gamma = 99\%$	14,913	14,913	3020	$R_{\text{sum}} = 533{,}820$

$O_{\text{sum}} = \sum_{j \in J} (1 - v_j) =$ total number of delayed orders.
$P_{\text{sum}} = \sum_{j \in J} n_j(1 - v_j) =$ total number of delayed products.
$Q_{\text{sum}} = \sum_{j \in J, t \in T: t > d_j} (t - d_j)w_{jt} =$ total delay.
$Q_{\text{max}} = \max_{j \in J, t \in T: t > d_j} (t - d_j)w_{jt} =$ maximum delay.
$R_{\text{sum}} = \sum_{j \in J} n_j r_j v_j + \sum_{j \in J, t \in T: t > d_j} n_j r_{jt} w_{jt} =$ total revenue.
($r_j = 1$ and $r_{jt} = 0.80 - 0.02(t - d_j)$; $t > d_j$, $j \in J$).

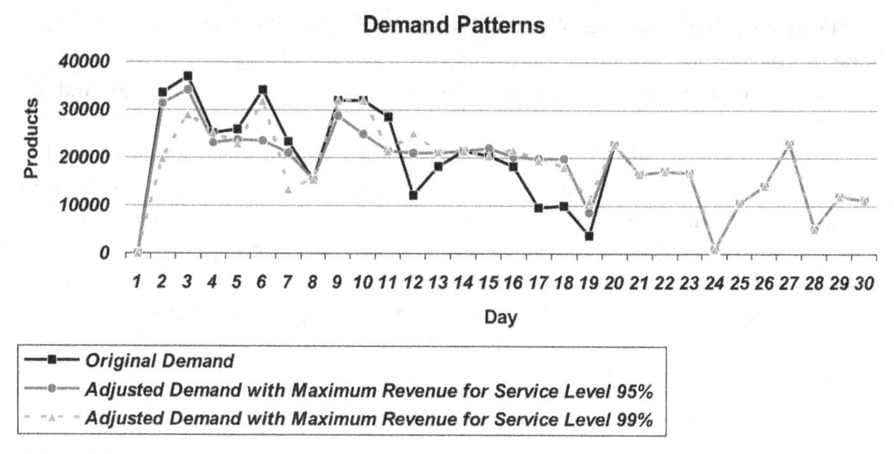

Figure 5.7 Ex post adjusted demand patterns with maximum revenue for different service levels.

demonstrates that both the total number of delayed orders and the total number of delayed products are smaller for the ex post solutions. The more demand-pattern information is available, that is, the longer is the batching interval, the better the solution results obtained.

The original demand pattern and the ex post adjusted demand patterns for various objective functions are compared in Figure 5.8. The corresponding critical load index Ψ (5.4) for the original and the ex post adjusted demands is shown in Figure 5.9, where $\Psi(d) \leq 1, \ \forall d \in T$ for the adjusted demand patterns. Figure 5.9 demonstrates that the primary objective of minimizing the number of delayed products leads to smaller values of Ψ at the beginning of the horizon for the adjusted demand, where $\Psi > 1$

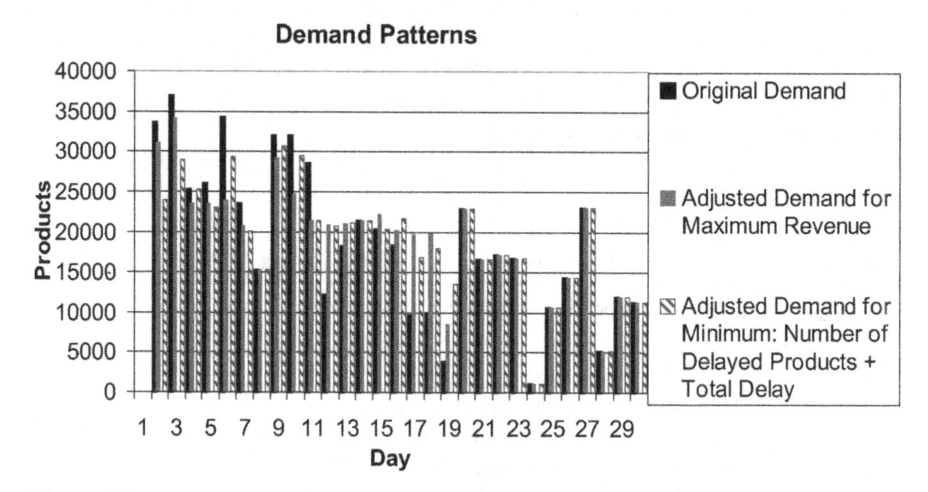

Figure 5.8 Original and ex post adjusted demand patterns over the entire planning horizon.

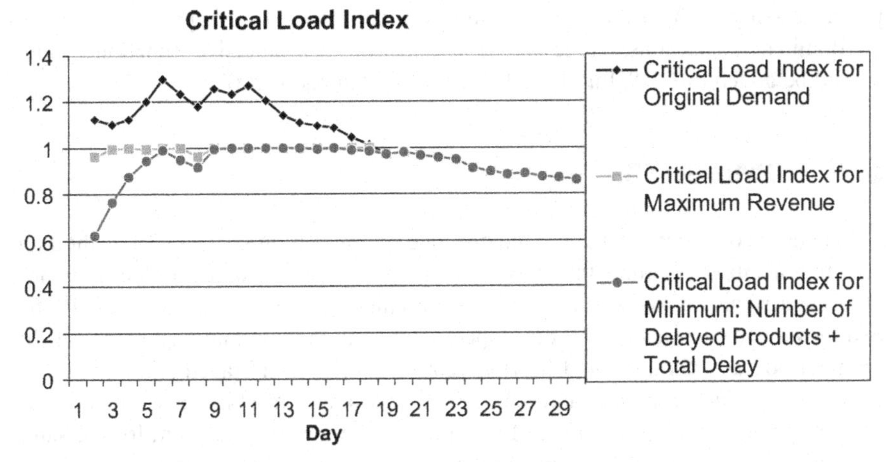

Figure 5.9 Critical load index Ψ (5.4) for original and ex post adjusted demand.

for the original demand. In contrast, the objective of maximizing total revenue leads to a smoother utilization of the capacity over the horizon.

The adjusted demand patterns and the corresponding solution values demonstrate that for a higher service level, more demand is reallocated to later periods; however, the number of delayed orders is reduced, which indicates that mainly large customer orders are selected for reallocation to achieve the required service level. The results indicate that the higher the service level required, the smaller the total number of delayed orders and the greater the total number of delayed products.

Table 5.4 also compares the weighted-sum and the lexicographic approach for the bi-objective problem formulations. In the example presented in Table 5.4, the optimal value $O_{\mathrm{sum}} = 8$ or $P_{\mathrm{sum}} = 39795$ for the primary objective function is identical for the two approaches, whereas the secondary objective functions Q_{sum}, $Q_{\max}$ are slightly greater for the lexicographic approach, since the optimal value of the primary objective can be achieved for alternative subsets of delayed orders.

Finally, the following simple example illustrates an attempt to find a subset of nondominated solutions to the bi-objective order acceptance and due date setting problem for the entire planning horizon. In the example $f_1 = O_{\mathrm{sum}}$, $f_2 \in \{Q_{\mathrm{sum}}, Q_{\max}\}$, and the nondominated solutions are determined by parameterizing the weighted-sum program *DDS* on $\lambda_1 = 0.1, 0.2, 0.3, 0.4, 0.5, 0.6, 0.7, 0.8, 0.9$ with $\lambda_2 = 1 - \lambda_1$.

For the objective function $\lambda_1 O_{\mathrm{sum}} + (1 - \lambda_1) Q_{\mathrm{sum}}$, only two nondominated solutions were found: $O_{\mathrm{sum}} = 11$, $Q_{\mathrm{sum}} = 68$ for $\lambda_1 = 0.1, 0.2, 0.3, 0.4, 0.5$, and $O_{\mathrm{sum}} = 8$, $Q_{\mathrm{sum}} = 71$ for $\lambda_1 = 0.6, 0.7, 0.8, 0.9$, and only one solution was found for the objective function $\lambda_1 O_{\mathrm{sum}} + (1 - \lambda_1) Q_{\max}$: $O_{\mathrm{sum}} = 11$, $Q_{\max} = 15$ for all $\lambda_1 = 0.1, 0.2, 0.3, 0.4, 0.5, 0.6, 0.7, 0.8, 0.9$.

Note, however, that the nondominated solution set of the bi-objective order acceptance and due date setting problem cannot be fully determined even by complete parameterizing on λ of the weighted-sum program *DDS*. To compute unsupported nondominated solutions, some upper bounds on the objective function values should be added to *DDS*, for example, Alves and Climaco (2007).

5.7 COMMENTS

The literature on order acceptance and due date setting is limited. An exact method for selecting a subset of orders that maximizes revenues for the static problem in which all order arrivals are known in advance is presented by Slotnick and Morton (1996), and Lewis and Slotnick (2002) developed a dynamic programming approach for the multiperiod case. Slotnick and Morton (2007) developed a branch-and-bound procedure for order acceptance and scheduling in a single machine environment and more recently, Rom and Slotnick (2009) proposed a genetic algorithm for the same problem. An MIP model for a quantity and due date quoting available to promise is presented by C.-Y. Chen et al. (2001). Hegedus and Hopp (2001) consider order delay costs that measure the positive difference between the quoted due date and the requested due date of an order.

The order acceptance strategies based on scheduling methods are presented by Wester et al. (1992) and Akkan (1997). In Wester et al. (1992) the decision whether or not to accept a new order depends on how much order tardiness it will introduce to the system. Akkan (1997) suggests accepting a new order if it can be included in the schedule such that it is completed by its due date, and without changing the schedule for already accepted orders. Ebben et al. (2005) developed a workload-based acceptance strategy in a job shop environment. Corti et al. (2006) propose a model supporting decision makers who have to verify the feasibility of customer-requested due dates. It adopts a capacity-driven approach to compare the capacity requested by both potential and already confirmed orders with the actual level of available capacity. Zorzini et al. (2008) investigate current practice supporting capacity and delivery lead-time management in the capital goods sector based on a sample of 15 Italian manufacturers and propose a model to formalize the decision process for setting due dates in the selected cases.

Another approach is order acceptance based on revenue management principles, for example, Harris and Pinder (1995), Bertrand and van Ooijen (2000), Geunes et al. (2006). Charnsirisaksul et al. (2004) studied an integrated order acceptance and scheduling problem on a single machine with the objective of maximizing the manufacturer's profit defined as revenue minus manufacturing, holding, and tardiness costs. A time-indexed MIP formulation was used to solve the problem in computational tests. The simultaneous order acceptance and scheduling decisions in a single machine make-to-order environment were also studied by Oğuz et al. (2010). They proposed an MIP model to maximize the total revenue from accepted orders, assuming that the revenue gained from an accepted order decreases linearly with the order's tardiness until its deadline. The model is realistic since it includes a release date, a due date, a deadline, and a sequence-dependent setup time for each order.

The material presented in this chapter is mostly based on the results published by Sawik (2009a). The IP approach proposed is capable of accepting orders and setting due dates in a make-to-order environment, either in a batch mode, where customer orders are collected over a specified time interval, or in a real-time mode, where a commitment due date is determined at the time of the customer order arrival. While the real-time mode is preferable for the customer, the batch mode offers the manufacturer more demand-pattern information, and the longer the batching interval, the larger the set of orders to optimize over. On the other hand, the computational effort required in real-time mode, where only a few newly arrived orders are considered at a time, is much less than for the batch mode, where a set of customer orders should be considered simultaneously.

The proposed monolithic, weighted-sum approach may outperform the hierarchical, lexicographic approach if multiple optima exist with the same value of the primary objective function, that is, if alternative minimal subsets of delayed and rejected orders exist. In this case, a smaller total or maximum delay may sometimes be achieved for the monolithic approach. The hierarchical approach, however, requires a much shorter CPU time to find the optimal solutions and hence seems to be more suitable for setting a due date for each newly arrived order in real-time mode. In particular, when the customer expects an immediate confirmation of the order acceptance (or rejection), where otherwise the potential customer can be lost, for example, in e-business. Furthermore, in make-to-order manufacturing the capacity evaluation at the customer enquiry stage is a critical issue and has a large impact on customer service and reliability of order fulfillment. Then, the introduced critical load index can help to quickly identify the system bottleneck and the overloaded periods.

In the proposed model various simplifying assumptions have been introduced. The model directly links customer orders with available capacity, whereas the other resources are assumed to be nonbinding, with material availability not being considered. However, by the addition of material availability constraints, the integer programming formulations can be easily enhanced to also account for a limited material availability.

In practice, a customer request for quotation may consist of the required quantities of several product types and the requested delivery dates. Then, a typical response to such a customer request should contain the quantity to be fulfilled, the date of delivery, and the price based on revenue management principles which may involve penalties associated with deviations from the customer-requested quantities and dates. The MIP approach can be enhanced to handle multiple product orders, with the pricing decisions based on both tactical factors such as estimated costs, as well as strategic factors, such as the value of a long-term relationship with a customer and the rejection costs. Finally, although the proposed approach is deterministic in nature, its usage on the rolling horizon basis allows for reactive decisions to be made in response to various disruptions in a supply chain.

EXERCISES

5.1 Formulate a weighted-sum mixed integer program for a triple-objective order acceptance and due date setting problem with the primary, secondary, and third objective of minimizing, respectively

(a) The number of rejected, delayed, and early customer orders.

(b) The number of rejected, delayed, and early products.

(c) The number of rejected orders, maximum delay of orders, and maximum earliness of orders.

5.2 Apply lexicographic optimization to formulate a sequence of three mixed integer programs for the triple-objective order acceptance and due date setting problems from Exercise 5.1.

5.3 Assume that the order acceptance and due date setting decisions are based on the triple-objective model of Exercise 5.1c. In model *SCO* for a nondelayed scheduling of customer orders, the objective function of minimizing the maximum earliness replaced is by minimization of maximum finished product inventory level, subject to additional output buffer capacity constraints. Formulate the modified mixed integer program *SCO*.

5.4 Which of the production system parameters in model *SCO* obtained in Exercise 5.3 would you select for changing their values, in order to ensure feasibility, if no feasible solution is found for some problem instance?

5.5 Modify model *SCO* for a nondelayed scheduling of customer orders to minimize the maximum level of finished product inventory, subject to output buffer capacity constraints, (Exercise 5.3).

(a) To minimize capacity of the output buffer required to complete all customer orders by their due dates.

(b) To minimize maximum tardiness of customer orders, subject to output buffer capacity constraints.

Chapter 6

Aggregate Production Scheduling in Make-to-Order Manufacturing

6.1 INTRODUCTION

This chapter deals with aggregate production scheduling in make-to-order manufacturing. One of the basic goals of aggregate production scheduling in make-to-order manufacturing is to maximize customer service level, that is, to maximize the fraction of customer orders filled on or before their due dates. A typical customer due date-related performance measure is minimization of the number of tardy orders. Simultaneously, increasing competition forces manufacturers to achieve low unit costs by utilizing renewable production resources (e.g., machines and people) highly and evenly and by minimizing the inventory. Both the input inventory of purchased materials waiting for processing in the system and the output inventory of finished products waiting for delivery to the customers should be minimized. These can be achieved by maximizing the number of orders assigned on due dates and for the early or tardy orders by minimizing, respectively, their earliness and tardiness with respect to the due dates. Minimization of the earliness of early orders reduces the output inventory of finished products waiting for delivery to the customers, whereas minimization of the tardiness of tardy orders reduces the input inventory of purchased materials waiting for processing.

The purpose of this chapter is to present time-indexed MIP formulations and a lexicographic approach to a bi- or multi-objective aggregate production scheduling in make-to-order manufacturing environment. The primary scheduling objective is to allocate a set of customer orders with various due dates among planning periods to minimize number of tardy orders and the secondary objectives are to level the total input and output inventory over a planning horizon or the aggregate production, the total capacity utilization or the machine assignments, with limited earliness and tardiness for the early and tardy orders, respectively. Two different measures of the

Scheduling in Supply Chains Using Mixed Integer Programming. By Tadeusz Sawik

variation of aggregate production, capacity utilization or machine assignment over the horizon are applied and compared. The basic MIP models are strengthened by the addition of some cutting constraints that are derived by relating the demand on required capacity to available capacity for each processing stage and each subset of orders with the same due date.

A close relation between minimizing the maximal inventory level and the maximum earliness of customer orders with respect to due dates is shown and used to simplify the inventory leveling problem. To reduce the required input inventory of purchased materials, the materials should be delivered as late as possible, that is, the order earliness with respect to due date should be as small as possible, while achieving the minimum number of tardy orders or, if possible, meeting all customer due dates. On the other hand, the smaller the earliness of customer orders, the smaller the output inventory of finished products completed before customer-required shipping dates and waiting for delivery to the customers. However, if for some customer orders the earliness is smaller than the minimal value of the maximum earliness, that is, order ready periods and due dates are closer to each other, then reallocation of orders to the earlier periods with surplus of capacity is restricted due to later material availability. As a result, the number of tardy orders may increase or even some orders may remain unscheduled during the planning horizon.

The following time-indexed IP and MIP models are presented in this chapter:

OA for assignment of customer orders to time periods to minimize number of tardy orders

PL for leveling of aggregate production

PLC for leveling of capacity utilization

IL for leveling of inventory

ILE for leveling of inventory by minimizing the maximum earliness, given the number of tardy orders

OA1 for assignment to time periods of single-period customer orders to minimize number of tardy orders

PL1 for leveling of aggregate production for single-period orders

PLC1 for leveling of capacity utilization for single-period orders

PLM1 for leveling of machine assignments for single-period orders

In addition, MIP models are provided to determine the minimum capacity of input, output, or central buffer required to attain the smallest number of tardy orders. The MIP approach has been applied to optimize aggregate production schedules in a flexible flow shop made up of several processing stages in series, with parallel identical machines, and a finite output buffer for holding completed products before delivery to the customers. Numerical examples modeled after a real-world make-to-order flexible assembly line in the electronics industry are provided and computational results are reported in Sections 6.4.3 and 6.5.2.

6.2 PROBLEM DESCRIPTION

The production system under study is a flexible flow shop (Fig. 6.1) that consists of m processing stages in series and input and output buffers of limited capacity, $B1$ and $B2$, respectively, for holding purchased materials waiting for processing in the system and for holding completed products waiting for delivery to the customers. Each stage $i \in I = \{1, \ldots, m\}$ is made up of $m_i \geq 1$ parallel identical machines. Typically, the capacity of input and output buffers is not large to limit material and finished product inventory and to limit early supplies of purchased materials before their processing dates and early completion of customer orders before the customer-required shipping dates.

In the system various types of products are produced in a make-to-order environment responding directly to customer orders. Let J be the set of customer orders. Each order $j \in J$ is described by a triple (a_j, d_j, n_j), where a_j is the order arrival date (e.g., the earliest period of material availability), d_j is the customer-requested due date (e.g., customer-required shipping date), and n_j is the size of order (quantity of ordered products). Denote by $J(d)$ the subset of orders with the same due date $d \in D$, where $D = \{d_j : j \in J\}$ is the set of distinct due dates of all customer orders.

Each order requires processing in various processing stages; however, some orders may bypass some stages. Let $J_i \subset J$ be the subset of orders that must be processed in stage i, and let $p_{ij} > 0$ be the processing time in stage i of each product in order $j \in J_i$.

The customer orders are processed and transferred among the stages in lots of various size and each lot is to be processed as a separate job. The size of production and transfer lots is a fixed parameter that depends only on product type and the capacity of containers used for product storage and transportation. If b_j is the size of production and transfer lot for order j, then the order quantity n_j is split across multiple lots of size b_j and at most one lot smaller than b_j. For each type of product, the lot size is determined so that processing of each lot as a separate job is fully completed in all stages of the multistage serial system in the same planning period, that is, $\sum_{i \in I} b_j p_{ij} \leq H \; \forall j \in J$, where H is the length of each planning period.

The planning horizon consists of h planning periods (e.g., working days). Let $T = \{1, \ldots, h\}$ be the set of planning periods and c_{it} the processing time available in period t on each machine in stage i. The amount c_{it} of time available on each machine in stage i in period t takes into account the flow shop configuration of the

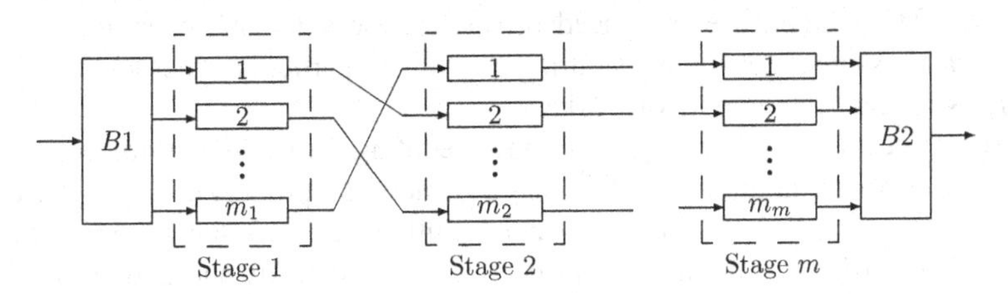

Figure 6.1 A flexible flow shop with finite input and output buffers.

production system and the transfer lot sizes. For each machine in stage i, c_{it} must take into account the time required for processing a single production lot at all upstream and downstream stages during the same planning period (see constraints (5.1) in Section 5.2). A correct estimation of c_{it} used for the medium-term production scheduling is crucial to ensure feasibility of daily machine schedules. A too large value of c_{it} may lead to infeasible daily schedules, that is, some orders assigned to the same planning period will not be completed during a single day horizon. On the other hand a too small value of c_{it} may result in significant underutilization of capacity. In addition the available processing capacity is assumed to be sufficient to schedule all the orders during the planning horizon, if all required materials are available at the beginning of the horizon.

The following two types of customer orders are considered (see Section 5.5):

1. Small (single-period) orders, where each order can be fully processed in a single time period, for example, during one day. The single-period orders are referred to as indivisible orders.
2. Large (multiperiod) orders, where each order cannot be completed in one period and must be split and processed in more than one time period. The multiperiod orders are referred to as divisible orders.

In practice, the two types of customer orders are simultaneously scheduled. Denote by $J1 \subseteq J$ the subset of indivisible orders. It is assumed that each customer order $j \in J1$ must be fully completed in exactly one planning period, and each customer order $j \in J \setminus J1$ must be completed in more than one consecutive planning periods. However, large orders that require more than one planning period for completion can be split into single-period suborders to be allocated among consecutive planning periods. The number of planning periods required to complete a multiperiod order $j \in J \setminus J1$ can be estimated as $\max_{i \in I, t \in T} \left\lceil \frac{n_j p_{ij}}{c_{it}} \right\rceil$, where $\lceil \cdot \rceil$ is the least integer not less than $\cdot$.

Processing of single-period orders does not require large intermediate buffers between the stages for holding semifinished products uncompleted during one planning period. On the other hand completion of each customer order during exactly one period (e.g., during one day) implies that downstream machines are idle early in the day and upstream machines are idle late in the day. In many make-to-order manufacturing environments, however, the production resources in addition to machines, also include the limited number of cross-trained workers required to attend the machines. The workers are even needed to supervise automatic operations and are dynamically assigned to machines during the day, in particular when the number of machines exceeds the number of workers.

The objective of aggregate production scheduling is to assign customer orders to planning periods to minimize number of tardy orders as a primary optimality criterion and to level aggregate production or capacity utilization over the planning horizon with limited earliness and tardiness, respectively, for the early and tardy orders, as a secondary criterion. While an explicit objective of the primary criterion is to maximize the customer service level, the implicit objective of the secondary criterion is to minimize unit production and inventory holding costs.

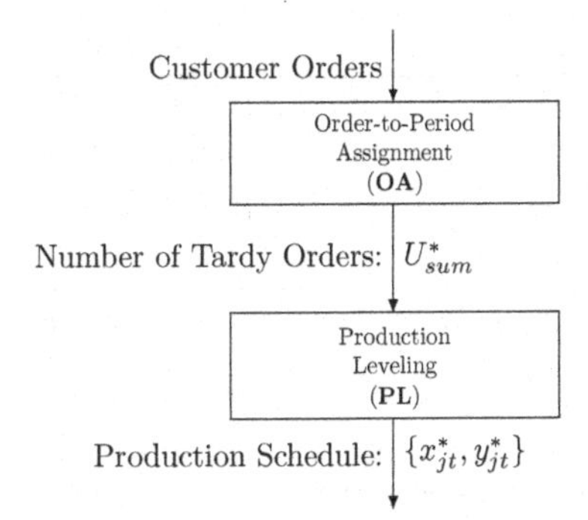

Figure 6.2 A lexicographic approach to bi-objective aggregate production scheduling.

A lexicographic approach will be applied, where first the customer orders are allocated among planning periods to find the minimum number of tardy orders and then the final production schedule is determined to level the aggregate production or the total capacity utilization over the horizon for a minimum number of tardy orders and limited earliness and tardiness of the early and tardy orders, respectively (see Fig. 6.2).

6.3 BI-OBJECTIVE SCHEDULING OF CUSTOMER ORDERS

In this section two time-indexed MIP models *OA* and *PL* (or *PLC*) are presented for a two-level approach to bi-objective aggregate production scheduling (for definition of decision variables, see Table 6.1).

The objective of the top-level problem *OA* is to determine an assignment of customer orders to planning periods over the horizon to minimize the number of tardy orders.

Given the subsets of tardy orders, the objective of the base-level problem *PL* (or *PLC*) is to allocate customer orders among planning periods to level the aggregate production (or the total capacity utilization) over the horizon.

The time-indexed MIP model *OA* for the optimal assignment of customer orders to planning periods over the planning horizon is shown below.

Model OA: *Order to Period Assignment to Minimize Number of Tardy Orders*

Minimize

$$U_{\text{sum}} = \sum_{j \in J} u_j \tag{6.1}$$

subject to

1. *Order-to-Period Assignment Constraints*:
   - each single-period customer order is assigned to exactly one planning period,

$$\sum_{t \in T:\, t \geq a_j} x_{jt} = 1; \; j \in J1 \tag{6.2}$$

Table 6.1 Notation: Aggregate Production Scheduling

Indices	
i	= processing stage, $i \in I = \{1, \ldots, m\}$
j	= customer order, $j \in J = \{1, \ldots, n\}$
t	= planning period, $t \in T = \{1, \ldots, h\}$
Input parameters	
a_j, d_j, n_j	= arrival date, due date, size of order j
b_j	= production and transfer lot size for order j
c_{it}	= processing time available in period t on each machine in stage i
m_i	= number of parallel identical machines in stage i
p_{ij}	= processing time in stage i of each product in order j
$B1, B2$	= input, output buffer capacity, respectively
D	= $\{d_j : j \in J\}$ set of distinct due dates of all customer orders
$J(d)$	= $\{j \in J : d_j = d\}$ subset of customer orders with identical due date d
J_i	= $\{j \in J : p_{ij} > 0\}$ subset of customer orders to be processed in stage i
H	= length of each planning period
Decision variables	
m_{it}	= number of machines selected for assignment in stage i in period t (machine selection variable)
r_{jt}	= 1, if material required for processing order j is available in period t; otherwise $r_{jt} = 0$ (material availability variable)
u_j	= 1, if order j is completed after due date; otherwise $u_j = 0$ (unit penalty for tardy orders)
x_{jt}	= 1, if order j is performed in period t; otherwise $x_{jt} = 0$ (order-to-period assignment variable)
$y_{jt} \in [0, 1]$	= fraction of customer order j to be processed in period t (order allocation variable)
Objective functions	
$E_{\text{sum}}, U_{\text{sum}}$	= number of early or on-time orders, tardy orders, respectively
$I_{\max}$	= maximum level of total (input and output) inventory
$M_{\min}, M_{\max}$	= minimum, maximum machine assignments, respectively
$P_{\min}, P_{\max}$	= minimum, maximum production level, respectively
$\tilde{P}_{\min}, \tilde{P}_{\max}$	= minimum, maximum capacity utilization, respectively

– each multi-period customer order is assigned to two or more consecutive planning periods,

$$x_{j\lfloor(t_1+t_2)/2\rfloor} \geq x_{jt_1} + x_{jt_2} - 1; \; j \in J \setminus J1, \, t_1, t_2 \in T: \quad a_j \leq t_1 < t_2 \leq h \tag{6.3}$$

2. *Order Allocation Constraints*:
   - each order must be completed,

$$\sum_{t \in T: t \geq a_j} y_{jt} = 1; \; j \in J \tag{6.4}$$

   - each indivisible order is completed in a single period,

$$x_{jt} = y_{jt}; \; j \in J1, \; t \in T: t \geq a_j \tag{6.5}$$

   - each divisible order is allocated among all the periods that are selected for its assignment,

$$x_{jt} \geq y_{jt}; \; j \in J \setminus J1, \; t \in T: t \geq a_j \tag{6.6}$$

   - the minimum portion of a divisible order alloted to one period is not less than the batch size,

$$y_{jt} \geq b_j x_{jt}/n_j; \; j \in J \setminus J1, \; t \in T: t \geq a_j \tag{6.7}$$

3. *Capacity Constraints*:
   - in every period the demand on capacity at each processing stage cannot be greater than the maximum available capacity in this period,

$$\sum_{j \in J_i} n_j p_{ij} y_{jt} \leq c_{it} m_i; \; i \in I, t \in T \tag{6.8}$$

4. *Input Buffer Capacity Constraints*:
   - in every period the total inventory of materials available for processing cannot exceed the input buffer capacity $B1$,

$$\sum_{j \in J: t \geq a_j} n_j - \sum_{j \in J, \tau \in T: a_j \leq \tau \leq t} n_j y_{j\tau} \leq B1; \; t \in T \tag{6.9}$$

5. *Output Buffer Capacity Constraints*:
   - in every period the total inventory of finished products completed before their due dates and waiting for shipping to customers cannot exceed the output buffer capacity $B2$,

$$\sum_{j \in J, \tau \in T: a_j \leq \tau \leq t < d_j} n_j y_{j\tau} \leq B2; \; t \in T \tag{6.10}$$

6. *Tardy Orders Constraints*:

– an indivisible tardy order is assigned after its due date,

$$u_j = \sum_{t \in T: t > d_j} x_{jt};\ j \in J1 \tag{6.11}$$

– a divisible tardy order is partly (or fully) assigned after its due date,

$$u_j \geq \sum_{t \in T: t > d_j} y_{jt};\ j \in J \setminus J1 \tag{6.12}$$

$$u_j \leq \sum_{t \in T: t > d_j} x_{jt};\ j \in J \setminus J1 \tag{6.13}$$

7. *Variable Nonnegativity and Integrality Conditions*:

$$u_j \in \{0, 1\};\ j \in J \tag{6.14}$$

$$x_{jt} \in \{0, 1\};\ j \in J, t \in T: t \geq a_j \tag{6.15}$$

$$y_{jt} \in [0, 1];\ j \in J, t \in T: t \geq a_j. \tag{6.16}$$

The objective function (6.1) represents the number of tardy orders to be minimized. The solution to *OA* determines the assignment of indivisible customer orders to single planning periods and the allocation of divisible orders among the consecutive planning periods.

In order to simplify the problem formulation, the input buffer capacity constraints (6.9) account for the inventory of product-specific parts only with no common parts for different product types considered. Furthermore, it is assumed that each product requires one unit of the corresponding product-specific part (e.g., one printed wiring board of a specific design per one electronic device of the corresponding type). As a result, for each customer order j the required quantity of product-specific parts equals the quantity of the ordered products n_j. The above assumptions can be easily relaxed by introducing different types of parts and unit requirements of each product for each part type (e.g., see Chapter 8).

If the set of multiperiod orders $j \in J \setminus J1$ can be partitioned into disjoint subsets $Jq,\ q = 1, \ldots, h$ of q-period customer order, such that $J = J1 \cup J2 \cdots \cup Jh$, then for every q, the order assignment constraints (6.3) can be replaced with the following pairs of constraints that will ensure that each q-period order is assigned to exactly q consecutive planning periods

$$\sum_{t \in T} x_{jt} = q;\ j \in Jq \tag{6.17}$$

$$x_{jt} + x_{jt+q} \leq 1;\ j \in Jq, t \in T: a_j \leq t \leq h - q. \tag{6.18}$$

The solution to *OA* determines the minimum number of tardy orders U^*_{sum}. Given the minimum number of tardy orders, the next optimization step is to find a production schedule such that the aggregate production is leveled over the planning horizon. The time-indexed MIP model *PL* for leveling the aggregate production over the planning horizon is shown below.

Model PL: *Leveling Aggregate Production*

Minimize

$$P_{max} \tag{6.19}$$

or

Maximize

$$P_{min} \tag{6.20}$$

subject to (6.2)–(6.16) and

1. *Order-to-Period Assignment Constraints*:
   – the number of tardy orders is at minimum

$$\sum_{j \in J} u_j = U^*_{sum} \tag{6.21}$$

2. *Production Leveling Constraints*:
   – if the objective (6.19) is selected, in every period the aggregate production cannot exceed the maximum production level to be minimized,

$$\sum_{j \in J} n_j y_{jt} \leq P_{max}; \quad t \in T \tag{6.22}$$

   – or if the objective (6.20) is selected, in every period the aggregate production must not be less than the minimum production level to be maximized,

$$\sum_{j \in J} n_j y_{jt} \geq P_{min}; \quad t \in T \tag{6.23}$$

3. *Variable Nonnegativity Conditions*:
   – if the objective (6.19) is selected

$$P_{max} \geq 0 \tag{6.24}$$

   – or if the objective (6.20) is selected

$$P_{min} \geq 0, \tag{6.25}$$

where P_{max} and P_{min} are the maximum and the minimum aggregate production level in a single planning period, respectively.

The objective function (6.19) tends to level the aggregate production from above, while leaving the lowest production volume close to zero, whereas (6.20) aims at leveling the aggregate production from below while the highest volume is constrained by the maximum available processing and buffer capacity.

The solution to *PL* determines the leveled production schedule, that is, the optimal allocation of customer orders among planning periods, $\{x^*_{jt}, y^*_{jt}\}$ such that the aggregate production is leveled over the horizon and the number of tardy orders is kept at a minimum.

If the objective of the base-level problem is leveling the total capacity utilization rather than the aggregate production, then the following modified MIP model *PLC* can be applied.

Model PLC: *Leveling Capacity Utilization*

Minimize

$$\tilde{P}_{\max} \tag{6.26}$$

or

Maximize

$$\tilde{P}_{\min} \tag{6.27}$$

subject to (6.2)–(6.16), (6.21), and

2′. *Capacity Leveling Constraints*:

- if the objective (6.26) is selected, in every period the total capacity utilization cannot exceed the maximum utilization to be minimized,

$$\sum_{i \in I, j \in J} n_j p_{ij} y_{jt} \Big/ \sum_{i \in I} c_{it} m_i \leq \tilde{P}_{\max}; \; t \in T \tag{6.28}$$

- or if the objective (6.27) is selected, in every period the total capacity utilization must not be less than the minimum utilization to be maximized,

$$\sum_{i \in I, j \in J} n_j p_{ij} y_{jt} \Big/ \sum_{i \in I} c_{it} m_i \geq \tilde{P}_{\min}; \; t \in T \tag{6.29}$$

3′. *Variable Nonnegativity Conditions*:

- if the objective (6.26) is selected

$$0 \leq \tilde{P}_{\max} \leq 1, \tag{6.30}$$

- or if the objective (6.27) is selected

$$0 \leq \tilde{P}_{\min} \leq 1, \tag{6.31}$$

where $\tilde{P}_{\max}$ and $\tilde{P}_{\min}$ are the maximum and minimum utilization of capacity, respectively.

In order to strengthen the MIP formulations, the following cumulative production and demand balancing constraint can be added to models *OA*, *PL* and *PLC*

Cumulative production and demand balancing constraint

- in every period cumulative production is not less than cumulative demand minus tardy demand

$$\sum_{j\in J,\tau\in T:a_j\le\tau\le t} n_j y_{j\tau} \ge \sum_{j\in J:d_j\le t} n_j(1-u_j);\ t\in T. \quad (6.32)$$

6.4 MULTI-OBJECTIVE SCHEDULING OF CUSTOMER ORDERS

In this section a triple-objective aggregate production scheduling is considered to assign customer orders to planning periods so as to minimize the number of tardy orders and the maximum level of the total (input and output) inventory or, equivalently, the maximum earliness of orders, respectively as primary and secondary optimality criteria, and to level aggregate production, capacity utilization or machine assignments over the planning horizon as an auxiliary criterion. An implicit objective is to achieve a high customer service level by meeting customer due dates, and a low unit production cost by leveling production or utilization of production resources and the inventory of purchased materials and finished products.

A lexicographic approach is applied, where the primary objective of maximizing customer service level is attained at the top level. At the top level the customer orders are allocated among planning periods to find the minimum number of tardy orders, then the maximum level of the total inventory or, equivalently, the maximum earliness of orders is minimized at the medium level and finally the aggregate production or the utilization of production resources are leveled over the horizon for the minimum number of tardy orders and the minimum value of the maximum earliness, see Figure 6.3.

In the top-level model *OA*, all customer orders are assumed to be available for processing at the beginning of the planning horizon, i.e., the order ready period (material availability period) is $a_j = 1, \forall j \in J$. As a result the solution to *OA* (Fig. 6.3) determines the smallest possible number of tardy orders U^*_{sum}.

Given the minimum number of tardy orders, the next optimization step is to minimize the maximum level of the total input and output inventory. The time-indexed MIP model *IL* for leveling of total inventory over the planning horizon is shown below.

Model IL: *Inventory Leveling*

Minimize

$$I_{\max} \quad (6.33)$$

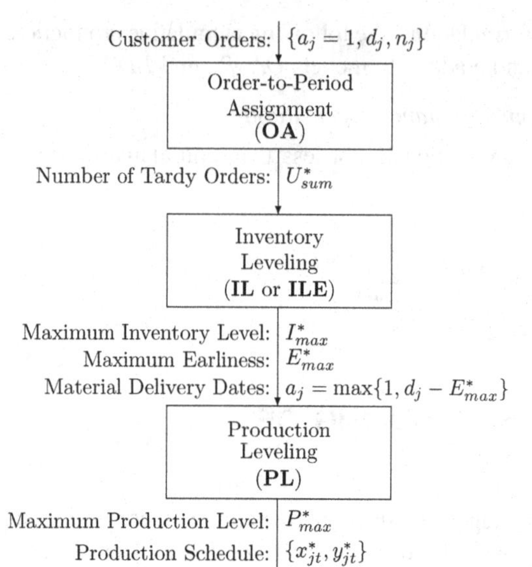

Figure 6.3 A lexicographic approach to multi-objective aggregate production scheduling.

subject to (6.2)–(6.16), (6.21), and

1. *Maximum Earliness Constraints*:

– for each early order j assigned to period $t < d_j$, its earliness $(d_j - t)$ cannot exceed the maximum earliness $E_{\max}$,

$$(d_j - t)x_{jt} \leq E_{\max};\ j \in J, t \in T: t \geq a_j \tag{6.34}$$

2. *Material Availability Constraints*:

– for each order j the required raw materials are available for processing $E_{\max}$ periods ahead of the order due date d_j (at the latest in period $d_j - 1$ if $E_{\max} = 1$),

$$r_{jt} \leq 1 + (t - d_j + E_{\max})/h;\ j \in J, t \in T: a_j \leq t \leq d_j - 1 \tag{6.35}$$

$$r_{jt} \geq (1 + t - d_j + E_{\max})/h;\ j \in J, t \in T: a_j \leq t \leq d_j - 1 \tag{6.36}$$

3. *Inventory Constraints*:

– in every period the total input inventory of raw materials and output inventory of finished products cannot exceed its maximum level $I_{\max}$ to be minimized,

$$\sum_{j \in J: a_j \leq t \leq d_j - 1} n_j r_{jt} + \sum_{j \in J: t \geq d_j} n_j - \sum_{j \in J, \tau \in T: a_j \leq \tau \leq t, t \geq d_j} n_j y_{j\tau} \leq I_{\max};\ t \in T \tag{6.37}$$

4. *Variable Nonnegativity and Integrality Conditions*:

$$r_{jt} \in \{0, 1\};\ j \in J, t \in T: a_j \leq t \leq d_j - 1 \tag{6.38}$$

$$E_{\max} \geq 1,\ integer \tag{6.39}$$

$$I_{\max} \geq 0 \tag{6.40}$$

The objective function (6.33) represents the maximum level of the total input and output inventory to be minimized, defined in the left-hand side of (6.37). Implicitly, (6.33) tends to level the total inventory over the planning horizon. The first two summation terms $\sum_{j\in J:a_j\leq t\leq d_j-1} n_j r_{jt} + \sum_{j\in J:t\geq d_j} n_j$ in the left-hand side of (6.37) represent the amount of product-specific materials supplied by period t. The third summation term $\sum_{j\in J,\tau\in T:a_j\leq\tau\leq t,\ t\geq d_j} n_j y_{j\tau}$ represents the amount of the finished products that had already been shipped to customers by period t.

Material availability constraints (6.35), (6.36) are formulated such that the material availability binary variable $r_{jt} = 1$ (material required for order j is available in period t), if $t \geq d_j - E_{\max}$, and $r_{jt} = 0$ (material required for order j is not available in period t), if $t \leq d_j - E_{\max} - 1$.

It should be noted that the actual input inventory of product-specific materials depends on the material supply schedule and may differ from the amount calculated in (6.37), where the materials are assumed to be supplied exactly $E_{\max}$ periods before each customer order due date. In contrast, (6.37) accounts for the actual inventory level of finished products.

The solution to the MIP model *IL* determines, for each order, its assignment period and from this the latest period of availability of the required materials. The corresponding earliness of each early order with respect to its due date is found such that the number of tardy orders remains at a minimum. As a result the maximum earliness $E_{\max}$ (the maximum length of the interval between order due date d_j and its ready period a_j, that is, the material availability period) is determined for all orders such that the maximum level $I_{\max}$ of the total input and output inventory is minimized and the number of tardy orders is at a minimum, U^*_{sum}.

Given the minimum number of tardy orders, and the minimum value of the maximum earliness of orders, the next optimization step is to solve the base-level problem *PL* to determine a leveled aggregate production over the planning horizon for a minimum number of tardy orders and the minimum total inventory level. In the base-level problem *PL*, each customer order is assumed to be available for processing $E_{\max}$ periods before its due date, that is, for each order j ready period (material delivery period) is $a_j = \max\{1, d_j - E_{\max}\}$.

The solution to *PL* determines the leveled production schedule, that is, the optimal allocation of customer orders among planning periods, $\{x^*_{jt}, y^*_{jt}\}$ such that the number of tardy orders and the maximum inventory level are kept at a minimum, and the aggregate production is leveled over the planning horizon.

6.4.1 Maximum Level of Total Inventory vs. Maximum Earliness of Customer Orders

This subsection shows that, for a given set of tardy orders, minimizing the maximal inventory of product-specific materials and finished products can be approximately achieved by minimizing the maximum earliness of customer orders.

The following formulae derived for $a_j = \max\{1, d_j - E_{\max}\}$ indicate that the smaller the maximum earliness $E_{\max}$ for customer orders, the later the required materials can be delivered by suppliers and the lower the output inventory of the finished products completed before due dates and waiting for delivery to customers.

- The cumulative supplies of product-specific materials by period t:

$$\sum_{j \in J: d_j - E_{\max} \leq t} n_j,$$

- The cumulative production by period t:

$$\sum_{j \in J, \tau \in T: d_j - E_{\max} \leq \tau \leq t} n_j y_{j\tau},$$

- The cumulative deliveries to customers by period t:

$$\sum_{j \in J, \tau \in T: d_j \leq t,\, d_j - E_{\max} \leq \tau \leq t} n_j y_{j\tau},$$

- The input inventory in period t equals the cumulative supplies by period t minus the cumulative production by period t:

$$\sum_{j \in J: d_j - E_{\max} \leq t} n_j \left(1 - \sum_{d_j - E_{\max} \leq \tau \leq t} y_{j\tau} \right),$$

- The output inventory in period t equals the cumulative production by period t minus the cumulative deliveries by period t:

$$\sum_{j \in J, \tau \in T: d_j - E_{\max} \leq \tau \leq t < d_j} n_j y_{j\tau},$$

- The total (input and output) inventory in period t equals the cumulative supplies by period t minus the cumulative deliveries by period t:

$$\sum_{j \in J: d_j \leq t + E_{\max}} n_j - \sum_{j \in J, \tau \in T: d_j \leq t,\, d_j - E_{\max} \leq \tau \leq t} n_j y_{j\tau}.$$

The last formula can be rewritten as

$$\sum_{j \in J: d_j \le t} n_j \left(1 - \sum_{d_j - E_{\max} \le \tau \le t} y_{j\tau} \right) + \sum_{j \in J: t+1 \le d_j \le t+E_{\max}} n_j \tag{6.41}$$

The first summation term in (6.41) is the inventory of product-specific materials for customer orders due by period t, and the second term is the inventory of product-specific materials and finished products of customer orders due after period t. The first term represents the input inventory in period t of product-specific materials for tardy orders and is greater than zero only if some customer orders are tardy, otherwise this term is equal to zero. The second term increases with the maximum earliness $E_{\max}$. Given the tardy orders, the total inventory increases with $E_{\max}$, that is, both the input inventory of product-specific materials and the output inventory of finished products can be reduced when ready periods and due dates of customer orders are closer.

Therefore, the maximum level $I_{\max}$ of the total input and output inventory can be implicitly minimized by minimizing the maximum earliness $E_{\max}$ of early orders, given the minimum number U^*_{sum} of tardy orders. As a consequence, a complex problem of minimizing the maximum inventory level $I_{\max}$ can be replaced by the much simpler problem of minimizing the maximum earliness $E_{\max}$ such that the minimum number of tardy orders U^*_{sum} is achieved. Accordingly, a complex MIP model *IL* can be replaced by the much simpler MIP model *ILE*, presented below.

In model *ILE* for each early order j assigned to period $t < d_j$, its earliness $(d_j - t)$ is determined, and the resulting maximum earliness over all early orders, $E_{\max} = \max_{j \in J, t \in T} (d_j - t) x_{jt}$, is directly minimized.

Model ILE: *Inventory Leveling by Minimizing the Maximum Earliness, Given the Number of Tardy Orders*

Minimize

$$E_{\max} \tag{6.42}$$

subject to (6.2)–(6.16), (6.21), (6.34), (6.39).

The objective (6.42) represents the maximum earliness of customer orders to be minimized or, equivalently, the maximum difference between the order due date and its ready period, that is, the latest period of material availability.

6.4.2 Finite Capacity of Input, Output, and Central Buffers

The minimum input and output buffer capacity, $B1_{\min}(E_{\max})$ and $B2_{\min}(E_{\max})$, respectively, required to begin processing of each order $j \in J$ at its ready period

$a_j = \max\{1, d_j - E_{\max}\}$ such that the smallest number of tardy orders is achieved can be determined as the optimal solutions to the following mixed integer programs:

$$B1_{\min}(E_{\max}) = \min\{B1 \geq 0: (6.2) - (6.9), (6.11) - (6.16), (6.21)\}, \quad (6.43)$$
$$B2_{\min}(E_{\max}) = \min\{B2 \geq 0: (6.2) - (6.8), (6.10) - (6.16), (6.21)\}. \quad (6.44)$$

Notice that the minimum capacity $B1_{\min}(E_{\max})$ (or $B2_{\min}(E_{\max})$) is determined assuming unlimited capacity of $B2$ (or $B1$) in (6.43) (or (6.44)), respectively.

If both the raw materials and the finished products are stored in a common central buffer, then the following central buffer capacity constraints should replace (6.9) and (6.10)

Central buffer capacity constraints:

- in every period the total inventory of raw materials and finished products stored in the central buffer cannot exceed its finite capacity BC

$$\sum_{j \in J: t \geq a_j} n_j - \sum_{j \in J, \tau \in T: a_j \leq \tau \leq t, t \geq d_j} n_j y_{j\tau} \leq BC; \; t \in T \quad (6.45)$$

The minimum capacity $BC_{\min}(E_{\max})$ of the central buffer required to begin processing of each order $j \in J$ at its ready period $a_j = \max\{1, d_j - E_{\max}\}$ such that the smallest number of tardy orders is achieved can be determined as the optimal solution to the following MIP model:

$$BC_{\min}(E_{\max}) = \min\{BC \geq 0: (6.2) - (6.8), (6.11) - (6.16), (6.21), (6.45)\}. \quad (6.46)$$

Note that the optimal capacity $BC^*_{\min}(E_{\max})$ of the central buffer is identical with the optimal value of the maximum inventory level $I^*_{\max}$ achieved for the same value of maximum earliness $E_{\max}$, that is, for ready periods $a_j = \max\{1, d_j - E_{\max}\}, \; j \in J$.

6.4.3 Computational examples

In this subsection numerical examples and some computational results are presented to illustrate possible applications of the proposed lexicographic approach with a triple of MIP models *OA*, *IL* or *ILE*, and *PL*. The examples are modeled after a real-world distribution center for high-tech products, where finished products are assembled for shipping to customers. The distribution center can be modeled as a flexible flow shop made up of six processing stages in series and parallel, with parallel machines (Fig. 6.4). In the distribution center 10 product types of three product groups are assembled. The processing stages are the following (see Fig. 5.3): material preparation stage, where all materials required for assembly of each product are prepared, postponement stage, where products for some orders are customized, three flashing/flexing stages in parallel, one for each group of products, where required software is

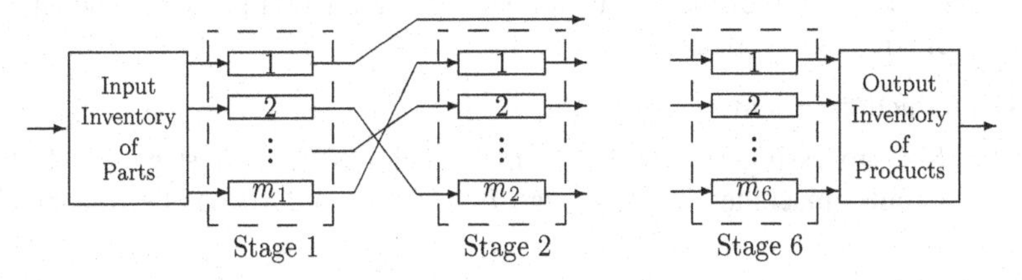

Figure 6.4 Distribution center: a flexible flow shop.

downloaded, and a packing stage, where products and required accessories are packed for shipping.

Customer orders require processing in at most four stages: material preparation stage, postponement stage, one flashing/flexing stage, and packing stage. However, some orders do not need postponement.

Customer orders are split into production lots of fixed sizes, each to be processed as a separate job. Each large (multiperiod) customer order must be completed in at most two planning periods (two days).

In the computational experiments four types of test problems are constructed with the following four regular patterns of demand:

1. Increasing, with demand skewed towards the end of the planning horizon.
2. Decreasing, with demand skewed towards the beginning of the planning horizon.
3. Unimodal, where demand peaks in the middle of the planning horizon and falls under available capacity in the first and last days of the horizon.
4. Bimodal, where demand peaks at the beginning and at the end of the planning horizon and slumps in mid-horizon.

Pattern 1 requires some orders to be completed earlier, for pattern 2 a majority of orders must be moved later in time, whereas patterns 3 and 4 require that orders are moved both early and late to reach feasibility.

For each demand pattern, the following two scenarios will be considered with different tightness measure *tcr*, (6.47) of the capacity constraint (6.8):

- Scenario I with medium tightness of capacity constraints: $tcr = 0.762$
- Scenario II with high tightness of capacity constraints: $tcr = 0.955$

where *tcr* (total capacity ratio) is defined below as the maximum over all processing stages of the total demand on capacity to total available capacity

$$tcr = \max_{i \in I} \left(\frac{\sum_{j \in J} n_j p_{ij}}{m_i \sum_{t \in T} c_{it}} \right). \tag{6.47}$$

A brief description of the production system, production process, products, and customer orders is given below.

1. **Production system**
   - Six processing stages: 10 parallel machines in each stage $i = 1, 2$; 20 parallel machines in each stage $i = 3, 4, 5$; and 10 parallel machines in stage $i = 6$.
2. **Products**
   - 10 product types of three product groups, each to be processed on a separate group of flashing/flexing machines
   - 100 customer orders, each consisting of several suborders (customer-required shipping volumes), known ahead of a monthly planning horizon. Every suborder has a different volume ranging from five to 6345 products (scenario I) or from five to 7930 products (scenario II). The total number of suborders is ranging from 669 to 816 depending on demand pattern and the capacity scenario. The total demand for all products is 429,685 and 537,995, respectively, for scenario I and II.
   - production (and transfer) lot sizes: 200,200,300,100,100,100,200,200,300, 100, respectively for product type 1,2,3,4,5,6,7,8,9,10.
3. **Processing times (in seconds) for product types:**

Product type/stage	1	2	3	4	5	6
1	20	0	120	0	0	15
2	20	0	140	0	0	15
3	10	0	160	0	0	10
4	15	5	0	120	0	15
5	15	10	0	140	0	15
6	10	5	0	160	0	10
7	15	10	0	180	0	15
8	20	5	0	0	120	15
9	15	0	0	0	140	10
10	15	0	0	0	160	10

4. **Planning horizon:** $h = 30$ days, each of length $H = 2 \times 9$ hours.

The number of two-period customer orders is not greater than 10 orders for each capacity scenario and demand pattern. Note that the suborders in the computational examples play the role of orders in the mathematical formulation. The primary objective is to determine an assignment of customer suborders over the planning horizon to minimize number of tardy suborders.

It is assumed that the processing time c_{it} available on each machine in stage i is the same in all periods, that is, $c_{it} = c_i, \ t \in T$. In the computational experiments the amount c_i of time available on each machine in stage i has been calculated as:

$$c_i = \underline{c}_i + \alpha(\overline{c}_i - \underline{c}_i),$$

where $\underline{c}_i$ and $\overline{c}_i$ are defined in (5.1) in Section 5.2, and $0 \leq \alpha \leq 1$.

The parameter α reflects the idle time of the machine waiting for the first production lot from upstream stages and the machine idle time during processing of the last production lot at downstream stages. The greater α and the greater the difference between the total processing times required for completing the maximum and the minimum production lot at all upstream and downstream stages, the greater can be the available capacity c_i of stage i. The value of parameter α has been determined experimentally to obtain a feasible machine schedule (for the machine scheduling approach, see Chapter 11) for the most highly loaded planning period, that is, the period with the highest demand on capacity required for the assigned customer orders. In the computational experiments parameter α ranged between 0.4 and 0.6.

The characteristics of MIP models *OA*, *IL*, *ILE*, and *PL* for the two capacity scenarios and various demand patterns and the solution results are summarized in Tables 6.2 to 6.9. The size of each mixed integer program is represented by the total number of variables, Var., number of binary variables, Bin., number of constraints, Cons., and number of nonzero elements in the constraint matrix, Nonz. The counts presented in the tables are taken from the models after presolving. The last column of each table presents the solution values U_{sum} for *OA*, I_{max} for *IL*, E_{max} for *ILE* and P_{max} for *PL*. The solution values I_{max} for *IL* and E_{max} for *ILE* are presented along with the corresponding associate value of E_{max} and I_{max} (in parentheses), respectively.

Table 6.2 Computational Results for Scenario I: Model *OA*

Demand pattern	Var.	Bin.	Cons.	Nonz.	U_{sum}
Increasing	49,759	24,880	28,696	217,772	0
Decreasing	49,813	24,907	28,750	223,519	2
Unimodal	43,253	21,627	25,052	191,630	0
Bimodal	41,089	20,545	23,833	182,233	0

Table 6.3 Computational Results for Scenario I: Model *IL*

Demand pattern	Var.	Bin.	Cons.	Nonz.	I_{max}, (E_{max})
Increasing	45,464	29,752	60,915	360,899	63,380, (2)
Decreasing	58,233	33,327	53,670	511,969	31,970, (1)
Unimodal	32,146	20,944	42,908	299,110	95,250, (3)
Bimodal	29,786	19,394	39,394	274,899	102,260, (3)

Table 6.4 Computational Results for Scenario I: Model *ILE*

Demand pattern	Var.	Bin.	Cons.	Nonz.	E_{max}, (I_{max})
Increasing	31,361	15,680	33,515	163,339	2, (63,380)
Decreasing	49,747	24,872	36,700	240,013	1, (33,280)
Unimodal	22,341	11,170	23,998	115,402	3, (95,250)
Bimodal	20,721	10,360	22,264	106,883	3, (102,260)

Table 6.5 Computational Results for Scenario I: Model *PL*

Demand pattern	Var.	Bin.	Cons.	Nonz.	P_{max}
Increasing	4791	2395	5127	59,932	18,495
Decreasing	32,845	16,421	20,000	429,639	20,685
Unimodal	5503	2751	5185	80,637	18,565
Bimodal	5231	2615	4941	77,577	17,855

Table 6.6 Computational Results for Scenario II: Model *OA*

Demand pattern	Var.	Bin.	Cons.	Nonz.	U_{sum}
Increasing	50,004	25,003	29,298	219,734	0
Decreasing	50,060	25,031	29,354	225,520	7
Unimodal	43,437	21,720	26,085	194,241	2
Bimodal	41,334	20,669	25,365	185,985	1

Table 6.7 Computational Results for Scenario II: Model *IL*

Demand pattern	Var.	Bin.	Cons.	Nonz.	I_{max}, (E_{max})
Increasing	45,674	29,890	61,389	363,060	175,355, (6)
Decreasing	58,592	33,531	54,949	942,010	95,635, (2)
Unimodal	53,282	31,533	55,586	489,773	146,785, (4)
Bimodal	30,105	19,583	41,856	28,1577	173,775, (5)

If cutting constraint (6.32) is applied, the CPU time can be reduced by up to 15%. The greater the number of tardy orders in the optimal solution, the more efficient the constraint. Tables 6.2 to 6.9 present computational results without application of cut (6.32), except for model *ILE* and scenario II with decreasing demand pattern in Table 6.8.

Table 6.8 Computational Results for Scenario II: Model *ILE*

Demand pattern	Var.	Bin.	Cons.	Nonz.	E_{max}, (I_{max})
Increasing	31,507	15,753	33,863	164,653	6, (175,355)
Decreasing	50,059	25,029	38,676	613,618	2, (97,395)[a]
Unimodal	43,438	21,720	36,630	462,309	4, (146,785)
Bimodal	20,985	10,494	24,403	150,177	5, (173,775)

[a]The optimality proven with cut (6.32)

Table 6.9 Computational Results for Scenario II: Model *PL*

Demand pattern	Var.	Bin.	Cons.	Nonz.	P_{max}
Increasing	10,813	5406	8248	146,203	17,935
Decreasing	34,581	17,290	21,822	459,781	22,655
Unimodal	27,848	13,925	17,834	349,848	20,245
Bimodal	7915	3959	7455	123,848	20,394

For the optimal values of maximum earliness E^*_{max}, various demand patterns and capacity scenario II, Figure 6.5 shows the aggregate production schedules, and Figure 6.6 shows the required input inventory of purchased materials and the output inventory of finished products. (For the total inventory, the sum of input and output inventories, see the corresponding charts for E^*_{max} in Fig. 6.8.)

Figure 6.5 shows that the aggregate production is best leveled over time for the increasing demand pattern, and Figure 6.6 indicates that the required material inventory and the finished product inventory are varying over time similarly, following or anticipating the demand pattern.

A comparison of the solution values I_{max} achieved for *IL* and *ILE* indicates that *ILE* generates the same optimal values for all demand patterns except for a slight difference for decreasing demand: $I^*_{max} = 31{,}970$ for *IL* versus $I_{max} = 33{,}280$ for *ILE*, for scenario I, and $I^*_{max} = 95{,}635$ for *IL* versus $I_{max} = 97{,}395$ for *ILE*, for scenario II. The difference is due to a different allocation of some orders among planning periods, in particular, a different tardiness of tardy orders.

For comparison, Table 6.10 presents the minimum capacity of the common buffer storage $BC_{min}(E_{max})$ for the input and output inventory as well as the minimum capacity of separate input buffer $B1_{min}(E_{max})$ and output buffer $B2_{min}(E_{max})$. The minimum capacities are obtained as solutions to the MIP models (6.43), (6.44), and (6.46), for optimal values of the maximum earliness E^*_{max} for each demand pattern. The solution results in Table 6.10 indicate that the minimum capacity of the central buffer $BC_{min}(E^*_{max})$ and the corresponding optimal value I^*_{max} of maximum inventory are identical for all demand patterns.

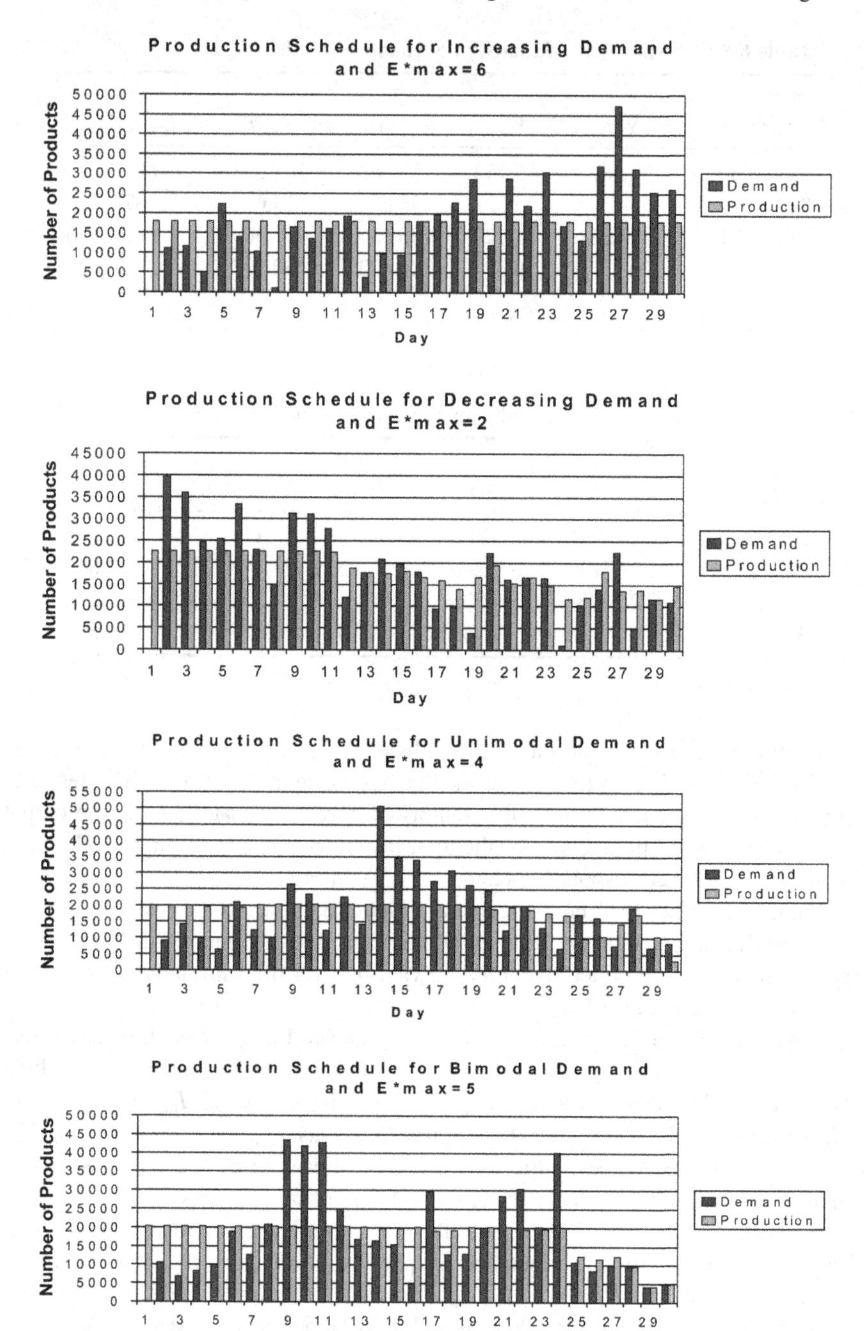

Figure 6.5 Leveled production schedules for scenario II and $E^*_{\max}$.

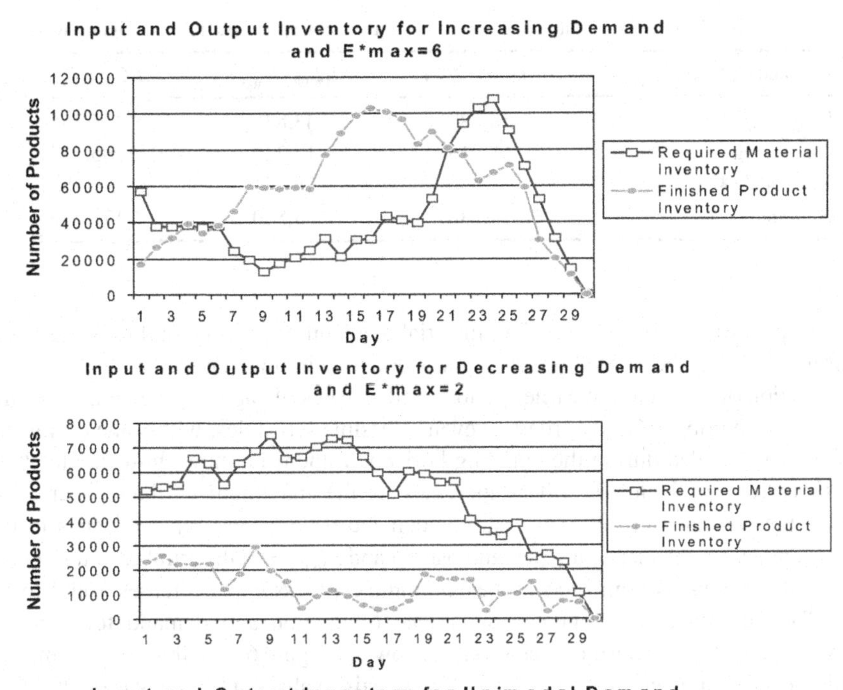

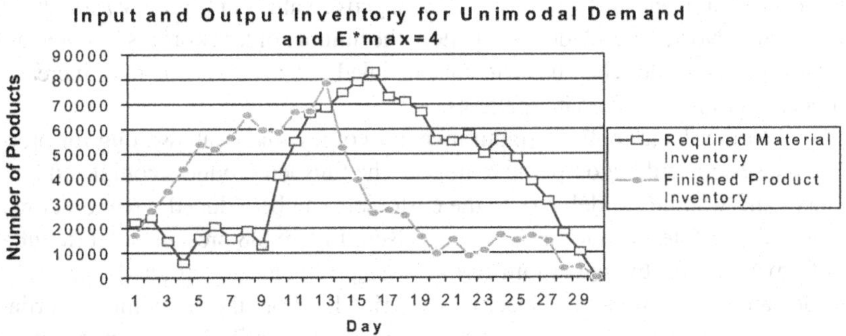

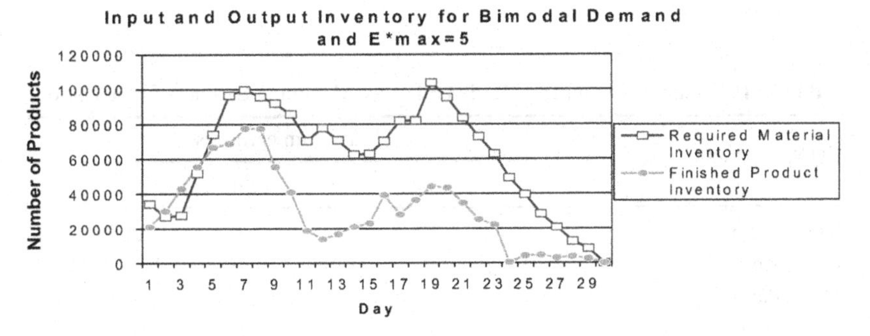

Figure 6.6 Input and output inventory for scenario II and $E^*_{\max}$.

Table 6.10 Minimum Capacity of Input, Output, and Central Buffers for Scenario II

Demand pattern/E^*_{max}	$B1_{min}(E^*_{max})$	$B2_{min}(E^*_{max})$	$BC_{min}(E^*_{max})$
Increasing/6	54,780	43,830	175,355
Decreasing/2	68,715	17,220	95,635
Unimodal/4	58,090	58,351	146,785
Bimodal/5	71,580	55,500	173,775

When order ready periods (i.e., material availability periods) and due dates are closer than E^*_{max}, the limited order earliness due to later material availability restricts reallocation of orders to the earlier periods with surplus of capacity, which may result in a greater number of tardy orders or even infeasible schedules, with some customer orders unscheduled during the planning horizon. Table 6.11 shows how the number of tardy (or unscheduled) orders increases as the maximum earliness decreases below the optimal value E^*_{max} for various demand patterns and capacity scenario II. For example, for the increasing demand pattern and $E^*_{max} = 6$, the number of unscheduled orders increases from 1 to 9 as the maximum earliness E_{max} decreases from 5 to 1.

The difference of cumulative aggregate production and demand for various demand patterns and capacity scenario II is shown in Figure 6.7 to illustrates examples with the maximum earliness $E_{max} = 1$. The negative values in Figure 6.7 indicate the tardy demand. Now, for each demand pattern the number of tardy orders has increased, in particular, infeasible schedules with unscheduled orders are obtained for increasing, unimodal, and bimodal demand patterns.

On the other hand, both the input inventory of raw materials waiting for processing in the system and the output inventory of the finished products completed before due dates and waiting for delivery to the customers can be reduced when order ready periods and due dates are closer. For comparison, Figure 6.8 presents total (input and output) inventory for the maximum earliness $E_{max} = 1$, $E_{max} = 10$, and E^*_{max} for various demand patterns and capacity scenario II. For the maximum earliness $E_{max} \geq E^*_{max}$, the total inventory is varying over time similarly to demand pattern,

Table 6.11 Minimum Number of Tardy Orders vs. Maximum Earliness for Scenario II

Demand pattern/E^*_{max}	Maximum earliness					
	1	2	3	4	5	6
Increasing/6	9[a]	4[a]	3[a]	1[a]	1[a]	0[b]
Decreasing/2	9	7[b]	–	–	–	–
Unimodal/4	11[a]	7	3	2[b]	–	–
Bimodal/5	12[a]	8	3	2	1[b]	–

[a]Number of unscheduled orders (no feasible solution).

[b]Minimum number of tardy orders for E^*_{max}.

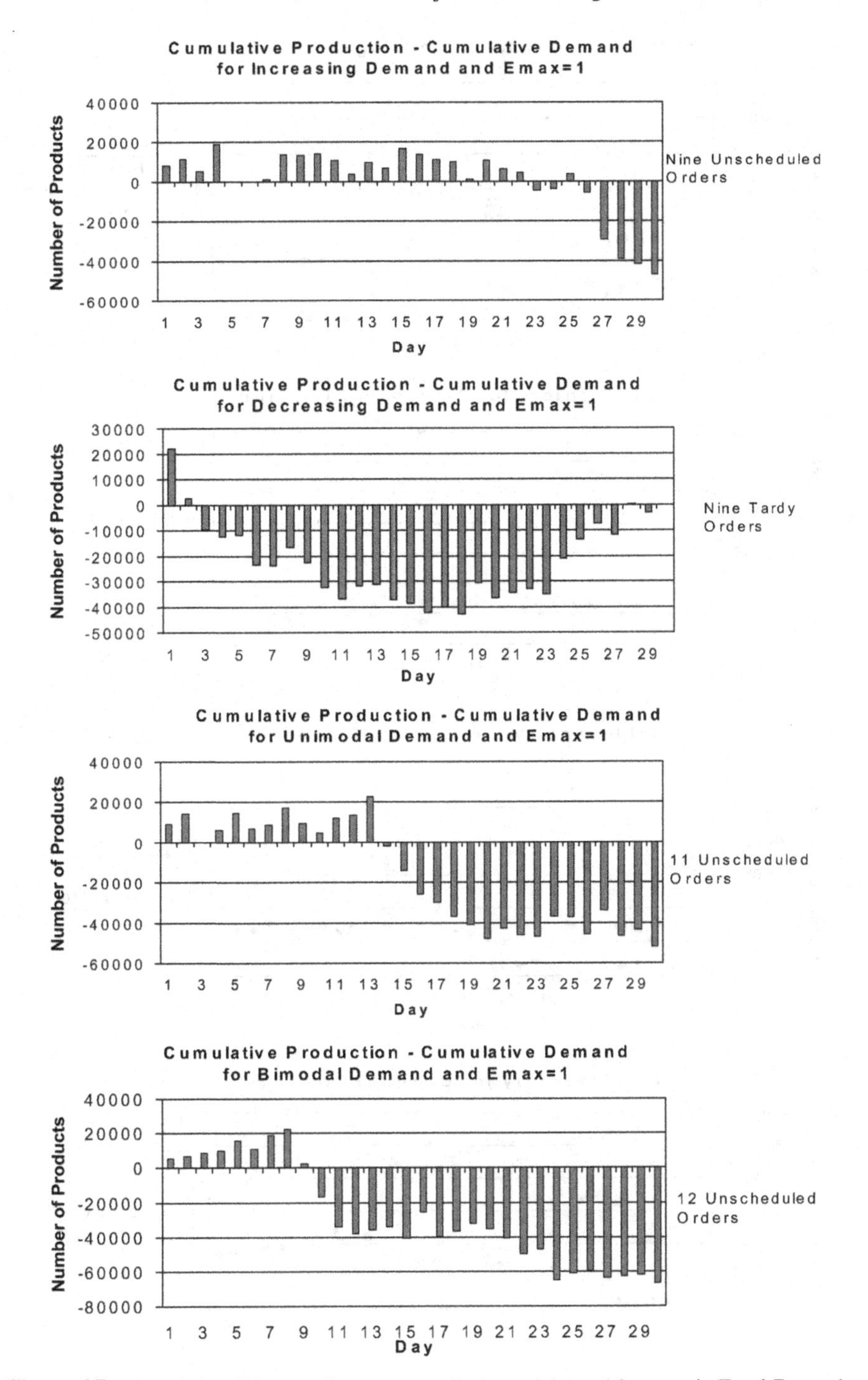

Figure 6.7 Cumulative difference of aggregate production and demand for scenario II and $E_{\max} = 1$.

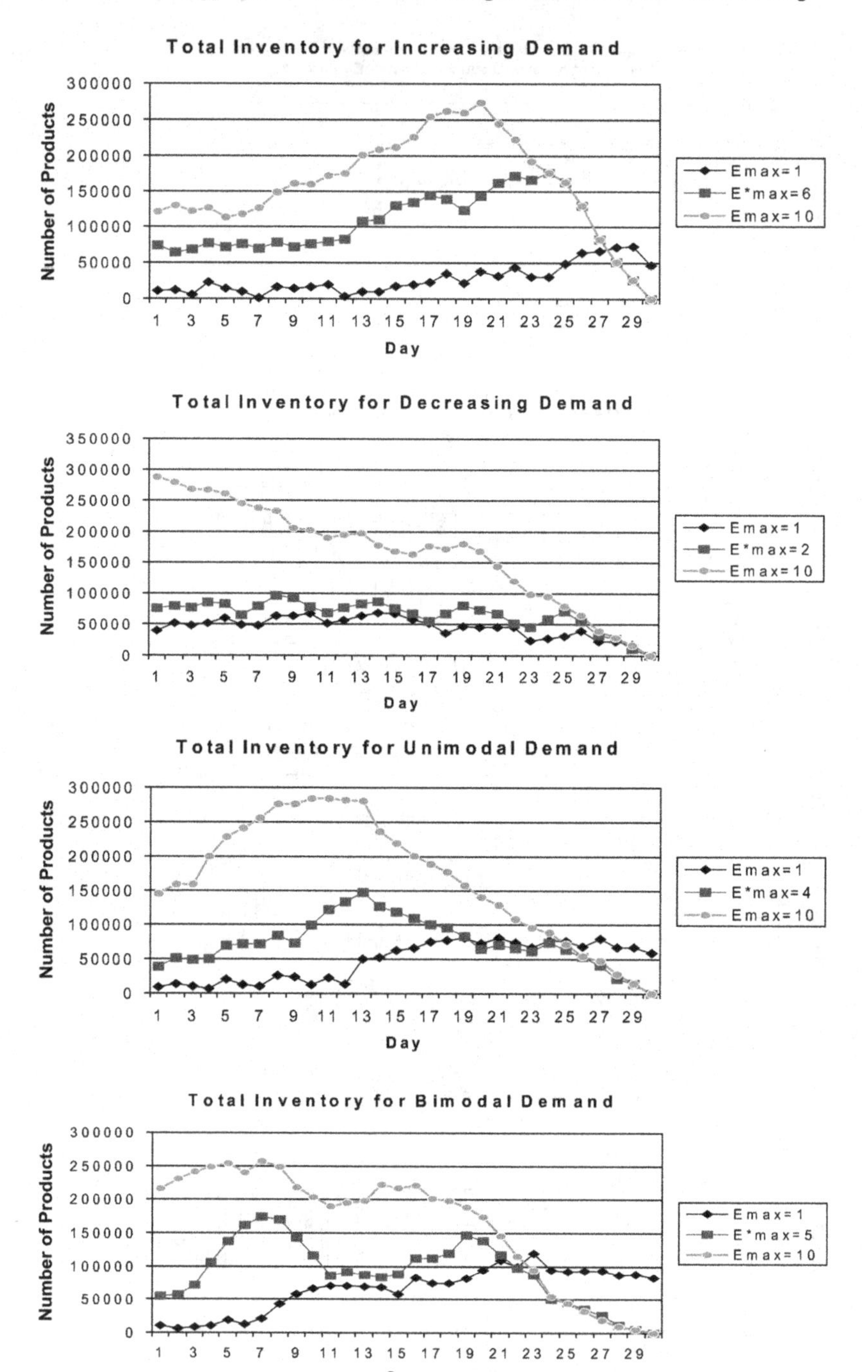

Figure 6.8 Total inventory for scenario II and different maximum earliness E_{max}.

and is best leveled over the planning horizon for E^*_{max}. For $E_{max} = 1$, the ending inventory level is greater than zero due to remaining materials for the unscheduled customer orders, except for the decreasing demand pattern.

The computational experiments were performed using the AMPL programming language and the CPLEX v.9.1 solver on a laptop with Pentium IV at 1.8 GHz and 1 GB RAM. The results indicate that the lexicographic approach is capable of finding proven optimal solutions in a reasonable CPU time (ranging from a few seconds to several hundred seconds) for large problems with typical patterns of demand that are encountered in industrial practice.

6.5 SCHEDULING OF SINGLE-PERIOD CUSTOMER ORDERS

This section presents IP and MIP models for scheduling single-period customer orders. If each customer order $j \in J$ can be fully completed in exactly one planning period, that is, $J = J1$, then the order allocation variables y_{it} and the unit penalties u_j are no longer required and model *OA* can be replaced with the following time-indexed binary program *OA1*.

Model OA1: *Assignment of Single-Period Orders to Time Periods to Minimize Number of Tardy Orders*

Minimize

$$U_{sum} = \sum_{j \in J, t \in T: t > d_j} x_{jt} \tag{6.48}$$

or equivalently

Maximize

$$E_{sum} = \sum_{j \in J, t \in T: t \leq d_j} x_{jt} \tag{6.49}$$

subject to (6.15) and

1. *Order-to-Period Assignment Constraints*:
   - each customer order is assigned to exactly one planning period,

$$\sum_{t \in T} x_{jt} = 1; \; j \in J \tag{6.50}$$

2. *Capacity Constraints*:
   - in every period the demand on capacity at each processing stage cannot be greater than the maximum available capacity in this period,

$$\sum_{j \in J_i} n_j p_{ij} x_{jt} \leq c_{it} m_i; \; i \in I, t \in T \tag{6.51}$$

3. *Input Buffer Capacity Constraints*:

– in every period the total inventory of materials available for processing cannot exceed the input buffer capacity $B1$,

$$\sum_{j \in J: t \geq a_j} n_j - \sum_{j \in J, \tau \in T: a_j \leq \tau \leq t} n_j x_{j\tau} \leq B1; \; t \in T \tag{6.52}$$

4. *Output Buffer Capacity Constraints*:

– in every period the total number of products completed before their due dates and waiting for shipping to customers cannot exceed the output buffer capacity $B2$,

$$\sum_{j \in J, \tau \in T: a_j \leq \tau \leq t < d_j} n_j x_{j\tau} \leq B2; \; t \in T. \tag{6.53}$$

The solution to *OA1* determines the minimum number of tardy orders U^*_{sum} as well as the minimum subset of tardy orders $JT \subset J$ and their assignment to planning periods $x_{j,d_j+\tau_j} = 1, j \in JT$, where $\tau_j \geq 1$ is the tardiness of order $j \in JT$. Similarly, the subset of early orders $JE \subset J$ is determined as well as their assignment to planning periods $x_{j,d_j-\tau_j} = 1, j \in JE$, where $\tau_j \geq 1$ is the earliness of order $j \in JE$. Implicitly, the subset $JD = J \setminus JE \cup JT$ of orders assigned on due dates is determined, i.e., such that $x_{j,d_j} = 1, j \in JD$.

The smaller the earliness for the early orders, the lower the inventory of the finished products waiting for delivery to customers. Similarly, the smaller the tardiness of the tardy orders, the lower the input inventory of purchased materials waiting for processing. Therefore, at the base-level problem the assignment of orders $j \in JD$ on due dates remains unchanged, and the earliness τ_j of the early orders $j \in JE$, and the tardiness τ_j of the tardy orders $j \in JT$ should not be increased.

Given the minimum number of tardy orders U^*_{sum}, the subsets of on due date orders, early orders, and tardy orders, the maximum earliness for early orders, and the maximum tardiness for tardy orders, the next optimization step is to find a production schedule such that the aggregate production is leveled over the planning horizon.

The MIP model *PL1* for the base-level problem can be formulated alternatively, either for minimum numbers of early and tardy orders (order assignment constraints 1a) or for limited order earliness and tardiness (order assignment constraints 1b).

Model PL1: *Leveling Aggregate Production for Single-Period Orders*

Minimize (6.19)

or

Maximize (6.20)

subject to (6.15), (6.24), (6.25), (6.51)–(6.53), and

1. *a. Order-to-Period Assignment Constraints*: (6.50) and
   – the number of tardy orders is at minimum

$$\sum_{j \in J: t > d_j} x_{jt} = U^*_{\text{sum}} \tag{6.54}$$

or

1. *b. Order-to-Period Assignment Constraints*:
   – on due date orders assignment constraints,

$$x_{jd_j} = 1; \; j \in JD \tag{6.55}$$

   – early orders assignment constraints (the earliness cannot be greater than τ_j),

$$\sum_{t \in T: d_j - \tau_j \le t \le d_j} x_{jt} = 1; \; j \in JE \tag{6.56}$$

   – tardy orders assignment constraints, (the tardiness cannot be greater than τ_j),

$$\sum_{t \in T: d_j < t \le d_j + \tau_j} x_{jt} = 1; \; j \in JT \tag{6.57}$$

2. *Production Leveling Constraints*:
   if the objective (6.19) is selected
   – in every period the aggregate production cannot exceed the maximum production level to be minimized,

$$\sum_{j \in J} n_j x_{jt} \le P_{\max}; \; t \in T \tag{6.58}$$

   or if the objective (6.20) is selected
   – in every period the aggregate production must not be less than the minimum production level to be maximized,

$$\sum_{j \in J} n_j x_{jt} \ge P_{\min}; \; t \in T. \tag{6.59}$$

The solution to *PL1* determines the leveled production schedule $\{x^*_{jt}\}$, that is, the optimal allocation of customer orders among planning periods with a leveled aggregate production over the horizon, such that number of tardy orders is at a minimum and their tardiness as well the earliness of the early orders are limited.

The corresponding machine assignments can be determined as below.

$$m_{it} = \left\lceil \frac{\sum_{j\in J} n_j p_{ij} x^*_{jt}}{c_{it}} \right\rceil; \; i \in I, t \in T \tag{6.60}$$

where m_{it} is the number of machines selected for assignment in stage i in period t.

If the objective of the base-level problem is leveling the total capacity utilization rather than the aggregate production, then the following modified MIP model *PLC1* can be applied.

Model PLC1: *Leveling Capacity Utilization for Single-Period Orders*

Minimize (6.26)

or

Maximize (6.27)

subject to (6.15), (6.30), (6.31), (6.51)–(6.57), and

2′. *Capacity Leveling Constraints*:

if the objective (6.26) is selected

– in every period the total capacity utilization cannot exceed the maximum utilization to be minimized,

$$\sum_{i\in I, j\in J} n_j p_{ij} x_{jt} \Big/ \sum_{i\in I} c_{it} m_i \le \tilde{P}_{\max}; \; t \in T \tag{6.61}$$

or if the objective (6.27) is selected

– in every period the total capacity utilization must not be less than the minimum utilization to be maximized,

$$\sum_{i\in I, j\in J} n_j p_{ij} x_{jt} \Big/ \sum_{i\in I} c_{it} m_i \ge \tilde{P}_{\min}; \; t \in T. \tag{6.62}$$

Alternatively, if the objective of the base-level problem is leveling the machine assignments rather than the aggregate production or capacity utilization, then the following IP model *PLM1* can be applied.

Model PLM1: *Leveling Machine Assignment for Single-Period Orders*

Minimize

$$M_{\max} \tag{6.63}$$

or

Maximize

$$M_{\min} \tag{6.64}$$

subject to (6.15), (6.51)–(6.57), and

1. *Machine Assignment Constraints*:
   - in every period the number of machines selected for assignment at each stage is not greater than the maximum number of available machines,
   - in every period the number of machines selected for assignment at each stage is not greater than the total number of assigned production lots,
   - in every period the demand on capacity at each processing stage cannot be greater than the total capacity of machines selected for assignment in this period,

$$m_{it} \leq m_i; \ \ i \in I, t \in T \tag{6.65}$$

$$m_{it} \leq \sum_{j \in J_i} \lceil n_j/b_j \rceil x_{jt}; \ \ i \in I, t \in T \tag{6.66}$$

$$\sum_{j \in J_i} n_j p_{ij} x_{jt} \leq c_{it} m_{it}; \ i \in I, t \in T \tag{6.67}$$

2. *Machine Assignment Leveling Constraints*:
   if the objective (6.63) is selected
   - in every period the total number of machines selected for assignment cannot exceed the maximum number of machine assignments to be minimized,

$$\sum_{i \in I} m_{it} \leq M_{\max}; \ \ t \in T \tag{6.68}$$

   - or if the objective (6.64) is selected, in every period the total number of machines selected for assignment cannot be less than the minimum number of machine assignments to be maximized,

$$\sum_{i \in I} m_{it} \geq M_{\min}; \ \ t \in T \tag{6.69}$$

3. *Variable Integrality Conditions*:

$$m_{it} \geq 0, \ integer; \ \ i \in I, t \in T \tag{6.70}$$

   if the objective (6.63) is selected

$$M_{\max} \geq 0, \ integer \tag{6.71}$$

or if the objective (6.64) is selected

$$M_{\min} \geq 0, \ integer, \tag{6.72}$$

where $M_{\max}$ and $M_{\min}$ are the maximum and minimum number of machine assignments in a single planning period, respectively.

The solution to *PLM1* determines the optimal production schedule $\{x^*_{jt}\}$, that is, the optimal allocation of customer orders among planning periods with a leveled machine assignment over the horizon, $\{m^*_{it}\}$, such that the number of tardy orders is at a minimum or alternatively their tardiness and earliness of the early orders are limited.

6.5.1 Cutting Constraints

In this subsection some cutting constraints on decision variables are derived by relating for each due date the local demand on required capacity to available capacity and the cumulative demand on required capacity to available cumulative capacity.

A necessary condition for problem *OA1* to have a feasible solution with all customer orders completed during the planning horizon is that for each processing stage $i \in I$ the total demand on capacity does not exceed total available capacity, that is, the total capacity ratio *tcr* (6.47) is not greater than one (see also (5.5) in Section 5.2.1)

$$tcr = \max_{i \in I} \left(\frac{\sum_{j \in J} n_j p_{ij}}{m_i \sum_{t \in T} c_{it}} \right) \leq 1. \tag{6.73}$$

Note that the available capacity c_{it} of each stage i takes into account processing time required for completing upstream and downstream operations, and hence is smaller than simply the available machine hours in period t (cf. (5.1)). Due to the discrete nature of customer orders, however, it is possible that some planning periods will not be filled exactly to their capacities. As a result the necessary condition (6.73) is not sufficient for all orders to be scheduled during the planning horizon.

In addition, the output buffer capacity should be sufficient for holding all early demand completed before due dates to reach feasibility of the production schedule. The minimum output buffer capacity $B2_{\min}$ required to complete all orders during the planning horizon (i.e., with no backlogging) can be determined as the optimal solution to the following mixed integer program:

$$B2_{\min} = \min\{B2 \geq 0\text{: } (6.15), (6.50), (6.51), (6.53)\}. \tag{6.74}$$

Furthermore, if no tardiness is allowed (i.e., $U_{\text{sum}} = 0$), the minimum output buffer capacity $B2^U_{\min} \geq B2_{\min}$ required to complete all orders on or before their corresponding due dates can be found by solving the following mixed integer program:

$$B2^U_{\min} = \min\left\{B2 \geq 0{:}(6.15), (6.50), (6.51), (6.53), \sum_{j\in J, t\in T: t>d_j} x_{jt} = 0\right\}. \quad (6.75)$$

For each due date $d \in D$ denote by Φ_d the local capacity ratio defined below.

$$\Phi_d = \max_{i\in I}(\varphi_{id});\ d \in D \quad (6.76)$$

where φ_{id} is the local capacity ratio for due date d with respect to processing stage i, that is, the ratio of local demand on capacity in stage i (i.e., the total processing time required to complete in stage i all the orders with due date d) to the total processing time available in stage i in planning period d

$$\varphi_{id} = \frac{\sum_{j\in J(d)} n_j p_{ij}}{c_{id} m_i};\ i \in I,\ d \in D. \quad (6.77)$$

Denote by D_0, D_1, the subset of locally satisfiable, locally unsatisfied due dates, respectively:

$$D_0 = \{d \in D{:}\ \Phi_d \leq 1\} \quad (6.78)$$

$$D_1 = \{d \in D{:}\ \Phi_d > 1\}. \quad (6.79)$$

For each locally satisfiable due date the entire demand due that planning period can be completed in that period. On the other hand, for each locally unsatisfied due date at least one order should be moved to another period to reduce local demand on capacity with respect to the overloaded processing stage, at least to the level of available capacity of that stage.

Cutting constraints on assignment of orders with locally unsatisfied due dates

$$\sum_{j\in J(d)\cap J_i, t\in T: t\neq d} x_{jt} \geq 1;\ i \in I,\ d \in D_1{:}\varphi_{id} > 1 \quad (6.80)$$

$$\sum_{j\in J(d)\cap J_i, t\in T: t\neq d} n_j p_{ij} x_{jt} \geq c_{id} m_i(\varphi_{id} - 1);\ i \in I,\ d \in D_1{:}\varphi_{id} > 1. \quad (6.81)$$

Constraint (6.80) ensures that, for each due date $d \in D_1$ and each overloaded processing stage i, at least one order locally unsatisfied with respect to that stage will not be completed on the due date. In addition, (6.81) guarantees that for each due date $d \in D_1$ and each overloaded processing stage i, the total work of orders locally unsatisfied with respect to that stage and moved to other periods is not less than the surplus of demand on capacity with respect to available capacity.

For each processing stage and each locally satisfied due date the work inflow of orders with other due dates less the work outflow of orders with this due date cannot exceed the slack capacity. Constraint (6.82) presented below ensures such a flow balance.

Flow balance cutting constraint for locally satisfied due dates

$$\sum_{d' \in D, j \in J(d'): d' \neq d} n_j p_{ij} x_{jd} - \sum_{j \in J(d), t \in T: t \neq d} n_j p_{ij} x_{jt} \leq c_{id} m_i (1 - \varphi_{id}); \; i \in I, \, d \in D_0. \tag{6.82}$$

For each due date $d \in D$ denote by Ψ_d the cumulative capacity ratio defined below (see also, critical load index (5.4) in Section 5.2.1).

$$\Psi_d = \max_{i \in I} (\psi_{id}); \; d \in D, \tag{6.83}$$

where ψ_{id} is the cumulative capacity ratio for due date d with respect to processing stage i

$$\psi_{id} = \max_{t \in T: t \leq d} \left(\frac{\sum_{d' \in D, j \in J(d'): t \leq a_j \leq d' \leq d} n_j p_{ij}}{m_i \sum_{\tau \in T: t \leq \tau \leq d} c_{i\tau}} \right); \; i \in I, \, d \in D. \tag{6.84}$$

If $\psi_{id} \leq 1$, then a due date $d \in D$ is globally satisfiable with respect to processing stage i, that is, for any period $t \leq d$, the cumulative demand on capacity in stage i of all the orders with due dates not greater than d and arrival dates not less than t (numerator in (6.84)) does not exceed the cumulative capacity available in this stage in periods t through d (denominator in (6.84)).

Denote by D^0, D^1, the subset of globally satisfiable, globally unsatisfied due dates, respectively:

$$D^0 = \{d \in D: \Psi_d \leq 1\} \tag{6.85}$$

$$D^1 = \{d \in D: \Psi_d > 1\}. \tag{6.86}$$

If all due dates are globally satisfiable (i.e., $D^1 = \emptyset$), and $B2 \geq B2^U_{\min}$, then the following constraints (6.87) and (6.88) guarantee nondelayed completion of all orders.

Order-to-period nondelayed assignment constraints

$$x_{jt} = 0; \; j \in J; \; t \in T: t > d_j, \; D^1 = \emptyset, \; B2 \geq B2^U_{\min} \tag{6.87}$$

$$\sum_{t=a_j}^{d_j} x_{jt} = 1; \; j \in J: D^1 = \emptyset, \; B2 \geq B2^U_{\min}. \tag{6.88}$$

The due dates both locally and globally unsatisfied are referred to as potentially unsatisfied due dates. In order to meet a potentially unsatisfied due date $d \in D_1 \cap D^1$, some orders with the due dates not greater than the earliest potentially unsatisfied due date $\underline{d} = \min\{d: d \in D_1 \cap D^1\}$ should be moved forward (delayed) to periods with slack capacity. In particular, for each stage i such that $\psi_{i\underline{d}} > 1$ at least one order with the due date not greater than $\underline{d}$ will be tardy and reallocated to some period $t > \underline{d}$ with slack capacity to reduce cumulative demand on capacity with respect to stage i in periods 1 through $\underline{d}$. Constraint (6.89) presented below ensures such a delayed order assignment.

Order-to-period delayed assignment constraints

$$\sum_{j \in J(d) \cap J_i, t \in D_0:\ d \leq \underline{d} < t,\ \psi_{i\underline{d}} > 1} x_{jt} \geq 1;\ i \in I. \qquad (6.89)$$

All the above cutting constraints can be added to models *OA1*, *PL1*, *PLC1*, and *PLM1* to reduce the computational effort required to find the optimal solution.

6.5.2 Computational Examples

In this subsection numerical examples and some computational results are presented to illustrate possible applications of the proposed lexicographic approach with a pair of integer programs *OA1* and *PL1* in the case of single-period orders. The examples are modeled after a real-world distribution center for high-tech products, where finished products are assembled for shipping to customers. The production system is a flexible flow shop (see Figs. 5.3 and 6.4) that consists of $m = 6$ processing stages in series and an output buffer of limited capacity $B2$ to hold completed products before delivery to the customers. A brief description of the production system, production process, products, and customer orders is given in Section 6.4.3. Now, the two-period orders are replaced with independent single-period suborders. In addition, the output buffer capacity is $B2 = 45{,}000$ products (scenario I) or $B2 = 76{,}000$ products (scenario II).

In the computational experiments the test problems described in Section 6.4.3 are applied, with the same demand patterns and capacity scenarios. For each capacity scenario and each demand pattern, Table 6.12 shows the minimum output buffer

Table 6.12 Minimum Capacity of the Output Buffer

Demand pattern	Scenario I		Scenario II	
	$B2_{\min}$	$B2^U_{\min}$	$B2_{\min}$	$B2^U_{\min}$
Increasing	24,150	24,720	75,040	75,690
Decreasing	0	x	0	x
Unimodal	0	41,200	37,850	x
Bimodal	0	39,000	25,770	x

$B2_{\min}$, $B2^U_{\min}$ - minimum buffer capacity required to complete all orders with no backlogging, with no tardiness, respectively.

x - insufficient processing capacity for completing all orders with no tardiness.

capacities $B2_{\min}$, (6.74), $B2^U_{\min}$, (6.75) required to complete all orders, respectively with no backlogging, with no tardiness. The table indicates that the output buffer capacity allows for completing all orders with no tardiness for three demand patterns in scenario I, and for one demand pattern in scenario II.

Note that the suborders in the computational examples play the role of orders in the mathematical formulation. Now, the problem objective is to determine an assignment of customer suborders over the planning horizon to minimize number of tardy suborders as a primary criterion and to level the aggregate production as a secondary criterion.

The characteristics of integer programs for the two capacity scenarios and various demand patterns and the solution results are summarized in Tables 6.13 and 6.14. The size of the integer program *OA1* (with the objective function (6.49) and the cutting constraints (6.80) to (6.82), (6.87) to (6.89)) and the mixed integer program *PL1*, is represented by the total number of variables, Var., number of binary variables, Bin., number of constraints, Cons., and number of nonzero elements in the constraint matrix, Nonz. The counts presented in the tables are taken from the models after presolving. The last column of each table presents the solution values E_{sum} and U_{sum} for *OA1* and $P_{\max}$ or $P_{\min}$ for *PL1*. The solution value $P_{\max}$ or $P_{\min}$ for *PL1* is presented along with the corresponding counter value of $P_{\min}$ or $P_{\min}$ (in parenthesis), respectively for the objective function (6.19) or (6.20).

In addition, Table 6.15 shows the solution results for scenario II obtained using the basic model *OA1* without any cutting constraint (notice that the increasing and decreasing demand patterns have identical numbers of variables and constraints). A comparison of the results in Tables 6.14 and 6.15 indicates that introduction of the cutting constraints has strengthened the formulation and may significantly reduce computation time required to find a proven optimal solution.

Table 6.13 Computational Results for Scenario I

Demand pattern/Model	Var.	Bin.	Cons.	Nonz.	Solution value
Increasing/*OA1*	15,856	15,856	1978	347,377	$E_{\text{sum}} = 812,\ U_{\text{sum}} = 0$
Increasing/*PL1*(6.19)[a]	5931	5930	1563	94,208	$(P_{\min} = 415),\ P_{\max} = 17{,}315$
Increasing/PL1(6.20)	5931	5930	1413	67458	$P_{\min} = 12{,}000,\ (P_{\max} = 20{,}075)$
Decreasing/*OA1*	24,360	24,360	1174	319,345	$E_{\text{sum}} = 809,\ U_{\text{sum}} = 3$
Decreasing/*PL1*(6.19)	1714	1713	799	19,071	$(P_{\min} = 3995),\ P_{\max} = 19{,}670$
Decreasing/*PL1*(6.20)	1714	1713	687	11,817	$P_{\min} = 7345,\ (P_{\max} = 19{,}940)$
Unimodal/*OA1*	11,275	11,275	1755	221,813	$E_{\text{sum}} = 704,\ U_{\text{sum}} = 0$
Unimodal/*PL1*(6.19)	3036	3035	1053	37,973	$(P_{\min} = 2750),\ P_{\max} = 20{,}830$
Unimodal/*PL1*(6.20)	3036	3035	1089	33,470	$P_{\min} = 6110,\ (P_{\max} = 22{,}060)$
Bimodal/*OA1*	10,437	10,437	1661	199,115	$E_{\text{sum}} = 671,\ U_{\text{sum}} = 0$
Bimodal/*PL1*(6.19)	3183	3182	1214	38,161	$(P_{\min} = 1135),\ P_{\max} = 19{,}605$
Bimodal/*PL1*(6.20)	3183	3182	1076	23,369	$P_{\min} = 3590,\ (P_{\max} = 21{,}125)$

[a]Objective function in *PL1*.

Table 6.14 Computational Results for Scenario II

Demand pattern/model	Var.	Bin.	Cons.	Nonz.	Solution value
Increasing/*OA1*	15,929	15,929	2005	353,387	$E_{\text{sum}} = 816,\ U_{\text{sum}} = 0$
Increasing/*PL1*(6.19)[a]	2498	2497	869	36,813	$(P_{\min} = 10{,}720),\ P_{\max} = 22{,}525$
Increasing/*PL1*(6.20)	2498	2497	779	26,422	$P_{\min} = 16{,}375,\ (P_{\max} = 22{,}605)$
Decreasing/*OA1*	24,480	24,480	1194	321,348	$E_{\text{sum}} = 798,\ U_{\text{sum}} = 18$
Decreasing/*PL1*(6.19)	1148	1147	586	10,846	$(P_{\min} = 14{,}970),\ P_{\max} = 21{,}185$
Decreasing/*PL1*(6.20)	1148	1147	502	7075	$P_{\min} = 16{,}630,\ (P_{\max} = 21{,}245)$
Unimodal/*OA1*	21,120	21,120	1091	312,791	$E_{\text{sum}} = 703,\ U_{\text{sum}} = 4$
Unimodal/*PL1*(6.19)	2920	2919	1167	37,499	$(P_{\min} = 11{,}275),\ P_{\max} = 19{,}865$
Unimodal/*PL1*(6.20)	2920	2919	1031	24,975	$P_{\min} = 15{,}250,\ (P_{\max} = 21{,}030)$
Bimodal/*OA1*	20,250	20,250	1053	293,129	$E_{\text{sum}} = 669,\ U_{\text{sum}} = 6$
Bimodal/*PL1*(6.19)	2346	2345	1134	26,262	$(P_{\min} = 10{,}295),\ P_{\max} = 19{,}345$
Bimodal/*PL1*(6.20)	2346	2345	988	16,237	$P_{\min} = 11{,}100,\ (P_{\max} = 20{,}580)$

[a]Objective function in *PL1*.

The demand distribution, the optimal assignment of customer orders over the monthly horizon, and the resulting aggregate production schedules for the two objective functions and the examples with the increasing demand pattern and scenario I with medium tightness of the capacity constraints are presented in Tables 6.16 to 6.18 and Figures 6.9 and 6.10. Tables 6.16 to 6.18 show detailed distribution and optimal assignment of all suborders (customer required shipping volumes) over the 30-day planning horizon. Figure 6.9 presents the distribution over the horizon of the aggregate demand and production for each product type, whereas in Figure 6.10 additionally the finished product inventory and machine assignments over the horizon are presented. The tables and figures show that objective functions (6.19) and (6.20) aim at leveling the aggregate production from above and below, respectively. However, the variations of production over the horizon is smaller for objective function (6.20).

Comparison (see Sawik, 2007c) of the solution values of *PL1* for various capacity scenarios and different demand patterns indicates that the objective function (6.20) leads to smaller fluctuations of the aggregate production over the planning horizon.

Table 6.15 Computational Results for Scenario II: Basic Integer Program *OA1*

Demand pattern	Var.	Bin.	Cons.	Nonz.	Solution value	GAP[a]
Increasing	24,480	24,480	1025	283,429	$E_{\text{sum}} = 815,\ U_{sum} = 1$	0.12
Decreasing	24,480	24,480	1025	194,366	$E_{\text{sum}} = 797,\ U_{\text{sum}} = 19$	0.22
Unimodal	21,210	21,210	916	195,723	$E_{\text{sum}} = 703,\ U_{\text{sum}} = 4$	–
Bimodal	20,250	20,250	884	184,874	$E_{\text{sum}} = 669,\ U_{\text{sum}} = 6$	–

[a]% GAP after 3600 seconds of CPU time.

Table 6.16 Distribution of Customer Orders for Increasing Demand Pattern: Scenario I

	1	2	3	4	5	6	7	8	9	10	11	12	13	14	15
1	0	1200	350	0	0	380	2800	0	900	0	420	1620	0	0	0
2	0	0	310	0	0	790	0	0	290	0	845	4105	0	0	0
3	0	0	0	0	0	1165	0	0	1330	0	785	0	0	0	0
4	0	0	0	0	0	420	0	0	555	0	155	0	0	0	0
5	0	0	0	0	0	240	0	0	205	0	620	0	0	0	0
6	0	0	0	0	0	220	0	0	730	0	400	0	0	0	0
7	0	0	0	0	0	640	0	0	10	0	0	0	0	0	0
8	0	0	0	0	0	790	0	0	145	0	480	0	0	0	0
9	0	0	0	0	0	0	0	0	770	0	0	0	0	0	0
10	0	0	0	0	0	0	0	0	380	0	0	0	0	0	0
11	0	0	0	1440	685	0	0	0	0	0	0	0	0	0	0
12	0	0	0	2555	540	0	0	0	0	0	0	0	0	0	0
13	0	0	0	0	1040	0	0	0	0	0	0	0	0	0	0
14	0	0	0	0	1540	0	0	0	0	0	0	0	0	0	0
15	0	0	0	0	1040	0	0	0	0	0	0	0	0	0	0
16	0	0	0	0	0	0	0	0	0	0	0	0	0	0	0
17	0	0	0	0	0	0	0	0	0	0	0	0	0	0	0
18	0	0	0	0	0	0	0	0	0	0	0	0	0	0	0
19	0	0	0	0	0	0	0	0	0	0	0	0	0	0	0
20	0	0	580	0	225	0	0	0	390	1460	520	2575	0	510	835
21	0	0	1300	0	170	0	0	0	190	895	285	2485	0	300	210
22	0	0	930	0	35	0	0	0	415	1355	85	0	0	550	425
23	0	0	1380	0	200	0	0	0	255	1210	720	0	0	595	740
24	0	0	0	0	90	0	0	0	405	70	550	0	0	395	15
25	0	0	0	0	65	0	0	0	315	0	600	0	0	645	255
26	0	0	0	0	125	0	0	0	35	0	210	0	0	80	130
27	0	0	0	0	0	0	0	0	170	0	120	0	0	200	700
28	0	0	0	0	0	0	0	0	415	0	530	0	0	500	0
29	0	0	0	0	0	0	0	0	150	0	355	0	0	0	0
30	0	0	0	0	0	0	0	0	565	0	0	0	0	0	0
31	0	0	0	0	1365	25	0	0	0	0	0	0	0	0	0
32	0	0	0	0	450	0	0	0	0	0	0	0	0	0	0
33	0	0	0	0	330	5	0	0	0	0	0	0	0	0	0
34	0	0	0	0	490	20	0	0	0	0	0	0	0	0	0
35	0	0	0	0	610	0	0	0	0	0	0	0	0	0	0
36	0	0	0	0	940	0	0	0	0	0	0	0	0	0	0
37	0	0	0	0	1735	0	0	0	0	0	0	0	0	0	0
38	0	0	0	0	0	0	0	0	0	0	0	0	0	0	0
39	0	0	0	0	0	0	0	0	0	0	0	0	0	0	0
40	0	0	0	0	0	0	0	0	0	0	0	0	0	0	0

16	17	18	19	20	21	22	23	24	25	26	27	28	29	30
1640	190	75	0	10	450	370	665	315	2650	1070	890	1505	0	340
2350	1180	495	0	355	5570	475	0	520	680	645	870	310	0	150
0	60	175	0	475	0	280	0	90	0	795	2870	670	0	70
0	890	600	0	145	0	60	0	0	0	775	0	840	0	130
0	1020	305	0	0	0	370	0	0	0	90	0	0	0	125
0	950	330	0	0	0	20	0	0	0	210	0	0	0	0
0	490	390	0	0	0	50	0	0	0	170	0	0	0	0
0	515	565	0	0	0	130	0	0	0	490	0	0	0	0
0	10	40	0	0	0	20	0	0	0	760	0	0	0	0
0	0	0	0	0	0	0	0	0	0	300	0	0	0	0
0	0	0	845	0	0	0	180	0	0	0	4800	80	780	2175
0	0	10	1080	0	0	0	1360	0	0	0	0	220	420	490
0	0	5	0	0	0	0	1465	0	0	0	0	250	75	3190
0	0	10	0	0	0	0	1535	0	0	0	0	265	50	0
0	0	10	0	0	0	0	580	0	0	0	0	210	2430	0
0	0	0	0	0	0	0	0	0	0	0	0	280	30	0
0	0	0	0	0	0	0	0	0	0	0	0	215	0	0
0	0	5	0	0	0	0	0	0	0	0	0	0	0	0
0	0	10	0	0	0	0	0	0	0	0	0	0	0	0
1290	1510	1375	3130	0	570	150	700	370	210	0	505	75	260	90
1270	365	90	0	0	585	125	1030	510	470	0	2305	325	575	740
450	250	980	0	0	720	820	740	445	210	0	355	490	155	660
165	0	1340	0	0	205	280	1440	0	105	0	1070	355	210	760
205	0	1170	0	0	65	60	310	0	620	0	130	90	225	0
0	0	695	0	0	200	610	650	0	490	0	750	200	435	0
0	0	0	0	0	550	335	245	0	745	0	0	660	470	0
0	0	0	0	0	700	0	0	0	700	0	0	585	80	0
0	0	0	0	0	585	0	0	0	690	0	0	530	400	0
0	0	0	0	0	255	0	0	0	735	0	0	0	95	0
0	0	0	0	0	0	0	0	0	0	0	0	0	0	0
0	20	80	10	55	110	0	0	70	365	1150	20	680	0	0
0	40	130	30	65	210	0	0	175	390	775	140	670	0	0
0	0	90	60	45	145	0	0	410	235	500	90	1360	0	0
0	0	110	40	135	220	0	0	130	60	15	620	1230	0	0
0	0	115	20	150	105	0	0	90	195	205	100	880	0	0
0	0	105	35	0	25	0	0	105	400	215	680	230	0	0
0	0	45	45	0	205	0	0	0	360	625	580	0	0	0
0	0	70	35	0	145	0	0	0	160	545	610	0	0	0
0	0	100	35	0	5	0	0	0	165	400	0	0	0	0
0	0	0	0	0	0	0	0	0	440	870	0	0	0	0

(*Continued*)

Table 6.16 *Continued*

	1	2	3	4	5	6	7	8	9	10	11	12	13	14	15
41	0	0	0	0	0	0	0	0	0	0	0	0	0	0	0
42	0	0	0	0	0	0	0	0	0	0	0	0	0	0	0
43	0	0	0	0	0	0	0	0	0	0	0	0	0	0	0
44	0	0	0	0	0	0	0	0	0	0	0	0	0	0	0
45	0	0	0	0	0	0	0	0	0	0	0	0	0	0	0
46	0	0	0	0	0	0	0	0	0	0	0	0	0	0	0
47	0	0	0	0	0	0	0	0	0	0	0	0	0	0	0
48	0	0	0	0	0	0	0	0	915	765	2405	525	10	945	390
49	0	0	0	0	0	0	0	0	590	970	985	510	185	1160	50
50	0	0	0	0	0	0	0	0	1000	925	765	340	60	840	770
51	0	0	0	0	0	0	0	0	100	1065	970	525	690	335	1055
52	0	0	0	0	0	0	0	0	850	0	0	780	590	80	695
53	0	0	0	0	0	0	0	0	0	0	0	10	425	125	90
54	0	0	0	0	0	0	0	0	0	0	0	35	695	490	350
55	0	0	0	0	0	0	0	0	0	0	0	675	0	0	780
56	0	0	0	0	0	0	0	0	0	0	0	835	0	0	0
57	0	0	0	0	0	0	0	0	0	0	0	335	0	0	0
58	0	0	0	0	0	225	365	0	0	755	0	0	0	0	0
59	0	0	0	0	0	1040	880	0	0	760	0	0	0	0	0
60	0	0	0	0	0	650	15	0	0	680	0	0	0	0	0
61	0	0	0	0	0	590	705	0	0	0	0	0	0	0	0
62	0	0	0	0	0	370	1035	0	0	0	0	0	0	0	0
63	0	0	0	0	0	80	950	0	0	0	0	0	0	0	0
64	0	0	0	0	0	570	60	0	0	0	0	0	0	0	0
65	0	0	0	0	0	210	560	0	0	0	0	0	0	0	0
66	0	0	0	0	0	270	605	0	0	0	0	0	0	0	0
67	0	0	0	0	0	155	0	0	0	0	0	0	0	0	0
68	0	0	0	0	0	0	0	0	0	0	0	0	0	0	0
69	0	215	1105	0	0	0	0	0	0	0	0	0	0	0	0
70	0	1700	980	0	0	0	0	0	0	0	0	0	0	0	0
71	0	750	190	0	0	0	0	0	0	0	0	0	0	0	0
72	0	710	640	0	0	0	0	0	0	0	0	0	0	0	0
73	0	255	0	0	0	0	0	0	0	0	0	0	0	0	0
74	0	165	0	0	0	0	0	0	0	0	0	0	0	0	0
75	0	515	0	0	0	0	0	0	0	0	0	0	0	0	0
76	0	25	0	0	0	0	0	0	0	0	0	0	0	0	0
77	0	845	0	0	0	0	0	0	0	0	0	0	0	0	0
78	0	0	0	0	0	0	0	0	0	0	0	0	0	0	0
79	0	0	0	0	0	0	0	0	0	0	0	0	0	0	0
80	0	0	0	0	900	500	0	0	0	0	0	0	0	0	0

16	17	18	19	20	21	22	23	24	25	26	27	28	29	30
0	25	0	0	0	0	0	0	0	0	0	0	0	0	0
0	20	0	0	0	0	0	0	0	0	0	0	0	0	0
0	370	0	0	0	0	0	0	0	0	0	0	0	0	0
0	240	0	0	0	0	0	0	0	0	0	0	0	0	0
0	260	0	0	0	0	0	0	0	0	0	0	0	0	0
0	195	0	0	0	0	0	0	0	0	0	0	0	0	0
0	210	0	0	0	0	0	0	0	0	0	0	0	0	0
750	1030	5200	1145	830	975	220	945	520	5130	390	2930	2150	1130	260
955	1230	0	2710	700	1020	535	155	1410	0	195	0	925	1195	1525
840	810	0	1075	1050	890	2790	945	1435	0	740	0	2020	335	2250
270	450	0	0	55	700	1910	695	480	0	1205	0	895	1090	625
720	535	0	0	730	1135	0	640	475	0	670	0	0	1060	0
380	50	0	0	770	0	0	795	480	0	105	0	0	640	0
65	385	0	0	285	0	0	1070	390	0	790	0	0	130	0
0	355	0	0	580	0	0	465	0	0	30	0	0	130	0
0	0	0	0	985	0	0	0	0	0	445	0	0	0	0
0	0	0	0	0	0	0	0	0	0	865	0	0	0	0
0	220	0	300	0	705	0	130	630	150	365	390	110	695	215
0	155	0	40	0	675	0	200	175	20	270	215	145	500	150
0	320	0	10	0	1080	0	510	650	350	320	50	120	0	120
0	120	0	365	0	0	0	245	0	100	60	50	10	0	265
0	345	0	105	0	0	0	435	0	510	190	40	100	0	405
0	400	0	295	0	0	0	270	0	365	185	240	90	0	5
0	90	0	535	0	0	0	0	0	280	260	95	60	0	310
0	5	0	0	0	0	0	0	0	265	0	30	90	0	320
0	320	0	0	0	0	0	0	0	0	0	420	50	0	50
0	0	0	0	0	0	0	0	0	0	0	40	30	0	0
0	0	0	0	0	0	0	0	0	0	0	275	170	0	0
0	35	0	0	10	530	835	750	10	0	70	105	90	135	0
0	160	0	0	335	0	805	1235	10	0	190	440	55	200	80
0	0	0	0	270	0	470	1340	5	0	350	315	195	240	215
0	0	0	0	315	0	965	0	0	0	325	0	175	80	55
0	0	0	0	475	0	985	0	0	0	390	0	145	0	205
0	0	0	0	110	0	140	0	5	0	140	0	0	0	170
0	0	0	0	0	0	575	0	10	0	110	0	0	0	130
0	0	0	0	0	0	0	0	5	0	460	0	0	0	195
0	0	0	0	0	0	0	0	0	0	40	0	0	0	0
0	0	0	0	0	0	0	0	0	0	165	0	0	0	0
0	0	0	0	0	0	0	0	0	0	325	0	0	0	0
0	0	0	0	130	20	0	265	170	0	575	160	0	640	325

(*Continued*)

Table 6.16 *Continued*

	1	2	3	4	5	6	7	8	9	10	11	12	13	14	15
81	0	0	0	0	975	365	0	0	0	0	0	0	0	0	0
82	0	0	0	0	2250	155	0	0	0	0	0	0	0	0	0
83	0	0	0	0	0	95	0	0	0	0	0	0	0	0	0
84	0	0	0	0	0	505	0	0	0	0	0	0	0	0	0
85	0	0	0	0	0	195	0	0	0	0	0	0	0	0	0
86	0	0	0	0	0	430	0	0	0	0	0	0	0	0	0
87	0	0	0	0	0	0	0	0	0	0	0	0	0	0	0
88	0	0	0	0	0	0	0	0	0	0	0	0	0	0	0
89	0	165	145	0	0	0	0	0	0	0	0	0	0	0	0
90	0	355	120	0	0	0	0	0	0	0	0	0	0	0	0
91	0	440	235	0	0	0	0	0	0	0	0	0	0	0	0
92	0	420	110	0	0	0	0	0	0	0	0	0	0	0	0
93	0	340	180	0	0	0	0	0	0	0	0	0	0	0	0
94	0	320	245	0	0	0	0	0	0	0	0	0	0	0	0
95	0	350	140	0	0	0	0	0	0	0	0	0	0	0	0
96	0	0	185	0	0	0	0	0	0	0	0	0	0	0	0
97	0	0	50	0	0	0	0	0	0	0	0	0	0	0	0
98	0	0	40	0	0	0	0	0	0	0	0	0	0	0	0
99	0	0	60	0	0	0	0	0	0	0	0	0	0	0	0
100	0	0	0	0	2065	0	270	800	1035	2465	90	2400	350	0	45

Table 6.17 Optimal Assignment of Customer Orders for min P_{max} Criterion

	1	2	3	4	5	6	7	8	9	10	11	12	13	14
1	350	1200	0	420	0	380	2800	190	900	0	0	1620	0	0
2	310	0	0	0	0	790	0	0	200	0	4950	0	0	0
3	0	0	0	0	0	1165	0	1330	176	0	785	0	0	0
4	0	0	0	0	0	420	0	556	0	0	155	0	0	0
5	0	0	0	0	0	240	0	0	205	0	620	0	0	305
6	0	0	0	0	0	220	0	0	730	0	400	0	0	330
7	0	0	0	0	0	640	0	0	10	0	0	0	490	0
8	0	0	0	0	400	790	0	0	145	0	0	0	0	0
9	0	0	0	0	0	0	0	0	770	40	0	0	0	0
10	0	0	0	0	0	0	0	0	380	0	0	0	0	0
11	0	0	0	1440	685	0	0	0	0	0	0	0	0	0
12	0	0	0	2555	540	0	0	0	0	0	0	0	0	0
13	0	0	5	0	1040	0	0	0	0	75	0	0	0	0
14	0	0	0	0	1540	0	0	0	0	0	0	0	0	0

16	17	18	19	20	21	22	23	24	25	26	27	28	29	30
0	0	0	0	0	385	0	0	85	0	1270	675	0	770	300
0	0	0	0	0	230	0	0	140	0	1805	630	0	285	290
0	0	0	0	0	100	0	0	5	0	270	465	0	1160	280
0	0	0	0	0	580	0	0	140	0	0	210	0	0	265
0	0	0	0	0	0	0	0	160	0	0	705	0	0	120
0	0	0	0	0	0	0	0	60	0	0	605	0	0	60
0	0	0	0	0	0	0	0	105	0	0	630	0	0	310
0	0	0	0	0	0	0	0	45	0	0	630	0	0	290
390	0	525	865	265	285	335	420	165	0	370	50	20	70	1665
175	0	80	0	130	205	370	290	400	0	230	60	120	15	0
340	0	390	0	0	110	230	1055	50	0	380	35	250	0	0
250	0	445	0	0	220	140	1025	645	0	435	50	0	20	0
385	0	110	0	0	220	310	0	680	0	0	25	0	75	0
390	0	260	0	0	260	0	0	0	0	260	10	0	40	0
55	0	0	0	0	280	0	0	245	0	490	10	0	25	0
0	0	410	0	0	0	0	0	0	0	80	0	0	15	0
0	0	405	0	0	0	0	0	0	0	0	0	0	35	0
0	0	350	0	0	0	0	0	0	0	0	0	0	75	0
0	0	0	0	0	0	0	0	0	0	0	0	0	0	0
900	0	380	5200	55	705	1795	180	370	35	240	3420	610	2800	165

15	16	17	18	19	20	21	22	23	24	25	26	27	28	29	30
1640	10	0	75	0	820	0	1070	665	315	2650	0	890	1505	0	340
2350	670	2150	0	0	355	5570	0	0	0	1325	870	0	310	0	0
0	625	280	0	0	0	0	0	0	2870	795	0	0	670	70	0
0	0	1490	0	0	335	0	0	0	0	0	775	0	840	0	0
0	0	1020	370	90	0	0	0	0	0	0	0	0	0	0	300
0	0	950	20	0	0	0	0	0	210	0	0	0	0	125	0
390	50	0	0	0	0	0	0	0	170	0	0	0	0	0	0
0	130	1080	0	0	0	0	0	0	0	0	490	0	0	0	0
0	30	0	0	0	0	0	0	0	0	760	0	0	0	0	0
0	0	0	0	0	0	0	0	0	0	0	300	0	0	0	0
0	0	0	0	845	180	0	0	0	0	0	0	4800	80	780	2175
420	0	0	1090	0	0	0	0	1360	0	0	0	0	220	0	490
0	0	0	0	0	0	0	0	1465	0	0	0	0	250	0	3190
0	0	0	1545	0	0	0	0	0	0	0	0	0	265	50	0

(*Continued*)

Table 6.17 *Continued*

	1	2	3	4	5	6	7	8	9	10	11	12	13	14
15	0	0	0	0	1040	0	0	0	0	0	0	0	0	0
16	0	0	0	0	0	0	0	0	0	0	0	0	0	0
17	0	0	0	0	0	0	0	0	0	0	0	0	0	0
18	0	0	0	0	0	0	0	0	0	0	0	0	0	0
19	0	0	0	0	0	0	0	0	0	0	0	0	0	0
20	500	0	0	0	225	0	0	0	910	1460	0	2575	90	1080
21	1300	300	0	0	455	0	0	90	315	895	210	2485	0	0
22	265	0	930	0	0	0	0	0	415	1355	85	0	975	0
23	1300	0	255	0	200	0	0	0	0	1930	0	740	0	595
24	90	0	15	0	0	0	0	60	405	465	615	0	205	0
25	0	0	0	0	65	0	0	0	915	0	0	255	645	0
26	0	0	0	0	125	0	210	0	35	0	0	0	0	325
27	0	0	0	0	0	0	0	0	170	120	0	0	0	900
28	0	0	0	415	0	0	0	0	0	0	530	0	500	0
29	0	0	0	0	0	0	0	0	150	0	355	0	0	255
30	0	0	0	0	0	0	0	0	565	0	0	0	0	0
31	0	0	0	20	1365	25	0	0	0	0	0	0	80	0
32	0	0	40	0	450	0	130	0	0	0	0	0	0	0
33	0	0	0	0	330	0	0	0	0	0	0	0	0	90
34	0	0	0	0	490	20	0	0	0	60	0	0	0	0
35	150	0	0	0	610	0	0	115	0	0	0	0	0	0
36	0	0	0	0	940	0	0	0	0	0	0	0	0	0
37	0	0	0	0	1780	0	0	0	0	0	0	0	0	0
38	0	160	0	0	0	0	0	0	0	0	0	0	0	70
39	0	0	0	100	0	0	0	0	0	0	0	0	0	0
40	0	0	0	0	0	0	0	0	0	0	0	0	0	0
41	0	0	0	0	25	0	0	0	0	0	0	0	0	0
42	0	0	0	0	0	0	0	0	20	0	0	0	0	0
43	0	0	0	0	0	0	0	0	0	0	0	0	0	0
44	0	0	0	0	0	0	0	0	0	0	0	0	0	0
45	0	0	0	0	0	0	0	0	0	0	0	0	0	260
46	0	0	0	0	195	0	0	0	0	0	0	0	0	0
47	0	0	0	0	0	0	0	0	0	0	0	0	0	0
48	0	0	0	0	0	0	0	0	915	765	2405	525	10	945
49	0	0	0	0	50	0	0	0	590	970	985	510	185	2115
50	0	0	0	0	0	0	0	0	1000	925	765	1110	60	1650
51	0	0	0	0	335	0	0	0	100	1065	1025	525	960	0
52	0	0	0	0	0	60	0	695	850	0	0	780	590	1255
53	0	0	175	0	0	0	0	0	0	0	0	100	425	0
54	0	0	0	0	0	0	0	0	0	490	0	35	1430	0

15	16	17	18	19	20	21	22	23	24	25	26	27	28	29	30
0	0	0	10	0	580	0	0	0	0	0	0	0	210	2430	0
0	0	0	0	30	0	280	0	0	0	0	0	0	0	0	0
0	0	0	0	0	215	0	0	0	0	0	0	0	0	0	0
0	0	0	5	0	0	0	0	0	0	0	0	0	0	0	0
0	0	0	10	0	0	0	0	0	0	0	0	0	0	0	0
835	2800	0	1745	3130	360	0	0	700	0	0	505	0	75	260	0
1635	585	0	0	0	0	0	1030	0	980	0	2305	0	325	575	740
980	450	0	0	155	720	0	1480	1185	210	0	0	355	490	0	0
0	165	0	1340	0	205	0	385	1440	0	0	0	1070	355	210	760
0	0	130	1170	0	0	0	0	310	0	620	0	0	315	0	0
0	0	695	0	0	0	850	610	0	0	1240	0	200	0	435	0
130	0	0	0	0	0	550	335	0	745	0	0	0	1130	0	0
0	0	0	0	0	700	0	700	80	585	0	0	0	0	0	0
0	0	585	0	0	0	0	0	0	0	690	0	0	530	400	0
0	0	0	0	0	95	0	0	0	735	0	0	0	0	0	0
0	0	0	0	0	0	0	0	0	0	0	0	0	0	0	0
0	85	0	110	0	0	0	0	70	1150	365	0	680	0	0	0
0	30	0	275	0	0	1165	140	0	175	0	0	0	670	0	0
0	0	0	250	410	0	235	0	0	0	90	500	0	1360	0	0
135	55	110	0	220	130	0	0	0	0	0	0	620	1230	0	0
0	0	0	125	0	190	0	0	0	1075	0	205	0	0	0	0
105	25	0	140	0	0	400	0	0	0	215	0	680	230	0	0
0	45	205	0	0	625	0	0	0	0	360	580	0	0	0	0
0	145	0	35	0	0	0	0	0	0	0	545	610	0	0	0
0	5	0	0	35	0	0	0	0	565	0	0	0	0	0	0
0	0	0	0	0	0	0	0	0	440	0	870	0	0	0	0
0	0	0	0	0	0	0	0	0	0	0	0	0	0	0	0
0	0	0	0	0	0	0	0	0	0	0	0	0	0	0	0
0	0	370	0	0	0	0	0	0	0	0	0	0	0	0	0
0	0	240	0	0	0	0	0	0	0	0	0	0	0	0	0
0	0	0	0	0	0	0	0	0	0	0	0	0	0	0	0
0	0	0	0	0	0	0	0	0	0	0	0	0	0	0	0
0	0	210	0	0	0	0	0	0	0	0	0	0	0	0	0
1140	5200	1030	1365	0	1775	975	0	0	520	5520	0	2930	2150	1130	260
0	0	2940	195	2710	700	1410	0	0	0	0	0	0	925	1195	1525
0	840	0	890	1075	1050	740	2790	945	1435	0	2020	0	335	0	2250
1055	0	1150	0	0	0	1205	2535	695	480	0	0	0	895	1090	0
0	0	0	730	1135	0	0	0	640	1145	0	0	0	0	1060	0
0	485	480	0	0	770	0	795	0	0	0	0	0	0	640	0
65	0	0	0	0	285	0	0	1070	390	0	790	0	0	130	0

(*Continued*)

Table 6.17 *Continued*

	1	2	3	4	5	6	7	8	9	10	11	12	13	14
55	0	0	0	0	0	0	0	0	0	0	0	675	0	780
56	0	0	0	0	0	0	0	0	0	0	0	835	445	0
57	0	0	0	0	0	0	0	0	0	0	0	335	0	0
58	0	0	0	0	0	225	305	220	0	1385	0	0	365	0
59	0	0	0	0	0	1195	880	0	500	1105	0	0	0	215
60	0	0	0	0	0	650	606	0	440	0	0	0	320	0
61	0	0	0	0	0	590	706	0	485	0	265	0	0	0
62	0	0	0	0	0	715	1035	105	0	0	0	510	100	0
63	0	0	0	0	0	88	960	0	670	185	0	295	0	0
64	0	0	0	0	0	570	60	0	0	260	90	0	535	310
65	0	0	0	0	0	210	560	0	5	0	0	0	0	0
66	0	0	0	0	0	270	505	0	0	0	0	0	320	0
67	0	0	0	0	0	155	0	0	0	0	0	0	0	0
68	0	0	0	0	0	0	0	0	0	0	0	0	0	0
69	0	215	1105	0	35	0	0	0	0	0	0	0	0	0
70	0	1700	980	0	0	0	0	0	160	0	0	0	190	0
71	0	750	190	0	0	0	0	270	0	0	0	0	350	0
72	0	710	840	0	0	0	0	0	0	0	0	0	965	0
73	0	256	0	0	0	0	0	0	0	0	0	0	0	0
74	0	166	0	0	0	0	0	0	0	0	0	0	0	0
75	0	515	0	0	0	0	0	0	0	0	0	0	0	0
76	0	25	0	0	0	0	0	0	0	0	0	0	0	0
77	0	845	0	0	0	0	0	0	0	0	0	0	0	0
78	0	0	0	0	0	0	0	0	0	0	0	0	0	0
79	0	0	0	0	0	0	0	0	0	0	0	0	0	0
80	0	0	0	0	900	600	0	0	0	0	0	0	0	0
81	0	0	0	0	975	365	0	0	0	0	0	0	0	0
82	0	0	0	0	2250	155	0	0	0	0	0	0	0	0
83	0	0	0	0	0	95	0	0	0	0	0	0	0	0
84	0	0	0	0	0	505	0	0	0	0	0	0	0	0
85	0	0	0	0	0	195	0	0	0	0	0	0	0	0
86	0	0	0	0	0	430	0	0	0	0	0	0	0	0
87	0	0	0	0	0	0	0	0	0	0	0	0	0	0
88	0	0	0	0	0	0	0	0	0	0	0	0	0	0
89	145	165	0	0	0	0	0	0	0	0	0	0	0	0
90	120	365	0	0	0	0	0	0	0	0	0	255	0	0
91	235	440	0	0	0	0	0	0	0	0	0	340	0	0
92	110	420	0	0	0	0	0	0	0	0	0	250	0	0
93	0	340	180	0	0	0	0	0	385	110	0	0	0	0
94	245	320	0	0	0	390	0	0	0	0	0	0	0	0

15	16	17	18	19	20	21	22	23	24	25	26	27	28	29	30
0	0	355	160	0	580	465	0	0	0	0	0	0	0	0	0
985	0	0	0	0	0	0	0	0	0	0	0	0	0	0	0
0	0	0	0	0	0	0	0	0	0	0	865	0	0	0	0
0	390	0	0	300	0	705	0	0	0	0	0	390	0	695	215
945	0	0	235	0	0	0	0	0	0	0	0	0	0	0	150
350	10	0	170	0	1080	0	0	510	650	0	0	0	0	0	0
245	60	0	160	0	0	0	0	0	0	0	0	0	0	0	0
625	0	0	40	0	0	0	0	0	0	0	0	0	0	0	405
0	95	0	365	0	0	0	0	240	0	0	0	0	0	0	0
0	155	280	0	0	0	0	0	0	0	0	0	0	0	0	0
0	30	0	90	0	0	0	265	0	0	0	0	0	0	0	320
0	0	0	50	0	0	0	0	0	0	0	50	420	0	0	0
0	40	0	30	0	0	0	0	0	0	0	0	0	0	0	0
0	0	0	0	0	0	0	0	0	0	0	0	275	170	0	0
0	265	530	155	835	750	0	0	0	0	0	0	0	0	0	0
805	145	0	0	335	0	0	0	1235	0	0	0	440	0	200	0
0	515	0	215	0	1340	0	0	0	0	0	0	315	0	0	0
0	315	0	0	0	0	0	0	0	0	325	0	0	255	0	55
0	390	475	0	0	0	0	985	0	0	0	0	0	350	0	0
0	175	0	250	0	0	140	0	0	0	0	0	0	0	0	0
110	0	0	10	0	0	0	575	0	0	0	0	0	0	0	130
0	5	0	0	0	0	0	460	0	0	0	0	0	195	0	0
0	0	0	40	0	0	0	0	0	0	0	0	0	0	0	0
0	0	0	0	0	0	0	0	0	0	165	0	0	0	0	0
0	0	0	0	0	0	0	0	0	0	0	325	0	0	0	0
0	20	0	290	0	265	170	0	575	0	0	0	0	0	0	0
0	85	0	0	0	385	675	0	0	0	0	1270	0	0	640	325
0	140	0	0	0	230	0	0	1805	0	0	0	0	0	770	300
0	105	0	0	0	270	0	0	0	0	0	0	630	0	285	290
0	0	0	0	0	0	580	0	0	0	0	0	465	0	1160	280
0	0	0	0	0	160	0	0	0	140	0	0	210	0	0	265
0	60	0	60	0	0	0	0	0	0	705	0	0	0	0	120
0	0	0	0	0	0	0	0	0	0	605	0	0	0	0	0
0	0	0	0	0	0	0	0	0	105	0	0	630	0	0	310
390	285	0	1390	0	1205	0	0	0	45	0	0	630	0	0	290
0	60	0	145	0	370	495	0	0	0	0	370	50	0	70	1665
0	265	0	550	0	0	0	0	0	400	230	0	0	120	0	0
0	70	445	0	140	220	0	0	1055	0	380	0	0	250	0	0
0	220	0	75	0	0	0	310	1025	1080	0	0	0	0	0	0
0	310	0	0	0	260	0	0	0	680	0	0	25	0	0	0

(Continued)

Table 6.17 *Continued*

	1	2	3	4	5	6	7	8	9	10	11	12	13	14
95	140	350	0	0	0	0	0	0	0	0	0	0	0	0
96	105	0	0	0	0	0	0	0	0	0	0	0	0	0
97	50	0	0	0	0	0	0	0	0	0	0	0	0	0
98	40	0	0	0	0	0	0	0	0	0	0	0	0	0
99	60	0	0	0	0	0	0	0	0	0	0	0	0	0
100	0	0	0	2110	0	0	270	800	1035	2465	90	2400	350	0

Table 6.18 Optimal Assignment of Customer Orders for max P_{min} Criterion

	1	2	3	4	5	6	7	8	9	10	11	12	13	14
1	420	1550	0	3180	0	10	190	0	900	1620	0	0	0	0
2	310	290	0	0	0	790	0	0	0	150	4950	0	0	520
3	0	0	0	0	0	1165	1330	0	0	0	785	0	265	60
4	155	0	0	0	0	420	555	0	0	0	0	0	0	0
5	0	0	0	0	0	240	0	0	205	0	620	0	0	90
6	0	0	0	0	0	220	730	0	0	0	400	0	330	0
7	0	0	0	0	10	640	0	0	0	0	0	0	930	0
8	0	0	480	0	0	790	0	0	145	0	0	0	130	0
9	0	0	0	0	0	0	0	0	790	0	50	0	0	0
10	0	0	0	0	0	0	0	380	0	0	0	0	0	0
11	0	0	685	1440	0	0	0	0	0	0	0	0	0	0
12	10	0	2555	540	0	0	0	0	0	0	0	0	0	0
13	0	0	0	1040	5	0	0	0	75	0	0	0	0	0
14	0	0	1540	0	0	0	0	0	0	0	0	0	0	0
15	10	0	0	0	1040	0	0	0	0	0	0	0	0	0
16	0	0	0	0	0	0	0	0	0	0	0	0	30	0
17	0	0	0	0	0	0	0	0	0	0	0	0	0	0
18	0	5	0	0	0	0	0	0	0	0	0	0	0	0
19	10	0	0	0	0	0	0	0	0	0	0	0	0	0
20	1325	0	0	0	0	0	0	1460	390	0	0	2665	570	510
21	1300	0	0	0	645	300	0	125	895	0	0	2485	210	365
22	0	250	930	0	35	0	0	0	415	1440	0	0	0	1955
23	1835	0	0	0	0	0	0	0	1210	0	720	0	0	1335
24	150	0	0	0	0	70	0	395	405	0	550	80	205	0
25	65	0	0	0	0	0	315	600	0	0	0	0	645	255
26	0	0	210	0	160	0	0	0	0	0	0	0	0	455
27	0	0	0	170	0	0	0	0	0	120	0	0	0	900
28	0	0	0	415	0	0	0	0	0	0	530	0	500	0
29	0	0	0	0	0	0	150	0	0	0	355	0	255	0

15	16	17	18	19	20	21	22	23	24	25	26	27	28	29	30
55	290	0	25	0	0	0	490	0	0	0	260	0	0	0	0
0	0	0	410	80	0	0	15	245	0	0	0	0	0	0	0
405	0	0	0	0	0	0	0	0	0	0	0	0	0	35	0
0	350	0	0	0	0	0	0	0	0	0	0	0	0	75	0
0	0	0	0	0	0	0	0	0	0	0	0	0	0	0	0
900	0	0	435	5200	0	705	2345	0	0	275	3420	0	610	2800	165

15	16	17	18	19	20	21	22	23	24	25	26	27	28	29	30
715	0	0	0	0	820	0	1070	665	315	2650	0	890	1505	340	0
350	1180	970	0	0	355	5570	0	0	645	680	0	870	310	0	0
475	0	0	0	0	0	0	280	0	3665	0	0	0	670	70	0
0	0	1490	0	0	145	0	60	0	0	0	775	0	840	130	0
305	0	1020	0	0	0	0	370	0	0	0	0	0	0	0	300
0	0	950	0	0	0	0	20	0	0	0	210	0	0	0	125
0	0	0	0	0	0	0	20	0	0	170	0	0	0	0	0
0	0	515	565	0	0	0	0	0	0	0	490	0	0	0	0
0	0	0	0	0	0	0	0	0	0	0	760	0	0	0	0
0	0	0	0	0	0	0	0	0	0	300	0	0	0	0	0
0	0	0	0	845	0	0	0	180	0	0	0	4800	80	780	2175
420	0	0	1080	0	0	0	0	1360	0	0	0	0	220	490	0
0	0	0	0	0	0	0	0	1465	0	0	0	0	250	0	3190
0	0	0	1545	0	0	0	0	0	0	0	0	0	265	50	0
0	0	0	0	0	580	0	0	0	0	0	0	0	210	2430	0
0	0	0	0	0	0	0	0	0	0	0	0	0	280	0	0
0	0	0	0	0	0	0	0	0	0	0	0	0	215	0	0
0	0	0	0	0	0	0	0	0	0	0	0	0	0	0	0
0	0	0	0	0	0	0	0	0	0	0	0	0	0	0	0
335	2800	0	1745	3130	0	0	150	700	0	715	0	0	75	260	0
360	585	0	0	0	0	0	1030	470	510	0	2630	0	0	575	740
0	450	0	0	0	720	0	1560	355	445	210	0	0	1150	155	0
0	165	0	1340	0	0	205	280	1440	0	105	0	1070	355	210	760
0	0	0	1170	0	0	0	0	310	0	620	130	0	90	225	0
0	0	695	0	0	0	200	610	650	0	1240	0	0	200	435	0
0	0	0	0	0	0	550	335	0	745	0	0	0	660	470	0
0	0	0	0	0	700	0	700	0	0	585	0	0	0	80	0
0	0	0	0	585	0	0	0	0	0	690	0	0	530	400	0
0	0	0	0	0	0	0	0	0	735	0	0	0	0	95	0

(Continued)

Table 6.18 *Continued*

	1	2	3	4	5	6	7	8	9	10	11	12	13	14
30	0	0	0	0	0	565	0	0	0	0	0	0	0	0
31	55	0	1365	0	0	25	0	0	0	0	20	0	0	80
32	0	0	130	65	450	0	0	40	0	0	0	0	0	0
33	5	0	0	0	330	0	0	0	45	0	0	0	0	90
34	0	490	60	15	20	0	0	0	0	0	0	40	0	135
35	265	0	0	610	0	0	0	0	0	20	0	105	0	0
36	0	105	0	0	940	0	0	0	0	0	0	35	0	105
37	0	45	0	0	1735	0	0	45	0	0	0	0	0	0
38	160	0	0	70	0	0	0	0	0	0	0	145	0	35
39	100	0	0	0	0	0	0	0	0	0	0	0	0	0
40	0	0	0	0	0	0	0	0	0	0	0	0	0	0
41	25	0	0	0	0	0	0	0	0	0	0	0	0	0
42	0	0	0	0	0	0	0	0	0	0	0	0	0	0
43	0	0	0	0	0	0	0	0	0	0	0	0	0	0
44	0	0	0	0	0	0	0	0	0	0	0	0	0	0
45	0	0	0	0	0	0	0	0	0	0	0	0	0	260
46	195	0	0	0	0	0	0	0	0	0	0	0	0	0
47	0	0	0	0	0	0	0	0	0	0	0	0	0	0
48	0	590	0	525	0	0	0	2405	915	765	0	0	955	0
49	1495	0	0	0	0	0	0	1020	0	0	1160	0	185	955
50	340	0	0	0	765	0	60	0	1000	925	0	0	840	1580
51	705	0	0	0	0	0	1065	0	0	970	0	525	745	0
52	0	0	0	0	0	850	695	0	0	0	0	780	590	1335
53	0	0	125	0	0	425	10	50	0	0	0	90	105	0
54	0	0	0	0	0	0	0	0	65	0	0	35	1045	490
55	0	0	0	0	0	0	0	0	0	130	30	675	780	0
56	0	0	0	0	0	0	985	0	0	445	0	0	0	0
57	0	0	0	0	0	335	0	0	0	0	0	0	0	0
58	0	0	0	365	0	445	0	0	760	755	475	0	150	0
59	0	0	0	1040	0	0	880	675	1615	0	270	215	145	215
60	0	0	650	0	0	15	680	320	0	670	0	170	130	0
61	0	0	0	0	590	0	1070	0	0	150	265	120	315	0
62	0	0	0	0	370	1035	0	105	780	100	550	190	0	0
63	0	0	0	0	0	80	950	295	405	90	0	185	270	0
64	0	0	0	0	0	720	0	535	370	355	0	280	0	0
65	0	0	0	0	0	300	560	0	0	0	5	30	0	0
66	605	0	0	0	270	0	50	0	0	0	0	0	0	0
67	0	0	0	0	155	0	0	0	0	0	70	0	0	0
68	0	0	0	0	0	0	0	0	0	0	0	0	0	0
69	35	1320	980	0	0	0	0	10	135	160	0	10	105	0

5	16	17	18	19	20	21	22	23	24	25	26	27	28	29	30
0	0	0	0	0	0	0	0	0	0	0	0	0	0	0	0
20	0	0	0	0	0	110	0	0	1220	365	0	0	680	0	0
30	0	0	210	0	0	775	0	0	175	390	0	140	670	0	0
0	0	0	60	0	0	145	0	0	645	0	500	90	1360	0	0
0	0	0	110	0	0	220	130	0	0	0	0	620	1230	0	0
0	0	0	0	0	0	0	0	195	970	0	305	0	0	0	0
25	0	0	0	0	0	0	0	0	0	615	0	680	230	0	0
0	0	0	205	0	0	0	0	0	580	360	625	0	0	0	0
0	0	0	0	0	0	0	0	0	0	0	545	610	0	0	0
5	0	0	0	35	0	0	0	0	165	0	400	0	0	0	0
0	0	0	0	0	0	0	0	0	0	440	870	0	0	0	0
0	0	0	0	0	0	0	0	0	0	0	0	0	0	0	0
20	0	0	0	0	0	0	0	0	0	0	0	0	0	0	0
0	0	370	0	0	0	0	0	0	0	0	0	0	0	0	0
0	0	240	0	0	0	0	0	0	0	0	0	0	0	0	0
0	0	0	0	0	0	0	0	0	0	0	0	0	0	0	0
0	0	0	0	0	0	0	0	0	0	0	0	0	0	0	0
0	0	210	0	0	0	0	0	0	0	0	0	0	0	0	0
360	5200	1030	1145	0	1775	975	0	0	520	5130	390	2930	2150	1130	260
0	0	2980	0	2710	700	1410	0	155	0	925	0	0	0	1195	1525
0	840	0	1075	890	1050	0	2790	945	2175	0	0	0	2355	0	2250
55	0	1150	0	0	0	0	3115	695	480	0	895	0	0	1090	625
0	0	0	730	1135	0	0	0	640	475	0	670	0	0	1060	0
0	860	0	0	0	770	0	795	0	0	0	0	0	0	640	0
385	0	0	0	285	0	0	0	1070	390	0	790	0	0	130	0
0	0	355	0	0	580	0	0	465	0	0	0	0	0	0	0
0	0	0	0	0	0	0	0	0	0	0	0	0	0	0	0
0	0	0	0	0	0	0	0	0	0	0	865	0	0	0	0
0	0	0	0	300	0	705	0	0	0	0	0	390	0	695	215
20	0	0	0	0	0	0	0	0	0	0	0	0	0	0	150
0	0	0	0	0	1080	0	510	0	650	0	0	0	0	0	0
0	0	0	0	0	0	0	0	0	0	0	0	0	0	0	0
0	0	0	0	0	0	0	0	0	0	0	0	0	0	0	405
0	0	0	0	365	0	0	0	0	0	0	0	240	0	0	0
0	0	0	0	0	0	0	0	0	0	0	0	0	0	0	0
0	0	0	0	0	0	0	0	0	0	265	0	0	0	0	320
320	0	0	0	0	0	0	0	0	0	0	420	0	50	0	0
0	0	0	0	0	0	0	0	0	0	0	0	0	0	0	0
0	0	0	0	0	0	0	0	0	0	0	0	275	170	0	0
0	0	0	530	0	835	0	0	750	0	0	0	0	0	0	0

(Continued)

Table 6.18 *Continued*

	1	2	3	4	5	6	7	8	9	10	11	12	13	14
70	0	1700	980	0	0	160	0	805	80	0	0	190	10	0
71	0	750	190	0	0	0	470	270	410	0	5	240	350	0
72	0	710	640	0	0	0	0	965	0	315	0	0	0	0
73	0	255	0	0	0	390	0	0	0	0	0	0	0	0
74	0	165	0	0	0	0	0	0	0	0	0	285	0	0
75	0	515	0	0	0	0	0	0	0	0	10	0	0	110
76	0	25	0	0	0	0	0	0	0	0	0	0	0	0
77	0	845	0	40	0	0	0	0	0	0	0	0	0	0
78	0	0	0	0	0	0	0	0	0	0	0	0	0	0
79	0	0	0	0	0	0	0	0	0	0	0	0	0	0
80	0	0	0	0	900	500	0	0	0	0	0	0	0	0
81	0	0	0	365	975	85	0	0	0	0	0	0	0	0
82	140	0	0	0	2405	0	0	0	0	0	0	0	0	0
83	95	0	0	0	0	0	0	0	0	0	5	0	0	0
84	505	0	0	0	0	0	0	0	0	0	0	0	0	0
85	0	0	0	0	195	0	0	0	0	0	0	0	0	0
86	0	0	0	0	0	430	0	60	0	0	0	0	0	0
87	0	0	0	0	0	0	0	0	0	0	0	0	0	0
88	0	0	0	0	0	0	0	0	0	0	0	0	0	0
89	145	165	0	0	0	0	20	0	0	0	0	0	0	0
90	120	355	175	0	0	0	80	0	0	0	0	0	60	0
91	0	440	235	0	0	340	0	35	0	0	50	0	0	0
92	110	420	0	0	0	0	0	20	0	0	50	0	250	0
93	0	340	180	0	0	0	0	220	0	385	75	0	0	110
94	245	320	0	0	0	390	10	40	0	0	0	0	0	0
95	420	350	0	0	10	0	0	0	0	0	0	25	0	55
96	185	0	0	0	0	0	0	0	0	0	0	0	0	0
97	50	0	0	0	0	0	0	0	0	0	0	0	0	0
98	350	0	40	0	0	0	0	0	0	0	0	0	0	0
99	60	0	0	0	0	0	0	0	0	0	0	0	0	0
100	0	0	0	2110	0	270	1150	1125	0	2465	0	2400	900	0

6.6 COMMENTS

The performance of aggregate production scheduling in customer-driven supply chains is evaluated by customer satisfaction and production costs (e.g., Silver et al., 1998). A typical measure of customer satisfaction is the customer service level, that is, the fraction of customer orders filled on or before their due dates, and a typical customer due date-related performance measure is the minimization of the number of tardy orders. For example, integer goal programming formulation for production

5	16	17	18	19	20	21	22	23	24	25	26	27	28	29	30
55	0	0	0	335	0	0	0	1235	0	0	0	440	0	200	0
0	0	0	0	0	1340	0	0	0	0	0	0	315	0	0	0
0	0	0	0	0	0	0	0	0	325	0	0	0	175	80	55
0	0	0	0	0	475	0	985	0	0	0	0	0	145	0	205
0	0	0	0	0	0	140	0	140	0	0	0	0	0	0	0
0	0	0	0	0	0	0	575	0	0	0	0	0	0	0	130
0	0	0	0	0	0	0	460	0	0	0	0	0	0	0	195
0	0	0	0	0	0	0	0	0	0	0	0	0	0	0	0
0	0	0	0	0	0	0	0	0	0	0	165	0	0	0	0
0	0	0	0	0	0	0	0	0	0	0	325	0	0	0	0
20	0	0	0	0	130	0	0	265	170	735	0	0	0	640	325
0	0	0	0	385	0	0	0	675	0	0	1270	0	0	770	300
0	0	0	0	0	0	230	0	1805	0	0	0	630	0	285	290
00	0	0	0	270	0	0	0	0	0	465	0	0	0	1160	280
0	0	0	0	0	0	580	0	140	0	0	0	210	0	0	265
0	0	0	0	0	160	0	0	0	0	0	0	705	0	0	120
0	0	0	0	0	0	0	0	0	60	0	0	605	0	0	0
0	0	0	0	0	0	0	0	0	105	0	630	0	0	0	310
0	0	0	0	0	0	0	0	0	45	0	0	630	0	0	290
55	0	0	1390	0	705	0	335	0	165	0	370	50	0	70	1665
15	0	0	0	0	130	205	370	290	400	0	230	0	120	0	0
0	0	0	390	0	230	110	0	1055	0	0	380	0	250	0	0
0	0	445	0	0	220	0	140	1025	645	0	435	0	0	0	0
0	0	0	0	0	0	0	310	0	680	0	0	25	0	0	0
0	0	0	260	0	260	0	0	0	0	0	260	0	0	0	0
0	0	0	0	0	0	0	0	0	245	490	0	0	0	0	0
0	0	0	410	0	0	0	0	0	0	0	80	0	0	15	0
405	0	0	0	0	0	0	0	0	0	0	0	0	0	35	0
0	0	0	0	0	0	0	0	0	0	0	0	0	0	75	0
0	0	0	0	0	0	0	0	0	0	0	0	0	0	0	0
55	0	0	380	5200	0	705	1795	180	370	35	3660	0	610	2800	165

scheduling with a due date-related criterion in make-to-order manufacturing is presented by Markland et al. (1990). A hierarchical approach and mixed integer programs for production scheduling in a make-to-order company are presented by Carravilla and Pinho de Sousa (1995). MIP formulations for aggregate production scheduling with various due date-related criteria are given in Shapiro (1993, 2001) and Sawik (2005a). A comprehensive description of MIP modeling and optimization approaches for solving different medium-term production planning problems, including examples of industrial applications, can be found in Pochet and Wolsey (2006).

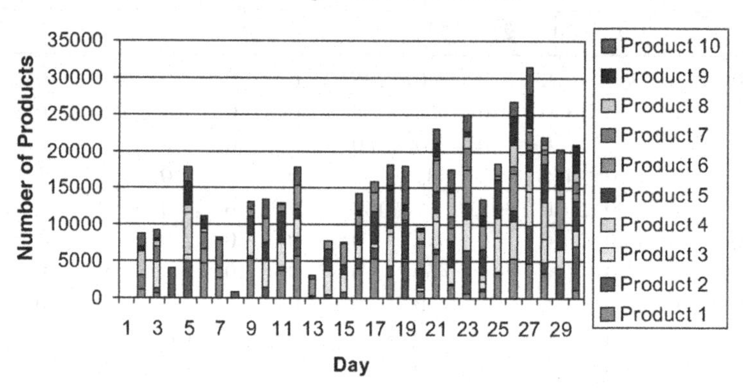

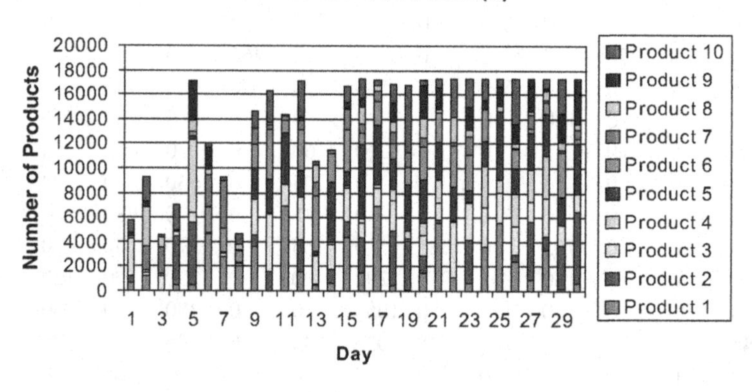

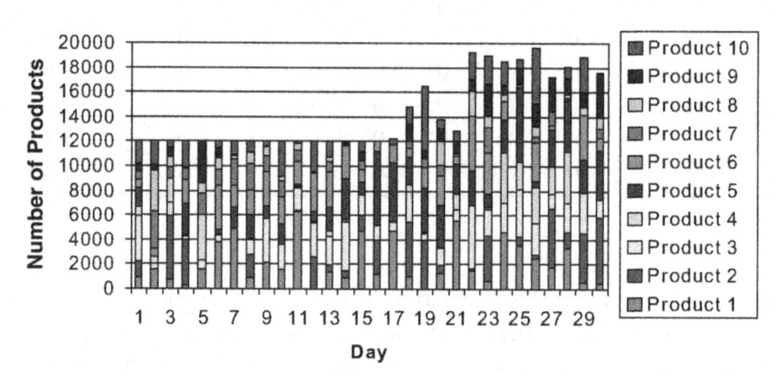

Figure 6.9 Demand pattern and production schedule: (a) min $P_{\max}$ criterion, (b) max $P_{\min}$ criterion.

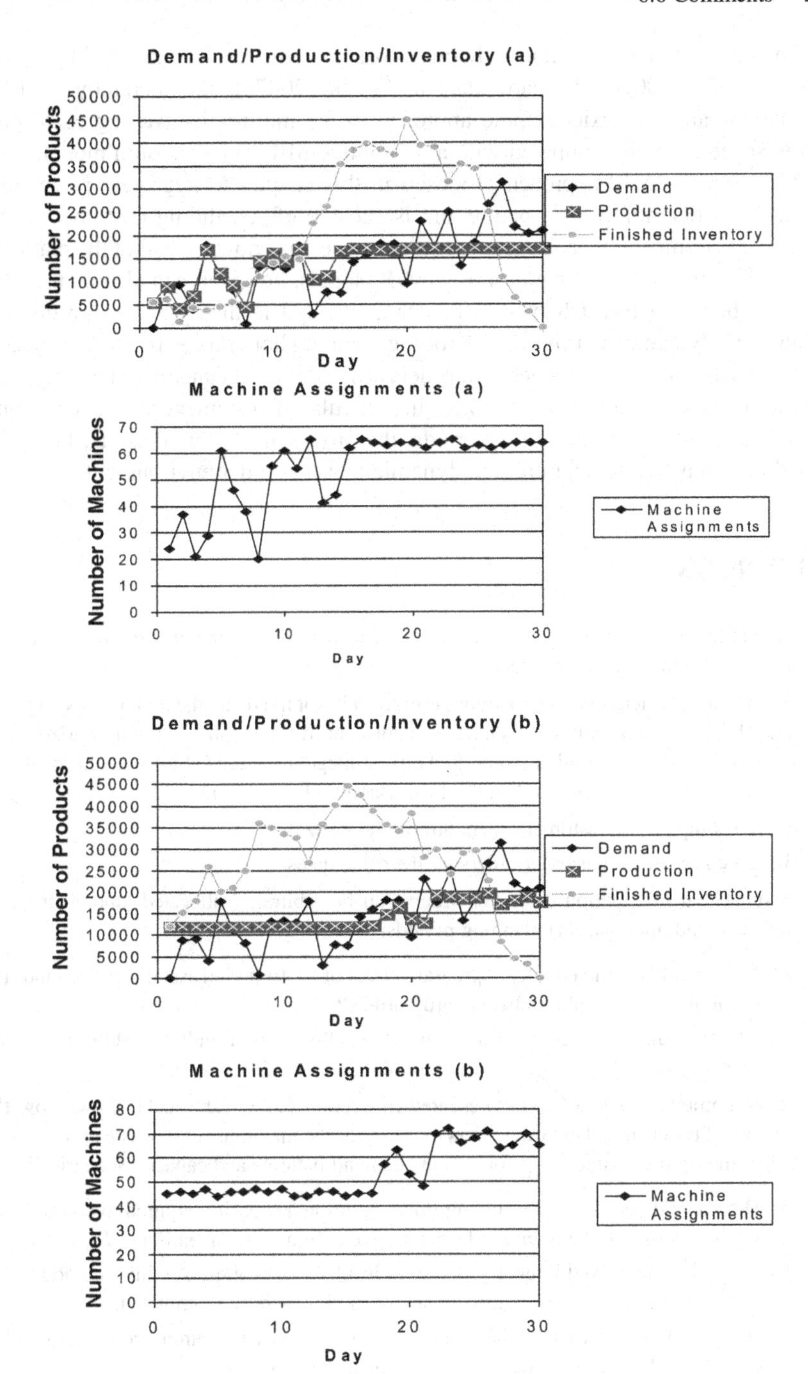

Figure 6.10 Production, inventory, and machine assignments: (a) min $P_{\max}$ criterion, (b) max $P_{\min}$ criterion.

The material presented in this chapter is based on the results published by Sawik (2007a, 2007b, 2007c). In particular, in Sawik (2007c), the hierarchy of MIP formulations and the lexicographic approach to the multi-objective aggregate production scheduling are compared with a monolithic MIP weighted-sum model.

Note that the MIP approach proposed in this chapter to aggregate production scheduling is deterministic in nature and is capable of scheduling customer orders in a static environment, where a set of customer orders is known ahead of the planning horizon. The approach and the proposed MIP models, however, can also be used for reactive scheduling (see Chapter 7) to iteratively update the aggregate production schedule in a dynamic environment. A rolling planning horizon can be used to successively solve the MIP models when new orders arrive or old, yet uncompleted orders are cancelled or modified during the horizon. In particular, if customer orders can be completed in at most two consecutive periods, the proposed MIP models can be easily applied for a reactive scheduling in a dynamic make-to-order environment.

EXERCISES

6.1 Modify the mixed integer program *OA* of order-to-period assignment to minimize the total number of tardy and early orders.

6.2 Assume that the length of the planning horizon h is not fixed and the total processing time available on each machine in each stage is constant over the entire planning horizon, that is, $c_{it} = c_i \forall t \in T$. Formulate a modified mixed integer program *OA* for scheduling of customer orders to minimize the length of the planning horizon and

(a) To complete all customer orders during the horizon.

(b) To complete all customer orders by their due dates.

6.3 Assume that multiperiod customer orders can be arbitrarily allocated among different (adjacent and not adjacent) planning periods.

(a) How should the mixed integer program *OA* of order-to-period assignment be modified to minimize the total number of tardy orders?

(b) What are the implications of noncontiguous allocation of multiperiod orders among planning periods on the modified formulations from Exercise 6.2?

6.4 In the computational examples presented in Section 6.4.3, Table 6.11 shows how the number of unscheduled or tardy orders increases as the maximum earliness $E_{\max}$ decreases below the optimal value $E^*_{\max}$, for various demand patterns and capacity scenario II.

(a) How would you determine the minimum additional capacity required to complete all the orders during the planning horizon, given the maximum earliness $E_{\max} < E^*_{\max}$?

(b) Formulate the mixed integer program to level the allocation of minimum additional capacity over the planning horizon such that all orders are completed.

(c) How would you determine the minimum length H of each planning period required to complete all customer orders during the planning horizon?

6.5 In the computational examples presented in Section 6.5.2, Table 6.12 indicates that for capacity scenario II and all demand patterns, except for the increasing one, there is not sufficient capacity to complete all customer orders by their due dates.

(**a**) How would you determine the minimum additional capacity required to complete all customer orders with no tardiness?

(**b**) Formulate the mixed integer program to level the allocation of minimum additional capacity over the planning horizon such that all customer orders are completed by their due dates.

(**c**) How would you determine the minimum length H of each planning period required to complete all customer orders with no tardiness?

Chapter 7

Reactive Aggregate Production Scheduling in Make-to-Order Manufacturing

7.1 INTRODUCTION

A customer-driven supply chain environment is dynamic in nature and the customers may modify, cancel, or add new orders during the planning horizon. As a result a predetermined aggregate production schedule may become inefficient and may need to be revised in reaction to unexpected changes of customer orders. A reactive scheduling algorithm should take into account the currently in progress production schedule and the remaining unmodified customer orders. The purpose of this chapter is to present MIP-based algorithms for reactive aggregate production scheduling in a dynamic, make-to-order manufacturing environment, where customers may modify or cancel their orders or place new orders during the planning horizon. The objective of the rescheduling algorithms is to update the current production schedule whenever the customer orders are modified or new orders arrive, subject to service level and inventory constraints. In the algorithms, different rescheduling policies are proposed, from a total reschedule of all remaining and unmodified customer orders to a nonreschedule of all such orders. In addition, a medium restrictive policy is considered for rescheduling only a subset of remaining customers orders awaiting for material supplies.

The following MIP models and MIP-based rescheduling algorithms are presented in this chapter:

OAS model for assignment of customer orders to time periods to maximize service level

OAS|E model for assignment of customer orders to time periods to maximize service level, subject to maximum earliness constraints

OAE|S model for assignment of customer orders to time periods to minimize maximum earliness, subject to service level constraints

Scheduling in Supply Chains Using Mixed Integer Programming. By Tadeusz Sawik

REALL reactive algorithm with rescheduling of all unmodified customer orders

REMAT reactive algorithm with rescheduling of all remaining customer order waiting for material supplies

RENON reactive algorithm with nonrescheduling of all unmodified customer orders.

Numerical examples modeled after a real-world scheduling/rescheduling of customer orders in electronics manufacturing are presented and some results of computational experiments are reported in Section 7.6.

7.2 PROBLEM DESCRIPTION

The production system under study is a flexible flow shop that consists of m processing stages in series and each stage $i \in I = \{1, \ldots, m\}$ is made up of $m_i \geq 1$ parallel identical machines. In the system various types of products are produced in a make-to-order environment responding directly to customer-requested orders. Let J be the set of customer orders that are known ahead of a planning horizon. Each order $j \in J$ is described by a triple (a_j, d_j, n_j), where a_j is the order arrival date (e.g., the earliest period of material availability), d_j is the customer-requested due date (e.g., customer-required shipping date), and n_j is the size of order (the quantity of ordered products of specified type). Each order requires processing in various processing stages; however, some orders may bypass some stages. Let $p_{ij} \geq 0$ be the processing time in stage i of each product in order $j \in J$. The orders are processed and transferred among the stages in lots of various size that depends on the ordered product type and let b_j be the size of production lot for order j.

The planning horizon consists of h planning periods (e.g., working days). Let $T = \{1, \ldots, h\}$ be the set of planning periods and c_{it} the processing time available in period t on each machine in stage i.

The following two types of customer orders are considered:

1. Single-period orders, where each order can be fully processed in a single time period, for example, during one day. The single-period orders are referred to as indivisible orders.
2. Two-period orders, where each order must be completed in two consecutive time periods, for example, during two days. The two-period orders are referred to as divisible orders.

In practice, the two types of customer orders are scheduled simultaneously. Denote by $J1 \subseteq J$, and $J2 \subseteq J$, respectively, the subset of single-period, and two-period orders. The approach proposed in this chapter, however, can be easily enhanced to allow for the inclusion of large orders that require more than two consecutive planning periods to complete.

A dynamic, make-to-order manufacturing environment is considered with a dynamic planning horizon used to successively update a production schedule when old, yet uncompleted customer orders are modified or new customer orders arrive during the horizon. The modifications of customer orders may include changes of order size, for example, increase, decrease, or cancellation, and/or changes of due dates, for example, postponement of delivery date, occurring during the planning horizon. The horizon can be progressively shifted to take into account modifications of the customer orders.

The objective of the medium-term aggregate production rescheduling is to assign/reassign customer orders to planning periods over a planning horizon to maximize the customer service level with limited input and output inventory.

7.3 MIXED INTEGER PROGRAMS FOR REACTIVE SCHEDULING

In this section MIP formulations are proposed for assignment/reassignment of customer orders over a medium-term planning horizon, to maximize service level, subject to the inventory constraints (for notation used, see Table 7.1).

Table 7.1 Notation: Initial Scheduling

Indices	
i	$=$ processing stage, $i \in I = \{1, \ldots, m\}$
j	$=$ customer order, $j \in J = \{1, \ldots, n\}$
l	$=$ product type, $l \in L$
t	$=$ planning period, $t \in T = \{1, \ldots, h\}$
Input parameters	
a_j, d_j, n_j	$=$ arrival date, due date, size of order j
b_j	$=$ production lot size for order j
c_{it}	$=$ processing time available in period t on each machine in stage i
m_i	$=$ number of parallel, identical machines in stage i
n	$=$ number of customer orders to be scheduled
p_{ij}	$=$ processing time in stage i of each product in order j
$J1 \subset J$	$=$ subset of single-period customer orders
$J2 \subset J$	$=$ subset of two-period customer orders
$J_l \subset J$	$=$ subset of customer orders for product type l
$\overline{E}$	$=$ upper limit on maximum earliness
$\overline{U}$	$=$ upper limit on number of tardy orders
Decision variables	
u_j	$=$ 1, if order j is completed after due date; otherwise $u_j = 0$ (unit penalty for tardy orders)
x_{jt}	$=$ 1, if order j is performed in period t; otherwise $x_{jt} = 0$ (order-to-period assignment variable)
$y_{jt} \in [0, 1]$	$=$ fraction of customer order j to be processed in period t (order allocation variable)
Objective functions	
$E_{\max}$	$=$ maximum earliness of orders
U_{sum}	$=$ number of tardy orders

7.3.1 Basic Model

The basic MIP model used to update the current production schedule whenever some customer orders are modified is presented below. The extent of required changes in the current schedule depends on the applied policy (see Section 7.4) and the changes in sizes and due dates of the modified customer orders. The updated schedule takes into account the current input inventory that is implicitly considered in the model by the upper bound $\overline{E}$ on the maximum earliness $E_{\max}$ of customer orders.

Model OAS|E: *Order to Period Assignment to Maximize Service Level, Subject to Maximum Earliness of Orders*

Maximize

$$1 - U_{\text{sum}}/n \quad (7.1)$$

or

Minimize

$$U_{\text{sum}} = \sum_{j \in J} u_j \quad (7.2)$$

subject to

1. *Order-to-Period Assignment Constraints*:

– each indivisible customer order is assigned to exactly one planning period,

$$\sum_{t \in T: t \geq a_j} x_{jt} = 1; \; j \in J1 \quad (7.3)$$

– each divisible customer order is assigned to at most two consecutive planning periods,

$$x_{jt} + x_{jt+1} \leq 2; \; j \in J2, t \in T\colon a_j \leq t \leq h-1 \quad (7.4)$$

$$x_{jt} + x_{jt'} \leq 1; \; j \in J2, t \in T, t' \in T\colon a_j \leq t \leq h-2, \; t' \geq t+2 \quad (7.5)$$

2. *Order Allocation Constraints*:

– each order must be completed,

$$\sum_{t \in T: t \geq a_j} y_{jt} = 1; \; j \in J \quad (7.6)$$

– each indivisible order is completed in a single period,

$$x_{jt} = y_{jt}; \; j \in J1, \; t \in T\colon t \geq a_j \quad (7.7)$$

– each divisible order is allocated among all the periods that are selected for its assignment,

$$x_{jt} \geq y_{jt}; \; j \in J2, \; t \in T\colon t \geq a_j \quad (7.8)$$

– the minimum portion of a divisible order allotted to one period is not less than the batch size,

$$y_{jt} \geq b_j x_{jt}/n_j;\ j \in J2,\ t \in T: t \geq a_j \tag{7.9}$$

3. *Tardy Orders Constraints*:
   – an indivisible tardy order is assigned after its due date,

$$u_j = \sum_{t \in T: t > d_j} x_{jt};\ j \in J1 \tag{7.10}$$

   – a divisible tardy order is partly (or fully) assigned after its due date,

$$u_j \geq \sum_{t \in T: t > d_j} y_{jt};\ j \in J2 \tag{7.11}$$

$$u_j \leq \sum_{t \in T: t > d_j} x_{jt};\ j \in J2 \tag{7.12}$$

4. *Capacity Constraints*:
   – in every period the demand on capacity at each processing stage cannot be greater than the maximum available capacity in this period,

$$\sum_{j \in J_i} n_j p_{ij} y_{jt} \leq c_{it} m_i;\ i \in I, t \in T \tag{7.13}$$

5. *Maximum Earliness Constraints*:
   – for each early order j assigned to period $t < d_j$, its earliness $(d_j - t)$ cannot exceed the maximum earliness $\overline{E}$

$$(d_j - t)x_{jt} \leq \overline{E};\ j \in J, t \in T: t \geq a_j \tag{7.14}$$

6. *Variable Nonnegativity and Integrality Conditions*:

$$u_j \in \{0, 1\};\ j \in J \tag{7.15}$$

$$x_{jt} \in \{0, 1\};\ j \in J, t \in T: t \geq a_j \tag{7.16}$$

$$y_{jt} \in [0, 1];\ j \in J, t \in T: t \geq a_j. \tag{7.17}$$

The objective function represents customer service level, that is, the fraction of nondelayed customer orders to be maximized (7.1) or equivalently the number of tardy orders to be minimized (7.2). The solution to MIP model $OAS|E$ determines the assignment of indivisible customer orders to single planning periods and the allocation of divisible orders among the pairs of consecutive planning periods such that the customer service level is maximized, subject to the limited maximum earliness of orders and implicitly, to the limited total input and output inventory level.

Model $OAS|E$ can be briefly rewritten as below.

$$\textbf{Model OAS|E:} \quad \max\{(7.1) : (7.2) - (7.17)\}. \tag{7.18}$$

7.3.2 Models for Hierarchical Initial Scheduling

In this subsection a hierarchical scheduling framework and two MIP models are presented to sequentially determine the beginning aggregate production schedule for the original customer orders known ahead of the planning horizon. The two MIP models are shown below.

Model OAS: *Order to Period Assignment to Maximize Service Level*

$$\textbf{Model OAS:} \quad \max\{(7.1) : (7.2) - (7.13), (7.15) - (7.17)\}, \tag{7.19}$$

where all materials are assumed to be available at the beginning of the planning horizon, that is, $a_j = 1$ for each order $j \in J$.

Model OAE|S: *Order to Period Assignment to Minimize Maximum Earliness, Subject to Service Level*

$$\begin{aligned} \textbf{Model OAE|S:} \quad & \min\{E_{\max}: (7.2) - (7.13), (7.15) - (7.17), \\ & U_{\text{sum}} \le \overline{U},\ (d_j - t)x_{jt} \le E_{\max};\ j \in J, t \in T : t \ge a_j, \\ & E_{\max} \ge 0\}, \end{aligned} \tag{7.20}$$

where $1 - \overline{U}/n$ is the solution value of (7.19).

The objective function of (7.20) implicitly limits the maximum level of the total input and output inventory over the planning horizon.

It is shown in Section 6.4.1 that for a given number of tardy orders (i.e., the required service level), the total inventory increases with $E_{\max}$, that is, both the input inventory of product-specific materials and the output inventory of finished products can be reduced when ready periods and due dates of customer orders are closer. Therefore, the maximum level of the total input and output inventory can be implicitly reduced by minimizing the maximum earliness $E_{\max}$ of early orders, given the minimum number U_{sum} of tardy orders.

In the above sequential decision-making framework, first the solution to *OAS* determines the maximum service level. Then the minimum value $E^*_{\max}$ of the maximum earliness $E_{\max}$ is found using model $OAE|S$ to implicitly limit total inventory, subject to service level constraints. The solution to $OAE|S$ determines the optimal allocation $\{x^*_{jt}, y^*_{jt}\}$ of customer orders among planning periods.

7.4 RESCHEDULING ALGORITHMS

In this section three different rescheduling algorithms are presented based on the proposed MIP models. Let t_{mod} be the first planning period immediately after a modification of customer orders. It is assumed that the customer orders completed before

t_{mod} or with due dates smaller than t_{mod} cannot be modified. In practice different rescheduling policies can be applied, from a total reschedule of all remaining customer orders, that is, reschedule of all unmodified orders that have been assigned to periods not less than t_{mod} (algorithm REALL) to a nonreschedule of all such orders (algorithm RENON). In addition to the above two extreme rescheduling policies, a medium restrictive algorithm REMAT is proposed for rescheduling of the remaining customer orders waiting for material supplies. For each order j, product-specific materials are assumed to be unavailable earlier than E^*_{max} periods ahead of the order due date d_j. Therefore, each order j cannot be assigned to periods earlier than $d_j - E^*_{\text{max}}$. In particular, in period t_{mod} product-specific materials are not available for orders due in periods greater than $t_{\text{mod}} + E^*_{\text{max}}$, and hence all such orders can be rescheduled. On the other hand, the unmodified orders with product-specific materials supplied by period t_{mod} (i.e., orders assigned to periods $t = t_{\text{mod}}, \ldots, t_{\text{mod}} + E^*_{\text{max}}$) are considered nonreschedulable in algorithm REMAT.

In all these algorithms the planning horizon is progressively shifted to take into account modifications of the customer orders (changes of order size and/or due date) occurring during the horizon. Table 7.2 presents the notation used in the rescheduling algorithms.

In all algorithms, first the set J_{old} of customer orders remaining for completion without modification is split into two disjoint subsets: J^S_{old} of reschedulable orders and J^N_{old} of fixed, nonreschedulable orders for which the assignment to planning periods cannot be changed. For example, in algorithm REMAT, the subset of nonreschedulable orders contains customer orders in J_{old} remaining for completion, such that have been assigned to periods in the subset $T^N_{\text{old}} = \{t_{\text{mod}}, \ldots, t_{\text{mod}} + E^*_{\text{max}}\}$ of remaining periods in $T_{\text{old}} = \{t_{\text{mod}}, \ldots, h\}$.

In what follows, let us denote by apostrophe (') the updated values of some parameters and decision variables after each modification of customer orders. For

Table 7.2 Notation: Rescheduling

Input parameters	
h'	= new planning horizon
t_{mod}	= the planning period immediately following modification of orders
J_{mod}	= set of modified orders
J_{old}	= subset of orders in J remaining for completion without modification
$J^N_{\text{old}}, J^S_{\text{old}}$	= subset of orders in J_{old}, respectively nonreschedulable, reschedulable
$T_{\text{new}} = \{h+1, \ldots, h'\}$	= set of new planning periods
$T_{\text{old}} = \{t_{\text{mod}}, \ldots, h\}$	= subset of remaining planning periods in T
T^N_{old}	= subset of periods in T_{old} with fixed assignment of orders in J_{old}
$J' = J_{\text{mod}} \cup J_{\text{old}}$	= updated set of orders
$T' = T_{\text{old}} \cup T_{\text{new}}$	= updated set of planning periods

Apostrophe (') denotes updated parameters after modification of customer orders.

example n'_j denotes the modified size of customer order $j \in J' = J_{\text{old}} \cup J_{\text{mod}}$, where $n'_j = n_j,\ j \in J_{\text{old}}$ and $n'_j \neq n_j,\ j \in J_{\text{mod}}$.

Algorithm REALL

Step 0. Split the set J_{old} of orders remaining for completion into two disjoint subsets: J^S_{old} of reschedulable orders and J^N_{old} of fixed, nonreschedulable orders.

$$J^N_{\text{old}} = \emptyset \tag{7.21}$$

$$J^S_{\text{old}} = J_{\text{old}} \tag{7.22}$$

Step 1. Determine the new planning horizon h' for the updated set J' of customer orders.

$$h_1 = \min\left\{h1\colon \max_{i \in I}\left(\frac{\sum_{j \in J'} n'_j p_{ij}}{m_i \sum_{t=t_{\text{mod}}}^{h1} c_{it}}\right) \leq 1\right\} \tag{7.23}$$

$$h_2 = \max_{j \in J'}(d_j) \tag{7.24}$$

If $\max\{h_1, h_2\} \leq h$, then set $h' = h$.
Otherwise set

$$h' = \max\{h_1, h_2\},\ T_{\text{new}} = \{h+1, \ldots, h'\} \quad \text{and} \quad T' = T_{\text{old}} \cup T_{\text{new}}.$$

Step 2. Do not change the assignment in period t_{mod} of partially completed, two-period orders in $J2$, that is,

$$y'_{j,t_{\text{mod}}} = y_{j,t_{\text{mod}}},\ j \in J2\colon x_{j,t_{\text{mod}}-1} = 1 \tag{7.25}$$

Step 3. Solve $OAS|E$, (7.18) for $\overline{E} = E^*_{\max}$ and subject to fixed assignment constraints from Step 2, to find a new schedule for the updated set J' of customer orders, updated set of planning periods T', and updated material availability periods

$$a'_j = \begin{cases} \max\{1, d_j - E^*_{\max}\} & \text{if } j \in J_{\text{old}} \\ \max\{t_{\text{mod}}, d_j - E^*_{\max}\} & \text{if } j \in J_{\text{mod}}. \end{cases} \tag{7.26}$$

In the algorithms REMAT and RENON presented below, Step 1 and Step 3 are identical with the corresponding steps of REALL.

Algorithm REMAT

Step 0. Split the set J_{old} of orders remaining for completion into two disjoint subsets: J_{old}^S of reschedulable orders and J_{old}^N of fixed, nonreschedulable orders.

$$J_{\text{old}}^N = \left\{ j \in J_{\text{old}} : \sum_{t_{\text{mod}} \le t \le t_{\text{mod}} + E_{\max}^*} x_{jt} = 1 \right\} \tag{7.27}$$

$$J_{\text{old}}^S = J_{\text{old}} \backslash J_{\text{old}}^N \tag{7.28}$$

Set $T_{\text{old}}^N = \{t_{\text{mod}}, \ldots, t_{\text{mod}} + E_{\max}^*\}$.

Step 2. Do not change the assignment of nonreschedulable orders $j \in J_{\text{old}}^N$, that is,

$$y'_{jt} = y_{jt},\ j \in J_{\text{old}}^N,\ t \in T_{\text{old}}^N \tag{7.29}$$

$$y'_{j,t_{\text{mod}}+E_{\max}^*+1} = y_{j,t_{\text{mod}}+E_{\max}^*+1},$$
$$j \in J_{\text{old}}^N \cap J2 : x_{j,t_{\text{mod}}+E_{\max}^*} = 1. \tag{7.30}$$

Algorithm RENON

Step 0. Split the set J_{old} of orders remaining for completion into two disjoint subsets: J_{old}^S of reschedulable orders and J_{old}^N of fixed, nonreschedulable orders.

$$J_{\text{old}}^N = J_{\text{old}} \tag{7.31}$$

$$J_{\text{old}}^S = \emptyset \tag{7.32}$$

Step 2. Do not change the assignment of all orders in J_{old}, that is,

$$y'_{jt} = y_{jt},\ j \in J_{\text{old}},\ t \in T_{\text{old}}. \tag{7.33}$$

The above MIP-based rescheduling algorithms differ in the extent of imposed schedule changes. In practice, the selection of a particular algorithm may depend on the impact of disruptions induced by rescheduling policy on the predetermined production schedule, currently in progress.

7.5 INPUT AND OUTPUT INVENTORY

In this section some formulae are derived for calculating the input inventory of raw materials and the output inventory of finished products. The input inventory of product-specific raw materials only is considered with no common materials for different product types taken into account. Furthermore, to make the calculations clearer it is assumed that each product requires one unit of the corresponding product-specific material (e.g., one printed wiring board of a specific design per one electronic device of the corresponding type). As a result, for each order j the required quantity of product-specific material equals the quantity of the ordered products n_j.

The original amount of product-specific materials required for customer orders $j \in J_{\text{mod}}$ such that $d_j - E^*_{\max} < t_{\text{mod}} \leq d_j$ and supplied before t_{mod} differs from the modified amount of those materials required after the order modification. As a result, the actual input inventory level in period t_{mod} may be higher or lower than the required level. For each product type $l \in L$, the shortage ($\Delta INP_l < 0$) or surplus ($\Delta INP_l > 0$) of product-specific material inventory in period $t_{\text{mod}} - 1$ with respect to the amount required for the modified orders $j \in J_{\text{mod}}$ is

$$\Delta INP_l = \sum_{j \in J_l \cap J_{\text{mod}}: d_j - E^*_{\max} < t_{\text{mod}} \leq d_j} (n'_j - n_j); \ l \in L \tag{7.34}$$

It is assumed that the shortage or surplus of product-specific materials is balanced with higher or lower supplies in period t_{mod}, respectively.

The input inventory $INP(t)$ of product-specific materials can be calculated as below.

$$INP(t) = INP(t_{\text{mod}} - 1) + \sum_{j \in J': t_{\text{mod}} \leq a'_j \leq t} n'_j - \sum_{j \in J_{\text{mod}}: d_j - E^*_{\max} < t_{\text{mod}} \leq d_j} n_j$$
$$- \sum_{j \in J', \tau \in T': a'_j \leq \tau \leq t} n'_j y'_{j\tau}; \ t \in T': t \geq t_{\text{mod}}, \tag{7.35}$$

where $INP(t_{\text{mod}} - 1)$ is the input inventory remaining in period $t_{\text{mod}} - 1$.

In (7.35), the input inventory $INP(t)$ in each period t is calculated as the difference between the amount of product-specific materials supplied by period t and the amount of these materials processed into finished products by this period. The first summation term with negative sign in the right-hand side of (7.35) balances in period t_{mod} the shortage or the surplus of product-specific materials supplied by period $t_{\text{mod}} - 1$.

Similarly, the output inventory $OUP(t)$ of finished products can be expressed as below.

$$OUP(t) = \sum_{d>t} OUP_d(t_{\text{mod}} - 1) + \sum_{j \in J', \tau \in T': a'_j \leq \tau \leq t < d_j} n'_j y'_{j\tau}; \ t \in T': t \geq t_{\text{mod}}, \tag{7.36}$$

where $OUP_d(t_{\text{mod}} - 1)$ is the output inventory of finished products remaining in period $t_{\text{mod}} - 1$, due in period $d > t_{\text{mod}}$.

In (7.36), the output inventory $OUP(t)$ in each period t is calculated as the amount of finished products processed by period t before the customer-required shipping dates.

The total inventory $TOT(t) = INP(t) + OUP(t)$ in each period t can be found by summing the corresponding right-hand sides of (7.35) and (7.36). In particular, the total input and output inventory $TOT(t_{\text{mod}} - 1)$ in the last period of previous schedule can be expressed as below.

$$TOT(t_{\text{mod}} - 1) = \sum_{j \in J: d_j \le t_{\text{mod}} - 1} n_j \left(1 - \sum_{a_j \le \tau \le t_{\text{mod}} - 1} y_{j\tau} \right) + \sum_{j \in J: t_{\text{mod}} \le d_j \le t_{\text{mod}} - 1 + E^*_{\max}} n_j. \tag{7.37}$$

The first summation term in (7.37) is the inventory of product-specific materials for customer orders due by period $t_{\text{mod}} - 1$, and the second term is the inventory of product-specific materials and finished products of customer orders due after period $t_{\text{mod}} - 1$, respectively, waiting for processing in the system and for shipping to customers. The first term represents the input inventory in period $t_{\text{mod}} - 1$ of product-specific materials for tardy orders and is greater than zero only if some customer orders are tardy, otherwise this term is equal to zero. The second term increases with the maximum earliness $E^*_{\max}$. Given the tardy orders, the total inventory in $t_{\text{mod}} - 1$ increases with $E^*_{\max}$, that is, both the input inventory of product-specific materials and the output inventory of finished products can be reduced when ready periods and due dates of customer orders are closer (see Section 6.4.1).

7.6 COMPUTATIONAL EXAMPLES

In this section numerical examples and some computational results are presented to illustrate possible applications of the proposed algorithms for reactive scheduling, based on the proposed MIP formulations. The examples are modeled after a real-world distribution center for high-tech products, where finished products are assembled for shipping to customers. The distribution center can be modeled as a flexible flow shop made up of six processing stages with parallel identical machines in each stage (see Figs. 5.3 and 6.4). A brief description of the production system, production process, products, and customer orders is given in Section 6.4.3. The beginning demand is the demand with increasing pattern, defined in Section 6.4.3 for scenario II, that is, is made up of 100 customer orders, each consisting of several suborders (customer-required shipping volumes). The total number of suborders is 816, and the beginning total demand for all products is 537,995. The initial planning horizon consists of $h = 30$ days, each of length $H = 2 \times 9$ hours.

Notice that the suborders in the computational examples play the role of orders in the mathematical formulation. Now, the problem objective is to assign/reassign customer suborders over the planning horizon to minimize number of tardy suborders as a measure of the customer service level, subject to maximum earliness constraints to limit the total inventory level.

In the computational experiments the following three scenarios with three modifications of customer orders during the planning horizon are considered:

Scenario A with total demand increased by 101,110 products

- 13 customer orders due in periods 8 to 30 are modified in period $t_{mod} = 6$. The resulting total change of demand is +70,200 products.
- 13 customer orders due in periods 15 to 30 are modified in period $t_{mod} = 14$. The resulting total change of demand is +15,950 products.
- 8 customer orders due in periods 26 to 30 are modified in period $t_{mod} = 24$. The resulting total change of demand is +14,960 products.

Scenario B with total demand decreased by 9330 products

- 11 customer orders due in periods 7 to 30 are modified in period $t_{mod} = 6$. The resulting total change of demand is −32,395 products.
- 14 customer orders due in periods 15 to 30 are modified in period $t_{mod} = 14$. The resulting total change of demand is +29,880 products.
- 7 customer orders due in periods 26 to 30 are modified in period $t_{mod} = 24$. The resulting total change of demand is −6815 products.

Scenario C with total demand unchanged

- Modifications of customer orders in period $t_{mod} = 6$ are the same as in the scenario with decreasing demand.
- Modifications of customer orders in period $t_{mod} = 14$ are the same as in the scenario with decreasing demand.
- 3 customer orders due in periods 26 to 30 are modified in period $t_{mod} = 24$. The resulting total change of demand is +2515 products.

The computational experiments are summarized in Tables 7.3, 7.4, and 7.5, respectively, for scenario A with increasing total demand, B with decreasing total demand, and C with unchanged total demand. The tables present solution results and the characteristics of MIP models *OAS* and $OAE|S$ (with $\overline{U} = 0$) for the initial scheduling of the beginning demand, $OAS|E$ (with $\overline{E} = 6$) for the three rescheduling algorithms REALL, REMAT, and RENON, and for the ex post scheduling (determined after all modifications are known) of the updated demand. The size of each MIP model is represented by the total number of variables, Var., number of binary variables, Bin., number of constraints, Cons., and number of nonzero elements in the constraint matrix, Nonz. The last column of each table presents the optimal solution values of U_{sum} for *OAS*, E_{max} for $OAE|S$, and U_{sum}, h' for $OAS|E$.

Since scenarios B and C are identical for planning periods $t < t_{mod} = 24$, Table 7.5 presents solution results only for the remaining periods $t \geq t_{mod}$.

Tables 7.3 to 7.5 indicate that the best results (the minimum number of tardy orders over the planning horizon and the smallest horizon length) are obtained for algorithm REALL, where total reschedule of all remaining customer orders is applied each time some orders are modified. In contrast, algorithm RENON, where the assignment of all remaining orders is not changed, produces the worst results. On the other

Table 7.3 Computational Results for Scenario A

Model/t_{mod}	Var.	Bin.	Cons.	Nonz.	Solution values[a]
Initial scheduling for beginning demand					
OAS	29,310	14,656	18,198	133,276	$U^*_{sum} = 0$
OAE|S	31,507	15,753	33,057	148,946	$E^*_{max} = 6$
REALL					
$OAS\|E = 6/t_{mod} = 6$	22,522	11,565	19,779	316,229	$U^*_{sum} = 2$, $h' = 31$
$OAS\|E = 6/t_{mod} = 14$	15,558	8013	13,116	195,823	$U^*_{sum} = 3$, $h' = 32$
$OAS\|E = 6/t_{mod} = 24$	4059	2112	3928	40,811	$U^*_{sum} = 5$, $h' = 33$
REMAT					
$OAS\|E = 6/t_{mod} = 6$	15,817	8145	11,237	186,527	$U^*_{sum} = 3$, $h' = 31$
$OAS\|E = 6/t_{mod} = 14$	7656	3959	5610	77,634	$U^*_{sum} = 6$, $h' = 32$
$OAS\|E = 6/t_{mod} = 24$	222	105	373	1898	$U^*_{sum} = 8$, $h' = 33$
RENON					
$OAS\|E = 6/t_{mod} = 6$	371	152	1281	6121	$U^*_{sum} = 8$, $h' = 35$
$OAS\|E = 6/t_{mod} = 14$	537	248	1625	8479	$U^*_{sum} = 10$, $h' = 35$
$OAS\|E = 6/t_{mod} = 24$	382	184	759	4344	$U^*_{sum} = 11$, $h' = 36$
Ex post scheduling for updated demand					
$OAS\|E = 6$	32,457	16,592	26,199	527,271	$U^*_{sum} = 2$, $h' = 32$

[a] U^*_{sum} = number of tardy orders, E^*_{max} = maximum earliness, h' = new planning horizon.

hand RENON requires the least, and REALL the greatest CPU time to find proven optimal schedules (see Sawik, 2007d).

The distribution of initial demand ahead of a monthly planning horizon, demand remaining and updated after each modification of orders, and the corresponding aggregate production schedules obtained for scenario A using scheduling/rescheduling algorithm REALL are shown in Figure 7.1.

For comparison, Figure 7.2 shows for each scenario the distribution over the horizon of the updated demand, optimal production schedule obtained ex post for the updated demand, and the concatenated production schedule for the entire planning horizon, constructed by using the REALL algorithm. The figure indicates that reactive scheduling by the REALL algorithm constructs production schedules very close to the optimal ex post schedule obtained when the updated demand is known for the entire horizon.

Finally, Figure 7.3 shows how the total inventory of product-specific materials and finished products varies over the horizon for each rescheduling algorithm, and each scenario with increasing, decreasing, and unchanged total demand. Again, the best results, that is, the lowest maximum inventory level is achieved for algorithm REALL, whereas RENON leads to the highest inventory level.

Table 7.4 Computational Results for Scenario B

Model/t_{mod}	Var.	Bin.	Cons.	Nonz.	Solution values[a]
Initial scheduling for beginning demand					
OAS	29,310	14,656	18,198	133,276	$U^*_{sum} = 0$
OAE|S	31,507	15,753	33,057	148,946	$E^*_{max} = 6$
REALL					
$OAS\|E = 6/t_{mod} = 6$	20,637	10,579	17,555	277,344	$U^*_{sum} = 0, h' = 30$
$OAS\|E = 6/t_{mod} = 14$	13,034	6726	12,069	151,584	$U^*_{sum} = 0, h' = 30$
$OAS\|E = 6/t_{mod} = 24$	2906	1520	3011	24,399	$U^*_{sum} = 0, h' = 30$
REMAT					
$OAS\|E = 6/t_{mod} = 6$	14,338	7386	9264	159,004	$U^*_{sum} = 0, h' = 30$
$OAS\|E = 6/t_{mod} = 14$	5898	3094	4305	54,439	$U^*_{sum} = 3, h' = 30$
$OAS\|E = 6/t_{mod} = 24$	64	25	92	382	$U^*_{sum} = 3, h' = 30$
RENON					
$OAS\|E = 6/t_{mod} = 6$	76	22	75	327	$U^*_{sum} = 1, h' = 30$
$OAS\|E = 6/t_{mod} = 14$	297	130	615	3624	$U^*_{sum} = 5, h' = 32$
$OAS\|E = 6/t_{mod} = 24$	111	50	153	835	$U^*_{sum} = 5, h' = 32$
Ex post scheduling for updated demand					
$OAS\|E = 6$	28,612	14,644	23,028	438,409	$U^*_{sum} = 0, h' = 30$

[a] U^*_{sum} = number of tardy orders, E^*_{max} = maximum earliness, h' = new planning horizon.

Table 7.5 Computational Results for Scenario C

Model/t_{mod}	Var.	Bin.	Cons.	Nonz.	Solution values[a]
REALL					
$OAS\|E = 6/t_{mod} = 24$	2798	1461	2914	23,512	$U^*_{sum} = 0, h' = 30$
REMAT					
$OAS\|E = 6/t_{mod} = 24$	39	12	41	159	$U^*_{sum} = 3, h' = 30$
RENON					
$OAS\|E = 6/t_{mod} = 24$	103	45	137	700	$U^*_{sum} = 5, h' = 32$
Ex post scheduling for updated demand					
$OAS\|E = 6$	28,654	14,665	23,073	438,760	$U^*_{sum} = 0, h' = 30$

[a] U^*_{sum} = number of tardy orders, E^*_{max} = maximum earliness, h' = new planning horizon.

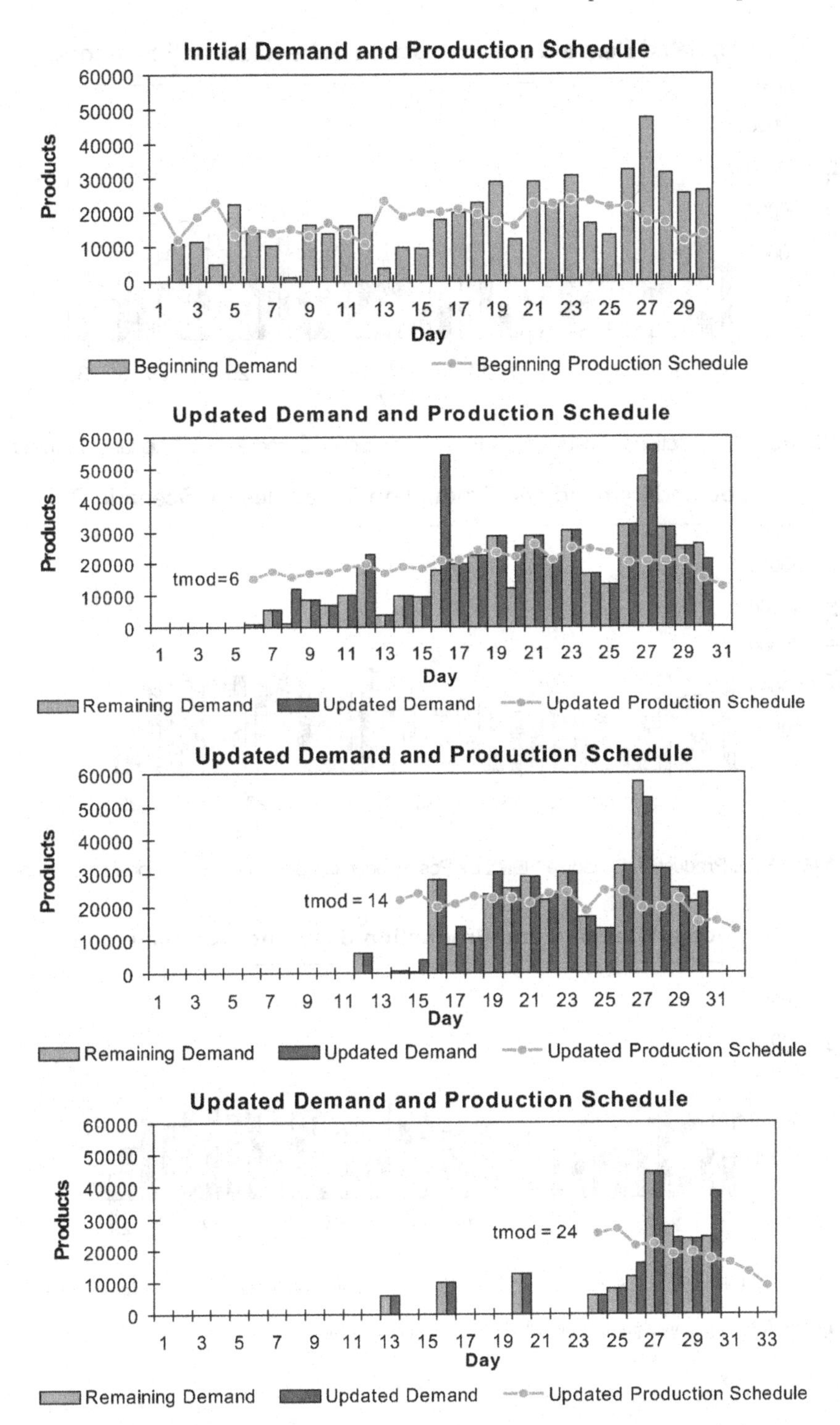

Figure 7.1 Demand and aggregate production schedule for algorithm REALL: scenario A.

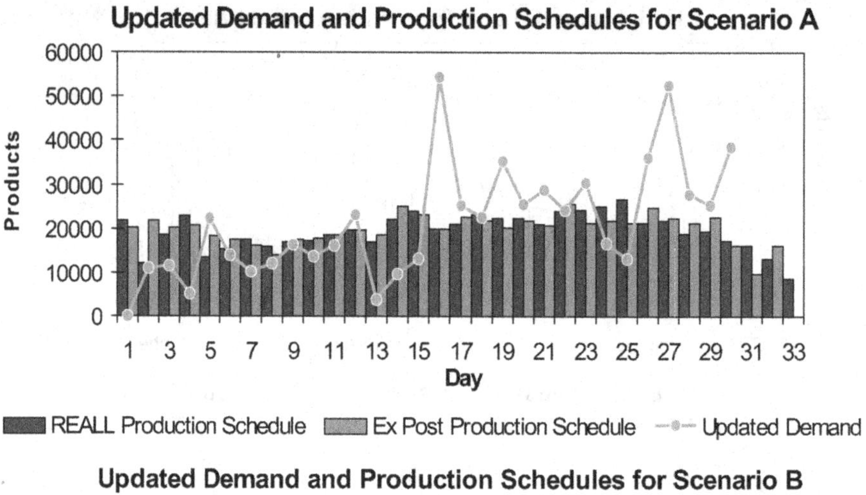

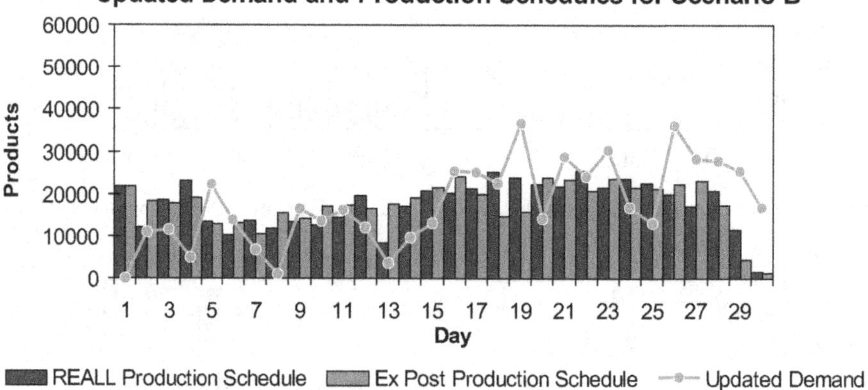

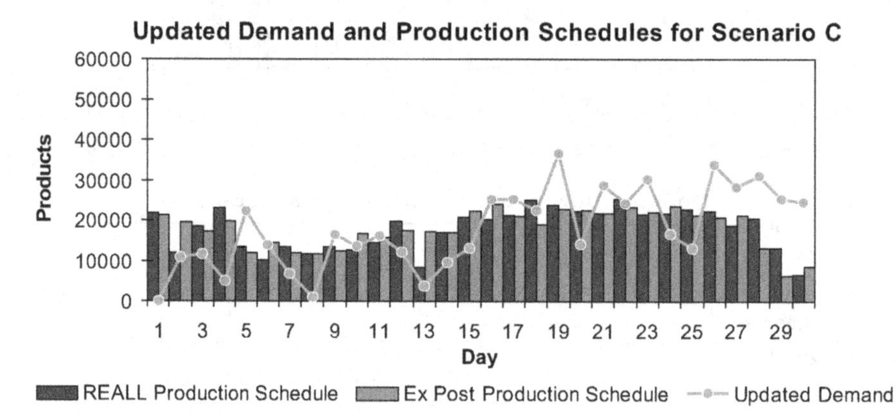

Figure 7.2 Aggregate production schedules for updated demand.

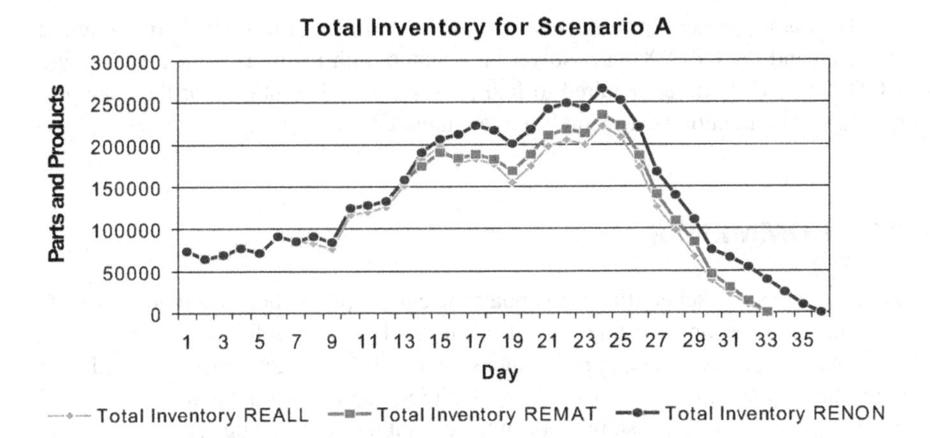

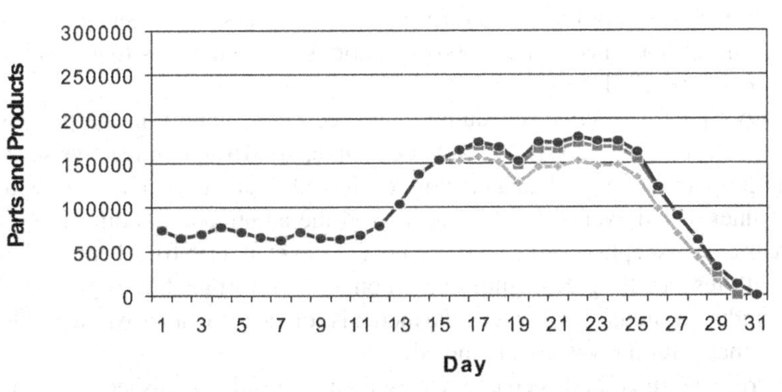

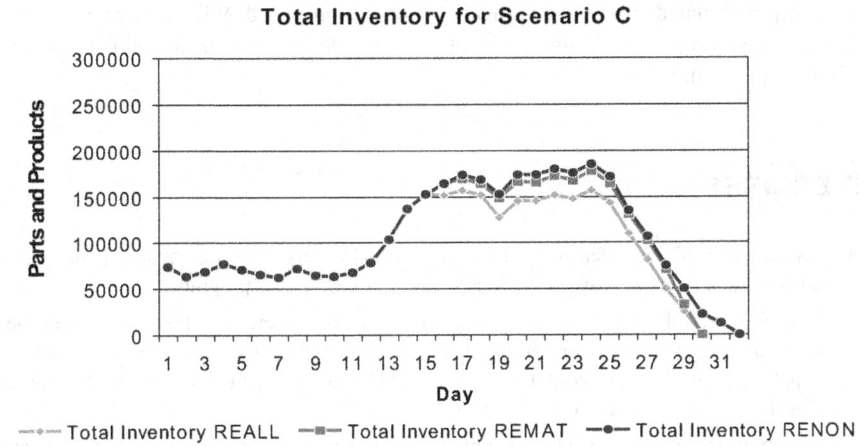

Figure 7.3 Total input and output inventory.

The computational experiments were performed using the AMPL programming language and the CPLEX v.9.1 solver on a laptop with Pentium IV at 1.8 GHz and 1 GB RAM. CPU times required to find proven optimal solutions for the examples ranged from a fraction of a second to a few hundred seconds.

7.7 COMMENTS

The aim of reactive scheduling is to update the current production schedule to provide an immediate response to unexpected events such as the addition or modification of customer orders or equipment breakdowns. In practice, reactive scheduling algorithms are applied for a dynamic rescheduling, and hence the algorithms should be fast and efficient. They should take into account the schedule currently in progress as well as the confirmed customer orders that are not affected by the unexpected events. A review of the literature (e.g., Vieira et al., 2003) indicates that research on reactive scheduling is mostly focused on heuristic approaches such as genetic algorithms (e.g., Rangsaritratsamee et al., 2004) or various AI techniques (e.g., Smith, 1995; J. Sun and Xue, 2001).

MIP approaches for reactive scheduling have been used mainly in the process industries (e.g., Röslof et al., 2001, 2002). For example, an MIP-based reactive scheduling of a large-scale industrial batch plant is considered by Janak et al. (2006), with two types of unexpected events: unit shutdown and the addition or modification of customer orders. An exception is the work by D.-Y. Liao et al. (1996), where a scheduling tool with rescheduling capabilities based on an MIP formulation is presented for the semiconductor industry. However, the model is solved by an approximate technique and optimal solution was not attempted.

The material presented in this chapter is based on the results published by Sawik (2007d), where various reactive scheduling policies and MIP-based rescheduling algorithms are proposed for a dynamic, make-to-order manufacturing environment. The computational results have indicated that the proposed MIP approach can be applied for reactive scheduling to iteratively update production schedules over a rolling planning horizon.

EXERCISES

7.1 Assume that all customer orders are single-period orders. Remove the unit penalty for tardy orders, u_j, and modify the corresponding mixed integer programs.

7.2 In Section 7.5, the input inventory of product-specific materials is calculated under the assumption that shortage or surplus of the inventory is balanced with higher or lower supplies, respectively, in period t_{mod}. How would you modify the proposed rescheduling algorithms, if the last assumption is removed?

7.3 Modify the formulae derived in Section 7.5 for the input and output inventory to account for both common and noncommon materials required for each product.

7.4 In the computational examples, Figure 7.2 indicates that the REALL rescheduling algorithm aims at leveling the aggregate production over the planning horizon, similarly to ex post scheduling. Why are both schedules similar?

7.5 Compare the total input and output inventory for different scenarios and rescheduling algorithms, illustrated in Figure 7.3, and explain why the total inventory in scenarios B and C is similar for both REMAT and RENON algorithms, whereas in scenario A the total inventories are different.

Chapter 8

Scheduling of Material Supplies in Make-to-Order Manufacturing

8.1 INTRODUCTION

An important issue of raw materials inventory management in make-to-order assembly is the problem of making ordering decisions for purchased components, subassemblies, or materials, to have them available when needed by the aggregate production schedule. This chapter deals with material ordering and supply scheduling in make-to-order assembly.

The aggregate production schedule (see Chapter 6) determines an assignment of customer orders to planning periods over a planning horizon and by this determines material requirements, varying over the horizon according to the allocation of customer orders among the planning periods. The objective of material ordering decisions is to coordinate material deliveries with the aggregate production schedule to meet the varying demand for materials and to minimize the total cost of material ordering, delivery, and inventory holding over the planning horizon.

Figure 8.1 shows a hierarchical framework for aggregate production scheduling and material ordering in a make-to-order assembly. First, at the upper level, the medium-term (e.g., monthly) aggregate production scheduling allocates customer orders among planning periods (e.g., days) to optimize some due date-related criterion. Then, at the lower level, material ordering determines delivery times and quantities to meet material requirements and to minimize total cost of ordering, delivery, and inventory holding of all materials.

The two most common ordering policies are:

- Stationary order quantity policy, where all replenishments of any given item are of the same quantity.
- Stationary order interval policy, where all orders of any given item are equally spaced in time.

Scheduling in Supply Chains Using Mixed Integer Programming. By Tadeusz Sawik

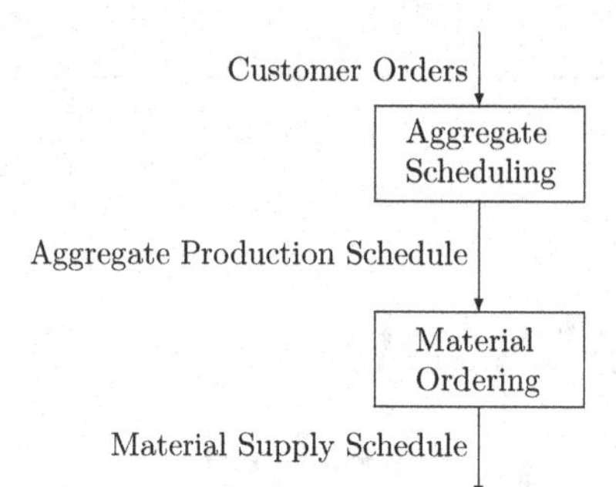

Figure 8.1 Aggregate production scheduling and material ordering.

The flexible ordering policy, where order intervals and/or order quantities are time-varying decision variables, is usually referred to as a nonstationary ordering policy.

This chapter provides the reader with MIP formulations and enumeration schemes for the two different approaches to optimal material ordering and supply scheduling:

- The cyclic approach, with different stationary order interval policy and different stationary order quantity policy for each material (the independent cycle ordering policy), or with the same stationary order interval for all materials and different stationary order quantities (the fixed cycle ordering policy).
- The flexible approach, with time-varying order intervals and order quantities (the nonstationary ordering policy).

The cyclic approach, in which the time between supplies is equal, ensures regular supplies of materials. However it may result in an excessive level of material inventory in some periods. On the other hand, the flexible approach means irregular supplies that can be better coordinated with the varying demand for materials required by an aggregate production schedule; however, it may incur higher ordering and delivery costs.

In the proposed ordering approaches, the order intervals and the order quantities are both assumed to be independent and stationary or time-varying decision variables that should be determined simultaneously to meet the varying demand and to minimize total ordering, delivery, and inventory holding cost.

The proposed MIP models and the enumeration scheme are enhanced to consider a finite input buffer capacity for storage of incoming materials or a safety lead time, and simple formulae are derived for the minimum buffer capacity, the maximum order interval, and the maximum safety lead time required for the existence of a feasible solution, when the buffer capacity is considered.

The following MIP and LP models with enumeration schemes are proposed to find the optimal schedule of material supplies for the two approaches:

FS for flexible scheduling of material supplies

CS for independent cyclic scheduling of material supplies

$CS_k(\tau)$ for cyclic scheduling of supplies of material k, with a fixed order interval τ

CSB$(\tau_1, \tau_2, \ldots, \tau_m)$ for independent cyclic scheduling, with a fixed order interval τ_k for each material $k = 1, \ldots, m$, and finite input buffer capacity

CSB(τ) for fixed cycle scheduling of supplies, with the same order interval τ for all materials, and finite input buffer capacity.

Section 8.4 presents computational examples modeled after a real-world make-to-order assembly in the electronics industry to illustrate possible applications of proposed MIP models for the flexible approach and enumeration scheme for the cyclic approach.

8.2 FLEXIBLE vs. CYCLIC MATERIAL SUPPLIES

In this section MIP models are proposed for the flexible and the cyclic approaches to material ordering and supply scheduling in a make-to-order assembly. Consider a production system where various types of products are assembled in a make-to-order environment using various types of purchased materials (for notation used, see Table 8.1). Let K be the set of m types of common and noncommon (product-specific) materials, L the set of n types of products, and a_{kl}, the unit requirement for material k of product l. The planning horizon (e.g., one month) consists of h planning periods (e.g., days). Let $T = \{1, \ldots, h\}$ be the set of planning periods.

The aggregate production schedule determined for the planning horizon assigns required production volumes of each product to each period, and let n_{lt} be the production of product $l \in L$ scheduled for period $t \in T$. The corresponding material requirement for each material k and each period t is $\sum_{l \in L} a_{kl} n_{lt}$. Denote by $M_{kt} = \sum_{l \in L} \sum_{t'=1}^{t} a_{kl} n_{lt'}$, the cumulative demand for material k in periods 1 through t and by $N_k = \sum_{l \in L: a_{kl}>0} \sum_{t \in T} n_{lt}$, the total production of products requiring material k. For each material $k \in K$, let p_k be the unit purchase price, o_k the fixed ordering and delivery cost, q_k the variable unit delivery cost, and φ_k the unit inventory holding cost, where $\varphi_k = \beta_k p_k$ is determined as the percentage β_k of the unit price p_k.

The problem objective is to determine supply times and quantities to meet material requirements (with no shortages allowed) and to minimize average cost per product of ordering, delivery, and inventory holding of all materials.

The MIP model *FS* for flexible scheduling of material supplies is formulated below.

Model FS: *Flexible Scheduling of Material Supplies*

Minimize average cost per product of ordering, delivery, and inventory holding

$$\sum_{k \in K} \sum_{t \in T} \left(o_k u_{kt} + q_k v_{kt} + \varphi_k \left(\sum_{t'=1}^{t} v_{kt'} - M_{kt} \right) \right) \Big/ N_k \tag{8.1}$$

subject to

1. *Material Requirement Constraints*:
   - for each material k and each period t the cumulative supplies cannot be less than the cumulative demand M_{kt};

$$\sum_{t'=1}^{t} v_{kt'} \geq M_{kt};\ k \in K, t \in T \tag{8.2}$$

2. *Material Supply Constraints*:
   - for each material k and each period t selected for supply, the supply quantity cannot be less than the minimum quantity $\underline{v}_k$ and cannot be greater than the total demand M_k;

$$\underline{v}_k u_{kt} \leq v_{kt} \leq M_k u_{kt};\ k \in K, t \in T \tag{8.3}$$

3. *Nonnegativity and Integrality Conditions*:

$$u_{kt} \in \{0, 1\};\quad k \in K, t \in T \tag{8.4}$$

$$v_{kt} \geq 0;\quad k \in K, t \in T. \tag{8.5}$$

Table 8.1 Notation: Scheduling of Material Supplies

Indices

k = material, $k \in K = \{1, \ldots, m\}$
l = product, $l \in L = \{1, \ldots, n\}$
t = planning period, $t \in T = \{1, \ldots, h\}$

Input parameters

a_{kl} = unit requirement for material k of product l
n_{lt} = production of product l scheduled for period t (aggregate production schedule)
o_k = the fixed ordering and delivery cost for material k
p_k = unit price for material k
q_k = the unit delivery cost for material k
β_k = the percentage of unit price used to determine the inventory holding cost for material k
φ_k = $\beta_k p_k$—the unit inventory holding cost for material k
$\underline{v}_k$ = minimum supply quantity of material k
$M_{kt} = \sum_{l \in L} \sum_{t'=1}^{t} a_{kl} n_{lt'}$—cumulative demand for material k in periods 1 through t
$M_k = \sum_{l \in L} \sum_{t \in T} a_{kl} n_{lt}$—total demand for material k
$N_k = \sum_{l \in L: a_{kl} > 0} \sum_{t \in T} n_{lt}$—total production of products that require material k

Decision variables

u_{kt} = 1, if supply of material k is scheduled for period t; otherwise $u_{kt} = 0$ (supply period selection variable—flexible approach)
$v_{kt} \geq 0$, the quantity of material k supplied in period t (supply quantity variable—flexible approach)
U_k = the stationary order interval for material k (cyclic approach)
V_k = the stationary order quantity for material k (cyclic approach)

Constraints (8.3) link the binary supply period variables and the continuous supply quantity variables for each material k and period t.

The independent cyclic scheduling problem is formulated below as a nonlinear mixed integer program *CS*.

Model CS: *Independent Cyclic Scheduling of Material Supplies*

Minimize average cost per product of ordering, delivery, and inventory holding

$$\sum_{k\in K}\left(o_k\lceil h/U_k\rceil + q_k\lceil h/U_k\rceil V_k + \varphi_k\sum_{t\in T}(\lceil t/U_k\rceil V_k - M_{kt})\right)\Big/N_k \tag{8.6}$$

subject to

1. *Material Requirement Constraints*:
   - for each material k and each period t, the cumulative supplies of V_k units every U_k periods cannot be less than the cumulative demand M_{kt};

$$\lceil t/U_k\rceil V_k \geq M_{kt};\ \ k\in K, t\in T, \tag{8.7}$$

   where $\lceil\cdot\rceil$ denotes the least integer not less than $\cdot$,
2. *Nonnegativity and Integrality Conditions*:

$$U_k \geq 0,\ integer;\ \ k\in K \tag{8.8}$$

$$V_k \geq \underline{v}_k;\ \ k\in K. \tag{8.9}$$

The objective (8.6) of problem *CS* is to minimize the average monthly cost per product of ordering, delivery, and inventory holding for all materials. The material requirement constraint (8.7) ensures that for each material k and each period t, the cumulative supplies in periods 1 through t ($\lceil t/U_k\rceil$ supplies, each of V_k units of material k), are not less than the cumulative demand M_{kt}.

Since the objective functions (8.1) and (8.6) are separable with regard to materials and the materials do not share a common resource (in models *FS* and *CS* there is no common resource constraint, for example, a finite input buffer capacity constraint), the optimal supply schedules can be found independently for each material.

For each material $k = 1,\ldots,m$ and each order interval $\tau = 1,\ldots,h$ define below the linear programming model $CS_k(\tau)$ of independent cyclic scheduling of supplies of material k.

Model CS$_k(\tau)$: *Cyclic Scheduling of Supplies of Material k, with a Fixed Order Interval τ*

Minimize

$$\left(o_k\lceil h/\tau\rceil + q_k\lceil h/\tau\rceil V_k + \varphi_k\sum_{t\in T}(\lceil t/\tau\rceil V_k - M_{kt})\right)\Big/N_k \tag{8.10}$$

subject to

1. *Material Requirement Constraints*:
   - for material k and each period t, the cumulative supplies of V_k units every τ periods cannot be less than the cumulative demand M_{kt},
   - the supply quantity of material k cannot be less than the minimum quantity $\underline{v}_k$,

$$\lceil t/\tau \rceil V_k \geq M_{kt};\ t \in T \tag{8.11}$$

$$V_k \geq \underline{v}_k . \tag{8.12}$$

The optimal solution value of $CS_k(\tau)$ is

$$\left(o_k \lceil h/\tau \rceil + q_k \lceil h/\tau \rceil V_k^\tau + \varphi_k \sum_{t \in T} (\lceil t/\tau \rceil V_k^\tau - M_{kt}) \right) \Big/ N_k \tag{8.13}$$

where

$$V_k^\tau = \max\left\{ \underline{v}_k ,\ \max_{t \in T} \lceil M_{kt} / \lceil t/\tau \rceil \rceil \right\}. \tag{8.14}$$

For each material k the optimal values U_k^* of the stationary order interval and V_k^* of the stationary order quantity are the solution of

$$\begin{aligned} &\min_{\tau \in T} \left(\frac{o_k \lceil h/\tau \rceil + q_k \lceil h/\tau \rceil V_k^\tau + \varphi_k \sum_{t \in T} (\lceil t/\tau \rceil V_k^\tau - M_{kt})}{N_k} \right) \\ &= \frac{o_k \lceil h/U_k^* \rceil + q_k \lceil h/U_k^* \rceil V_k^* + \varphi_k \sum_{t \in T} (\lceil t/U_k^* \rceil V_k^* - M_{kt})}{N_k} \end{aligned} \tag{8.15}$$

where

$$V_k^* = \max\left\{ \underline{v}_k ,\ \max_{t \in T} \lceil M_{kt} / \lceil t/U_k^* \rceil \rceil \right\}. \tag{8.16}$$

The optimal solution (8.16) can easily be found by enumeration for all values $\tau = 1, \ldots, h$ of the order interval U_k, that is, for each value τ of the order interval, the best value V_k^* of the order quantity is calculated, and the minimum among all V_k^* determines the optimal solution.

8.3 MODEL ENHANCEMENTS

In this section three practical extensions of the material ordering and supply scheduling problems and the MIP models are presented. The first extension accounts for limited storage space for the input materials, and the second one deals with the safety lead time for material supplies. Finally, supplies of common and noncommon (product-specific) materials are distinguished.

In make-to-order assembly the aggregate production schedule already accounts for any resource restrictions such as finite storage capacity for the input materials. Therefore, in the basic formulations presented in this chapter, the finite input buffer is not considered. Nevertheless, the formulations are enhanced in this section to account for the limited input buffer capacity. Furthermore, the order intervals are assumed to already include the delivery lead times so that the supplies meet delivery requirements for materials of the aggregate production schedule. As a result, order intervals are assumed to be identical with material supply intervals. In this section, however, introduction of additional safety lead time is considered.

8.3.1 Finite Input Buffer

In the MIP models *FS* and *CS*, the aggregate production schedule is assumed to already account for a finite capacity of an input storage for the purchased materials. If, nevertheless, a finite capacity input storage should be considered in the ordering problem, then the objective of the supply scheduling problem, also called a staggering problem, is to time phase (stagger) the arrivals of materials so as to satisfy the input buffer finite capacity constraint. To this end, the following common resource constraints (8.17) and (8.18) should be introduced, respectively, in models *FS* and *CS*:

$$\sum_{k \in I} b_k \left(\sum_{t'=1}^{t} v_{kt'} - M_{kt} \right) \leq B; \; t \in T \tag{8.17}$$

$$\sum_{k \in I} b_k (\lceil t/U_k \rceil V_k - M_{kt}) \leq B; \; t \in T, \tag{8.18}$$

where b_k is the space occupied by one unit of material k and B denotes the storage space of the input buffer.

This constraint requires careful scheduling of arrival of different items to limit the peak usage of the storage resource and the ordering problems with the common resource constraints (8.17) or (8.18) are no longer separable with regard to materials. In particular, the optimal solution to nonlinear MIP model *CS* with the input buffer capacity constraint (8.18) can be found by enumeration of solutions to the following linear programming model $CSB(\tau_1, \tau_2, \ldots, \tau_m)$ for all values of $\tau_k = 1, \ldots, h$ of the order interval U_k for each material $k = 1, \ldots, m$.

Model CSB($\tau_1, \tau_2, \ldots, \tau_m$): *Independent Cyclic Scheduling, with a Fixed Order Interval τ_k for each Material $k = 1, \ldots, m$, and Finite Input Buffer Capacity*

Minimize

$$\sum_{k \in K} \left(o_k \lceil h/\tau_k \rceil + q_k \lceil h/\tau_k \rceil V_k + \varphi_k \sum_{t \in T} (\lceil t/\tau_k \rceil V_k - M_{kt}) \right) \Big/ N_k \tag{8.19}$$

subject to (8.9) and

1. *Material Requirement Constraints*:
   - for each material k and each period t, the cumulative supplies of V_k units every τ_k periods cannot be less than the cumulative demand M_{kt},

$$\lceil t/\tau_k \rceil V_k \geq M_{kt};\ \ k \in K, t \in T \tag{8.20}$$

2. *Input Buffer Capacity Constraints*:
   - in each period, the total space required for storing purchased materials waiting for processing cannot exceed the input buffer finite storage space,

$$\sum_{k \in K} b_k(\lceil t/\tau_k \rceil V_k - M_{kt}) \leq B;\ \ t \in T. \tag{8.21}$$

The computational effort required to find optimal solution based on enumeration can be significant for large problems (the number of problems $CSB(\tau_1, \tau_2, \ldots, \tau_m)$ to be solved is h^m). Therefore, instead of seeking the independent cycle solution with different order intervals U_k for each material $k \in K$ one may more easily find a fixed cycle solution with the same order interval $U_k = \tau$ for all materials $k \in K$. The MIP model $CSB(\tau)$ for fixed cycle ordering with the same order interval τ is presented below.

Model CSB(τ): *Fixed Cycle Scheduling of Supplies, with the Same Order Interval τ for All Materials, and Finite Input Buffer Capacity*

Minimize

$$\sum_{k \in K} \left(o_k \lceil h/\tau \rceil + q_k \lceil h/\tau \rceil V_k + \varphi_k \sum_{t \in T} (\lceil t/\tau \rceil V_k - M_{kt}) \right) \Big/ N_k \tag{8.22}$$

subject to

1. *Material Requirement Constraints*:
   - the supply quantity of each material k cannot be less than the optimal order quantity V_k^τ for the order interval τ,

$$V_k \geq V_k^\tau;\ \ k \in K \tag{8.23}$$

2. *Input Buffer Capacity Constraints*:
   - in each period, the total space required for storing purchased materials waiting for processing cannot exceed the input buffer finite storage space,

$$\sum_{k \in K} b_k(\lceil t/\tau \rceil V_k - M_{kt}) \leq B;\ \ t \in T, \tag{8.24}$$

where V_k^τ is defined in (8.14).

Notice that the longer the fixed cycle τ, the greater is the order quantity V_k^τ and the greater is the required capacity B of the input buffer. Given the capacity B, the optimal fixed cycle U^* can be found by enumeration for $1 \leq \tau \leq \tau_{\max}(B)$, where the maximum order interval is

$$\tau_{\max}(B) = \max\left\{\tau \in T: \sum_{k \in K} b_k(\lceil t/\tau \rceil V_k^\tau - M_{kt}) \leq B;\ t \in T\right\}. \tag{8.25}$$

The fixed cycle supply scheduling problem $CSB(\tau)$ has a feasible solution only if the input buffer capacity B is not less than the minimum capacity $B_{\min}$ required for a unit order interval $\tau = 1$, that is,

$$B_{\min} = \max_{t \in T} \sum_{k \in K} b_k(t \max\{\underline{v}_k,\ \max_{t \in T} \lceil M_{kt}/t \rceil\} - M_{kt}). \tag{8.26}$$

8.3.2 Safety Lead Time

In the MIP models *FS* and *CS*, the supply period and the order interval, respectively, are assumed to already include the delivery lead time so that the supplies meet delivery requirements for materials of the aggregate production schedule. However, if some material k should be ordered to arrive a fixed safety lead time of s_k periods in advance of the aggregate production schedule, then the material requirement constraints (8.2) and (8.7) should be replaced with the following constraints (8.27) and (8.28), respectively, in *FS* and *CS* models:

$$\sum_{t'=1}^{t} v_{kt'} \geq M_{k,t+s_k};\ k \in K, t \in T : t \leq h - s_k, \tag{8.27}$$

$$\lceil t/U_k \rceil V_k \geq M_{k,t+s_k};\ k \in K, t \in T : t \leq h - s_k. \tag{8.28}$$

Accordingly, M_{kt} in (8.16) is replaced with $M_{k,t+s_k}$, and the optimal solution (8.16) is

$$V_k^* = \max\left\{\underline{v}_k,\ \max_{1 \leq t \leq h - s_k} \lceil M_{k,t+s_k}/\lceil t/U_k^* \rceil \rceil\right\}. \tag{8.29}$$

The supplies of some materials in advance of the aggregate production schedule may increase the level of material inventory to be held in an input buffer storage. If the storage capacity is limited, then the following constraint on selection of safety lead times s_k must be satisfied

$$\max_{t \in T} \sum_{k \in K} b_k(M_{k,t+s_k} - M_{kt}) \leq B. \tag{8.30}$$

In particular, if all safety lead times are identical, that is, $s_k = s \; \forall k \in K$, then the maximum lead time $s_{\max}$ can be determined as follows

$$s_{\max} = \max\left\{ s \in T: \max_{1 \le t \le h-s} \sum_{k \in K} b_k (M_{k,t+s} - M_{kt}) \le B \right\}. \tag{8.31}$$

8.3.3 Common and Product-Specific Materials

The MIP models can be easily enhanced to simultaneously account for both common and noncommon (product-specific) materials. To this end, the additional decision variables should be introduced to control the supplies of product-specific materials:

$u_{klt} = 1$, if supply of product-specific material k for product l is scheduled for period t; otherwise $u_{klt} = 0$ (product-specific material supply period selection variable—flexible approach)

$v_{klt} \ge 0$, the quantity of product-specific material k for product l supplied in period t (product-specific material supply quantity variable—flexible approach)

U_{kl} = the stationary order interval for product-specific material k for product l (cyclic approach)

V_{kl} = the stationary order quantity for product-specific material k for product l (cyclic approach).

Furthermore, the problem formulations should be enhanced by the addition of appropriate new product-specific material constraints and new product-specific material cost terms in the objective functions.

8.4 COMPUTATIONAL EXAMPLES

In this section computational examples are presented to illustrate possible applications of the proposed MIP approach and enumeration schemes. The examples are modeled after a real-world distribution center for high-tech products, where finished products are assembled for shipping to customers.

A brief description of the materials, products and the customer orders is given below.

1. $m = 5$, common and product-specific material types (accessories, batteries, manuals/packaging, plastics, and other).

   The fixed ordering and delivery costs

   $$o_1 = 400, \; o_2 = 400, \; o_3 = 100, \; o_4 = 400, \; o_5 = 300.$$

   The unit purchase price

   $$p_1 = 1.65, \; p_2 = 4.52, \; p_3 = 0.62, \; p_4 = 1.17, \; p_5 = 1.49.$$

   The unit delivery costs

   $$q_1 = 0.10, \; q_2 = 0.10, \; q_3 = 0.01, \; q_4 = 0.05, \; q_5 = 0.05.$$

The percentage of purchase price $\beta_k = 0.16/365,\ k = 1, 2, 3, 4, 5$ and the corresponding unit inventory holding costs $\beta_k p_k$ are

$$\varphi_1 = 0.00072,\ \varphi_2 = 0.00198,\ \varphi_3 = 0.00027,\ \varphi_4 = 0.00051,\ \varphi_5 = 0.00065.$$

The minimum supply quantity $\underline{v}_k = 1000,\ k = 1, 2, 3, 4, 5.$

2. $n = 11$, product types (cellular phones of different types).

3. The unit requirements for common materials

$$[a_{kl}] = \begin{bmatrix} 11111111100 \\ 00011111100 \\ 44444444433 \\ 11122222211 \\ 11111111100 \end{bmatrix}.$$

4. The unit requirements for product-specific materials

$$[a_{kl}] = \begin{bmatrix} 11111111122 \\ 11100000011 \\ 11111111122 \\ 11100000011 \\ 00000000011 \end{bmatrix}.$$

5. $h = 30$, planning periods (days).

In the example, all cost parameters for both common and product-specific materials of the same type are identical; however, the unit requirements and the corresponding material demands are different.

The time-varying demand for both common and product-specific materials is generated by the aggregate production schedule $n_{lt},\ l \in L,\ t \in T$ for the 30-day planning horizon and 11 product types, shown in Table 8.2 and in Figure 8.2.

For the flexible approach, the optimal supplies v_{kt} and v_{klt}, respectively, of common and product-specific materials are shown in Tables 8.3 and 8.4.

For the independent cyclic approach, the optimal order intervals U_k, U_{kl} and order quantities V_k, V_{kl}, respectively, for common, product-specific materials are shown in Tables 8.5 and 8.6. The tables show that common materials are on average ordered every week, whereas product-specific materials are supplied once or twice per month. The results indicate that for both the flexible and the cyclic ordering policies, the optimal solution requires a single supply of product-specific materials at the beginning of the planning horizon, whereas common materials are supplied during the entire horizon. The inventory of common and product-specific materials for the two approaches are compared in Figure 8.3. The figure indicates that material inventory

Table 8.2 Aggregate Production Schedule n_{lt}

t/l	1	2	3	4	5	6	7	8	9	10	11
1	0	1565	0	4019	0	0	0	0	0	0	0
2	1800	0	142	0	0	0	0	8096	0	3600	8882
3	3002	0	6557	921	0	0	0	4107	0	2302	7365
4	5298	6000	344	0	0	1600	2007	0	3525	0	3700
5	3183	5735	914	3960	0	0	0	0	2824	0	5185
6	7000	0	0	100	0	0	7011	0	3251	0	4949
7	3678	0	0	0	0	1899	8382	0	0	0	8405
8	7097	0	10,979	0	0	8974	0	0	0	0	1208
9	4900	0	12,293	0	0	11,936	0	0	0	0	638
10	6000	0	2700	0	0	5130	0	0	0	1758	8181
11	8128	0	5751	0	1573	8205	1575	0	0	0	1323
12	3414	0	11,679	105	0	11,334	852	55	0	651	1500
13	206	2900	1787	196	396	8770	1848	0	0	1300	7800
14	2383	0	9164	37	0	9553	715	1699	0	919	3688
15	9040	0	583	528	0	1795	1018	0	0	173	9360
16	1836	23	995	70	0	405	2691	7782	872	725	7200
17	1528	7446	931	63	31	3827	2043	3770	783	555	2405

18	1324	60	4326	1125	0	8769	1361	1858	349	3677	4141
19	3980	0	3310	3336	0	7387	2911	413	0	3738	1211
20	3763	0	0	6606	0	2436	0	714	0	198	8410
21	0	7200	3143	3331	0	1578	0	0	545	2400	5133
22	1512	0	3989	6376	0	1137	265	0	0	1545	8450
23	4313	5092	5534	141	0	8367	673	0	400	1078	893
24	3458	3268	2000	2444	0	7800	840	99	1052	2694	1951
25	2328	4790	6379	893	0	7700	825	0	2714	0	648
26	3062	0	0	944	0	8197	2500	3206	3186	3401	802
27	1339	0	7700	1421	0	4400	2388	1300	7100	400	422
28	1728	2300	5000	2384	0	9000	1503	1000	0	626	2696
29	0	1621	4400	0	0	8599	1800	1000	4300	573	4200
30	1700	0	3400	0	0	7000	2798	1596	3397	2500	3449
Total monthly production of product types											
$\sum$	97,000	48,000	11,4000	39,000	2000	155,798	46,006	36,695	36,695	34,813	124,195

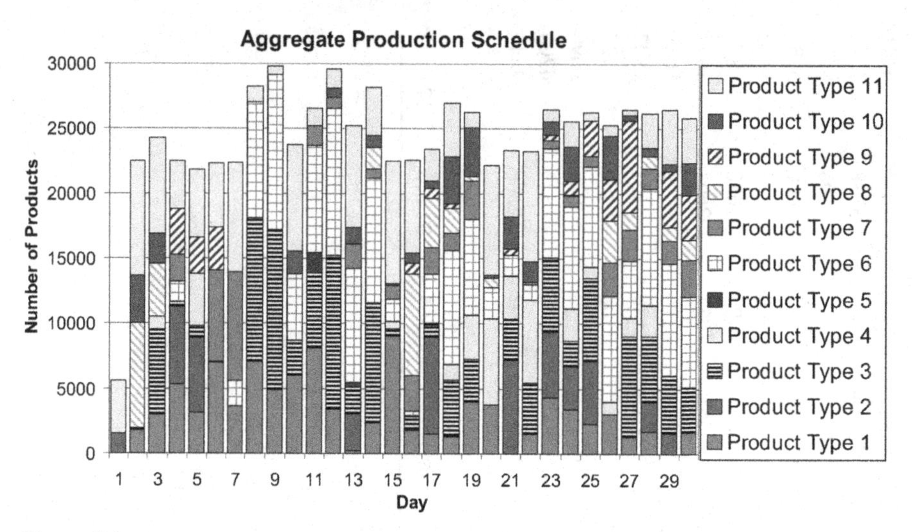

Figure 8.2 Aggregate monthly production schedule.

for the flexible ordering policy is much smaller than that for the cyclic ordering and so are the corresponding material inventory holding costs.

Average material costs per product for each type of product for the optimal flexible and cyclic supplies are compared in Figure 8.4. In addition, the results are

Table 8.3 Supplies of Common Materials v_{kt}: Flexible Approach

t/k	1	2	3	4	5
1	166,929	24,275	187,283	160,365	166,929
4	0	0	333,561	0	0
5	0	64,820	0	0	0
7	0	0	0	178,387	0
8	0	0	230,254	0	0
10	0	0	190,034	0	0
11	68,774	0	0	0	68,774
12	0	38,901	396,401	178,768	0
14	179,235	0	0	0	246,920
16	0	72,834	273,181	0	0
17	0	0	0	280,316	0
19	0	0	265,882	0	0
22	0	0	284,873	0	0
23	0	51,981	0	0	0
24	157,859	0	0	0	0
25	0	0	306,507	247,766	0
27	0	60,986	0	0	90,174
28	0	0	300,236	0	0

Table 8.4 Supplies of Product-Specific Materials v_{klt}: Flexible Approach

k,l \ t	1	2	3	..													....												..	30
1,1	0	97,000	0	0	0	0	0	0	0	0	0	0	0	0	0	0	0	0	0	0	0	0	0	0	0	0	0	0	0	0
1,2	48,000	0	0	0	0	0	0	0	0	0	0	0	0	0	0	0	0	0	0	0	0	0	0	0	0	0	0	0	0	0
1,3	0	72,455	0	0	0	0	0	0	0	0	0	0	0	0	0	0	0	0	0	0	41,545	0	0	0	0	0	0	0	0	0
1,4	39,000	0	0	0	0	0	0	0	0	0	0	0	0	0	0	0	0	0	0	0	0	0	0	0	0	0	0	0	0	0
1,5	0	0	0	0	0	0	0	0	0	0	2000	0	0	0	0	0	0	0	0	0	0	0	0	0	0	0	0	0	0	0
1,6	0	0	0	93,598	0	0	0	0	0	0	0	0	0	0	0	0	0	0	0	0	0	62,200	0	0	0	0	0	0	0	0
1,7	0	0	0	46,006	0	0	0	0	0	0	0	0	0	0	0	0	0	0	0	0	0	0	0	0	0	0	0	0	0	0
1,8	0	36,695	0	0	0	0	0	0	0	0	0	0	0	0	0	0	0	0	0	0	0	0	0	0	0	0	0	0	0	0
1,9	0	0	0	34,298	0	0	0	0	0	0	0	0	0	0	0	0	0	0	0	0	0	0	0	0	0	0	0	0	0	0
1,10	0	69,626	0	0	0	0	0	0	0	0	0	0	0	0	0	0	0	0	0	0	0	0	0	0	0	0	0	0	0	0
1,11	0	102,672	0	0	0	0	0	0	0	0	0	0	145,718	0	0	0	0	0	0	0	0	0	0	0	0	0	0	0	0	0
2,1	0	35,958	0	0	0	0	0	0	0	43,114	0	0	0	0	0	0	0	0	0	0	0	0	17,928	0	0	0	0	0	0	0
2,2	16,223	0	0	0	0	0	0	0	0	0	0	0	0	0	0	0	31,777	0	0	0	0	0	0	0	0	0	0	0	0	0
2,3	0	7957	0	0	0	0	0	71,630	0	0	0	0	0	0	0	0	0	0	0	0	0	0	34,413	0	0	0	0	0	0	0
2,10	0	10,530	0	0	0	0	0	0	0	0	0	0	0	0	24,283	0	0	0	0	0	0	0	0	0	0	0	0	0	0	0
2,11	0	51,336	0	0	0	0	0	0	0	0	0	0	35,805	0	0	0	0	0	0	37,054	0	0	0	0	0	0	0	0	0	0
3,1	0	56,089	0	0	0	0	0	0	0	0	0	0	0	0	40,911	0	0	0	0	0	0	0	0	0	0	0	0	0	0	0
3,2	16,223	0	0	0	0	0	0	0	0	0	0	0	0	0	0	0	31,777	0	0	0	0	0	0	0	0	0	0	0	0	0
3,3	0	72,455	0	0	0	0	0	0	0	0	0	0	0	0	0	0	0	0	0	0	41,545	0	0	0	0	0	0	0	0	0
3,4	11,124	0	0	0	0	0	0	0	0	0	0	0	0	0	0	0	0	0	27,876	0	0	0	0	0	0	0	0	0	0	0
3,5	0	0	0	0	0	0	0	0	0	0	2000	0	0	0	0	0	0	0	0	0	0	0	0	0	0	0	0	0	0	0
3,6	0	0	0	29,539	0	0	0	0	0	0	65,196	0	0	0	0	0	0	0	0	0	0	0	61,063	0	0	0	0	0	0	0
3,7	0	0	0	46,006	0	0	0	0	0	0	0	0	0	0	0	0	0	0	0	0	0	0	0	0	0	0	0	0	0	0
3,8	0	36,695	0	0	0	0	0	0	0	0	0	0	0	0	0	0	0	0	0	0	0	0	0	0	0	0	0	0	0	0
3,9	0	0	0	12,549	0	0	0	0	0	0	0	0	0	0	0	0	0	0	0	0	0	0	0	21,749	0	0	0	0	0	0
3,10	0	15,320	0	0	0	0	0	0	0	0	54,306	0	0	0	0	0	0	0	0	0	0	0	0	0	0	0	0	0	0	0
3,11	0	97,026	0	0	0	0	0	0	0	0	74,834	0	0	0	0	0	0	0	76,530	0	0	0	0	0	0	0	0	0	0	0
4,1	0	97,000	0	0	0	0	0	0	0	0	0	0	0	0	0	0	0	0	0	0	0	0	0	0	0	0	0	0	0	0
4,2	48,000	0	0	0	0	0	0	0	0	0	0	0	0	0	0	0	0	0	0	0	0	0	0	0	0	0	0	0	0	0
4,3	0	114,000	0	0	0	0	0	0	0	0	0	0	0	0	0	0	0	0	0	0	0	0	0	0	0	0	0	0	0	0
4,10	0	34,813	0	0	0	0	0	0	0	0	0	0	0	0	0	0	0	0	0	0	0	0	0	0	0	0	0	0	0	0
4,11	0	124,195	0	0	0	0	0	0	0	0	0	0	0	0	0	0	0	0	0	0	0	0	0	0	0	0	0	0	0	0
5,10	0	34,813	0	0	0	0	0	0	0	0	0	0	0	0	0	0	0	0	0	0	0	0	0	0	0	0	0	0	0	0
5,11	0	62,824	0	0	0	0	0	0	0	0	0	0	0	0	61,371	0	0	0	0	0	0	0	0	0	0	0	0	0	0	0

Table 8.5 Optimal Ordering Policy for Common Materials: Cyclic Approach

Material k	Order interval U_k	Order quantity V_k
1	6	114,560
2	8	78,450
3	3	276,822
4	6	209,121
5	6	114,560

Table 8.6 Optimal Ordering Policy for Product-Specific Materials: Cyclic Approach

	Order interval U_k, order quantity V_k for material k				
Product l	$k = 1$	$k = 2$	$k = 3$	$k = 4$	$k = 5$
1	30,97000	30,97000	30,97000	30,97000	–
2	30,48000	20,24000	30,48000	30,48000	–
3	30,114000	30,114000	10,39680	30,114000	–
4	30,39000	–	19,19500	–	–
5	30,2000	–	30,2000	–	–
6	14,77899	–	14,77899	–	–
7	30,46006	–	30,46006	–	–
8	30,36695	–	30,36695	–	–
9	30,34298	–	22,17149	–	–
10	17,34813	17,17407	17,34813	30,34813	30,34813
11	30,248390	30,124195	30,248390	30,124195	30,124195

compared with average material costs for monthly supplies, that is, cyclic supplies every 30 days. The average cost per product for the independent cycle policy is 0.57, and the total monthly cost is 418,204. For the flexible policy the costs are 0.56, and 411,266, respectively. For a single per month supply policy, the corresponding costs are 0.61, and 446,114.

Comparison of the three ordering policies indicates that for the example data, the best is the flexible policy, and a simple independent cycle policy produces only slightly higher costs. The worst is a single supply policy as it incurs the highest material inventory carrying costs.

For the example problem, the size of the MIP model *FS* for the flexible approach was as follows: the total number of variables, 2,329, number of binary variables, 1,159, number of constraints, 3,409, and number of nonzero coefficients in the constraint matrix, 22,570. The CPU time required to find the proven optimal solution (using the AMPL programming language and the CPLEX v.9.1 solver on a laptop with Pentium IV at 1.8 GHz and 1 GB RAM) for the flexible approach was not greater than a few hundred seconds and was negligible (fraction of a second) for the cyclic approach based on enumeration.

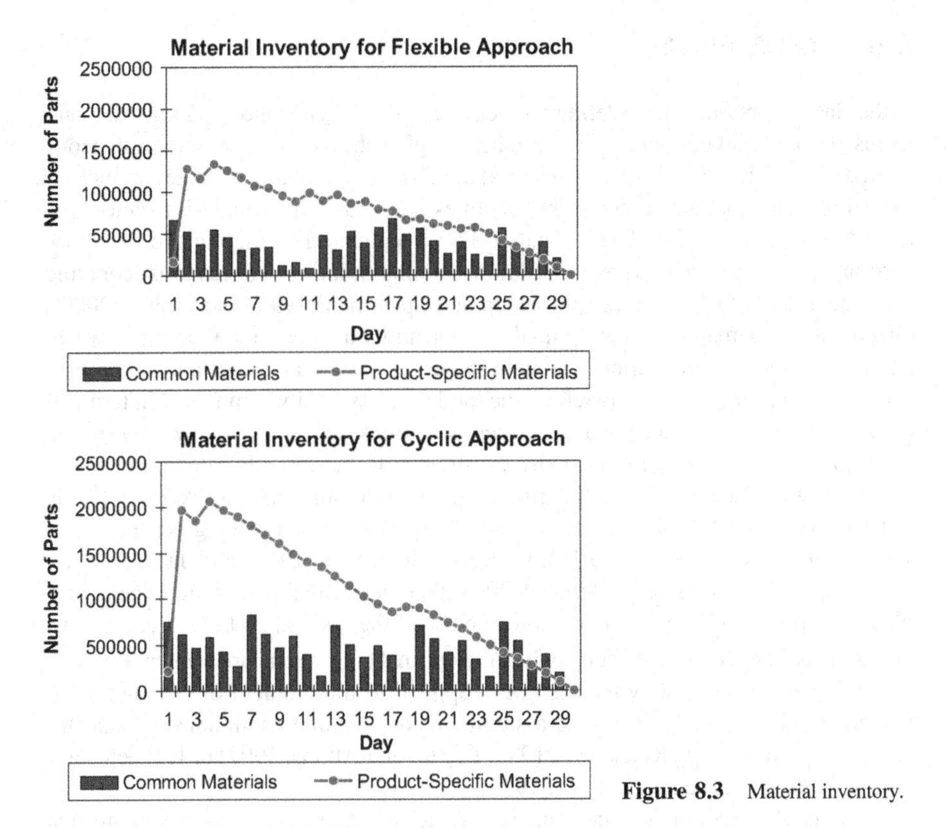

Figure 8.3 Material inventory.

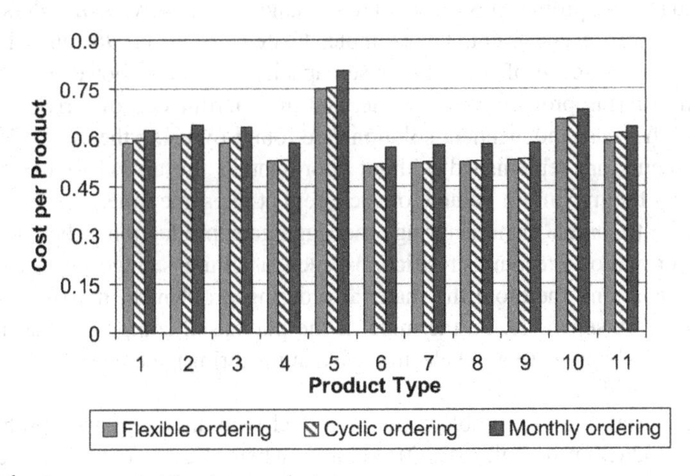

Figure 8.4 Average material cost per product.

8.5 COMMENTS

In the classic version of the ordering problem for multi-item inventory systems, demand rate is constant and continuous for each item, replenishment of stock is instantaneous upon delivery, shortages are not allowed, and the unit prices charged by the supplier for the items are independent of the order quantities. The problem objective is to determine an ordering policy to find the best trade-off between material ordering and inventory carrying costs. The problem has a well-known optimal solution, referred to as economic order quantity (EOQ), for example, Hopp and Spearman (1996) and Zipkin (2000). Given that the demand and the optimal order quantity are constant, the time between any two consecutive deliveries of an item (order interval) is also constant and proportional to the order quantity. Therefore, the models are typically formulated in terms of order intervals rather than order quantities (see Muckstadt and Roundy, 1993). The order interval for each item is usually constrained to be not only integer but also a power of two of a base planning period (e.g., an hour, shift, day, week, etc.), that is order intervals are $1, 2, 4, 8, \ldots, 2^k$ times as long as the base planning period, where k is a nonnegative integer, see Maxwell and Muckstadt (1985) and Jackson et al. (1988). It is well known (e.g., Bertazzi, 2003) that the optimal cost of the discrete problem with power-of-two restriction is within 6% of the optimal cost of the continuous problem and the worst case occurs when the optimal continuous order interval is $\sqrt{2}$.

A power-of-two policy has also been applied to the problem of ordering components used in assembly systems, assuming a constant and continuous demand for each final product (e.g., Muckstadt, 1985; D. Sun and Atkins, 1997) or for each component (see Federgruen and Zheng, 1995).

The classic economic order quantity model has been enhanced assuming that there is a single resource constraint (e.g., a limited storage space of a warehouse) that limits maximal inventory level over time (Rosenblatt and Rothblum, 1990). This constraint requires careful scheduling of arrival of different items to limit the peak usage of the resource. The objective of the supply scheduling problem (also called a staggering problem) is to time phase (stagger) the arrivals of orders so as to satisfy the resource constraint, for example, Gallego et al. (1996) and Hall et al. (1998b). In the absence of the warehouse capacity constraint, however, the classic multi-item ordering problem is separable, and the coordination of arrivals is not an issue. In such a case independent solutions are found by using the classic EOQ formula and items are replenished without coordinating the arrival of the orders to avoid storing the maximum volume of each item at the same time.

In make-to order manufacturing, the aggregate production schedule already accounts for various resource restrictions such as finite storage capacity for the input materials, and therefore the material ordering problem in most cases can be considered to be separable. Furthermore, if the purchased components do not need a large space for storage (e.g., in electronics manufacturing), the input buffer capacity can be considered unlimited.

The material ordering problem can be linked with another classic problem, the economic production quantity (EPQ) problem, where the objective is to determine production lot size to find the best trade-off between manufacturing setup and finished product inventory holding costs. The EOQ and EPQ problems can be considered simultaneously within a supply chain framework; see Goyal and Deshmukh (1992). The integrated EOQ and EPQ models are introduced with the objective of finding a joint economic lot size to simultaneously minimize total cost of material ordering and holding and manufacturing setup and finished products holding. However, the models are based on various simplifying assumptions such as a single finished product with a constant or a piece-wise linear demand pattern; for example, W. Lee (2005). Therefore, the models cannot be directly applied in a complex, multiproduct make-to-order manufacturing environment, with an arbitrary demand pattern for different finished products.

The material presented in this chapter is an extension of the results published by Sawik (2005b), where cyclic (stationary) and flexible (nonstationary) approaches for optimal material ordering in a make-to-order assembly and the corresponding MIP formulations have been compared. The cyclic approach, more common in practice, results in regular supplies: however, in most cases it incurs higher costs of material inventory holding (e.g., Güder and Zydiak 2000). In addition, the independent cycle approach with a common resource constraint (finite input buffer capacity) results in a nonlinear mixed integer program to be solved by enumeration that requires a large computational effort. In contrast, the flexible approach, similar to just-in-time procurement, enables material deliveries to be better coordinated with the aggregate production schedule and can be more easily accommodated for a common resource constraint (e.g., Güder et al., 1995). It leads to a higher synchronization of material supplies with the aggregate production schedule and as a result to a lower total cost. However, the just-in-time procurement system with the flexible approach requires a close cooperation between producer and material supplier.

EXERCISES

8.1 Reformulate model *FS* for flexible scheduling and model *CS* for independent cyclic scheduling of material supplies to account for the additional cost of material shortages, when supplies of some materials are delayed.

8.2 Generalize model *FS* for flexible scheduling of material supplies in the case of multiple suppliers with different ordering and delivery costs and different prices.

8.3 Assuming that for each supplier the maximum supply quantity and the total number of supplies over the planning horizon are limited, add the additional constraints to the model formulated in Exercise 8.2.

8.4 Suppose that common materials and product-specific materials are ordered from different suppliers using different ordering policies. What policy would you select to best coordinate supplies of materials from different suppliers?

8.5 Compare solution results for the flexible and the cyclic approach, presented in Section 8.4 (Tables 8.4 and 8.6, and Fig. 8.3) and explain why for both approaches, the supplies of product-specific materials are concentrated at the beginning of the planning horizon.

Chapter 9

Selection of Static Supply Portfolio in Supply Chains with Risks

9.1 INTRODUCTION

This chapter deals with a static (single-period) supplier selection and order allocation in a customer-driven supply chain with risks, that is with the allocation of orders for parts among the selected suppliers with no timing decisions, in contrast to the dynamic (multiperiod) supply portfolio, which allocates orders among the selected suppliers over a time horizon (see Chapter 10).

Supply chain risk management addresses two risk levels: operational risks and disruption risks. Operational risks are the inherent uncertainties arising from the problems of coordinating supply and demand, such as uncertain customer demand, uncertain supply, and uncertain cost. Disruption risks are major disruptions to normal activities caused by natural and man-made disasters, such as earthquakes, floods, hurricanes, equipment breakdowns, labor strikes, terrorist attacks, etc. In most cases, the business impact associated with disruption risks is much greater than that of operational risks. For example, Toyota production lines were shut down for two weeks when their sole supplier of the brake-fluid proportioning valves had a big fire in 1977 (Norrman and Jansson, 2004). The Taiwan earthquake of September 1999 created huge losses for many electronics companies supplied with components by Taiwanese manufacturers, for example, Apple lost many customer orders due to supply shortage of DRAM chips (Sheffi, 2005). Another example is the Philips microchip plant fire of March 2000 in New Mexico that resulted in 400 million euros in lost sales by a major cell phone producer, Ericsson (Norrman and Jansson, 2004).

In a make-to-order environment, customer-oriented manufacturers should be prepared to produce varieties of products to meet the different customer needs. Each product is typically composed of many common and non-common (custom) parts that can be sourced from different approved suppliers with different supply capacity. An important issue is how to best allocate the orders for parts among various

Scheduling in Supply Chains Using Mixed Integer Programming. By Tadeusz Sawik

part suppliers to fulfill all customer orders for products and to achieve a high customer service level at a low cost and, in addition, to mitigate the impact of supply chain risks. Since common parts can be efficiently managed by material requirement planning methods, this chapter focuses on custom parts that can be critical in make-to-order manufacturing. For custom-engineered products no inventory of custom parts can be kept on hand. Instead, the custom parts need to be requisitioned with each customer order and hence the custom parts inventory need not be considered. Furthermore, in stochastic supply settings, supplier selection allows the producer to decide whether it should cooperate with low-cost yet risky suppliers instead of more expensive but possibly more reliable suppliers. A common risk-neutral objective of minimizing expected cost is therefore influenced by uncertainty and risk. As a result, new non-risk-neutral objectives of minimizing the number of outcomes that could occur above an acceptable cost level are observed in practice. Given a set of customer orders for products, the decisionmaker needs to decide from which supplier to purchase custom parts required for each customer order. The above decisions are based on price, quality (defect rate), and reliability (on-time delivery) criteria that may conflict with each other, for example, the supplier offering the lowest price may not have the best quality or the supplier with the best quality may not deliver on time. To reduce the fixed ordering (transaction) costs the number of suppliers and the total number of orders should be minimized. On the other hand, the selection of more suppliers sometimes may divert the risk of unreliable supplies.

The problem of single-period allocation of orders for custom parts among suppliers in a make-to-order manufacturing environment with risks will be formulated as a single- or multi-objective binary or mixed integer program. The following MIP or IP models are presented:

SP for selection of static supply portfolio without discount to minimize expected cost under operational risks

SPD for selection of static supply portfolio with discount to minimize expected cost under operational risks

SP_E for selection of static supply portfolio to minimize expected cost under disruption risks

SP_CV for selection of static supply portfolio to minimize expected worst cost under disruption risks

WSP weighted-sum program for selection of bi-objective static supply portfolio to minimize weighted sum of expected cost and expected worst-case cost under disruption risks

CSP Chebyshev program for selection of bi-objective static supply portfolio to minimize weighted distance from a reference expected cost and expected worst-case cost under disruption risks

The *SP* and *SPD* models for selection of supply portfolio under conditions of operational risk associated with uncertain quality and reliability of supplies incorporate risk constraints. These constraints control the risk of defective or unreliable supplies by limiting the maximum number of delivery patterns (combinations of suppliers' delivery dates) for which the average defect rate or late delivery rate may exceed the maximum acceptable rates. In addition, in the *SPD* model the quantity or business volume discounts offered by the suppliers are considered.

In the *SP_CV*, *WSP*, and *CSP* models for selection of supply portfolio in the presence of supply chain disruption risks, the two percentile measures of risk popular in financial engineering, value-at-risk (VaR) and conditional value-at-risk (CVaR), are applied to control the risk of supply disruptions. The two different types of disruption scenarios are considered: scenarios with independent local disruptions of each supplier and scenarios with local and global disruptions that may result in disruption of all suppliers simultaneously. The resulting scenario-based optimization problem under uncertainty is formulated as a single- or bi-objective mixed integer program that can be solved using commercially available software for MIP. The proposed MIP models provide the decisionmaker with a simple tool for evaluating the relationship between expected and worst-case costs. For a finite number of scenarios, CVaR allows the evaluation of worst-case costs and shaping of the resulting cost distribution through optimal supplier selection and order allocation decisions, that is, the selection of an optimal supply portfolio.

The objective of the risk-neutral models *SP*, *SPD*, and *SP_E* is to minimize the expected costs, whereas in risk-averse models *SP_CV*, *WSP*, and *CSP*, the nonrisk-neutral objective of minimizing the expected worst-case costs is applied.

The proposed MIP models are illustrated with a set of computational examples in Sections 9.4 and 9.8.

9.2 SELECTION OF A SUPPLY PORTFOLIO WITHOUT DISCOUNT UNDER OPERATIONAL RISKS

In this section selection of a single-period supply portfolio in a nondiscount environment in the presence of operational risks associated with uncertain quality and reliability of supplies is considered and the MIP model for determining the static supply portfolio without discount is developed.

In the supply chain under consideration various types of products are assembled by a single producer to satisfy customer orders, using custom parts purchased from multiple suppliers (for notation used, see Table 9.1). Each supplier can provide the producer with custom parts for all customer orders. However, the suppliers have different limited capacity and, in addition, differ in price and quality of offered parts and in reliability of on-time delivery of parts. Let $I = \{1, \ldots, m\}$ be the set of m suppliers and $J = \{1, \ldots, n\}$ the set of n customer orders for the products, known ahead of time. Each order $j \in J$ is described by the quantity n_j of required custom parts and requested delivery date, where the latter need not be explicitly considered when selecting a supplier. Each supplier is assumed to have sufficient capacity to complete manufacturing and to deliver the ordered parts to the producer by the requested dates. All parts ordered from a supplier are shipped together, with a single shipment at one of a series of fixed delivery dates. The parts are dispatched to the producer at the

Table 9.1 Notation: Supply Portfolio without Discount

Indices

i = supplier, $i \in I = \{1, \ldots, m\}$
j = customer order, $j \in J = \{1, \ldots, n\}$
s = delivery pattern, $s \in S = \{1, \ldots, \sigma\}$

Input parameters

c_i = capacity of supplier i
n_j = number of parts to be purchased for customer order j
$N = \sum_{j \in J} n_j$—total demand for parts
o_i = cost of ordering parts from supplier i
p_{ij} = price of part for customer order j purchased from supplier i
q_{is} = expected defect rate of supplier i for delivery date in pattern s
r_{is} = expected late delivery rate of supplier i for delivery date in pattern s
$\overline{q}$ = the largest acceptable average defect rate of supplies
$\overline{r}$ = the largest acceptable average late delivery rate of supplies
$\overline{v}$ = the maximum allowed number of delivery patterns with the average defect rate or average late delivery rate of supplies greater than $\overline{q}$ or $\overline{r}$, respectively

Problem variables

v_s = 1, if for delivery pattern s the average defect rate or the average late delivery rate of supplies for the selected portfolio is, respectively, greater than $\overline{q}$ or greater than $\overline{r}$, otherwise $v_s = 0$ (portfolio selection variable)
x_i = the fraction of total demand for parts ordered from supplier i (order allocation variable)
y_i = 1, if an order for parts is placed on supplier i; otherwise $y_i = 0$ (supplier selection variable)
z_{ij} = 1, if parts required for customer order j are ordered from supplier i; otherwise $z_{ij} = 0$ (customer order assignment variable)

earliest fixed delivery date after the completion time of their manufacturing. Hence, for each supplier the delivery date and the corresponding reliability of supply depend on the completion time of manufacturing the ordered parts, which is unknown to the producer when the supplier selection decision is made. Likewise, the quality of supply may depend on the completion time. When the suppliers are selected, however, the risk of defective and unreliable supplies can be considered using past observations. Since different suppliers may complete manufacturing of ordered parts at different times, and then deliver the parts at different dates, a different risk can be associated with each combination of suppliers' delivery dates.

Let us call each combination of m fixed delivery dates, one delivery date for each supplier, a delivery pattern. Each delivery pattern must be feasible with respect to requested delivery dates. The total number of all feasible combinations of m fixed delivery dates consists of σ delivery patterns and let $S = \{1, \ldots, \sigma\}$ be the index set of all feasible delivery patterns. The probability that is assigned to the occurrence of each delivery pattern is identical and equals $1/\sigma$.

Let c_i be the capacity of supplier $i \in I$, o_i—cost of ordering parts from supplier $i \in I$, p_{ij}—purchasing price of part for customer order $j \in J$ from supplier $i \in I$, and q_{is}, r_{is}—respectively, the expected defect rate, the expected late delivery rate of supplier $i \in I$ for delivery date in pattern $s \in S$. The rates q_{is} and r_{is} are based on past observations.

We assume that the risk of defective or unreliable supplies from the selected suppliers can be measured by the number of delivery patterns for which the average defect rate or the late delivery rate of supplies are unacceptable.

The decisionmaker needs to decide from which supplier to purchase custom parts required for each customer order to achieve a low unit cost and high quality and reliability of supplies.

The static supply portfolio is defined as

$$(x_1, \ldots, x_m),$$

where

$$\sum_{i \in I} x_i = 1$$

and $0 \leq x_i \leq 1$ is the fraction of the total demand for parts ordered from supplier i (for definition of problem variables, see Table 9.1).

When deciding on a static portfolio of suppliers it is assumed that the orders for all parts are simultaneously placed on selected suppliers (e.g., at time 0), and each supplier delivers all the ordered parts at the earliest possible delivery date. Therefore, the allocation of orders for parts among the suppliers is not combined with the allocation of orders among the planning periods. Nevertheless, given past observations, the static portfolio should be checked over the time horizon against the risk of too low quality of purchased parts (too high defect rate) and too low reliability of supplies (too high late delivery rate), where both the quality and the reliability are randomly varying over time. The model incorporates risk constraints, where the risk of defective or unreliable supplies is controlled by the maximum number of delivery patterns (combinations of suppliers' delivery dates) for which the average defect rate or late delivery rate may exceed the maximum acceptable rates. The maximum number of delivery patterns and the corresponding maximum rates represent, respectively, the confidence level and the targeted rates above which a risk-averse decisionmaker wants to limit the number of outcomes.

Denote by $\overline{q}$, $\overline{r}$, and $\overline{v}$ the maximal acceptable defect rate of the portfolio, the maximal acceptable late delivery rate of the portfolio, and the maximum number of delivery patterns for which the average defect rate or the average late delivery rate of the portfolio can be above the threshold $\overline{q}$ or $\overline{r}$, respectively. The parameters $\overline{q}$, $\overline{r}$, and $\overline{v}$ are fixed by the decisionmaker to control the risk of defective and unreliable supplies.

We assume that the decisionmaker is willing to accept only portfolios for which the number of delivery patterns with the average defect rate greater then the threshold $\overline{q}$ or with the average late delivery rate greater than the threshold $\overline{r}$ is not greater than $\overline{v}$.

The performance of the static portfolio can be measured by the following two criteria:

$$f_1(y, z) = \sum_{i\in I} o_i y_i / N + \sum_{i\in I} \sum_{j\in J} n_j p_{ij} z_{ij} / N, \tag{9.1}$$

$$f_2(x) = \sum_{i\in I} \sum_{s\in S} (q_{is} + r_{is}) x_i / \sigma, \tag{9.2}$$

where $f_1(y, z)$ is the average ordering and purchasing cost per part, and $f_2(x)$ is the average defect and late delivery rate.

The overall performance can be measured by the sum $F(y, z)$ of average cost per part of ordering, purchasing, defects, and delays, in which the cost of a defective or delayed part is assumed to be identical with its price:

$$\begin{aligned} F(y, z) = & \sum_{i\in I} \left(o_i y_i + \sum_{j\in J} n_j p_{ij} z_{ij} \right) / N \\ & + \sum_{i\in I} \left(\sum_{s\in S} (q_{is} + r_{is}) / \sigma \right) \sum_{j\in J} n_j p_{ij} z_{ij} / N. \end{aligned} \tag{9.3}$$

The second summation term in (9.3) can be interpreted as the purchasing cost of additional parts required to compensate for defective and delayed parts. Clearly, the assumption that the cost of a defective or delayed part is identical with its price may be too relaxed in practice. The lack of all required parts may lead to unfulfilled customer orders and then the resulting shortage cost can be much higher than the purchasing cost of additional parts. On the other hand, parts not passing some quality acceptance level may not be paid for, and parts delivered late may be paid for at a reduced price.

The binary program *SP* for the single-objective supplier selection problem in a nondiscount environment is formulated below.

Model SP: *Selection of Supply Portfolio Without Discount under Operational Risks*

Minimize (9.3) subject to

1. *Order Assignment Constraints*:
   - for each customer order required parts are supplied by exactly one supplier,
   - for each supplier the total quantity of ordered parts cannot exceed its capacity,

$$\sum_{i\in I} z_{ij} = 1; \quad j \in J \tag{9.4}$$

$$\sum_{j\in J} n_j z_{ij} \le c_i; \quad i \in I \tag{9.5}$$

2. *Portfolio Selection Constraints*:
   - the portfolio definition constraint (note that the order allocation variable x_i is an auxiliary variable determined by z_{ij}),

- for each delivery pattern s, the portfolio with average defect rate greater than $\overline{q}$ is not acceptable,
- for each delivery pattern s, the portfolio with average late delivery rate greater than $\overline{r}$ is not acceptable,
- the portfolio can be unacceptable for at most $\overline{v}$ delivery patterns,
- parts are ordered from supplier i, if at least one customer order is assigned to supplier i,

$$x_i = \sum_{j \in J} n_j z_{ij}/N; \quad i \in I \tag{9.6}$$

$$v_s \geq \frac{\sum_{i \in I} q_{is} x_i - \overline{q}}{1 - \overline{q}}; \quad s \in S \tag{9.7}$$

$$v_s \geq \frac{\sum_{i \in I} r_{is} x_i - \overline{r}}{1 - \overline{r}}; \quad s \in S \tag{9.8}$$

$$\sum_{s \in S} v_s \leq \overline{v} \tag{9.9}$$

$$x_i \leq \left(\frac{C_i}{N}\right) y_i; \quad i \in I \tag{9.10}$$

$$y_i \leq \sum_{j \in J} z_{ij}; \quad i \in I \tag{9.11}$$

3. *Nonnegativity and Integrality Conditions*:

$$v_s \in \{0, 1\}; \quad s \in S \tag{9.12}$$

$$x_i \geq 0; \quad i \in I \tag{9.13}$$

$$y_i \in \{0, 1\}; \quad i \in I \tag{9.14}$$

$$z_{ij} \in \{0, 1\}; \quad i \in I, \quad j \in J. \tag{9.15}$$

Constraints (9.7) and (9.8) prevent the choice of portfolios whose average defect rate or whose average late delivery rate is above the fixed threshold $\overline{q}$, (9.7) or $\overline{r}$, (9.8), respectively. For each delivery pattern s such that the average defect rate $\sum_{i \in I} q_{is} x_i$ is greater than the largest acceptable rate $\overline{q}$, (9.7) or the average late delivery rate $\sum_{i \in I} r_{is} x_i$ is greater than the largest acceptable rate $\overline{r}$, (9.8), $v_s = 1$. Then all the delivery patterns with the average defect rate or with the average late delivery rate above the threshold are summed up in (9.9). If the result is greater than $\overline{v}$, then the portfolio is infeasible.

In view of risk constraints (9.7) to (9.9) and under the assumption of identical probability $1/\sigma$ that each delivery pattern is realized, $\overline{q}$ and $\overline{r}$ can be considered as the targeted average defect and late delivery rates based on the α-percentile of these rates, where $\alpha = (1 - \overline{v}/\sigma)$ represents the confidence level for the average rates across all delivery patterns. In other words, we allow $100(1 - \alpha)\% = 100(\overline{v}/\sigma)\%$ of the delivery patterns to exceed $\overline{q}$ and/or $\overline{r}$.

If different probability π_s is assigned to each delivery pattern $s \in S$, then constraint (9.9) should be replaced by

$$\sum_{s \in S} \pi_s v_s \leq 1 - \alpha, \qquad (9.9')$$

where the confidence level $\alpha \in (0, 1)$ is fixed by the decisionmaker to control the risk of defective and unreliable supplies, and the higher the confidence level, the more risk-averse is the selected supply portfolio.

Notice that $\overline{v} = 0$ implies $\sum_{i \in I} q_{is} x_i \leq \overline{q}$ and $\sum_{i \in I} r_{is} x_i \leq \overline{r} \; \forall \, s \in S$, that is, the selected portfolio must be acceptable for each delivery pattern. In particular, for the stationary expected defect rates $q_{is} = q_i \; \forall \, i \in I,\, s \in S$ and the stationary expected late delivery rates $r_{is} = r_i \; \forall \, i \in I,\, s \in S$, $\overline{v}$ can take on either value 0 or σ, since the selected portfolio can be either acceptable or unacceptable over the entire set of all possible delivery patterns. Hence, in the stationary case variable v can be eliminated from the model.

Since the continuous allocation variables x_i are auxiliary variables defined by the binary assignment variables z_{ij}, x_i can be replaced by z_{ij} using relation (9.6), and hence the *SP* model is a pure binary program.

9.3 SELECTION OF SUPPLY PORTFOLIO WITH DISCOUNT UNDER OPERATIONAL RISKS

In a make-to-order environment, in which custom parts are typically ordered in small lot sizes, a supplier may sometimes offer a discount that depends on total value of sales volume (business volume) or on total quantity of ordered parts. In the context of a business volume discount the quantity or variety of purchased parts does not affect the offered price, while for the quantity discount the price does not depend on the total "dollar" amount of the sales volume.

In this section the presence of price discounts offered by suppliers is considered; first, selection of a supplier with business volume discount and then selection of a supplier with quantity discount, based on the total value and the total quantity, respectively, of the order (for notation used, see Table 9.2).

Assume that each supplier i offers cumulative (all-units) price breaks having g_i discount intervals

$$(\beta_{i0}, \beta_{i1}], \ldots, (\beta_{i,g_i-1}, \beta_{i,g_i}]$$

according to total business volume (or total quantity), where β_{ik} is the upper limit on *kth* ($k \in K_i = \{1, \ldots, g_i\}$) business volume (or quantity) discount interval $(\beta_{ik-1}, \beta_{ik}]$ and $\beta_{i0} = 0 \forall i$.

Let $0 < a_{ik} < 1$ be the discount rate (percentage of discount) associated with interval k of supplier i. If total business volume (total quantity) from supplier i falls on interval k, then the price of each part for customer order j is $(1 - a_{ik})p_{ij}$.

Table 9.2 Notation: Supply Portfolio with Discount

Indices

k = discount interval of supplier i, $k \in K_i = \{1, \ldots, g_i\}$

Input parameters

K_i = set of discount intervals of supplier i

a_{ik} = price discount rate associated with discount interval k of supplier i

β_{ik} = upper limit on discount interval k of supplier i, $0 = \beta_{i0} \leq \beta_{i1} \leq \cdots \leq \beta_{ig_i}$

Problem variables

u_{ik} = 1, if total business volume (or total quantity) from supplier i falls on the discount interval k; otherwise $u_{ik} = 0$

Z_{ijk} = 1, if parts required for customer order j are ordered from supplier i and total business volume (or total quantity) of parts purchased from this supplier falls on the discount interval k; otherwise $Z_{ijk} = 0$

The following discount interval selection variable needs to be added to enhance the supplier selection and order allocation problem *SP* for the discount environment:

$u_{ik} = 1$, if total business volume (or total quantity) from supplier i falls on the discount interval k; otherwise $u_{ik} = 0$.

Furthermore, the customer order assignment variable z_{ij} needs to be modified as below.

$Z_{ijk} = 1$, if parts required for customer order j are ordered from supplier i and total business volume (or total quantity) of parts purchased from this supplier falls on the discount interval k; otherwise $Z_{ijk} = 0$.

The total business volume or total quantity of parts purchased from supplier i is $\sum_{j \in J} \sum_{k \in K_i} n_j p_{ij} Z_{ijk}$ or $\sum_{j \in J} \sum_{k \in K_i} n_j Z_{ijk}$, respectively. Notice that the total business volume or the total quantity purchased from each supplier falls on exactly one discount interval of this supplier.

Now, the average purchasing and ordering cost per part can be expressed as follows:

$$f_1(y, Z) = \sum_{i \in I} o_i y_i / N + \sum_{i \in I} \sum_{j \in J} \sum_{k \in K_i} (1 - a_{ik}) n_j p_{ij} Z_{ijk} / N. \tag{9.16}$$

The overall performance of the static portfolio with discount can be measured by the sum $F(y, Z)$ of the average cost per part of ordering, purchasing, defects, and delays, in which the cost of a defective or delayed part is assumed to be identical with its regular price:

$$F(y, Z) = \sum_{i \in I} \left(o_i y_i + \sum_{j \in J} \sum_{k \in K_i} (1 - a_{ik}) n_j p_{ij} Z_{ijk} \right) / N$$

$$+ \sum_{i \in I} \left(\sum_{s \in S} (q_{is} + r_{is}) / \sigma \right) \sum_{j \in J} \sum_{k \in K_i} n_j p_{ij} Z_{ijk} / N. \quad (9.17)$$

The binary program *SPD* for the single-objective supplier selection problem with a business volume discount or quantity discount is formulated below.

Model SPD: *Selection of Supply Portfolio with Discount under Operational Risks*

Minimize (9.17) subject to (9.7)–(9.10), (9.12)–(9.14) and

1. *Order Assignment Constraints*:

- the parts required for each customer order are supplied by exactly one supplier in one discount interval,
- for each supplier the total quantity of ordered parts cannot exceed its capacity,

$$\sum_{i \in I} \sum_{k \in K_i} Z_{ijk} = 1; \quad j \in J \quad (9.18)$$

$$\sum_{j \in J} \sum_{k \in K_i} n_j Z_{ijk} \le c_i; \quad i \in I \quad (9.19)$$

2. *Portfolio Selection Constraints*:

- the portfolio definition constraint (note that the order allocation variable x_i is an auxiliary variable determined by Z_{ijk}),
- an order for parts is placed on supplier i, if for at least one customer order the required parts are ordered from supplier i,

$$x_i = \sum_{j \in J} \sum_{k \in K_i} n_j Z_{ijk} / N; \quad i \in I \quad (9.20)$$

$$y_i \le \sum_{j \in J} \sum_{k \in K_i} Z_{ijk}; \quad i \in I \quad (9.21)$$

3. *Discount Constraints*:

- total business volume purchased from each selected supplier can be exactly in one discount interval

$$(\beta_{i,k-1} + 1) u_{ik} \le \sum_{j \in J} n_j p_{ij} Z_{ijk} \le \beta_{ik} u_{ik}; \quad i \in I, k \in K_i, \quad (9.22)$$

for a business volume discount environment,

or

- total quantity purchased from each selected supplier can be exactly in one discount interval

$$(\beta_{i,k-1} + 1)u_{ik} \leq \sum_{j \in J} n_j Z_{ijk} \leq \beta_{ik} u_{ik}; \quad i \in I,\ k \in K_i, \tag{9.23}$$

for a quantity discount environment,

- parts are ordered from supplier i, if supplier i is selected and assigned at least one customer order,

$$\sum_{k \in K_i} u_{ik} = y_i; \quad i \in I \tag{9.24}$$

$$Z_{ijk} \leq u_{ik}; \quad i \in I, j \in J, k \in K_i \tag{9.25}$$

4. *Integrality Conditions*:

$$u_{ik} \in \{0, 1\}; \quad i \in I, k \in K_i \tag{9.26}$$

$$Z_{ijk} \in \{0, 1\}; \quad i \in I, j \in J, k \in K_i. \tag{9.27}$$

9.4 COMPUTATIONAL EXAMPLES

In this section some computational examples are presented to illustrate possible applications of the proposed MIP approach for supplier selection of a static supply portfolio in a nondiscount or discount environment under operational risks. The following parameters have been used for the example problems:

- σ, the number of delivery patterns, was equal to 30.
- m, the number of suppliers, was equal to 20.
- n, the number of customer orders, was equal to 100.
- o_i, the cost of ordering parts from supplier i, was equal to 5000 for each supplier i.
- p_{ij}, the unit price of parts required for each customer order j, purchased from each supplier i was uniformly distributed over [10,15] (i.e., drawn from U[10;15]) and reduced by the factor $(1 - \max_{s \in S}(q_{is} + r_{is}))$ to get a lower price for parts from the suppliers with higher defect and late delivery rates.
- q_{is} and r_{is}, the expected defect rate and the expected late delivery rate of each supplier i and delivery date in each pattern s, was uniformly distributed over (0,0.08) and (0,0.10), respectively.
- $\bar{q}$ and $\bar{r}$, the largest acceptable average defect rate and late delivery rate, were equal to 0.05 and 0.06, respectively.

- $\overline{v}$, the maximum allowed number of delivery patterns with the average defect or late delivery rates greater than $\overline{q} = 0.05$ or $\overline{r} = 0.06$, respectively, was equal to 0, 1, or 2.
- n_j, the numbers of required parts for each customer order, were integers uniformly distributed over [100, 5000].
- c_i, the capacity of each supplier i, was equal to $\lceil 2\sum_{j\in J} n_j/m \rceil$ ($\lceil \cdot \rceil$ denotes the smallest integer not less than $\cdot$), that is, the total capacity of all suppliers was equal to the double total demand for parts.
- quantity discount: $g_i = 3$ discount intervals for each supplier i, with the upper limits $\beta_{ik} = \lceil kc_i/g_i \rceil$; $i \in I$, $k = 1, \ldots, g_i$.
- business volume discount: $g_i = 3$ discount intervals for each supplier i, with the upper limits $\beta_{ik} = \lceil k\bar{p}_i c_i/g_i \rceil$; $i \in I$, $k = 1, \ldots, g_i$, where $\bar{p}_i = \sum_{j\in J} p_{ij}/n$ is the average unit price of parts from supplier i.
- a_{ik}, discount coefficients, were equal to $0.05(k-1)$; $i \in I$, $k = 1, \ldots, g_i$, that is, the maximum percentage discount is 10%.

The computational results for the three risk levels represented by the maximum number of delivery patterns $\overline{v} = 1, 2, 3$ for which the average defect rate or late delivery rate of supplies can be unacceptable are summarized in Table 9.3. The size of the MIP models *SP* and *SPD*, respectively, for the supplier selection without and with discount (business volume or quantity) environment are represented by the total number of variables, Var., number of binary variables, Bin., and number of constraints, Cons. The last two columns of the table present the solution values and CPU time in seconds

Table 9.3 Computational Results

$\overline{v}$	Var.	Bin.	Cons.	$F\,(f_1, f_2)^a$, No. of suppliers^b	CPU^c
Nondiscount environment (model *SP*)					
0	2040	2020	241	9.108 (9.366, 0.090), 13	17
1	2070	2050	242	9.027 (9.321, 0.087), 13	7
2	2070	2050	242	9.004 (9.283, 0.089), 12	15
Business volume discount (model *SPD*)					
0	6120	6100	6381	8.335 (7.592, 0.090), 12	2191
1	6130	6110	6382	8.239 (7.528, 0.087), 12	1265
2	6130	6110	6382	8.207 (7.494, 0.087), 11	292
Quantity discount (model *SPD*)					
0	6120	6100	6381	8.392 (7.580, 0.090), 12	234
1	6130	6110	6382	8.226 (7.515, 0.087), 13	380
2	6130	6110	6382	8.204 (7.495, 0.087), 12	140

[a] $F(f_1, f_2)$ = the average cost (the average ordering and purchasing cost per part, the average defect and late delivery rate), respectively.

[b] The number of selected suppliers.

[c] CPU seconds for proving optimality on a PC Pentium IV, 1.8 GHz, 1 GB RAM/CPLEX 11.

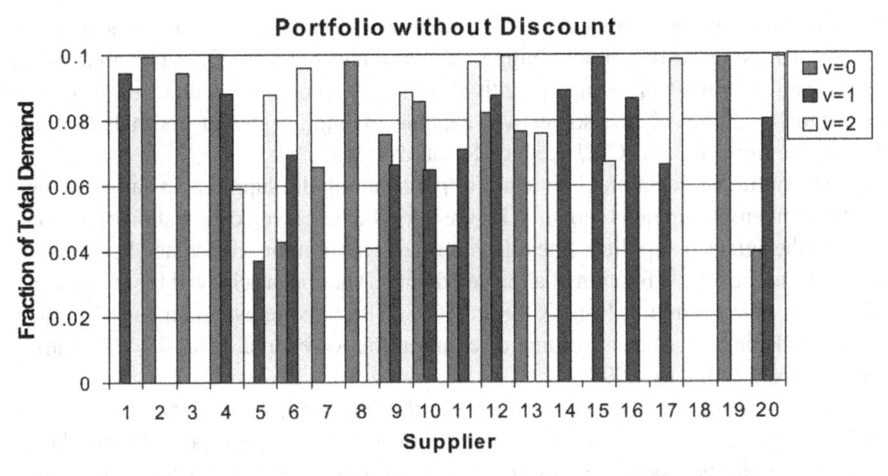

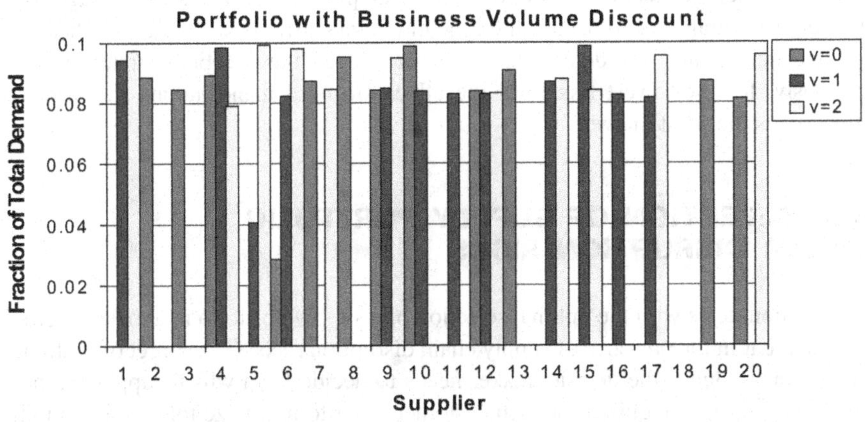

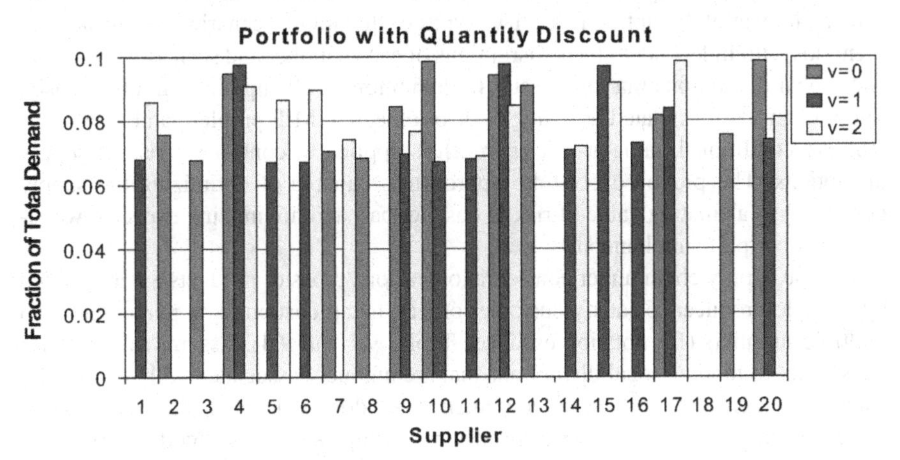

Figure 9.1 Optimal supply portfolios.

required to prove optimality of the solution. The computational experiments were performed using AMPL programming language and the CPLEX v.11 solver (with the default settings) on a laptop with Pentium IV processor running at 1.8 GHz and with 1 GB RAM The solver was capable of finding proven optimal solutions within the limit of 3600 CPU seconds for all examples.

The optimal allocation of demand for parts among the suppliers in a nondiscount or discount environment is shown in Figure 9.1. The best leveled allocation of demand among the selected suppliers (the allocation with a minimum difference between the highest and the lowest fraction of allocated demand) has been achieved for the quantity discount, with x_i ranging from 0.068 to 0.098. The most uneven demand allocation was found for the business volume discount at the lowest risk level ($\overline{v} = 0$), with x_i ranging from 0.028 to 0.098.

The computational results demonstrate that the greater the allowed number of delivery patterns $\overline{v}$ with the average defect and late delivery rates above the thresholds (i.e., the higher the risk of defective or delayed supplies), the lower is the average cost. The largest average cost is achieved for the lowest risk level ($\overline{v} = 0$), that is, where the average defect and late delivery rates never exceed the acceptable rates. Simultaneously, for the lowest risk level the allocation of demand among the selected suppliers is the most uneven.

9.5 SELECTION OF SUPPLY PORTFOLIO UNDER DISRUPTION RISKS

This section deals with the optimal selection of a supply portfolio in a make-to-order environment in the presence of supply chain disruption risks. Given a set of customer orders for products, the decisionmaker needs to decide from which supplier to purchase custom parts required for each customer order to minimize total cost and mitigate the impact of disruption risks. Two types of disruption scenarios are considered: scenarios with independent local disruptions of each supplier and scenarios with local and global disruptions that may result in disruption of all suppliers simultaneously. The problem is formulated as a single- or bi-objective MIP problem and a value-at-risk and conditional value-at-risk approach is applied to control the risk of supply disruptions. The proposed portfolio approach is capable of optimizing the supply portfolio by calculating value-at-risk of cost per part and minimizing expected worst-case cost per part simultaneously.

In the supply chain under consideration various types of products are assembled by a single producer to satisfy customer orders, using custom parts purchased from multiple suppliers (for notation used, see Tables 9.1 and 9.4). Assume that supplies are subject to random local disruptions that are uniquely associated with a particular supplier, which may arise from equipment breakdowns, local labor strikes, bankruptcy, from local natural disasters such as earthquakes, fires, floods, hurricanes, etc. Let $\delta_i \in \{0, 1\}$ denote the binary random variable of operational state of supplier i, where $\delta_i = 1$, if supplier i operates without disruptions, otherwise $\delta_i = 0$. Denote by π_i the local disruption probability for supplier i, that is, the parts ordered from supplier

Table 9.4 Notation: Supply Portfolio under Disruption Risks

Indices

s = disruption scenario, $s \in S = \{1, \ldots, \sigma\}$

Input parameters

b_j = the per unit shortage cost for customer order j

q_i = expected defect rate of supplier i

α = confidence level

π_i = the local disruption probability for supplier i

π^* = the global disruption probability for all suppliers

Problem variables

$\mathcal{T}_s$ = the amount by which cost of portfolio in scenario s exceeds VaR (tail cost)

VaR = the targeted cost of portfolio based on the α-percentile of costs, that is, in $100\alpha\%$ of scenarios, the outcome cannot exceed VaR (value-at-risk)

z_{ij} = the fraction of parts required for customer order j ordered from supplier i (customer order allocation variable)

i are delivered without disruptions with probability $(1 - \pi_i)$, or not at all with probability π_i. Let P_s be the probability that disruption scenario s is realized, where each scenario $s \in S$ is comprised of a unique subset $I_s \subset I$ of suppliers who deliver parts without disruptions, and $S = \{1, \ldots, \sigma\}$ is the index set of all scenarios (note that there are a total of $\sigma = 2^m$ potential scenarios). The probability of disruption scenario s in the presence of independent local disaster events is

$$P_s = \prod_{i \in I_s} (1 - \pi_i) \cdot \prod_{i \notin I_s} \pi_i.$$

Assume that in addition to independent local disruptions of each supplier, there are potential global disasters that may result in disruption of all suppliers simultaneously. For example, such global events may include economic crisis, terrorist attack, widespread labor strike in a transportation sector, etc. Although the probability of such events usually is very low, their consequences may be very high. Let π^* be the probability of simultaneous global disruption of all suppliers due to some disaster. The global disaster and the local disasters at each supplier are assumed to be independent events; therefore; the probability P_s^* of each disruption scenario $s \in S$ under the risks of both type of events is

$$P_s^* = \begin{cases} (1 - \pi^*)P_s & \text{if } I_s \neq \emptyset \\ \pi^* + (1 - \pi^*) \prod_{i \in I} \pi_i & \text{if } I_s = \emptyset. \end{cases}$$

Note that if the probability of global disruption $\pi^* = 0$, then the probability P_s^* reduces to P_s for independent local disaster events.

The producer does not need to pay for ordered and undelivered parts. However, it is charged with a much higher cost of unfulfilled customer orders for products, caused by the shortage of parts, undelivered due to supply disruptions. Let b_j be the per unit cost of unfulfilled customer order j.

The decisionmaker needs to decide from which suppliers to purchase custom parts required for each customer order to achieve a minimum cost of ordering, purchasing, defects, and shortages and to mitigate the impact of disruption risks by minimizing the potential worst-case cost.

9.6 SINGLE-OBJECTIVE SUPPLY PORTFOLIO UNDER DISRUPTION RISKS

In this section two MIP models are proposed for a single-period supplier selection and order allocation problem, that is, for determining a static supply portfolio (for definition of problem variables, see Table 9.4).

When deciding on a static supply portfolio it is assumed that the orders for all parts are placed simultaneously on selected suppliers (e.g., at time 0), and each supplier delivers all the ordered parts at the earliest possible delivery date. Therefore, the allocation of orders for parts among the suppliers is not combined with the allocation of orders among the planning periods. Nevertheless, the static portfolio should be checked against the risk of supply disruptions across all potential disruption scenarios.

9.6.1 Value-at-Risk vs. Conditional Value-at-Risk

This subsection briefly defines and compares VaR and CVaR, the two popular percentile measures of risk; see Sarykalin et al. (2008).

Let $F_W(u) = Prob\{W \leq u\}$ be the cumulative distribution function of a random variable W representing cost. The VaR of W with the confidence level $\alpha \in [0, 1)$ is defined as a lower α-percentile of the random variable W

$$VaR_\alpha(W) = \min\{u\colon F_W(u) \geq \alpha\}.$$

VaR represents the maximum cost associated with a specified confidence level of outcomes (i.e., the likelihood that a given portfolio's costs will not exceed the amount defined as VaR). However, VaR does not account for properties of the cost distribution beyond the confidence level and hence does not explain the magnitude of the cost when the VaR limit is exceeded.

On the other hand, CVaR focuses on the tail of the cost distribution, that is, on outcomes with the highest cost. Assuming that $F_W(u)$ is a continuous distribution function, the CVaR of W with the confidence level $\alpha \in [0, 1)$, $CVaR_\alpha(W)$, equals the expectation of W subject to $W \geq VaR_\alpha(W)$. However, in the general case $CVaR_\alpha(W)$ is not equal to an average of outcomes greater than $VaR_\alpha(W)$ and is defined as the mean of the generalized α-tail distribution

$$CVaR_\alpha(W) = \int_{-\infty}^{\infty} u dF_W^\alpha(u),$$

where

$$F_W^\alpha(u) = \begin{cases} 0 & \text{if } u < VaR_\alpha(W) \\ \dfrac{F_W(u) - \alpha}{1 - \alpha} & \text{if } u \geq VaR_\alpha(W). \end{cases}$$

$CVaR_\alpha(W)$ can be considered as a generalization of the expected value, when $\alpha = 0$ they are equivalent. On the other hand, the higher the confidence level α, the closer are values of $VaR_\alpha(W)$ and $CVaR_\alpha(W)$.

Alternatively, $CVaR_\alpha(W)$ can be defined as the weighted average of $VaR_\alpha(W)$ and the conditional expectation of W subject to $W > VaR_\alpha(W)$, for example, Sarykalin et al. (2008).

When the distribution function has a vertical jump at $VaR_\alpha(W)$ (the probability interval of confidence level α with the same VaR), a probability "atom" is said to be present at $VaR_\alpha(W)$. For example, when the distribution is modeled by scenarios, the probability measure is concentrated in finitely many points and the corresponding distribution function is a step function (constant between jumps) with jumps at those points. Since $CVaR_\alpha(W)$ is obtained by averaging a fractional number of scenarios, the $VaR_\alpha(W)$ atom can be split. When $F_W(VaR_\alpha(W)) > \alpha$, then probability $1 - F_W(VaR_\alpha(W))$ of the cost interval $[VaR_\alpha(W), \infty)$ is smaller than $1 - \alpha$.

Note that if $F_W(VaR_\alpha(W)) = 1$, so that $VaR_\alpha(W)$ is the highest cost that may occur, then $CVaR_\alpha(W) = VaR_\alpha(W)$.

Summarizing (see also Fig. 9.2) the above definitions (from now on, VaR and CVaR will be denoted without subscript α):

- Value-at-risk (VaR) at a $100\alpha\%$ confidence level is the targeted cost of the portfolio such that for $100\alpha\%$ of the scenarios, the outcome will not exceed VaR. In other words, VaR is a decision variable based on the α-percentile of costs, that is, in $100(1 - \alpha)\%$ of the scenarios, the outcome may exceed VaR.
- Conditional value-at-risk (CVaR) at a $100\alpha\%$ confidence level is the expected cost of the portfolio in the worst $100(1 - \alpha)\%$ of the cases. In other words, we allow $100(1 - \alpha)\%$ of the outcomes to exceed VaR, and the mean value of these outcomes is represented by CVaR.

In other words, VaR is the acceptable cost level above which we want to minimize the number of outcomes, and CVaR considers those portfolio outcomes where costs exceed VaR. Since VaR and CVaR measure different parts of the cost distribution, VaR may be better for optimizing portfolios when good models for tails are not available; otherwise CVaR may be preferred.

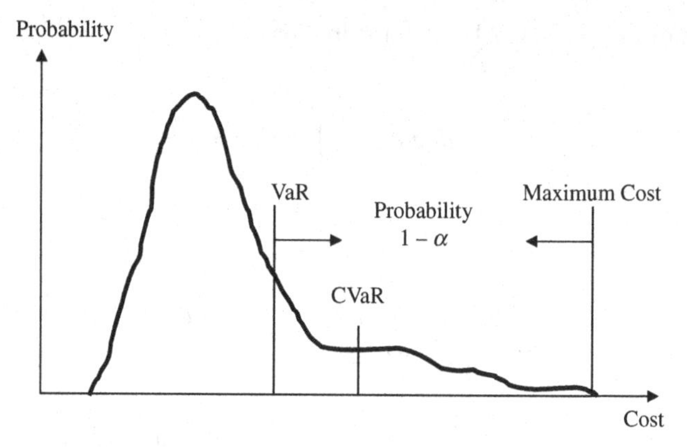

Figure 9.2 VaR versus CVaR.

9.6.2 Minimization of Expected Costs

The overall quality of the supply portfolio can be measured by the expected cost per part, $E(x)$, (9.28), of ordering, $\sum_{i\in I} o_i y_i/N$, purchasing and defects, $\sum_{i\in I}\sum_{j\in J} n_j p_{ij}(1+q_i)z_{ij}/N$, in which the cost of a defective part is assumed to be identical with its price, and shortage of parts due to supply disruptions (cost of unfulfilled orders less cost of nondelivered parts), $\sum_{s\in S}\sum_{i\notin I_s}\sum_{j\in J} P_s n_j(b_j - p_{ij}(1+q_i))z_{ij}/N$.

The MIP model *SP_E* for selection of a supply portfolio is formulated below. In the proposed model, the portfolio will be optimized by minimizing expected cost per part $E(x)$, (9.28).

Model SP_E: *Selection of a Supply Portfolio to Minimize Expected Cost per Part under Disruption Risks*

Minimize expected cost per part

$$E(x) = \sum_{i\in I} o_i y_i/N + \sum_{i\in I}\sum_{j\in J} n_j p_{ij}(1+q_i)z_{ij}/N + \sum_{s\in S}\sum_{i\notin I_s}\sum_{j\in J} P_s n_j(b_j - p_{ij}(1+q_i))z_{ij}/N \qquad (9.28)$$

subject to

1. *Supplier Selection and Order Allocation Constraints*:
   - for each customer order all required parts must be provided by selected suppliers,

- for each selected supplier the total quantity of ordered parts cannot exceed its capacity,
- at least one customer order should be assigned to each selected supplier,

$$\sum_{i \in I} z_{ij} = 1; \quad j \in J \tag{9.29}$$

$$\sum_{j \in J} n_j z_{ij} \le c_i y_i; \quad i \in I \tag{9.30}$$

$$y_i \le \sum_{j \in J} z_{ij}; \quad i \in I \tag{9.31}$$

2. *Nonnegativity and Integrality Conditions:*

$$y_i \in \{0, 1\}; \quad i \in I \tag{9.32}$$

$$z_{ij} \in [0, 1]; \quad i \in I, j \in J. \tag{9.33}$$

The resulting supply portfolio $(x_1, \ldots, x_m)$ is determined by the customer order allocation variables z_{ij}

$$x_i = \sum_{j \in J} n_j z_{ij} / N; \quad i \in I. \tag{9.34}$$

Note that unlike the *SP* and *SPD* models, in the proposed model the parts required for each customer order are assumed to be partially provided by one or more suppliers and the customer order allocation variable z_{ij} represents the fraction of all parts required for order j provided by supplier i. In some practical cases, all custom parts of the same type that are required for a customer order are purchased from a single supplier. Then the corresponding continuous allocation variable z_{ij} should be redefined as a binary assignment variable denoting whether or not all parts required for order j are provided by supplier i. If all z_{ij} are defined to be binary variables, then *SP_E* becomes a pure binary program, similar to the *SP* and *SPD* models, with the auxiliary continuous variables x_i replaced by binary variables z_{ij}.

The *SP_E* model will be used to compare the risk-neutral results with those obtained by applying a risk aversive decision-making model described in the next subsection.

9.6.3 Minimization of Expected Worst-Case Costs

In the supply portfolio selection under disruption risks, the confidence level α is fixed by the decisionmaker to control the risk of losses due to supply disruptions. We assume that the decisionmaker is willing to accept only portfolios for which the total probability of scenarios with costs greater than VaR is not greater than $1 - \alpha$.

Furthermore, in the model presented below, a risk-averse decisionmaker wants to minimize the expected worst-case costs exceeding VaR.

Define $\mathcal{T}_s$ as the tail cost for scenario s, where tail cost is defined as the amount by which costs in scenario s exceed VaR.

The portfolio will be optimized by calculating VaR and minimizing CVaR simultaneously. By measuring CVaR, the magnitude of the tail costs is considered to achieve a more accurate estimate of the risks of minimizing cost. When using CVaR to minimize worst-case costs, CVaR is always not less than VaR. In the proposed model CVaR is represented by an auxiliary function (9.35) introduced by Rockafellar and Uryasev (2000). The MIP model *SP_CV* for selection of a static supply portfolio to reduce the risk of high costs due to supply disruptions is formulated below.

Model SP_CV: *Selection of Supply Portfolio to Minimize Expected Worst-Case Cost per Part under Disruption Risks*

Minimize expected worst-case cost (CVaR)

$$C(x) = VaR + (1-\alpha)^{-1}\sum_{s\in S} P_s\mathcal{T}_s \tag{9.35}$$

subject to

1. *Supplier Selection and Order Allocation Constraints*: (9.29) to (9.31)
2. *Risk Constraints*:
   - the tail cost for scenario s is defined as the nonnegative amount by which cost per part in scenario s exceeds VaR,

$$\mathcal{T}_s \geq \sum_{i\in I} o_i y_i/N + \sum_{i\in I}\sum_{j\in J} n_j p_{ij}(1+q_{is})z_{ij}/N + \sum_{i\notin I_s}\sum_{j\in J} n_j(b_j - p_{ij}(1+q_i))z_{ij}/N - VaR; \quad s\in S \tag{9.36}$$

3. *Nonnegativity and Integrality Conditions*: (9.32), (9.33), and

$$\mathcal{T}_s \geq 0; \quad s\in S. \tag{9.37}$$

Note that as $\mathcal{T}_s$ is constrained of being positive, the model tries to decrease VaR and hence positively impact the objective function. However, large reduction in VaR may result in more scenarios with positive tail costs.

If for some customer order j all required parts must be supplied by a single supplier, then the corresponding nonnegative allocation variables z_{ij}, $i\in I$ should be redefined to be binary assignment variables, as for the *SP_E* model.

Notice that the number of variables and constraints in the proposed mixed integer program grows linearly in the number $\sigma = 2^m$ of scenarios considered and hence exponentially in the number m of suppliers. For example, a 10-supplier problem has approximately 2000 decision variables and 3000 constraints, while for a 20-supplier problem, this increases to over 2,000,000 variables and 3,000,000 constraints. Even the construction of problems of this size becomes intractable, for example, Chahar and Taaffe (2009).

9.7 BI-OBJECTIVE SUPPLY PORTFOLIO UNDER DISRUPTION RISKS

In the single-objective approach the supply portfolio is selected by minimizing either the expected cost per part, $E(x)$, (9.28) or the expected worst-case cost per part, $C(x)$, (9.35). In this section the two cost functions are considered simultaneously, and a bi-objective selection of the supply portfolio is presented aimed at minimizing both objective functions to balance expected costs with the risk tolerance. This trade-off model is known as the mean-risk model (e.g., Ogryczak and Ruszczynski, 2002), formulated as the optimization of a composite objective consisting of the expected cost and the CVaR as a risk measure.

The nondominated solution set of the bi-objective supply portfolio can be found by the parameterization on λ the weighted-sum program *WSP* or the reference point based program *CSP* presented below. The scalarizing MIP models are based on *SP_CV* model with the addition of objective (9.28) of the *SP_E* model.

Model WSP: *Bi-Objective Selection of Supply Portfolio to Minimize Weighted Sum of per Unit Expected Cost and Expected Worst-Case Cost under Disruption Risks*

Minimize

$$\lambda E(x) + (1 - \lambda)C(x) \tag{9.38}$$

where $0 \leq \lambda \leq 1$, subject to (9.28)–(9.33), (9.35)–(9.37).

In the next model, let $(\underline{E}, \underline{C})$ be a reference point in the criteria space such that $\underline{E} < E(x)$ and $\underline{C} < C(x)$ for all feasible supply portfolios x.

Model CSP: *Bi-Objective Selection of Supply Portfolio to Minimize the Augmented Weighted Chebyshev Metric under Disruption Risks*

Minimize

$$\gamma + \varepsilon(E(x) + C(x)) \tag{9.39}$$

where ε is a small positive value,
subject to (9.28)–(9.33), (9.35)–(9.37), and

1. *Weighted Distance Constraints:*
   - the weighted distance in the criteria space of each solution value from the reference point cannot be greater than the maximum distance γ to be minimized,

$$\lambda(E(x) - \underline{E}) \leq \gamma \tag{9.40}$$

$$(1 - \lambda)(C(x) - \underline{C}) \leq \gamma \tag{9.41}$$

$$\gamma \geq 0, \tag{9.42}$$

where $0 \leq \lambda \leq 1$.

The Chebyshev program *CSP* is based on the augmented $\lambda -$ weighted Chebyshev metric $\min\{\lambda|E(x) - \underline{E}| + \varepsilon(E(x) + C(x)), (1 - \lambda)|C(x) - \underline{C}| + \varepsilon(E(x) + C(x))\}$.

It should be pointed out that for mixed integer programs, there may be portions of the nondominated set (nearby weakly nondominated solution) that both above approaches are unable to compute, even if the complete parameterization on λ is attempted, and very small ε is considered, for example, Steuer (1986).

9.8 COMPUTATIONAL EXAMPLES

In this section some computational examples are presented to illustrate possible applications of the proposed MIP approach for selection of a supply portfolio under disruption risks. The following parameters have been used for the example problems:

- m, the number of suppliers, was equal to 10 and the number $\sigma = 2^m$ of disruption scenarios, was equal to 1024.
- n, the number of customer orders, was equal to 50.
- n_j, the numbers of required parts for each customer order, were integers uniformly distributed over [100, 500], that is, generated from a U[100;500] distribution.
- c_i, the capacity of each supplier i, was equal to $\lceil 2\sum_{j \in J} n_j/m \rceil$, that is, the total capacity of all suppliers was equal to double the total demand for parts.
- b_j, the per unit shortage cost for customer order j, was equal to 100 for all orders j.
- o_i, the cost of ordering parts from supplier i, was equal to 500 for each supplier i.
- q_i, the expected defect rate of each supplier i, was exponentially distributed, ranging from 0.0003 to 0.03.
- p_{ij}, the unit price of parts required for each customer order j purchased from each supplier i, was uniformly distributed over [10,15] (i.e., drawn from

U[10;15]) and reduced by the factor $(1 - q_i)$ to get a lower price for parts from suppliers with a higher defect rate.

- π_i, the local disruption probability was uniformly distributed over [0,0.06], that is, the disruption probabilities were drawn independently from U[0;0.06].
- π^*, the global disruption probability was equal to 0.01.
- α, the confidence level, was equal to 0.50, 0.75, 0.90, 0.95, or 0.99.

For the test examples, the total demand for parts is $N = \sum_{j \in J} n_j = 14{,}750$ parts.

Solution results for the scenarios with local and with local and global disruptions for the risk-neutral model *SP_E* are shown in Table 9.5, and for the risk-averse model *SP_CV* with different confidence levels, in Tables 9.6 and 9.7. The size of the MIP models is represented by the total number of variables, Var., number of binary variables, Bin., number of constraints, Cons, and number of nonzero coefficients in the constraint matrix, Nonz. The tables also present CPU time in seconds required to prove optimality of the solution and the probability $1 - F(VaR)$ of outcomes with worst-case cost above VaR. The tables demonstrate that the number of selected suppliers increases with the confidence level α, which indicates that the impact of disruption risks is mitigated by diversification of the supply portfolio. Note that VaR becomes smaller than expected cost when $\alpha = 0.50$ and $\alpha = 0.75$.

Table 9.5 Solution Results for the *SP_E* Model

Disruption scenario	Var.	Bin.	Cons.	Nonz.	Mean cost, no. of suppliers	CPU[a]
Local	520	10	81	2040	12.30, 5	0.01
Local & global	520	10	81	2040	13.18, 5	0.02

[a]CPU seconds for proving optimality on a PC Pentium IV, 1.8 GHz, 1 GB RAM/CPLEX 11.

Table 9.6 Solution Results for the *SP_CV* Model: Local Disruptions

Confidence level α	0.50	0.75	0.90	0.95	0.99
Var. = 1545, Bin. = 10, Cons. = 1115, Nonz. = 526,338[a]					
CPU[b]	1.75	2.17	5.23	3.17	8.91
Expected cost	12.30	12.30	13.11	13.04	13.45
CVaR	14.26	18.18	23.70	26.51	32.16
VaR	10.35	10.35	19.70	22.30	28.32
$1 - F(VaR)$	0.151	0.124	0.043	0.035	0.005
No. of suppliers selected	5	5	9	10	10

[a]Var. = number of variables, Bin. = number of binary variables, Cons. = number of constraints, Nonz. = number of nonzero coefficients.

[b]CPU seconds for proving optimality on a PC Pentium IV, 1.8 GHz, 1 GB RAM/CPLEX 11.

Table 9.7 Solution Results for the *SP_CV* Model: Local and Global Disruptions

Confidence level α	0.50	0.75	0.90	0.95	0.99
Var. = 1545, Bin. = 10, Cons. = 1115, Nonz. = 526,338[a]					
CPU[b]	1.00	1.10	4.09	5.15	5.63
Expected cost	13.18	13.18	14.06	13.96	15.89
CVaR	16.02	21.69	31.67	42.05	100.17
VaR	10.35	10.35	20.29	23.19	100.17
$1 - F(VaR)$	0.113	0.113	0.041	0.032	0
No. of suppliers selected	5	5	9	10	5

[a]Var. = number of variables, Bin. = number of binary variables, Cons. = number of constraints, Nonz. = number of nonzero coefficients.

[b]CPU seconds for proving optimality on a PC Pentium IV, 1.8 GHz, 1 GB RAM/CPLEX 11.

The optimal supply portfolios for *SP_E* model and different types of disruption scenarios are shown in Figure 9.3. In addition, the figure presents for each supplier i, the expected defect rate q_i, the average unit price $\sum_{j\in J} p_{ij}/n$, and for each type of scenario the disruption probabilities π_i and $\pi^* + (1 - \pi^*)\pi_i$. The portfolios are identical for both types of disruption scenario with the total demand equally allocated among five suppliers with the lowest disruption rates.

In the computational experiments the confidence level α is set at five levels of 0.5, 0.75, 0.90, 0.95, and 0.99, which means that the focus is on minimizing the highest 50%, 25%, 10%, 5%, and 1% of all scenario outcomes, that is, costs per part. For the *SP_CV* model and scenarios with local disruptions, Figure 9.4 shows the probability mass functions of optimal cost per part and the supply portfolios for the three confidence levels: 0.5, 0.90, and 0.99. When α increases, a more risk-averse decision-making focuses on a smaller set of outcomes and the number of selected suppliers also is increasing to mitigate the impact of disruptions risks by diversification of the supply portfolio.

Figure 9.5 shows the probability mass functions and the cumulative distribution functions for the optimal portfolios with different confidence levels for the scenarios with local and global disruptions. In addition, the optimal portfolios are presented. Figure 9.5 indicates that the mass function of cost per part is concentrated in a few points and the resulting cumulative distribution is a discontinuous step function with jumps at those points. Such results are typical for the scenario-based optimization under uncertainty, where the probability measure is concentrated in finitely many points. The resulting discontinuity (vertical jumps) of the distribution function leads to probability intervals of confidence level α with the same VaR. The discrete distributions of cost per part for the optimal portfolios with four different confidence levels and the corresponding probabilities concentrated at each level of cost are also presented in Table 9.8. The table shows that the probabilities are concentrated at 6, 10, 10, 10 points, respectively for the confidence level $\alpha = 0.5, 0.9, 0.989, 0.99$. In the examples, a large probability atom is concentrated at the highest cost. As a consequence, a slight increase of the confidence level from $\alpha = 0.989$ to $\alpha = 0.99$ results

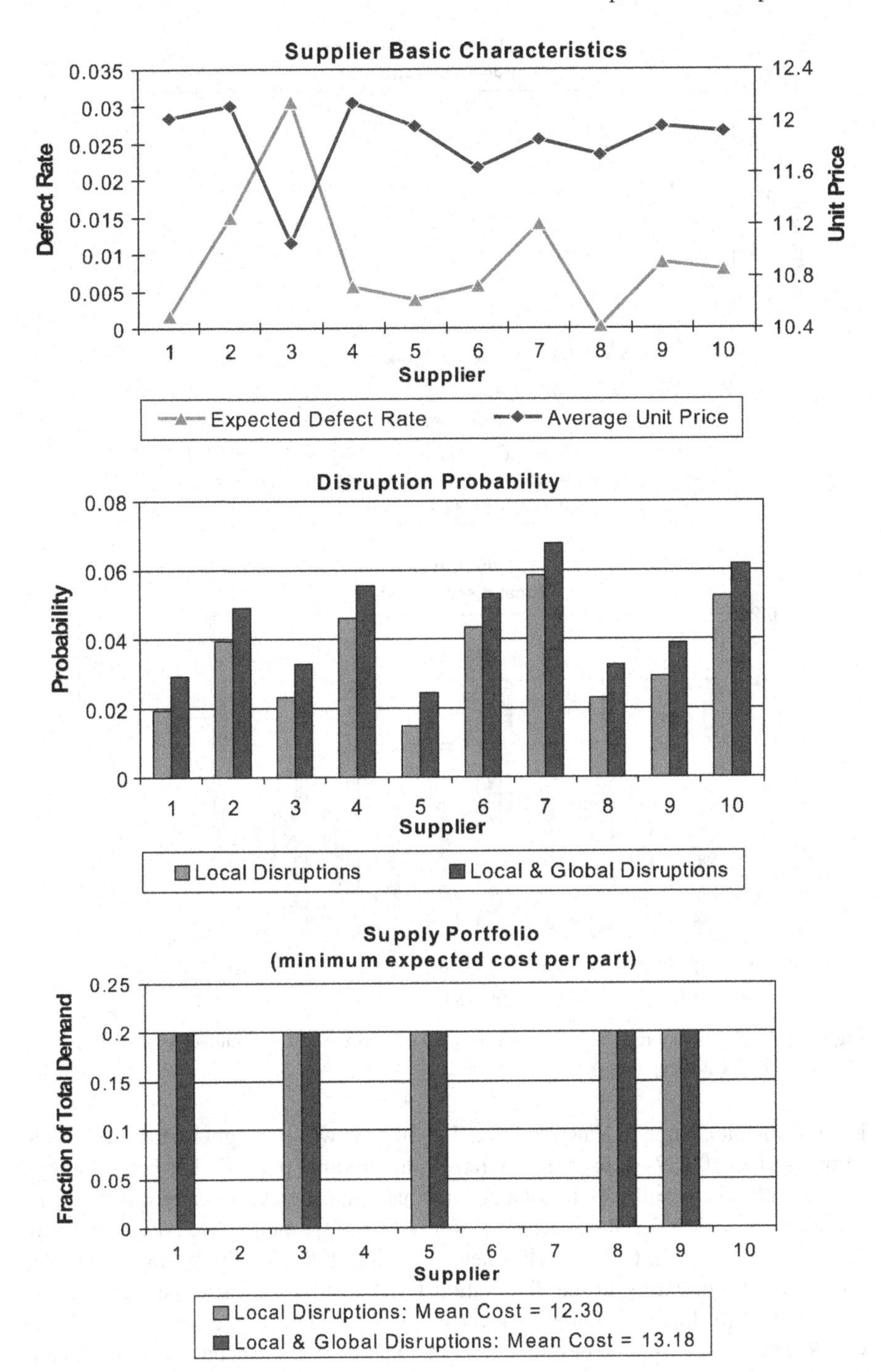

Figure 9.3 Optimal supply portfolio for the *SP_E* model and different types of disruption scenarios.

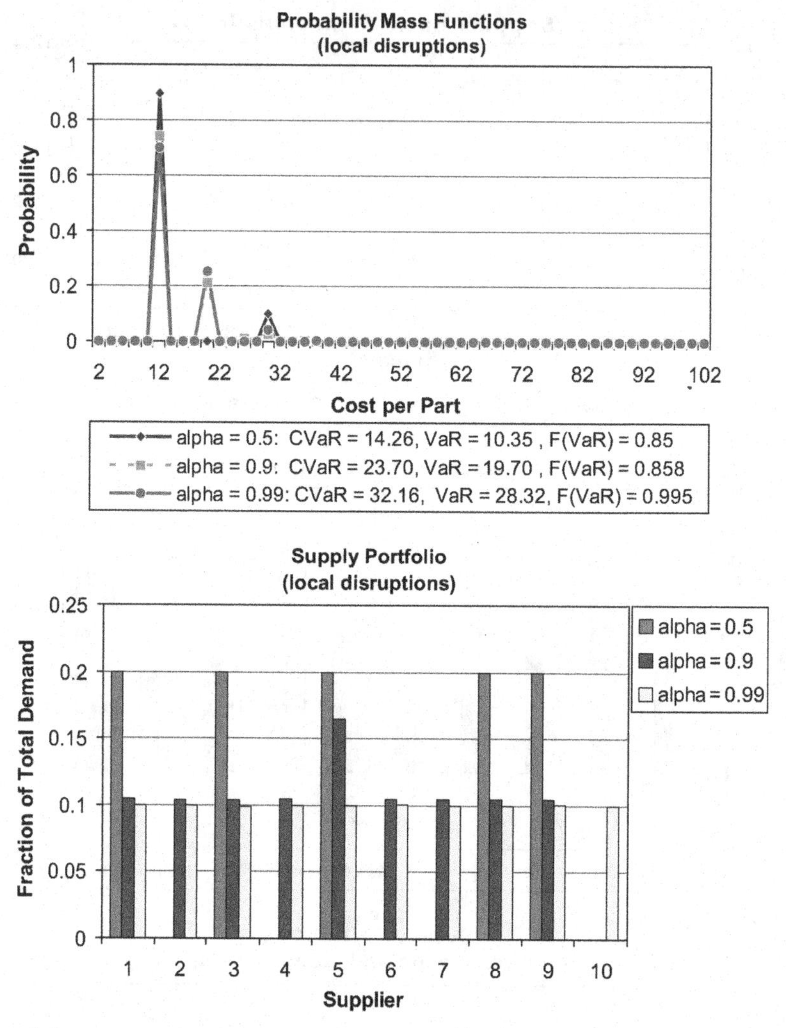

Figure 9.4 Probability mass functions and supply portfolios for different confidence levels in the *SP_CV* model: local disruptions.

in a significant change in VaR from 37.33 to 100.17, while a slight increase of CVaR from 94.81 to 100.17 is observed. Moreover, the optimal portfolio has been changed significantly; for $\alpha = 0.989$ the total demand has been equally allocated among all 10 suppliers ($x_i = 0.1$ for all i), whereas for $\alpha = 0.99$ among five suppliers only ($x_i = 0.2$ for $i = 1, 2, 6, 7, 8$). This degree of instability of the optimal portfolio due to the discontinuity in the distribution function may be distressing in practice, when a slightly higher confidence level is required. Despite the limited change in CVaR, the above results demonstrate that the well-known misbehavior in the dependence of VaR and optimal portfolio on the confidence level can as well be encountered when CVaR is applied as a risk measure.

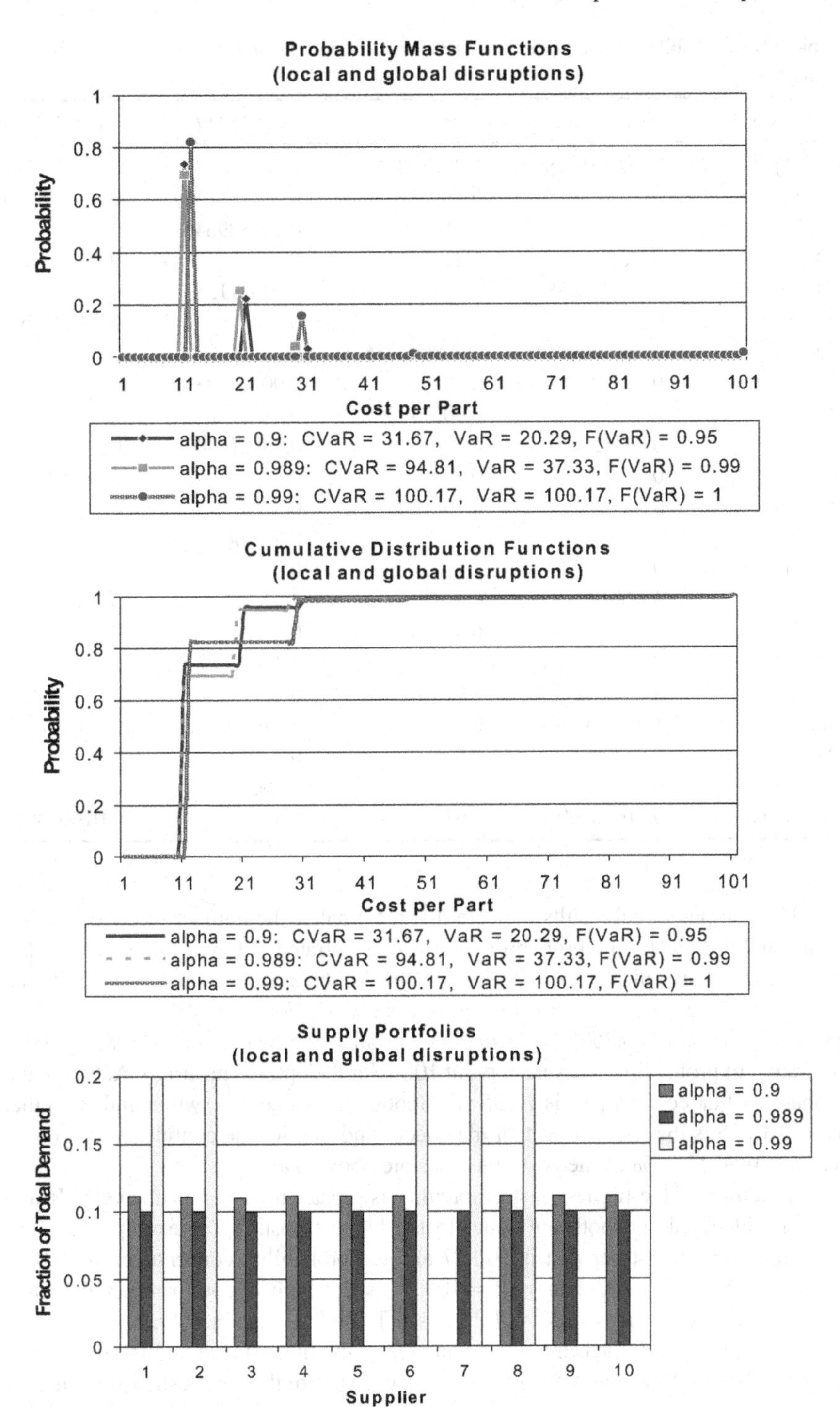

Figure 9.5 Probability mass functions, cost distributions, and supply portfolios for different confidence levels in the *SP_CV* model: local and global disruptions.

Table 9.8 Probability of Cost per Part for Optimal Supply Portfolios: Local and Global Disruptions

Cost interval	$\alpha = 0.5$	$\alpha = 0.9$	$\alpha = 0.989$	$\alpha = 0.99$
[10, 11)	0.886661618	0.736768317	0.693881575	0
[11, 12)	0	0	0	0.821176973
[19, 20)	0	0	0.251829646	0
[20, 21)	0	0.221857039	0	0
[28, 29)	0.098887038	0	0.040311352	0
[29, 30)	0	0	0	0.156904596
[30, 31)	0	0.029090556	0	0
[37, 38)	0	0	0.003744839	0
[40, 41)	0	0.002178297	0	0
[46, 47)	0.004355651	0	0.000223405	0
[47, 48)	0	0	0	0.011507829
[50, 51)	0	0.000102579	0	0
[55, 56)	0	0	8.94e-06	0
[60, 61)	0	3.15e-06	0	0
[64, 65)	9.47e-05	0	2.43e-07	0.0004038
[70, 71)	0	6.30e-08	0	0
[73, 74)	0	0	4.41e-09	0
[80, 81)	0	7.91e-10	0	0
[82, 83)	1.01e-06	0	5.14e-11	6.76e-06
[90, 91)	0	5.66e-12	0	0
[91, 92)	0	0	3.46e-13	0
[100, 101)	0.010000004	0.01	0.01	0.010000043

The computational results indicate that the smaller the number of concentration points and the greater the probability atoms concentrated at those points, the greater can be the positive difference $F(VaR) - \alpha$, that is, the smaller than $1 - \alpha$ can be the probability of outcomes with cost higher than VaR. For example (see Table 9.7), for $\alpha = 0.5$, $VaR = 10.35$ and $F(VaR) = 0.88666 > 0.5$, which indicates a high concentration of probability measure at point 10.35 for the optimal portfolio. Actually, the probability that cost per part is 10.35 is 0.88666 (see Table 9.8), which indicates that $VaR = 10.35$ is the lowest cost that may occur and that for the confidence level $\alpha =$ 0.5, less than 11.33% of the cost outcomes are above VaR.

Moreover, if the highest cost probability is greater than $1 - \alpha$, then CVaR and VaR are identical and both are equal to the highest cost. In the example for $\alpha =$ 0.99, the highest cost per part is 100.17 and the probability concentrated at 100.17 is $0.01000004 > 1 - \alpha$, then $VaR = 100.17$ is the highest cost per part that may occur and hence $CVaR = VaR = 100.17$ (see Table 9.8 and Fig. 9.5).

If the probability measure is concentrated at the highest cost and is greater than $1 - \alpha$, so that CVaR and VaR are identical with the highest cost, then for a higher confidence level α, a smaller number of suppliers are selected, which indicates that

Table 9.9 Solution Results for the *SP_CV* Model with Binary Assignment Variables z_{ij}

Confidence level α	0.50	0.75	0.90	0.95	0.99
Var. = 1545, Bin. = 510, Cons. = 1105, Nonz. = 526,328[a]					
Local disruptions					
Expected cost	12.34	12.35	13.16	13.06	13.46
CVaR	14.33	18.27	23.78	26.56	32.28
VaR	10.35	10.38	20.01	22.29	28.50
$1 - F(VaR)$	0.144	0.144	0.083	0.052	0.001
No. of suppliers selected	6	6	9	10	10
Local and global disruptions					
Expected cost	13.22	13.25	14.07	13.94	15.74
CVaR	16.08	21.80	31.75	42.08	100.20
VaR	10.36	10.40	20.16	23.21	100.20
$1 - F(VaR)$	0.155	0.155	0.094	0.053	0
No. of suppliers selected	6	6	9	10	6

[a]Var. = number of variables, Bin. = number of binary variables, Cons. = number of constraints, Nonz. = number of nonzero coefficients.

diversification of the supply portfolio is no longer necessary. For instance, for scenarios with local and global disruptions, the optimal portfolio selected for $\alpha = 0.99$ consists of five suppliers only, the same number as that for a much lower α (cf. Table 9.7, Fig. 9.5).

All the above observations indicate that the probability of disruption of a supply is a key determinant in the decision of allocation of demand among suppliers. In a risk-averse model, an order for delivery of parts from a particular supplier is selected more on the likelihood of no disruption in supply than on its purchasing cost or defect rate.

Finally, Table 9.9 presents solution results for the *SP_CV* model with binary assignment variables z_{ij}, that is, when for each customer order, the required custom parts must be provided by a single supplier only. Comparison of the results shown in Table 9.9 with the corresponding results presented in Tables 9.6 and 9.7 for continuous allocation variables z_{ij}, indicates that in the former case both the expected cost per part and CVaR were slightly higher and, in addition, for a low α the number of selected suppliers was greater. Such results were expected, since the *SP_CV* model with the continuous allocation variables is a partial LP relaxation of that model with the binary assignment variables.

For the bi-objective approach, the subsets of nondominated solutions were computed by parameterization on $\lambda \in \{0.01, 0.10, 0.25, 0.50, 0.75, 0.90, 0.99\}$ the weighted-sum program *WSP* and the Chebyshev program *CSP*. The subsets of nondominated solutions that were found using the two approaches are similar and for the selected seven levels of weight λ, from three to five different nondominated solutions were found using each approach. In total, five and six different solutions were found: (Mean Cost, CVaR) = (12.30, 29.14), (12.41, 26.77), (12.59, 24.98),

(12.84, 24.05), (13.06, 23.70) and (Mean Cost, CVaR) = (13.18, 36.32), (13.33, 33.97), (13.49, 32.65), (13.84, 31.84), (13.86, 31.82), (14.06, 31.67), respectively, for scenarios with local and with local and global disruptions. The trade-off between the expected cost and the expected worst-case cost is clearly shown in Figure 9.6, where the two convex efficient frontiers of mean cost—CVaR models with $\alpha = 0.9$ are presented. The results emphasize the effect of varying cost/risk preference of the decisionmaker.

Note that solutions to single-objective models *SP_E* and *SP_CV* are equivalent to the nondominated solutions of the weighted-sum program *WSP* with $\lambda = 1$ and $\lambda = 0$, respectively.

The computational experiments were performed using the AMPL programming language and the CPLEX v.11 solver (with the default settings) on a laptop with Pentium IV processor running at 1.8 GHz and with 1 GB RAM. The CPLEX solver was capable of finding proven optimal solutions within CPU seconds for all examples.

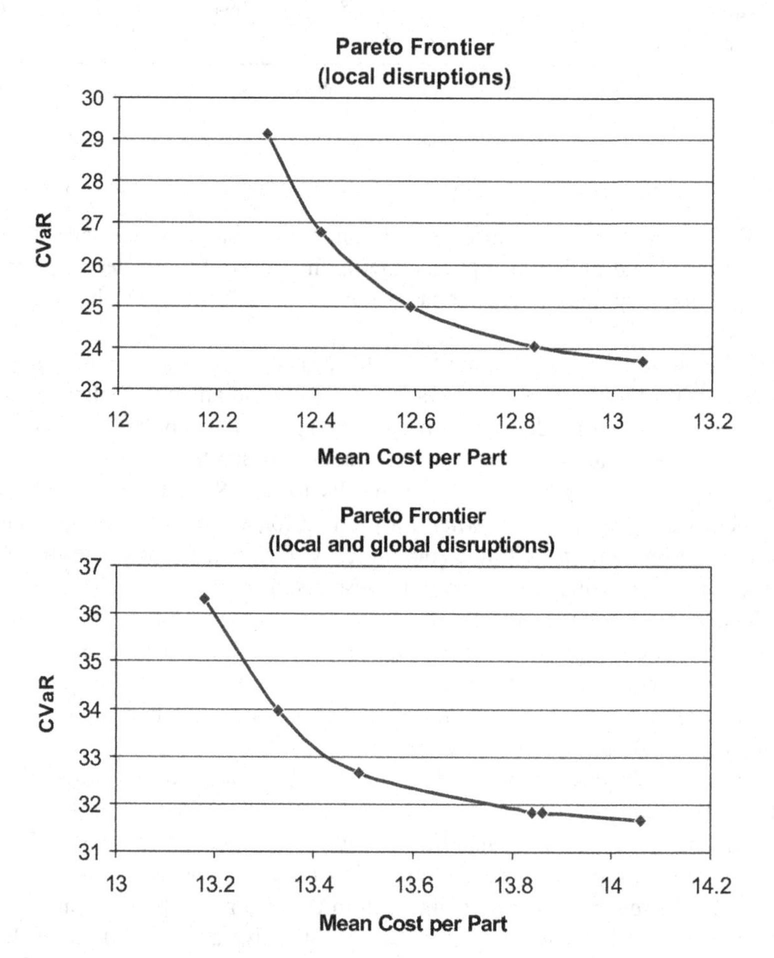

Figure 9.6 Efficient frontiers of Mean cost - CVaR models: $\alpha = 0.9$.

9.9 COMMENTS

Supply chain risk management has been extensively studied over the past decade. In practice four basic approaches can be applied to mitigate the impact of supply chain risks (e.g., Kleindorfer and Saad, 2005; Tang, 2006; Cohen and Kunreuther 2007): supply management, demand management, product management, and information management. In particular, to ensure efficient supply of materials along the customer driven supply chain, supply management deals with selection of supply portfolio, that is, supplier selection and order allocation under uncertain quality of supplied materials and reliability of on-time delivery.

The supplier selection is a complex decision-making problem which includes both quantitative and qualitative factors and one of the disadvantages of the mathematical programming methods is their failure to account for qualitative factors that may affect suppliers' performance. In order to consider both quantitative and qualitative factors some researchers propose hybrid approaches that combine different methods. For example, Sanayei et al. (2008) propose an integration of multi-attribute utility theory and linear programming, first to rate and choose the best suppliers and then to find optimal allocation of order quantities among the selected suppliers to maximize total additive utility. The combined method allows both quantitative and qualitative factors under risk and uncertainty to be considered as well as to account for the probabilistic nature of supplier performance. Another integrated approach that combines analytic network process and multi-objective mixed integer programming is proposed by Demirtas and Ustun (2008) and Ustun and Demirtas (2008). First, the potential suppliers are evaluated according to 14 criteria that are involved in the four clusters: benefits, opportunities, costs, and risks, to calculate the priorities of each supplier. Then, the optimum quantities are allocated among selected suppliers to maximize total value of purchasing (using the calculated priorities) and to minimize the total cost and total defect rate. However, the disadvantages of the integrated methods usually may affect the performance of hybrid approaches.

In spite of the importance of supplier selection and order allocation problems, the decision making is not sufficiently addressed in the literature (for a recent review, see Aissaoui et al. (2007), in particular for make-to-order manufacturing environments, for example, Murthy et al. (2004), Sawik (2005b), Yue et al. (2010). Basically, the authors distinguish between single and multiple item models and supplier selection with single or multiple sourcing, where each supplier can fully meet all requirements or none of the suppliers is able to satisfy the total requirements, respectively.

The vast majority of the decision models are mathematical programming models either single objective, for example, Kasilingam and Lee (1996), Basnet and Leung (2005), Jayaraman et al. (2006), or multiple objectives, for example, Weber and Current (1993), Xia and Wu (2007), Demirtas and Ustun (2008), Ustun and Demirtas (2008). For a recent literature review of the multicriteria decision-making approaches for supplier evaluation and selection, see Ho et al. (2010).

Although, the supplier selection problem is stochastic in nature, research seldom considers uncertainty and risk. For example, chance-constrained programming models were developed by Kasilingam and Lee (1996) to account for stochastic demand and by Wu and Olson (2008) to consider expected losses from quality acceptance inspection or late delivery. Furthermore, most models assume that the supply capacity is unlimited or known. However, unexpected machine breakdowns could affect the supply capacity. Relatively little work has been done in the area of uncertain supply capacity. Parlar and Perry (1996) present a continuous time model in which the availability of each of the m suppliers is uncertain because of disruptions such as equipment breakdown. By considering the case that each supplier is either "on" or "off," there are 2^m possible number of states for the whole system. For each of these 2^m states, they analyze a state-specific (s, Q) ordering policy so that the buyer would order Q units when the on-hand inventory reaches s. Berger et al. (2004) consider risks associated with a supplier network, which include catastrophic super events that affect all suppliers, as well as unique events that impact only one single supplier, and then present a decision tree-based model to help determine the optimal number of suppliers needed for the buying firm. Following the same decision framework, Ruiz-Torres and Farzad (2007) consider unequal failure probabilities for all the suppliers. Berger and Zeng (2006) study the optimal supply size under a number of scenarios that are determined by various financial loss functions, the operating cost functions and the probabilities of all the suppliers being down. The impacts of supply disruption risks on the choice between single and dual sourcing methods in a two-stage supply chain with a non-stationary and price-sensitive demand is considered by Yu et al. (2009). Li and Zabinsky (2009) developed a two-stage stochastic programming model and a chance-constrained programming model to determine a minimal set of suppliers and optimal order quantities with consideration of business volume discounts and with no time dimension. Both models include several objectives and strive to balance a small number of suppliers with the risk of not being able to meet demand. The stochastic programming model is scenario based and uses penalty coefficients, whereas the chance-constrained programming model assumes a probability distribution and constrains the probability of not meeting demand.

The material presented in this chapter is based on the results discussed in Sawik (2010a, 2011a). In Sawik (2010a) a portfolio approach is proposed for the problem of a single-period allocation of orders for custom parts among suppliers in a make-to-order manufacturing environment with operational risks. The problem is formulated as a single- or multi-objective MIP problem with the risk of defective or unreliable supplies controlled by the maximum number of delivery patterns (combinations of suppliers' delivery dates) for which the average defect rate or late delivery rate can be unacceptable. In Sawik (2011a), the portfolio approach has been enhanced to consider a single-period supplier selection and order allocation in a make-to-order environment in the presence of supply chain disruption risks. To mitigate the risk of supply disruptions, the two popular percentile measures of risk, value-at-risk (VaR) and conditional value-at-risk (CVaR), have been applied.

VaR and CVaR have been widely used in financial engineering in the field of portfolio management (e.g., Sarykalin et al., 2008). CVaR is used in conjunction with VaR and is applied for estimating the risk with nonsymmetric cost distributions. Uryasev (2000) and Rockafellar and Uryasev (2000, 2002) introduced a new approach to select a portfolio with the reduced risk of high losses. The portfolio is optimized by calculating VaR and minimizing CVaR simultaneously. For example, this approach has been applied to solution of the newsvendor problem (e.g., Gotoh and Takano, 2007) or to risk-averse selection of orders, where the approach is combined with a scenario-based method (Chahar and Taaffe, 2009).

The probability measure for the scenario-based optimization under uncertainty is concentrated in finitely many points and the resulting discrete distribution of cost may have a different effect on the optimal portfolio. In particular, the well-known misbehavior in the dependence of VaR on the confidence level can as well be encountered when CVaR is applied as a risk measure. For instance, if a large probability atom is concentrated at some cost, a slight increase of the confidence level may result in a significant change in VaR as well as in the optimal portfolio, while a slight increase of CVaR may occur. Such an instability of the optimal portfolio due to the discontinuity in the distribution function may be distressing in practice, when a sligthly higher confidence level is required. On the other hand, the computational results indicate that the smaller the number of concentration points and the greater the probability atoms concentrated at those points, the greater can be the positive difference $F(VaR) - \alpha$, that is, the smaller than $1 - \alpha$ can be the probability of outcomes with cost higher than VaR.

The computational experiments also indicate that the probability of disruption of a supply is a key determinant in the decision of allocation of demand among the suppliers. In a risk-averse model, an order for delivery of parts from a particular supplier is selected more on the likelihood of no disruption of supply than on its purchasing cost or defect rate. The suppliers associated with the highest disruption rates are rarely selected. In most cases the number of selected suppliers increases with the confidence level α, which indicates that the impact of disruption risks is mitigated by diversification of the supply portfolio.

In the models proposed in this chapter various simplifying assumptions have been introduced. For example, it has been assumed that each supplier is capable of manufacturing all required part types. In a more general setting, each supplier may only be prepared to manufacture a subset of part types and provide with the parts the corresponding subset of customer orders. Similarly, the scenarios with local and global disruptions can be relaxed for the situation with local and semiglobal disruptions, where there are "semiglobal" events that affect only a subset of all suppliers simultaneously, for example, when a disaster event affects only some restricted geographical area. In the *SPD* model, the quantity discount offered by the supplier has been assumed to be based on the total quantity of all ordered parts. In practice, independent quantity discounts may be offered for the individual part types. On the other hand, the *SP_E* and *SP_CV* models do not account for any type of discounts; however, they can also be enhanced for a discount environment.

EXERCISES

9.1 Assuming that each supplier i is capable of providing with parts only a subset $J_i \subset J$ of all customer orders, introduce appropriate changes in the models presented in this chapter.

9.2 Modify model *SPD* for the selection of supply portfolio with quantity discount offered on the total quantity of all ordered parts to account for the independent quantity discounts offered for individual part types.

9.3 Generalize models *SP_E* and *SP_CV* for selection of static supply portfolio under disruption risks to account for:

(**a**) the total quantity discount for all ordered parts.

(**b**) the independent quantity discounts for individual part types.

(**c**) the total business volume discount.

9.4 Define the probability for disruption scenarios with local and semiglobal disruptions that affect only a subset of all suppliers simultaneously.

9.5 In the computational examples in Section 9.8, Figure 9.3 indicates that the optimal supply portfolios with minimum expected cost are identical for both local and local and global disruptions. Why are the portfolios identical?

Chapter 10

Selection of a Dynamic Supply Portfolio in Supply Chains with Risks

10.1 INTRODUCTION

In this chapter the portfolio approach and the static MIP formulations presented in Chapter 9 are enhanced for dynamic (multiperiod) supplier selection and order allocation decisions in a make-to-order environment in the presence of both the low probability and high impact supply disruptions and the high probability and low impact supply delays. Given a set of customer orders for finished products, the decisionmaker needs to decide from which supplier and when to purchase product-specific parts required for each customer order to meet customer-requested due dates at a low cost and to mitigate the impact of supply chain risks. The selection of suppliers and the allocation of orders over time is based on price and quality of purchased parts, and reliability of supplies. For the selection of a dynamic supply portfolio, a time-indexed MIP approach is proposed to incorporate risk that uses conditional value-at-risk via scenario analysis. In the scenario analysis, the low probability and high impact supply disruptions are combined with the high probability and low impact supply delays. In addition to general scenarios of supplies with on-time, delayed, or disrupted deliveries, the two extreme cases of supplies are considered for which disruptions were combined either with on-time or with longest delay deliveries.

The following time-indexed MIP models are presented in this chapter:

TDN_E for selection of a dynamic supply portfolio to minimize expected cost for TDN (on-time, delayed or never) supplies

TDN_CV for selection of a dynamic supply portfolio to minimize expected worst-case cost for TDN (on-time, delayed or never) supplies

TON_E for selection of a dynamic supply portfolio to minimize expected cost for TON (on-time-or-never) supplies

Scheduling in Supply Chains Using Mixed Integer Programming. By Tadeusz Sawik

TON_CV for selection of a dynamic supply portfolio to minimize expected worst-case cost for TON (on-time-or-never) supplies

DON_E for selection of a dynamic supply portfolio to minimize expected cost for DON (longest delay-or-never) supplies

DON_CV for selection of a dynamic supply portfolio to minimize expected worst-case cost for DON (longest delay-or-never) supplies

The objective of the risk-neutral models *TDN_E*, *TON_E* and *DON_E* is to minimize the expected costs, whereas in risk-averse models *TDN_CV*, *TON_CV* and *DON_CV*, the nonrisk-neutral objective of minimizing the expected worst-case costs is applied.

The material presented in this chapter shows that the MIP approach combined with percentile measures of risk, VaR and CVaR, and scenario analysis, is capable of solving a hard discrete stochastic optimization problem of selection of a dynamic supply portfolio under disruption and delay risks. The proposed models provide the decisionmaker with a simple tool for evaluating the relationship between expected and worst-case costs. This will be illustrated in Section 10.6 with a set of simple computational examples.

10.2 MULTIPERIOD SUPPLIER SELECTION AND ORDER ALLOCATION

The approach proposed in this paper was originally developed for a very common type of supply chain, with a producer of a single product who obtains raw materials from several different suppliers with limited supply capacity to meet customer orders by customer-requested due dates. However, the approach is also applicable to the case of a single producer who assembles different types of products using product-specific parts purchased from multiple suppliers (for notation used, see Table 10.1). In order to simplify further considerations it is assumed that for each product type, one product-specific part type (e.g., a critical custom part type) needs to be supplied in required amounts by custom parts manufacturers. For example, in the electronics industry a producer of different electronic devices needs to be supplied by electronics manufacturers with printed wiring boards of different device-specific design. The last assumption can be easily relaxed to consider supplies of different product-specific part types required for each product type.

Let $I = \{1, \ldots, m\}$ be the set of m suppliers and $J = \{1, \ldots, n\}$ the set of n customer orders for the finished products, known ahead of time and not varying over the time period. Each order $j \in J$ is described by the quantity n_j of required single-type custom parts required for the ordered product type, and denote by e_j and f_j, respectively, the earliest and the latest delivery date of ordered parts to ensure meeting the customer-requested due date for finished products. The planning horizon consists of h planning periods (e.g., days or weeks) and let $T = \{1, \ldots, h\}$ be the set of planning periods.

Each supplier can provide the producer with custom parts for all customer orders. However, the suppliers may have different capacity and, in addition, may differ in price and quality of purchased parts and in reliability of delivery. Let c_{it} be the capacity

Table 10.1 Notation: Dynamic Supply Portfolio

Indices	
i	= supplier, $i \in I = \{1, \ldots, m\}$
j	= customer order, $j \in J = \{1, \ldots, n\}$
l	= delivery scenario, $l \in L$
s	= disruption scenario, $s \in S$
t	= planning period, $t \in T = \{1, \ldots, h\}$
Input parameters	
a_j	= per unit penalty cost of delayed customer order j caused by delayed delivery of required parts
b_j	= per unit penalty cost of unfulfilled customer order j caused by shortage of required parts
c_{it}	= capacity of supplier i in period t
e_j	= the earliest delivery date of parts for customer order j
f_j	= the latest delivery date of parts for customer order j
n_j	= number of parts to be purchased for customer order j
o_{it}	= cost of ordering parts from supplier i in period t
p_{ijt}	= unit price of part for customer order j purchased from supplier i in period t
q_{it}	= expected defect rate of supplier i in period t
r_{it}	= expected delay rate of supplier i in period t
w_{it}	= expected disruption rate of supplier i in period t
$\overline{q}$	= the largest acceptable average defect rate of supplies
$\overline{r}$	= the largest acceptable average delay rate of supplies
$\overline{w}$	= the largest acceptable average disruption rate of supplies
α	= confidence level
π_{it}	= disruption probability for supplier i and period t
$\rho_{it}(\theta)$	= probability that delivery from supplier i scheduled for period t is delayed by θ periods, where $\theta \in \{0, \ldots, h-t+1\}$

of supplier i in period t, o_{it} the cost of ordering parts from supplier i in period t, p_{ijt} the per unit price of product-specific part for customer order j purchased from supplier i in period t, and q_{it} the expected defect rate of supplier i in period t.

Assume that supplies are subject to random disruptions. Although the probability of supply disruption usually is very low, its financial consequences may be very high. In addition to low probability and high impact disruptions, the supplies are also subject to high probability and low impact supply delays. Unlike supply disruptions indicating that ordered parts are not delivered at all during the planning horizon, delayed supplies indicate late delivery of required parts within the horizon.

Note that on-time delivery scheduled for period t occurs in that period, whereas the late delivery may occur in one of the remaining periods $t+1, \ldots, h$, that is, to deliver the ordered parts by the end of the planning horizon the delay cannot be longer than $h-t$ periods. By convention, deliveries delayed until a dummy period $t = h+1$ represent disruptions of supplies, that is, no delivery of ordered parts.

Summarizing, it is assumed that the longest delay of each delivery scheduled for period t cannot be greater than $h - t$ periods, whereas the infeasible delay of $h - t + 1$ periods (i.e., delivery in a dummy period $t = h + 1$), by convention represents a disruption, that is, no delivery.

Let us call an on-time, delayed, or disrupted delivery, a TDN (on-time, delayed or never) supply. For convenience, denote the total number of potential supplies in all periods by $M = hm$. A particular realization of TDN supplies can be represented by a delivery scenario $l \in L$ (L is the index set of delivery scenarios) in the form of an integer vector $\Delta^l \in \{0, \ldots, h\}^M$, where $\Delta^l_{it} \in \{0, \ldots, h - t + 1\}$ represents that delivery of parts from supplier $i \in I$ scheduled for period $t \in T$ is either on time, if $\Delta^l_{it} = 0$, is delayed by Δ^l_{it} periods, if $1 \leq \Delta^l_{it} \leq h - t$ or is disrupted, if $\Delta^l_{it} = h - t + 1$.

The probability that scenario l occurs is denoted by P_l, and let π_{it} be the probability of disruption for supplier i in period t, that is, the parts ordered from supplier i and scheduled for period t are delivered (on time or delayed) within the planning horizon with probability $(1 - \pi_{it})$, or not at all with probability π_{it}.

Denote by $\rho_{it}(\theta)$ the probability that delivery from supplier i scheduled for period t is delayed by θ periods, where $\theta \in \{0, \ldots, h - t + 1\}$ and

$$\sum_{\theta=0}^{h-t+1} \rho_{it}(\theta) = 1; i \in I, t \in T. \tag{10.1}$$

In particular, $\rho_{it}(h - t + 1) = \pi_{it}$ is the probability of disruption for supplier i and period t. As a consequence, the last formula can be rewritten as

$$\sum_{\theta=0}^{h-t} \rho_{it}(\theta) = 1 - \pi_{it}; i \in I, t \in T. \tag{10.2}$$

The delays Δ_{it} are assumed to be statistically independent, discrete multivariate random variables, where delay $\Delta_{it} = h - t + 1$, by convention represents a disruption of delivery. Therefore, the probability P_l of each delivery scenario $l \in L$ is given by

$$\begin{aligned} P_l &= \prod_{i \in I, t \in T} \rho_{it}(\Delta^l_{it}) \\ &= \left(\prod_{i \in I, t \in T: \Delta^l_{it} = h-t+1} \pi_{it} \right) \left(\prod_{i \in I, t \in T: 0 \leq \Delta^l_{it} \leq h-t} \rho_{it}(\Delta^l_{it}) \right); l \in L, \end{aligned} \tag{10.3}$$

where the right-hand-side product is factorized into the disruption and the delay probabilities.

For each supplier i, deliveries of parts can be scheduled for each period $t = 1, 2, \ldots, h - 1, h$, and the corresponding maximum possible delays are $\Delta_{it} = h - 1, h - 2, \ldots, 1, 0$. On the other hand, all those deliveries can be disrupted. Therefore, the maximum total delay of all possible deliveries from each supplier is $\sum_{t \in T} (h - t) = h(h - 1)/2$, and the maximum total number of disruptions is h.

Let r_{it} be the expected delay rate of supplier i in period t, that is, the ratio of the expected delay of supply (i, t) to the maximum total delay, and let w_{it} be the expected disruption rate of supplier i in period t, that is, the ratio of the total probability of scenarios with disruption of supply (i, t) to the maximum number of disruptions. The expected delay rate and disruption rate are defined below.

$$r_{it} = \sum_{l \in L: 1 \leq \Delta_{it}^{l} \leq h-t} P_l \Delta_{it}^{l} / (h(h-1)/2); \; i \in I, t \in T \tag{10.4}$$

$$w_{it} = \sum_{l \in L: \Delta_{it}^{l} = h-t+1} P_l (\Delta_{it}^{l} - h + t)/h = \sum_{l \in L: \rho_{it}(\Delta_{it}^{l}) = \pi_{it}} P_l / h; \; i \in I, t \in T \tag{10.5}$$

where $h(h-1)/2$ and h represent, respectively, the maximum total delay and the maximum number of disruptions, of all deliveries from each supplier.

Note that the above scenarios do not allow for deliveries beyond the planning horizon. The longest delay of $h - t$ periods of each delivery scheduled for period t depends on t and allows for delivery in the last period, h, at the latest. It is possible to consider also the other scenarios that allow for delayed deliveries beyond the planning horizon. For example, an alternative scenario could be defined with the same maximum possible delay of $\Delta_{\max}$ periods for all deliveries scheduled over the planning horizon. Then, for every scenario $l \in L$ and all $i \in I$ and $t \in T$, any feasible delay Δ_{it}^{l} should satisfy the inequality $1 \leq \Delta_{it}^{l} \leq \Delta_{\max}$. As a result, the latest delivery could occur in period $t = h + \Delta_{\max}$, while any later deliveries would not be allowed.

10.3 SELECTION OF A DYNAMIC SUPPLY PORTFOLIO TO MINIMIZE EXPECTED COSTS

In this section an MIP model is proposed for selection of a risk-neutral multiperiod supply portfolio in the presence of supply delays and disruptions. The decisionmaker needs to decide what quantities of various custom parts to order, from which supplier, and in which periods to minimize expected costs. Hence, the decisionmaker needs to select a dynamic supply portfolio, that is, the allocation of orders for parts among the suppliers and among the planning periods.

A dynamic supply portfolio is defined as

$$\{x_{it} : i \in I, t \in T\},$$

where

$$\sum_{i \in I} \sum_{t \in T} x_{it} = 1$$

and $0 \leq x_{it} \leq 1$ is the fraction of the total demand for parts ordered from supplier i for period t (for definition of problem variables, see Table 10.2).

The dynamic supply portfolio should be checked over the time horizon against the highest acceptable defect rate, delay rate, and disruption rate of supplies. Denote

Table 10.2 Decision Variables

VaR	= the targeted cost of the portfolio based on the α-percentile of costs, that is, in $100\alpha\%$ of scenarios, the outcome cannot exceed VaR (value-at-risk)
$\mathcal{T}_l$	= the amount by which the cost of the portfolio in scenario l exceeds VaR (tail cost)
x_{it}	= the fraction of total demand for parts ordered from supplier i to be delivered in period t (order allocation variable)
y_{it}	= 1, if an order for parts to be delivered in period t is placed on supplier i; otherwise $y_{it} = 0$ (supplier selection variables)
z_{ijt}	= the fraction of total demand for parts required for customer order j to be delivered from supplier i in period t (customer order allocation variable)

by $\overline{q}$, $\overline{r}$, and $\overline{w}$ the maximal acceptable average defect rate, delay rate, and disruption rate, respectively. The supply delays and disruptions may result in the shortage of required parts and the corresponding delay and shortage penalty costs of delayed or unfulfilled customer orders should be incorporated into the model. Clearly, the producer does not need to pay for ordered and defective or undelivered parts, whereas parts delivered late may be paid for at a reduced price. However, the producer can be charged with a much higher penalty cost of delayed or unfulfilled customer orders for products, caused by the shortage of required parts due to defective, delayed, or undelivered parts.

Let a_j be the per unit and per period penalty cost of delayed customer order j caused by the late delivery of required custom parts, where deliveries not later than f_j are not penalized. Similarly, let b_j be the per unit cost of unfulfilled (lost) customer order j, caused by supply disruptions.

For each delivery scenario l and each customer order j with $f_j < h$ to be supplied with parts by supplier i in period t (i.e., with $z_{ijt} > 0$), the penalty cost of delivery delayed by $1 \leq \Delta_{it}^l \leq h - t$ periods is given by

$$a_j \max\{0, t + \Delta_{it}^l - f_j\} n_j,$$

that is, only deliveries later than f_j are penalized.

The resulting total expected delay penalty cost over all delivery scenarios, given an order allocation vector z, is then given by

$$\sum_{l \in L} P_l \left(\sum_{i \in I} \sum_{t \in T: 1 \leq \Delta_{it}^l \leq h-t} \sum_{j \in J: f_j < h, f_j < t + \Delta_{it}^l} a_j (t + \Delta_{it}^l - f_j) n_j z_{ijt} \right).$$

Note that if $f_j = h$, then no delivery within the planning horizon can be delayed and penalized. However, any delay beyond $f_j = h$ by convention is a disruption, and hence penalized with the cost of unfulfilled customer order.

The total expected penalty cost of unfulfilled customer orders due to shortage of required parts caused by supply disruptions (i.e., deliveries delayed until period $h + 1$) is

$$\sum_{l \in L} P_l \left(\sum_{i \in I} \sum_{t \in T: \Delta_{it}^l = h-t+1} \sum_{j \in J} (b_j - p_{ijt}(1 + q_{it})) n_j z_{ijt} \right),$$

where the producer does not need to pay for ordered and undelivered parts and hence the purchasing costs of those parts are deducted from the cost of unfulfilled customer orders.

In risk-neutral operating conditions the overall quality of the supply portfolio can be measured by the expected cost per part of

ordering, $\sum_{i\in I}\sum_{t\in T} o_{it}y_{it}/N$

purchasing, $\sum_{i\in I}\sum_{j\in J}\sum_{t\in T} n_j p_{ijt} z_{ijt}/N$

defects (where the penalty cost of a defective part is assumed to be identical with its regular price), $\sum_{i\in I}\sum_{j\in J}\sum_{t\in T} n_j q_{it} p_{ijt} z_{ijt}/N$

delay penalty, $\sum_{l\in L} P_l\Big(\sum_{i\in I}\sum_{t\in T:1\le\Delta_{it}^l\le h-t}\sum_{j\in J:f_j<h, f_j<t+\Delta_{it}^l} a_j(t+\Delta_{it}^l-f_j)\, n_j z_{ijt}\Big)\Big/N$

disruption penalty, $\sum_{l\in L} P_l\Big(\sum_{i\in I}\sum_{t\in T:\Delta_{it}^l=h-t+1}\sum_{j\in J}(b_j-p_{ijt}(1+q_{it}))\, n_j z_{ijt}\Big)\Big/N$

where $N=\sum_{j\in J} n_j$ is the total demand for parts.

Note that in make-to-order manufacturing no inventory of custom parts (e.g., printed wiring boards of product-specific design) can be kept on hand and the parts are requisitioned with each customer order. In addition, custom parts required for each customer order j are to be delivered within a time window $[e_j, f_j]$ derived from the customer-requested due date to further reduce any inventory holding cost. The delivery cannot be earlier than the earliest date e_j and the required parts are assumed to be processed as soon as they are delivered. As a result no overage costs are considered. On the other hand, the required parts can be delivered late, beyond the latest date f_j, or not delivered at all. Then delay or disruption penalty costs represent the underage costs.

The time-indexed MIP model *TDN_E* for optimal selection of a dynamic supply portfolio to minimize the expected cost per part for TDN supplies is presented below.

Model TDN_E: *Selection of a Dynamic Supply Portfolio to Minimize Expected Cost per Part: TDN Supplies*

Minimize expected cost per part

$$E(x)=\sum_{i\in I}\sum_{t\in T} o_{it}y_{it}/N+\sum_{i\in I}\sum_{j\in J}\sum_{t\in T} p_{ijt}(1+q_{it})n_j z_{ijt}/N$$

$$+\sum_{l\in L} P_l\left(\sum_{i\in I}\ \sum_{t\in T:1\le\Delta_{it}^l\le h-t}\ \sum_{j\in J:f_j<h, f_j<t+\Delta_{it}^l} a_j(t+\Delta_{it}^l-f_j)n_j z_{ijt}\right)\Bigg/N$$

$$+\sum_{l\in L} P_l\left(\sum_{i\in I}\ \sum_{t\in T:\Delta_{it}^l=h-t+1}\ \sum_{j\in J}(b_j-p_{ijt}(1+q_{it}))n_j z_{ijt}\right)\Bigg/N \qquad (10.6)$$

subject to

1. *Order Allocation Constraints*:
   - for each customer order the required parts must be delivered by selected suppliers not earlier than the earliest and not later than the latest delivery date,
   - for each selected supplier the amount of parts delivered in each period cannot exceed the supplier available capacity,

$$\sum_{i \in I} \sum_{t \in T: e_j \leq t \leq f_j} z_{ijt} = 1; \; j \in J \tag{10.7}$$

$$\sum_{j \in J} n_j z_{ijt} \leq c_{it} y_{it}; \; i \in I, t \in T \tag{10.8}$$

2. *Portfolio Selection Constraints*:
   - the portfolio definition constraint,
   - the supplier i is selected for delivery of parts in period t if at least one customer order is assigned to supplier i in period t,
   - the average defect rate of the portfolio cannot be greater than $\overline{q}$,
   - the average delay rate of the portfolio cannot be greater than $\overline{r}$,
   - the average disruption rate of the portfolio cannot be greater than $\overline{w}$,

$$x_{it} = \sum_{j \in J} n_j z_{ijt} \Big/ \sum_{j \in J} n_j; \; i \in I, t \in T \tag{10.9}$$

$$y_{it} \geq x_{it}; \; i \in I, t \in T \tag{10.10}$$

$$y_{it} \leq \sum_{j \in J} z_{ijt}; \; i \in I, t \in T \tag{10.11}$$

$$\sum_{i \in I} \sum_{t \in T} q_{it} x_{it} \leq \overline{q} \tag{10.12}$$

$$\sum_{i \in I} \sum_{t \in T} r_{it} x_{it} \leq \overline{r} \tag{10.13}$$

$$\sum_{i \in I} \sum_{t \in T} w_{it} x_{it} \leq \overline{w} \tag{10.14}$$

3. *Nonnegativity and Integrality Conditions*:

$$x_{it} \geq 0; \; i \in I, t \in T \tag{10.15}$$

$$y_{it} \in \{0, 1\}; \; i \in I, t \in T \tag{10.16}$$

$$z_{ijt} \in [0, 1]; \; i \in I, j \in J, t \in T. \tag{10.17}$$

While the objective function (10.6) aims at minimizing the expected costs caused by defects, delays, and disruptions, constraints (10.12) to (10.14) limit their average rates. Note that x_{it} is an auxiliary variable determined by z_{ijt} and can be removed from the model.

If the ordering costs were negligible, the binary variables y_{it} could be eliminated and then the MIP model *TDN_E* reduces to an LP program. On the other hand, if for each customer order the order for required parts cannot be allocated among different suppliers and all custom parts should be provided by a single delivery from one supplier only, then the continuous allocation variables z_{ijt} should be redefined as binary assignment variables denoting whether or not all parts required for order j are provided by supplier i in period t (cf. variables z_{ij} in Section 9.2).

10.4 SELECTION OF A DYNAMIC SUPPLY PORTFOLIO TO MINIMIZE EXPECTED WORST-CASE COSTS

In this section the MIP model *TDN_CV* is presented for the selection of a dynamic risk-averse supply portfolio to reduce the risk of high costs.

Let $\alpha \in (0, 1)$ represent the confidence level for the cost distribution across all scenarios. The following two percentile risk measures will be considered when selecting a dynamic supply portfolio (see Section 9.6.1):

- Value-at-risk (VaR) at a $100\alpha\%$ confidence level, that is, the targeted cost of the portfolio such that for $100\alpha\%$ of the scenarios, the outcome will not exceed VaR.
- Conditional value-at-risk (CVaR) at a $100\alpha\%$ confidence level, that is, the expected cost of the portfolio in the worst $100(1 - \alpha)\%$ of the cases.

In the dynamic supply portfolio selection problem discussed in this section α is assumed to be fixed by a risk-averse decisionmaker to control the risk of losses due to supply delays and disruptions. The decisionmaker is willing to accept only portfolios for which the total probability of scenarios with costs greater than VaR is not greater than $1 - \alpha$. Furthermore, the decisionmaker wants to minimize the worst-case costs exceeding VaR.

The portfolio will be optimized by calculating VaR and minimizing CVaR simultaneously.

The time-indexed MIP model *TDN_CV* for optimal selection of a dynamic supply portfolio to minimize the expected worst-case cost per part for TDN supplies is formulated below.

Model TDN_CV: *Selection of a Dynamic Supply Portfolio to Minimize Expected Worst-Case Cost per Part: TDN Supplies*

Minimize expected worst-case cost per part (CVaR)

$$C(x) = \text{VaR} + (1 - \alpha)^{-1} \sum_{l \in L} P_l T_l \tag{10.18}$$

subject to

1. *Order Allocation and Portfolio Selection Constraints*: (10.7) to (10.14)
2. *Risk Constraints*:
   - the tail cost for scenario l is defined as the nonnegative amount by which cost per part in scenario l exceeds VaR,

$$\mathcal{T}_l \geq \sum_{i\in I}\sum_{t\in T} o_{it}y_{it}/N + \sum_{i\in I}\sum_{j\in J}\sum_{t\in T} p_{ijt}(1+q_{it})n_j z_{ijt}/N$$
$$+\sum_{i\in I}\sum_{t\in T:1\leq\Delta_{it}^l\leq h-t}\sum_{j\in J:f_j<h,\, f_j<t+\Delta_{it}^l} a_j(t+\Delta_{it}^l - f_j)n_j z_{ijt}/N$$
$$+\sum_{i\in I}\sum_{t\in T:\Delta_{it}^l=h-t+1}\sum_{j\in J}(b_j - p_{ijt}(1+q_{it}))n_j z_{ijt}/N - \text{VaR};\ l\in L \tag{10.19}$$

3. *Nonnegativity and Integrality Conditions*: (10.15) to (10.17) and

$$\mathcal{T}_l \geq 0;\ l\in L. \tag{10.20}$$

Note that the variable VaR does not need to be constrained of being nonnegative. As $\mathcal{T}_l$ is constrained of being positive, the model tries to decrease VaR and hence positively impact the objective function. However, large reduction in VaR may result in more scenarios with positive tail costs.

10.5 SUPPLY PORTFOLIO FOR BEST-CASE AND WORST-CASE TDN SUPPLIES

In this section, the following two extreme cases of TDN supplies are considered:

- Best-case TDN supplies, such that the ordered parts are delivered on time or not at all. The best-case TDN supplies will be called TON (on-time-or-never) supplies.
- Worst-case TDN supplies, such that the ordered parts are delivered with the longest delay (i.e., in the last period $t = h$ of the planning horizon) or not at all. The worst-case TDN supplies will be called DON (longest delay-or-never) supplies.

In the TON or DON supplies a particular realization of supplies is represented by a disruption scenario $s \in S$ (S is the index set of all scenarios) in the form of a binary vector $\delta^s \in \{0, 1\}^M$, ($M = hm$), where $\delta_{it}^s = 1$ represents that supplier $i \in I$ in period $t \in T$ operates without disruptions, otherwise $\delta_{it}^s = 0$. The probability that this scenario occurs is denoted by $\mathcal{P}_s$ and is given by (see Section 9.5)

$$\mathcal{P}_s = \prod_{i\in I,t\in T:\delta_{it}^s=0} \pi_{it}\cdot \prod_{i\in I,t\in T:\delta_{it}^s=1}(1-\pi_{it});\ k\in\{1,\ldots,2^M\}, \tag{10.21}$$

where π_{it} is the probability of disruption for supplier i in period t.

10.5.1 Minimization of Expected Costs

In this subsection two MIP models *TON_E* and *DON_E* are presented for the selection of an optimal risk-neutral supply portfolio for best-case TDN supplies with no delays, and worst-case TDN supplies with longest delays, respectively.

For each disruption scenario s, the total shortage of required parts faced by the producer, given an order allocation vector z, is

$$\sum_{i \in I} \sum_{j \in J} \sum_{t \in T} n_j z_{ijt}(1 - \delta_{it}^s),$$

and the corresponding penalty cost of unfulfilled customer orders is

$$\sum_{i \in I} \sum_{j \in J} \sum_{t \in T} b_j n_j z_{ijt}(1 - \delta_{it}^s).$$

The time-indexed MIP model *TON_E* for an optimal selection of a dynamic supply portfolio to minimize expected cost per part in the case of "on-time-or-never" supplies is presented below.

Model TON_E: *Selection of a Dynamic Supply Portfolio to Minimize Expected Cost per Part: TON Supplies*

Minimize expected cost per part

$$\underline{E}(x) = \sum_{i \in I} \sum_{t \in T} o_{it} y_{it}/N + \sum_{i \in I} \sum_{j \in J} \sum_{t \in T} p_{ijt}(1 + q_{it}) n_j z_{ijt}/N$$
$$+ \sum_{i \in I} \sum_{j \in J} \sum_{s \in S} \sum_{t \in T} P_s(b_j - p_{ijt}(1 + q_{it})) n_j z_{ijt}(1 - \delta_{it}^s)/N \qquad (10.22)$$

subject to (10.7)–(10.17).

In the DON supplies, the longest delays $(h - t)$ occur for all (i, t) and each disruption scenario s with $\delta_{it}^s = 1$ and for all j with $z_{ijt} > 0$ and $f_j < h$. For each disruption scenario the corresponding worst-case delay penalty cost, given an order allocation vector z, is then given by

$$\sum_{i \in I} \sum_{t \in T} \sum_{j \in J: f_j < h} a_j(h - f_j) n_j z_{ijt} \delta_{it}^s.$$

The time-indexed MIP model *DON_E* for an optimal selection of a dynamic supply portfolio to minimize expected cost per part in the case of "longest delay-or-never" supplies is presented below.

Model DON_E: *Selection of a Dynamic Supply Portfolio to Minimize Expected Cost per Part: DON Supplies*

Minimize expected cost per part

$$\overline{E}(x) = \sum_{i \in I} \sum_{t \in T} o_{it} y_{it}/N + \sum_{i \in I} \sum_{j \in J} \sum_{t \in T} p_{ijt}(1 + q_{it}) n_j z_{ijt}/N$$

$$+ \sum_{i \in I} \sum_{j \in J: f_j < h} \sum_{s \in S} \sum_{t \in T} \mathcal{P}_s a_j (h - f_j) n_j z_{ijt} \delta_{it}^s / N$$

$$+ \sum_{i \in I} \sum_{j \in J} \sum_{s \in S} \sum_{t \in T} \mathcal{P}_s (b_j - p_{ijt}(1 + q_{it})) n_j z_{ijt} (1 - \delta_{it}^s)/N \quad (10.23)$$

subject to (10.7)–(10.17).

Note that the total number $|S|$ of disruption scenarios is 2^M, $(M = hm)$, while the number $|L|$ of possible delivery scenarios for TDN supplies is bounded by $2^M (h!)^m$ ($| \cdot |$ denotes the power of a set $\cdot$). This bound can be derived as follows. Since on-time or delayed supplies occur only for supplies (i, t) without disruptions in the disruption scenario s (i.e., for (i, t) with $\delta_{it}^s = 1$), and the delay for such supplies may take on $h - t + 1$ possible values $0, 1, \ldots, h - t$, while for the supplies with disruption (for (i, t) with $\delta_{it}^s = 0$) the delay is $h - t + 1$, that is, it takes on only one value, each disruption scenario $s \in S$ is associated with $\prod_{t \in T} (h - t + 1)^{\left(\sum_{i \in I} \delta_{it}^s\right)}$ combinations of the delays, that is, the delivery scenarios. Then, the total number of the delivery scenarios $l \in L$ is $|L| = \sum_{k=1}^{2^M} \prod_{t \in T} (h - t + 1)^{\left(\sum_{i \in I} \delta_{it}^s\right)} < 2^M (h!)^m$.

10.5.1.1 Equivalent Single-Period Supply Portfolio

The multiperiod models *TON_E, DON_E* for m suppliers and h planning periods can be transformed into equivalent single-period models for $M = hm$ suppliers. To this end, each pair of indices (i, t), $i \in I$, $t \in T$ should be replaced with a single index $\iota \in \mathcal{I}$, where $\mathcal{I} = \{1, \ldots, M\}$ represents the set of equivalent M single-period suppliers. In the MIP models *TON_E* and *DON_E*, variables x_{it}, y_{it}, z_{ijt} are replaced with x_ι, y_ι, $z_{\iota j}$. Then, the order allocation constraint (10.7) should be replaced with

$$\sum_{\iota \in \mathcal{I}_j} z_{\iota j} = 1; \; j \in J, \quad (10.7')$$

where $\mathcal{I}_j$ is the subset of equivalent single-period suppliers that are capable of delivering parts required for customer order j and it represents the subset $\{(i, t){:}\ i \in I,\ e_j \le t \le f_j\}$ of pairs (i, t) feasible for customer order j.

10.5.2 Minimization of Expected Worst-Case Costs

In this subsection two MIP models *TON_CV* and *DON_CV* are presented for the selection of an optimal risk-averse supply portfolio for best-case TDN supplies with no delays, and worst-case TDN supplies with longest delays, respectively.

The time-indexed MIP model *TON_CV* for TON supplies with on-time or no deliveries is presented below, where $\underline{\mathcal{T}}_s$ denotes the tail cost for the best-case supplies associated with disruption scenario s.

Model TON_CV: *Selection of a Dynamic Supply Portfolio to Minimize Expected Worst-Case Cost per Part: TON Supplies*

Minimize expected worst-case cost per part (CVaR)

$$\underline{C}(x) = \text{VaR} + (1-\alpha)^{-1} \sum_{s \in S} \mathcal{P}_s \underline{\mathcal{T}}_s \tag{10.24}$$

subject to

1. *Order Allocation and Portfolio Selection Constraints*: (10.7) to (10.17)
2. *Risk Constraints*:
   - the tail cost for scenario s is defined as the nonnegative amount by which cost per part in scenario s exceeds VaR,

$$\underline{\mathcal{T}}_s \geq \sum_{i \in I} \sum_{t \in T} o_{it} y_{it}/N + \sum_{i \in I} \sum_{j \in J} \sum_{t \in T} p_{ijt}(1+q_{it}) n_j z_{ijt}/N$$
$$+ \sum_{i \in I} \sum_{j \in J} \sum_{t \in T} (b_j - p_{ijt}(1+q_{it})) n_j z_{ijt}(1-\delta_{it}^s)/N - \text{VaR}; \;\; s \in S \tag{10.25}$$

$$\underline{\mathcal{T}}_s \geq 0; \;\; s \in S. \tag{10.26}$$

Finally, the *DON_CV* model is formulated for the worst-case TDN supplies with longest delays, such that ordered parts are delivered either in the last period $t = h$ of the planning horizon or not at all. $\overline{\mathcal{T}}_s$ denotes the tail cost for the worst-case supplies associated with disruption scenario s.

Model DON_CV: *Selection of a Dynamic Supply Portfolio to Minimize Expected Worst-Case Cost per Part: DON Supplies*

Minimize expected worst-case cost per part (CVaR)

$$\overline{C}(x) = \text{VaR} + (1-\alpha)^{-1} \sum_{s \in S} \mathcal{P}_s \overline{\mathcal{T}}_s \tag{10.27}$$

subject to

1. *Order allocation and portfolio selection constraints*: (10.7) to (10.17)
2. *Risk constraints*:

– the tail cost for scenario s is defined as the nonnegative amount by which cost per part in scenario s exceeds VaR,

$$\begin{aligned}\overline{T}_s \geq & \sum_{i\in I}\sum_{t\in T} o_{it}y_{it}/N + \sum_{i\in I}\sum_{j\in J}\sum_{t\in T} p_{ijt}(1+q_{it})n_j z_{ijt}/N \\ & + \sum_{i\in I}\sum_{t\in T}\sum_{j\in J: f_j<h} a_j(h-f_j)n_j z_{ijt}\delta_{it}^s/N \\ & + \sum_{i\in I}\sum_{j\in J}\sum_{t\in T}(b_j - p_{ijt}(1+q_{it}))n_j z_{ijt}(1-\delta_{it}^s)/N - \text{VaR}; \quad s\in S\end{aligned} \tag{10.28}$$

$$\overline{T}_s \geq 0; \quad s \in S. \tag{10.29}$$

10.6 COMPUTATIONAL EXAMPLES

In this section some computational examples are presented to illustrate possible applications of the proposed MIP approach for selection of a dynamic supply portfolio in the presence of delay and disruption risks.

In the delivery scenarios, for each supplier i and period t, the delivery scheduled for that period is assumed to occur either in period t or in period h or not at all, that is, TDN supplies of ordered parts may occur either on time or with the longest delay $h - t$ or never. Therefore, for each supplier and each period there are three possible delivery states.

The following parameters have been used for the example problems:

- h and m, the number of planning periods and the number of suppliers were equal to 3, and hence $M = hm = 9$. The corresponding number, $|L| = 3^M$ of delivery scenarios was equal to 19,683.
- n, the number of customer orders, was equal to 50.
- a_j, delay penalty cost per period and per unit of customer order j, was equal to 10 for all orders j.
- b_j, shortage penalty cost per unit of customer order j, was equal to 100 for all orders j.
- n_j, the number of required parts for each customer order j, was integer uniformly distributed over [100,500], that is, generated from a U[100;500] distribution.
- c_{it}, the capacity of each supplier i in each period t, was equal to $\lceil 2\sum_{j\in J} n_j/m \rceil$, that is, in each period t the total capacity of all suppliers was equal to double the total demand for parts.
- e_j, the earliest delivery date of parts required for customer order j, was equal to 1 for all orders j.

- f_j, the latest delivery date of parts required for customer order j, was integer uniformly distributed over $[e_j, h]$, that is, generated from a U[1;3] distribution.
- o_{it}, the cost of ordering parts from supplier i in period t, was equal to 500 for each supplier i and period t.
- q_{it}, the expected defect rate of each supplier i in each period t, was exponentially distributed.
- $\overline{q}$ the largest acceptable average defect rate of supplies, was equal to 0.05.
- p_{ijt}, the unit price of parts required for each customer order j purchased from each supplier i in each period t, was uniformly distributed over [10,15] (i.e., drawn from U[10;15]) and reduced by the factor $(1 - q_{it})$ to get a lower price for parts from the suppliers with a higher defect rate in period t.
- π_{it}, the disruption probability for each supplier i in each period t was uniformly distributed over [0,0.05], that is, the disruption probabilities were drawn independently from U[0;0.05]. The corresponding probability of on-time and delayed delivery, was equal to $(1 - \pi_{it})/3$ and $2(1 - \pi_{it})/3$, respectively.
- $\overline{r}$ the largest acceptable average delay rate of supplies was equal to 0.25.
- $\overline{w}$ the largest acceptable average disruption rate of supplies was equal to 0.02.
- α, the confidence level, was equal to 0.50, 0.75, 0.90, 0.95 or 0.99.

For the test examples, the total demand for parts is $N = \sum_{j \in J} n_j = 14{,}750$ parts.

In the disruption scenarios considered for best-case TDN supplies with no delays, and worst-case TDN supplies with longest delays, for each supplier i and each period t there are two possible states: delivery (on time for TON supplies or in period h for DON supplies) with probability $(1 - \pi_{it})$ or no delivery, that is, disruption, with probability π_{it}. The corresponding number, $|S| = 2^M$, of disruption scenarios, for $h = m = 3$ was equal to 512.

The computational results are summarized in Tables 10.3 and 10.4, respectively, for the risk-neutral models *TON_E, TDN_E, DON_E* and the risk-averse models *TON_CV, TDN_CV, DON_CV.* The size of the MIP models is represented by the total number of variables, Var., number of binary variables, Bin., number of constraints, Cons, and number of nonzero coefficients in the constraint matrix, Nonz. For the risk-averse models, Table 10.4 shows the associated value of the expected cost calculated for the optimal risk-averse portfolio. The tables also present CPU time in seconds required to prove optimality of the solution and for the risk-averse models, the probability $1 - F(\text{VaR})$ of outcomes with worst-case cost above VaR. For TDN supplies the probabilities are slightly less than the corresponding values of $1 - \alpha$. The experiments with the two extreme cases show that for TON and DON supplies, however, the probability measure is more concentrated in finitely many points than for general TDN supplies. Large positive differences $F(\text{VaR}) - \alpha$ for TON and DON supplies indicate a large concentration of probability measure at the corresponding levels of cost. The greater the positive difference $F(\text{VaR}) - \alpha$, that is, the smaller than $1 - \alpha$ is the probability of outcomes with cost higher than VaR, the greater probability measures concentrated at those points. As a result, for optimal risk-averse

Table 10.3 Solution Results for Risk-Neutral Models

Model	Var.	Bin.	Cons.	Nonz.	Expected cost	CPU[a]
TDN_E	468	9	90	1746	19.27	0.02
TON_E	468	9	89	1740	12.39	0.02
DON_E	468	9	90	1746	22.69	0.02

Var. = number of variables, Bin. = number of binary variables, Cons. = number of constraints, Nonz.= number of nonzero coefficients.

[a]CPU seconds for proving optimality on a PC Pentium IV, 1.8 GHz, 1 GB RAM/CPLEX 11.

portfolios with TON and DON supplies, the probability of outcomes with cost higher than VaR is much smaller than $1 - \alpha$, given a confidence level α. For example, for TON supplies and $\alpha = 0.5$, VaR = 10.87 and $1 - F(\text{VaR}) = 0.066 < 0.5$, which indicates a high concentration of probability measure at point 10.87 for the optimal

Table 10.4 Solution Results for Risk-Averse Models

Confidence level α	0.50	0.75	0.90	0.95	0.99
***TDN_CV* model, $\|L\|$ = 19,683 delivery scenarios**					
Var. = 20,152, Bin. = 9, Cons. = 19,771, Nonz. = 9,075,594					
CPU[a]	3530	3600	1765	871	840
Expected cost	19.24	19.35	19.69	19.69	19.72
VaR	18.04	20.71	26.85	29.04	35.55
CVaR	22.42	25.39	30.12	32.39	38.32
$1 - F(\text{VaR})$	0.474	0.223	0.098	0.046	0.010
***TON_CV* model, $\|S\|$ = 512 disruption scenarios**					
Var. = 981, Bin. = 9, Cons. = 601, Nonz. = 237,772					
CPU[a]	0.7	0.6	0.8	0.8	1
Expected cost	12.41	12.45	13.34	13.27	13.34
VaR	10.87	10.92	20.86	21.23	30.79
CVaR	13.96	17.03	23.25	25.61	32.20
$1 - F(\text{VaR})$	0.066	0.066	0.023	0.023	0.005
***DON_CV* model, $\|S\|$ = 512 disruption scenarios**					
Var. = 981, Bin. = 9, Cons. = 600, Nonz. = 237,763					
CPU[a]	5	5	5	7	5
Expected cost	22.36	22.53	23.05	22.89	23.05
VaR	20.93	21.18	29.73	30.42	38.55
CVaR	23.80	26.57	31.85	33.90	39.80
$1 - F(\text{VaR})$	0.176	0.066	0.023	0.023	0.005

Var. = number of variables, Bin. = number of binary variables, Cons. = number of constraints, Nonz.= number of nonzero coefficients.

[a]CPU seconds for proving optimality on a PC Pentium IV, 1.8 GHz, 1 GB RAM/CPLEX 11.

portfolio. Actually, the probability that cost per part is 10.87 is 0.934, which indicates that VaR $= 10.87$ is the lowest cost that may occur and that for the confidence level $\alpha = 0.5$, at most 6.6% of the cost outcomes are above VaR.

The optimal supply portfolios for TDN supplies are presented in Figure 10.1. The figure also presents the expected disruption rate w_{it}, expected delay rate r_{it} and the expected defect rate q_{it}. In the computational experiments suppliers with higher expected defect rates q_{it}, that are known ahead of time, have lower prices p_{ijt}. The prices, however, are independent of disruption rates. On the other hand, to ensure that the impact of disruption probabilities π_{it} on the allocation of orders among suppliers is independent of suppliers' capacities, all capacities c_{it} are identical.

It is observed that the supplies (i, t) associated with the lowest expected disruption rates w_{it} are allotted the largest fractions x_{it} of total demand. The results indicate that the probability of disruption of a supply is a key determinant in the decision of dynamic allocation of demand among the suppliers. In a risk-averse model, a supply is selected more on its likelihood of nondisruption than on its expected delay or defect rate. Note that cost of lost customer orders, b_j, is set to be much higher than the corresponding cost, a_j, of delayed orders, which is typical for industrial practice.

Figure 10.1 also compares the optimal risk-neutral and risk-averse portfolios. For the risk-averse model the ordered quantities are less than those determined by a risk-neutral model that minimizes the expected cost. The more risk-averse the supply portfolio is, that is, the higher the confidence level α, the more balanced is the allocation of demand among the selected suppliers.

10.6.1 Scenarios with at Most One or with Consecutive Disruptions

In addition to the general delivery scenario, the following two special subsets of the general scenario have also been considered:

- scenarios with at most one disruption of each supplier over the horizon
- scenarios with consecutive disruptions (no delivery) of each supplier over the horizon since its first disruption

In the special scenarios it is assumed that for each supplier the supply disruption may occur either at most once over the planning horizon or in all successive periods that follow the disruption period. The latter case represents a permanent disruption of supplies from a particular supplier beginning with the first such event until the end of the horizon.

The total number of TDN supply scenarios with at most one disruption of each supplier is

$$|L'| = \left(\sum_{k_1,k_2,k_3:k_1 \le 1, k_1+k_2+k_3=3} \binom{3}{k_1, k_2, k_3} \right)^3 = 20^3 = 8000,$$

where k_1, k_2, and k_3 represent, respectively, the number of disruptions, the number of delayed deliveries, and the number of on-time deliveries of each supplier over the horizon.

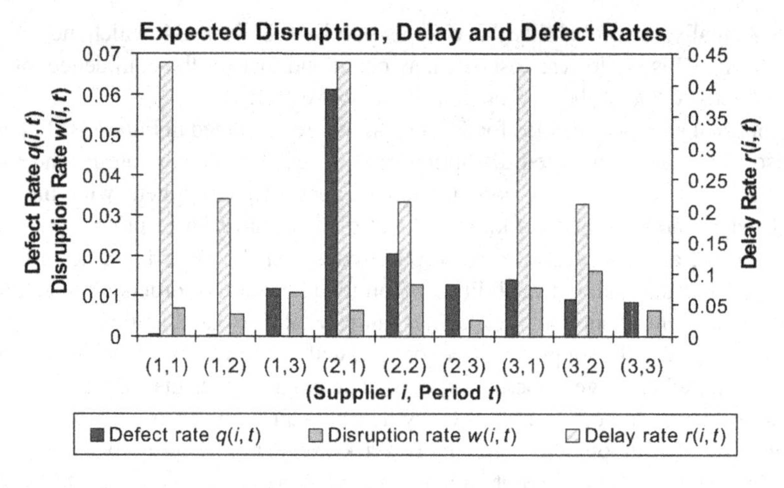

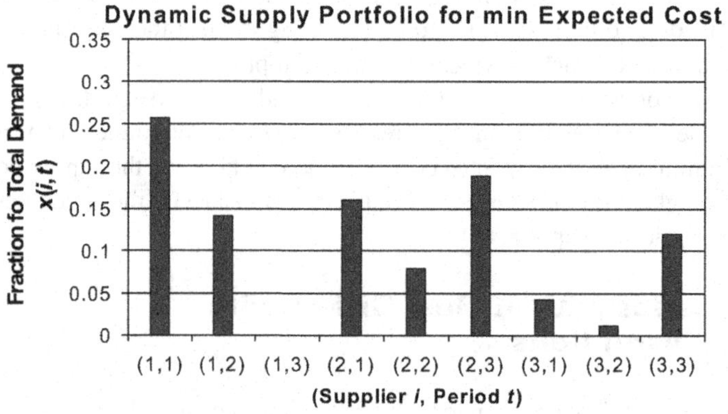

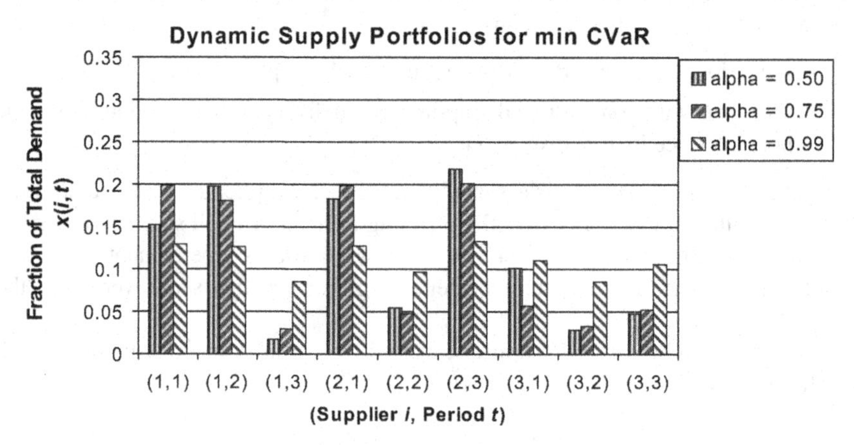

Figure 10.1 Optimal supply portfolios for TDN supplies.

The multinomial coefficient is defined as

$$\binom{k}{k_1, k_2, \ldots, k_r} = \frac{k!}{k_1!k_2!\cdots k_r!},$$

where $k_1 + k_2 + \cdots + k_r = k$.

The total number of TDN supply scenarios with consecutive disruptions of each supplier can be found by simple enumeration of all such scenarios and is $(15)^3 = 3375$.

The probabilities P_l of realization of special delivery scenarios $l \in L' \subset L$ do not sum to one and need to be adjusted. The adjusted probability P'_l is assumed to be the sum of its original probability P_l (10.3) and the adjustment term proportional to its contribution to the total probability of all special scenarios

$$P'_l = P_l + \left(P_l \Big/ \sum_{l \in L'} P_l\right)\left(1 - \sum_{l \in L'} P_l\right); \ l \in L'.$$

The solution results for the two special scenarios are summarized in Table 10.5, while Figure 10.2 presents examples of optimal risk-averse portfolios for the confidence level $\alpha = 0.9$.

Table 10.5 shows that all solution values for the scenario with consecutive disruptions are smaller than the corresponding values for the scenario with at most

Table 10.5 Solution Results for the *TDN_CV* Model: Special Scenarios

Confidence level α	0.50	0.75	0.90	0.95	0.99
$\lvert L' \rvert = 8000$ scenarios with at most one disruption of each supplier					
Var. = 8469, Bin. = 9, Cons. = 8090, Nonz. = 3,689,746					
CPU[a]	430	576	47	45	54
Expected cost	19.36	19.31	19.60	19.61	19.60
VaR	18.46	20.71	26.65	28.52	33.82
CVaR	22.36	25.11	29.55	31.42	36.84
$1 - F$(VaR)	0.46	0.22	0.098	0.042	0.009
$\lvert L' \rvert = 3375$ scenarios with consecutive disruptions of each supplier					
Var. = 3844, Bin. = 9, Cons. = 3464, Nonz. = 1,557,612					
CPU[a]	65	60	62	71	14
Expected cost	17.85	17.93	17.98	17.99	18.15
VaR	17.75	19.05	21.08	21.09	25.87
CVaR	19.76	21.02	22.58	24.08	28.59
$1 - F$(VaR)	0.45	0.25	0.09	0.048	0.009

Var. = number of variables, Bin. = number of binary variables, Cons. = number of constraints, Nonz.= number of nonzero coefficients.

[a]CPU seconds for proving optimality on a PC Pentium IV, 1.8 GHz, 1 GB RAM/CPLEX 11.

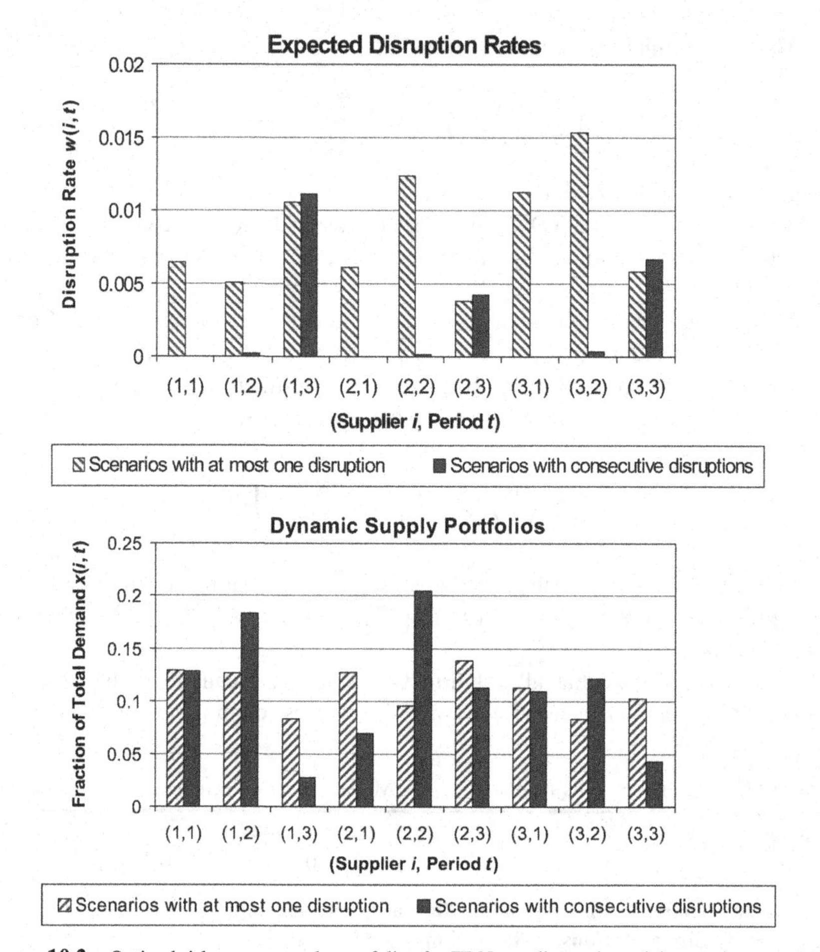

Figure 10.2 Optimal risk-averse supply portfolios for TDN supplies and special scenarios: $\alpha = 0.9$.

one disruption over the horizon. Furthermore, the solution values for the latter scenario are similar to the corresponding values for the general scenario (see Table 10.4), and so are the corresponding optimal portfolios (see Fig. 10.1). It is once again observed that the supplies associated with lowest expected disruption rates w_{it} are allotted the largest fractions x_{it} of the total demand. This is clearly shown in Figure 10.2, for the scenario with consecutive disruptions, where the expected disruption rates are negligible in the first two periods and large in the last period. As a result most of the demand is allotted among the first two periods.

The computational results point out that the less distinguishable are expected disruption rates w_{it} for different supplies (i, t), the more leveled over the supplies is the optimal portfolio, which is demonstrated in Figure 10.1 for the general scenario and Figure 10.2 for the scenario with at most one disruption over the horizon.

For TON and DON supplies, Figure 10.3 presents the optimal risk-neutral and risk-averse (for confidence level $\alpha = 0.5$ and $\alpha = 0.9$) supply portfolios. The risk-

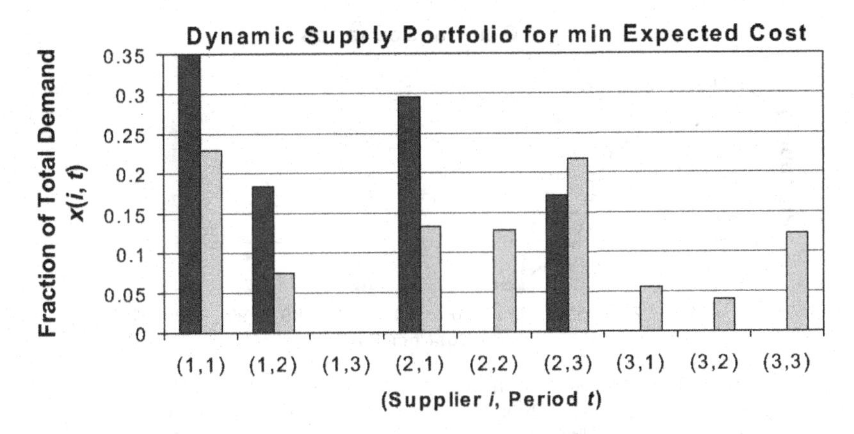

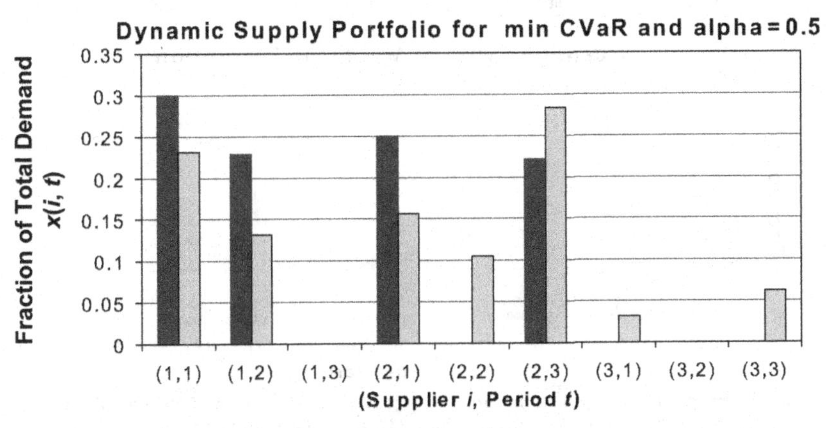

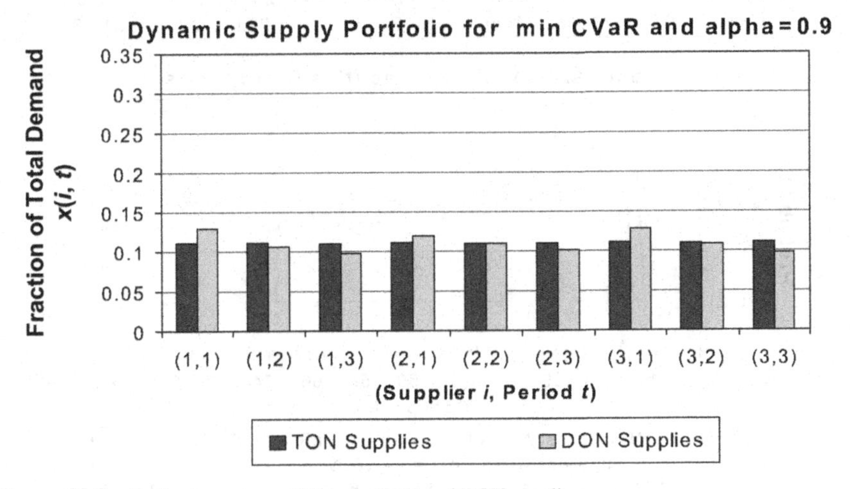

Figure 10.3 Optimal supply portfolios for TON and DON supplies.

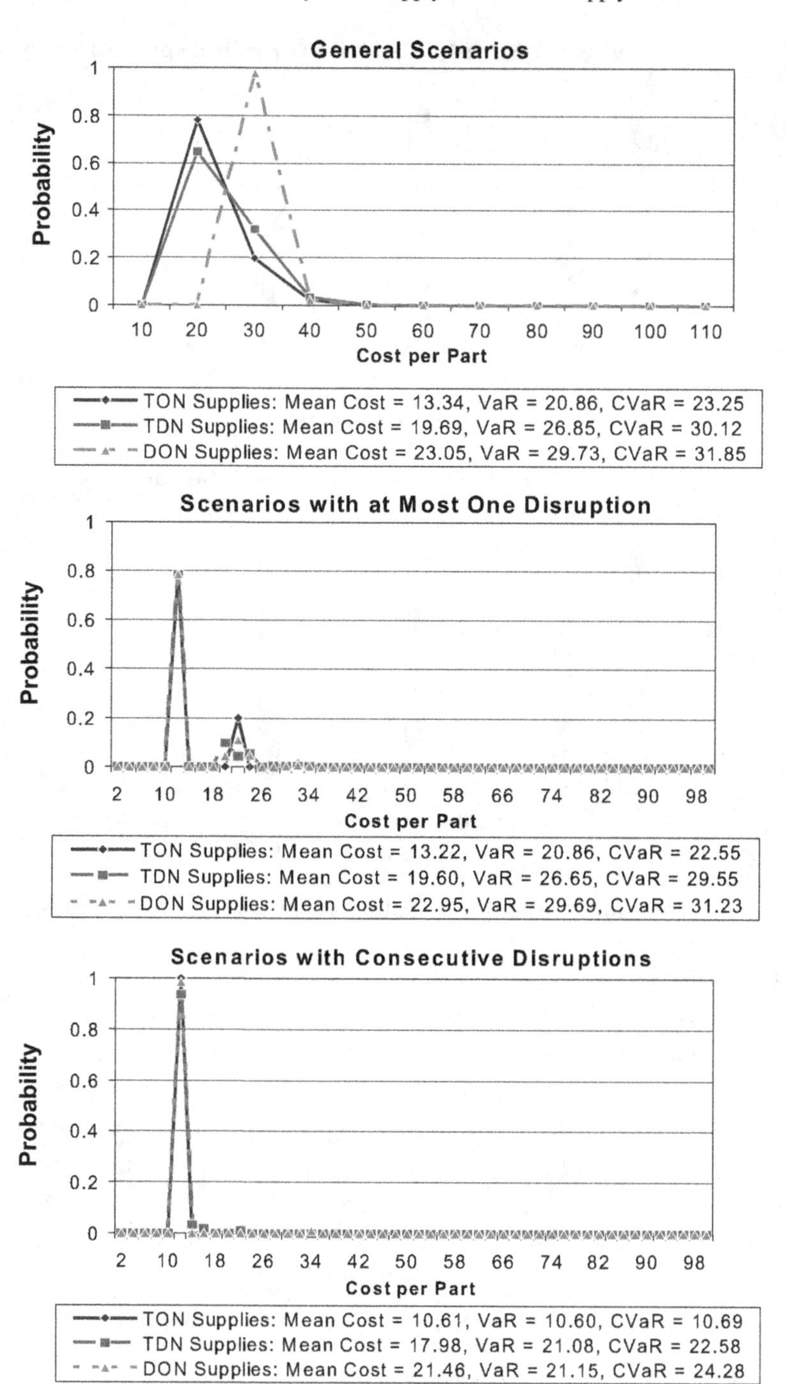

Figure 10.4 Risk-averse cost distributions for different scenarios and confidence level $\alpha = 0.9$.

averse portfolios for $\alpha = 0.5$ are different for TON and DON supplies and are very close to the corresponding risk-neutral portfolios that minimize the expected cost. The portfolios for TON supplies (subject to disruption risks only) consist of supplies (1,1), (1,2), (2,1), and (2,3) with the lowest expected disruption rates, while for DON supplies (subject to both disruption and delay risks) the portfolios are more diversified. In contrast, for a higher confidence level $\alpha = 0.9$ the risk-averse portfolios are balanced and are very close each other, which indicates that for both TON and DON supplies a greater diversification was required to mitigate the impact of risks.

In the computational experiments the confidence level α is set at five levels of 0.5, 0.75, 0.90, 0.95, and 0.99, which means that focus is on minimizing the highest 50%, 25%, 10%, 5%, and 1% of all scenario outcomes, that is, costs per part. When α increases, a more risk-averse decision making focuses on a smaller set of outcomes and the number of selected suppliers also is increasing to mitigate the impact of disruptions risks by diversification of the supply portfolio.

The optimal risk-averse cost distributions (probability mass functions) for confidence level $\alpha = 0.9$ and different scenarios are shown in Figure 10.4. The probability mass function of cost per part is concentrated in a few points, which is typical for the scenario-based optimization under uncertainty, where the probability measure is concentrated in finitely many points. Comparison of the optimal cost distributions indicates that for very low probability scenarios with consecutive disruptions, the outcomes are clustered around the expected cost values, while for higher probability scenarios with at most one disruption over the horizon, the outcomes are slightly skewed towards the higher costs. In particular, for TON supplies and scenarios with consecutive disruptions all outcomes are concentrated around the expected cost value.

The computational experiments were performed using the AMPL programming language and the CPLEX v.11 solver (with the default settings) on a laptop with a Pentium IV processor running at 1.8 GHz and with 1 GB RAM. The CPLEX solver was capable of finding proven optimal solutions within CPU seconds for all examples.

10.7 COMMENTS

The literature on multiperiod supplier selection and order allocation models is very limited. For example, Basnet and Leung (2005) proposed an MIP formulation for multiproduct, multiperiod lot sizing with supplier selection in a deterministic environment, Ustun and Demirtas (2008) presented an integration of analytic network processes and a deterministic multiperiod, multi-objective MIP formulation to consider both tangible and intangible factors for choosing the best suppliers and determine the optimum quantities ordered from the selected suppliers.

The material presented in this chapter is based on the results discussed in the paper by Sawik (2011b) on dynamic (multiperiod) supplier selection and order allocation in the presence of supply chain disruption and delay risks. The computational experiments presented in Sawik (2011b) indicate that the probability of disruption to a supply is a key determinant in the decision of dynamic allocation of demand among the suppliers. In a risk-averse model, an order for delivery of parts from a particular supplier in some period is selected more on the likelihood of nondisruption of supply than on its expected delay or defect rate. The supplies associated with the lowest expected disruption rates are allotted the largest fractions of the total demand and the less distinguishable are expected disruption rates for different suppliers and periods, the more leveled over the horizon and the suppliers is the optimal portfolio.

The experiments demonstrate that the more risk-averse the supply portfolio is, that is, the higher the confidence level α, the more leveled is the allocation of demand among the selected suppliers. Furthermore, the number of selected suppliers increases with the confidence level, which indicates that the impact of disruption and delay risks is mitigated by diversification of the supply portfolio.

The experiments also show that in a risk-neutral environment the more costly suppliers are rarely selected, the portfolio can be highly unbalanced over the horizon, and the suppliers and the ordered quantities are greater than those determined by the risk-averse solutions.

In the models proposed in this chapter each supplier is assumed to be capable of manufacturing all required part types. In a more general setting, each supplier may only be prepared to manufacture a subset of part types and provide with the parts the corresponding subset of customer orders.

In practice, the proposed MIP models can be applied only for a small problem as the size of the mixed integer programs increases rapidly with the number of suppliers m and the length of the planning horizon h. As each scenario specifies a supply realization for all suppliers in all periods, and the total number of potential supplies in all periods is hm, the number of potential scenarios as well as the number of variables and constraints in the MIP models grow exponentially in hm. For a greater number of suppliers and planning periods and the number of scenarios considered the problem of running out of memory may arise when the branch and cut tree becomes so large that insufficient memory is available to solve LP subproblems by an MIP solver.

EXERCISES

10.1 Assuming that each supplier i is capable of providing with parts only a subset $J_i \subset J$ of all customer orders, introduce appropriate changes in the models presented in this chapter.

10.2 Generalize models *TDN_E* and *TDN_CV* for selection of a dynamic supply portfolio under disruption and delay risks to account for:

(**a**) the total quantity discount for all ordered parts.

(**b**) the independent quantity discounts for individual part types.

(**c**) the total business volume discount.

10.3 Modify the probability for delivery scenarios (10.3) to account for the local and global disruptions that may affect all suppliers simultaneously

(**a**) over the entire planning horizon.

(**b**) during a fixed time interval.

10.4 Modify the probability for disruption scenarios (10.21) to account for local and semiglobal disruptions that affect only a subset of all suppliers simultaneously

(**a**) over the entire planning horizon.

(**b**) during a fixed time interval.

10.5 In the computational examples in Section 10.6, Figure 10.3 indicates that the optimal risk-averse supply portfolios for TON and DON supplies and a high confidence level α are balanced over the set of suppliers and the planning horizon and are very close each other, while for a small α the portfolios are unbalanced and different. Why do the optimal portfolios have such properties?

Part Three

Coordinated Scheduling in Supply Chains

Chapter 11

Hierarchical Integration of Medium- and Short-Term Scheduling

11.1 INTRODUCTION

This chapter deals with intertemporal coordination of medium-term, aggregate production scheduling and short-term, detailed machine assignment and scheduling in a customer-driven supply chain, and presents a hierarchical decision-making framework with IP and MIP models. In the medium-term aggregate production scheduling, a set of customer orders for different products and with various due dates is allocated among planning periods so that each order is fully completed in exactly one period, and machines are selected for assignment in every period. The medium-term scheduling objective is to minimize, in decreasing order of importance, the number of unscheduled (rejected), tardy, and early orders and to level machine assignments over a planning horizon. Then the customer orders assigned to one planning periods are split into production lots, each to be processed as a separate job. The objective of the short-term machine assignment and scheduling is to determine an assignment of production lots to machines to minimize the number of operators required to attend the machines as a primary optimality criterion and to minimize makespan as a secondary criterion. The IP model for medium-term production scheduling is strengthened by the addition of cutting constraints that are derived by relating the demand on required capacity to available capacity for each processing stage and each subset of orders with the same due date. Similarly, the MIP model for short-term machine scheduling is also strengthened by the addition of cutting constraints derived for batch scheduling.

A general hierarchical framework for production scheduling in a make-to order manufacturing environment is shown in Figure 11.1. First, at the top level, the medium-term (e.g., monthly) aggregate production scheduling allocates customer

Scheduling in Supply Chains Using Mixed Integer Programming. By Tadeusz Sawik

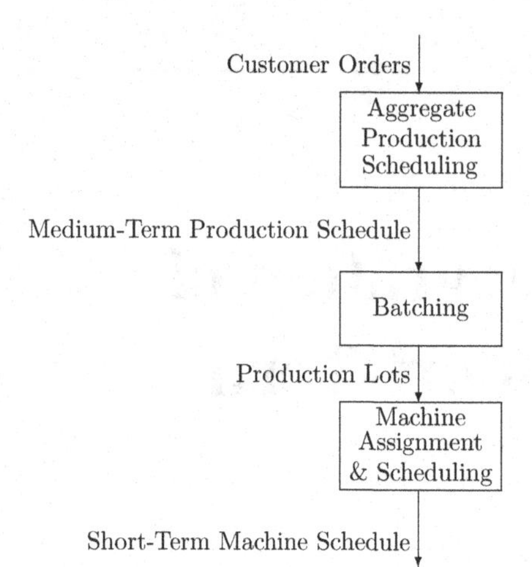

Figure 11.1 Hierarchical production scheduling.

orders among planning periods (e.g., days) to optimize some due date-related criterion. Then, at the medium level, the short-term (e.g., daily) batching (also called lot splitting or lot streaming) divides customer orders into production lots, each to be processed as a separate job, to best utilize the production resources. Finally, at the base level, the detailed machine assignment and scheduling translates a subset of production lots into short-term (e.g., daily) machine scheduling to maximize throughput or equivalently to minimize makespan.

The proposed hierarchical decision-making framework integrates the medium-term aggregate production scheduling (see Chapter 6) and the short-term detailed machine scheduling (see Chapters 1 to 4), with both maximization of the customer service level and best utilization of the production resources integrated in the objective functions.

The following IP and MIP models are presented in this chapter:

MPS for medium-term aggregate production scheduling

MA for short-term machine assignment

MS/A for short-term machine scheduling with prefixed machine assignment

The proposed formulations are applied to hierarchical scheduling of a flexible flow shop with multicapacity machines in a make-to-order manufacturing environment. Numerical examples modeled after a real-world make-to-order flexible assembly line (FAL) for high-tech products and computational results are provided in Section 11.5.

11.2 PROBLEM DESCRIPTION

The production system under study is a flexible flow shop that consists of m processing stages in series (see Fig. 11.2) and an output buffer of limited capacity B to hold

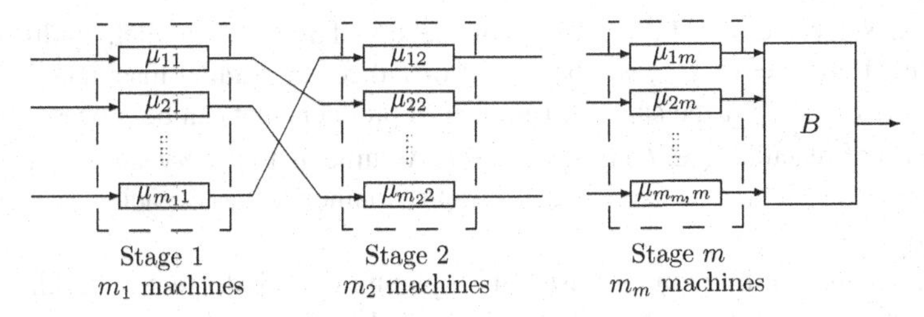

Figure 11.2 A flexible flow shop with multicapacity machines and finite output buffer.

completed products before delivery to the customer (for notation used, see Table 11.1). Typically, the capacity B of the output buffer is not large to keep low inventory costs and to limit early completion of customer orders before the customer-required shipping dates (see Section 6.4.2).

Table 11.1 Notation: Medium-Term Production Scheduling

Indices

g	= parallel machine in stage i, $g \in G_i = \{1, \ldots, m_i\}$
i	= processing stage, $i \in I = \{1, \ldots, m\}$
j	= customer order, $j \in J$
t	= planning period, $t \in T = \{1, \ldots, h\}$

Input parameters

a_j, d_j, n_j	= arrival date, due date, size of order j
b_j	= production and transfer lot size for order j
c_{it}	= processing time available in period t on each machine in stage i
p_{ij}	= processing time in stage i of each product in order j
μ_{gi}	= capacity of machine g in stage i (number of products that can be processed simultaneously)
B	= output buffer capacity
D	= $\{d_j: j \in J\}$ set of distinct due dates of all customer orders
H	= length of each planning period
J_i	= $\{j \in J: p_{ij} > 0\}$ subset of customer orders to be processed in stage i
$J(d)$	= $\{j \in J: d_j = d\}$ subset of customer orders with identical due date d
T^*	= $T \cup \{h + 1\}$, where $h + 1$ is a dummy period with infinite capacity

Decision variables

x_{jt}	= 1, if order j is performed in period t; otherwise $x_{jt} = 0$ (order-to-period assignment variable)
w_{git}	= 1, if machine $g \in G_i$ in stage $i \in I$ is selected for assignment in period t; otherwise $w_{git} = 0$ (machine selection variable)
$M_{\max}$	= maximum number of machines selected for assignment in a single planning period

Each stage $i \in I = \{1, \ldots, m\}$ is made up of $m_i \geq 1$ parallel, multicapacity machines. Let $G_i = \{1, \ldots, m_i\}$ be the set of indices of parallel machines in stage i and denote by $\mu_{gi} \geq 1$ the capacity (number of products that can be processed simultaneously) of machine $g \in G_i$ in stage $i \in I$. Assume that in every stage i the parallel machines $g \in G_i$ are ordered according to nonincreasing capacity μ_{gi}, that is, $\mu_{1i} \geq \mu_{2i} \geq \cdots \geq \mu_{m_i,i}$.

The planning horizon consists of h planning periods (e.g., working days) of equal length H (e.g., hours or minutes). Let $T = \{1, \ldots, h\}$ be the set of planning periods.

In the system various types of products are produced in a make-to-order environment responding directly to customer orders. Let J be the set of customer orders, each to be fully completed in exactly one planning period. Each order $j \in J$ is described by a triple (a_j, d_j, n_j), where a_j is the order ready period (e.g., the earliest period of material availability), d_j is the customer-requested due date (e.g., customer-required shipping date), and n_j is the quantity of ordered products. The assumption of single-period orders, which simplifies the mathematical formulations, can be easily relaxed (see Sections 5.5, 6.2, and 7.2) to allow for the inclusion of multiperiod orders.

Each order requires processing in various processing stages; however, some orders may bypass some stages. Let $J_i \subset J$ be the subset of orders that must be processed in stage i, and let $p_{ij} > 0$ be the processing time in stage i of each product in order $j \in J_i$.

The orders are processed and transferred among the stages in lots of various size and each lot is to be processed as a separate job. Let b_j be the size of the production and transportation lot for order j. The size of production and transportation lots is a fixed parameter and not a decision variable to be optimized in a framework of a lot-sizing model. Setups and machine changeovers between different product types are negligible, and the lot size b_j depends only on product type and the capacity of containers used for product storage and transportation. For each type of product, parameter b_j is determined so that processing of each lot b_j as a separate job is fully completed in all stages of the flexible flow shop in exactly one planning period, that is, $\sum_{i \in I} p_{ij} \lceil b_j / \mu_{im_i} \rceil < H$.

Processing of products in some stages additionally requires human operators and the number of required operators per machine depends on the type of product.

The objective of the medium-term aggregate production scheduling is to assign customer orders to planning periods and to select machines for assignment in every period so as to minimize, in decreasing order of importance, number of unscheduled, tardy, and early orders and to level machine assignments over a planning horizon. Leveling machine assignments and, in turn, balancing the number of people required for machine attendance, results in lower operational costs.

For each planning period, given the set of production lots of all customer orders assigned to this period and the set of machines selected for assignment in this period, the objective of short-term machine assignment and scheduling is to determine an assignment of the production lots to machines and a detailed processing schedule to minimize the number of people as a primary optimality criterion and to minimize makespan as a secondary criterion.

11.3 MEDIUM-TERM PRODUCTION SCHEDULING

In this section the monolithic, time-indexed IP model is presented for the multi-objective, medium-term aggregate production scheduling, where each customer order must be fully completed in a single planning period (e.g., during one day).

Due to the discrete nature of orders it is unlikely that any time period will be filled exactly to its capacity. As a result, some orders may be left unscheduled during the planning horizon. The orders that cannot be scheduled in periods 1 through h due to insufficient capacity are assigned at a significant penalty to a dummy planning period $h^* = h + 1$ with infinite capacity. Let $T^* = \{1, \dots, h, h + 1\}$ be the enlarged set of planning periods with a dummy period $h^* = h + 1$ included.

A weighted-sum approach is applied, where the quad-objective production scheduling problem is reduced to a single-objective problem by introducing the weights λ_1, λ_2, λ_3, and λ_4 representing the relative importance of the four objective functions to be minimized, respectively, the number of unscheduled orders, tardy orders, early orders, and the maximum machine assignments in a single planning period (to level machine assignments over the horizon).

The time-indexed IP model *MPS* for medium-term production scheduling is presented below (for definition of decision variables, see Table 11.1).

Model MPS: *Medium-Term Production Scheduling*

Minimize

$$\lambda_1 \sum_{j \in J} x_{jh+1} + \lambda_2 \sum_{j \in J: t > d_j} x_{jt} + \lambda_3 \sum_{j \in J: t < d_j} x_{jt} + \lambda_4 M_{\max} \qquad (11.1)$$

where $\lambda_1 > \lambda_2 > \lambda_3 \geq \lambda_4 \geq 0$ and $\lambda_1 + \lambda_2 + \lambda_3 + \lambda_4 = 1$,

subject to

1. *Order-to-Period Assignment Constraints*:
   - each customer order is assigned to exactly one planning period,

$$\sum_{t \in T^*} x_{jt} = 1; \; j \in J \qquad (11.2)$$

2. *Machine Assignment Constraints*:
   - in every stage the machines are selected for assignment in the order of nonincreasing capacity,
   - in every period the number of machines selected for assignment in each stage is not greater than the maximum number of assigned transfer lots,
   - in every period the number of machines selected for assignment cannot exceed the maximum number of machine assignments to be minimized,

– in every period demand on capacity in each processing stage cannot be greater than the total capacity of machines selected for assignment in this period,

$$w_{g+1it} \leq w_{git};\ i \in I, g \in G_i, t \in T: g < m_i \tag{11.3}$$

$$\sum_{g \in G_i} w_{git} \leq \sum_{j \in J_i} \lceil n_j/b_j \rceil x_{jt};\ i \in I, t \in T \tag{11.4}$$

$$\sum_{i \in I} \sum_{g \in G_i} w_{git} \leq M_{\max};\ t \in T \tag{11.5}$$

$$\sum_{j \in J_i} n_j p_{ij} x_{jt} \leq c_{it} \sum_{g \in G_i} \mu_{gi} w_{git};\ i \in I, t \in T \tag{11.6}$$

3. *Output Buffer Capacity Constraints*:
   – in every period the total number of products completed before their due dates and waiting for shipping to customers cannot exceed the output buffer capacity,

$$\sum_{j \in J} \sum_{\tau \in T: a_j \leq \tau \leq t < d_j} n_j x_{j\tau} \leq B;\ t \in T \tag{11.7}$$

4. *Variable Integrality Conditions*:

$$x_{jt} \in \{0, 1\};\ j \in J, t \in T^*: t \geq a_j \tag{11.8}$$

$$w_{git} \in \{0, 1\};\ i \in I, g \in G_i, t \in T \tag{11.9}$$

$$M_{\max} \geq 0,\ integer, \tag{11.10}$$

where $\lceil \cdot \rceil$ is the least integer not less than $\cdot$, ($\lceil n_j/b_j \rceil$ denotes the number of production lots of customer order j).

The objective of the medium-term aggregate scheduling problem is to minimize the weighted sum of unscheduled, tardy, and early orders and to level machine assignments over a planning horizon. In the objective function (11.1) unscheduled orders are penalized at a higher rate than tardy orders and tardy orders are penalized at a higher rate than early orders and maximum machine assignments, that is, $\lambda_1 > \lambda_2 > \lambda_3 \geq \lambda_4$.

The left-hand side of capacity constraint (11.6) denotes the total processing time required for completing, in stage i, all customer orders assigned to period t, whereas the right-hand side of (11.6) is the total available processing time on all machines selected for assignment in stage i in period t.

Note that, if all machines have unit capacity, that is, each machine can process at most one product at a time, then $\mu_{gi} = 1$ for all $i \in I$, $g \in G_i$ and $\sum_{g \in G_i} \mu_{gi} = m_i$, and represents the number of parallel identical machines in stage $i \in I$.

The cutting constraints (11.3) and (11.4) reduce the computational effort required to find the proven optimal solution to the integer program *MPS*.

The amount of time available on each machine in stage i in period t, c_{it}, must take into account the flow shop configuration of the production system and the transfer lot sizes; c_{it} is bounded as follows (see constraints (5.1) in Section 5.2):

$$\begin{aligned}
H - & \max_{j\in J_i: a_j \le t \le d_j} \left(\sum_{i'\in I: i'<i} \lceil b_j/\mu_{1i'} \rceil p_{i'j} \right) \\
- & \max_{j\in J_i: a_j \le t \le d_j} \left(\sum_{i'\in I: i'>i} \lceil b_j/\mu_{1i'} \rceil p_{i'j} \right) \\
& \le c_{it} \le \\
H - & \min_{j\in J_i: a_j \le t \le d_j} \left(\sum_{i'\in I: i'<i} \lceil b_j/\mu_{1i'} \rceil p_{i'j} \right) \\
- & \min_{j\in J_i: a_j \le t \le d_j} \left(\sum_{i'\in I: i'>i} \lceil b_j/\mu_{1i'} \rceil p_{i'j} \right).
\end{aligned} \tag{11.11}$$

For each machine in stage i, c_{it} accounts for the shortest processing time required for completing a single production lot on machines with the highest capacity, at all upstream and downstream stages during the same planning period.

A correct estimation of c_{it} used for the medium-term aggregate scheduling level is crucial to ensure the feasibility of the short-term detailed machine assignment and scheduling. A too large value of c_{it} may lead to infeasible daily schedules, that is, some customer orders assigned to the same planning period will not be completed during a single-day scheduling horizon. On the other hand, a too small value of c_{it} may result in significant underutilization of processing capacity.

11.3.1 Cutting Constraints for Multicapacity Machines

The cutting constraints presented in this subsection are simple extensions for the system with multicapacity machines of the cuts derived in Section 6.5.1 for the case of unit-capacity machines.

A necessary condition for problem *MPS* to have a feasible solution with all customer orders completed during the planning horizon is that, for each processing stage $i \in I$, the total demand on capacity does not exceed total capacity available (cf. (6.73)), that is,

$$\frac{\sum_{j\in J} n_j p_{ij}}{\sum_{t\in T} c_{it} \sum_{g\in G_i} \mu_{gi}} \le 1; \forall i \in I. \tag{11.12}$$

If the customer orders were arbitrarily divisible and could be completed during more than one planning period, all orders were ready for completing at the beginning of the horizon and the output buffer capacity was unlimited to allow for any earliness in completing the orders, then the feasibility condition (11.12) would be sufficient. Due to the discrete nature of customer orders, however, it is possible that some time periods will not be filled exactly to their capacities. As a result the necessary condition (11.12) is not sufficient for all orders to be scheduled during the planning horizon. Nevertheless, the IP model *MPS* can be strengthened by relating for each due date the local demand on required capacity to available capacity and the cumulative demand on required capacity to available cumulative capacity (see Section 6.5.1).

For each due date $d \in D$, the local capacity ratio Φ_d is defined in (6.76), and the local capacity ratio for due date d with respect to processing stage i, φ_{id}, is

$$\varphi_{id} = \frac{\sum_{j \in J(d)} n_j p_{ij}}{c_{id} \sum_{g \in G_i} \mu_{gi}}; \; i \in I, \, d \in D. \tag{11.13}$$

The subset of locally satisfiable, locally unsatisfied due dates, D_0, D_1, are defined in (6.78), (6.79), respectively.

For each processing stage and each locally unsatisfied due date at least one order should be moved to another period to reduce local demand on capacity with respect to the overloaded stage, at least to the level of available capacity. Hence, some orders with due dates in periods with the local demand on capacity exceeding available capacity are potentially subject to earliness or tardiness.

Cutting constraints on assignment of orders with locally unsatisfied due dates

$$\sum_{j \in J(d) \cap J_i, t \in T: t \neq d} x_{jt} \geq 1; \; d \in D_1, \, i \in I: \varphi_{id} > 1 \tag{11.14}$$

$$\sum_{j \in J(d) \cap J_i, t \in T: t \neq d} n_j p_{ij} x_{jt} \geq (\varphi_{id} - 1) c_{id} \sum_{g \in G_i} \mu_{gi};$$

$$d \in D_1, \, i \in I: \varphi_{id} > 1. \tag{11.15}$$

Furthermore, for each locally satisfied due date and each processing stage with surplus of capacity, the work inflow of orders with other due dates less the work outflow of orders with this due date cannot exceed the slack capacity.

Flow balance cutting constraint for locally satisfied due dates

$$\sum_{d' \in D, j \in J(d') \cap J_i: d' \neq d} n_j p_{ij} x_{jd} - \sum_{j \in J(d) \cap J_i, t \in T: t \neq d} n_j p_{ij} x_{jt} \leq$$

$$(1 - \varphi_{id}) c_{id} \sum_{g \in G_i} \mu_{gi}; \; d \in D_0, \, i \in I. \tag{11.16}$$

Constraint (11.16) ensures that, for each locally satisfied due date and each processing stage, the total processing time of the orders with other due dates and assigned to this period is not greater than the slack capacity plus the total processing time of the orders moved early or late from this period.

For each due date $d \in D$, the cumulative capacity ratio Ψ_d is defined in (6.83) (see also critical load index (5.4) in Section 5.2.1), and the cumulative capacity ratio for due date d with respect to processing stage i, ψ_{id}, is

$$\psi_{id} = \max_{t \in T: t \leq d} \left(\frac{\sum_{d' \in D, j \in J(d'): t \leq a_j \leq d' \leq d} n_j p_{ij}}{\sum_{\tau \in T: t \leq \tau \leq d} c_{i\tau} \sum_{g \in G_i} \mu_{gi}} \right); i \in I, d \in D. \quad (11.17)$$

The subset of globally satisfiable, globally unsatisfied due dates, D^0, D^1, are defined in (6.85), (6.86) respectively.

If all due dates are globally satisfiable (i.e., $D^1 = \emptyset$) and, in addition, the output buffer capacity B is not less than the minimum capacity $B_{\min}$ (cf. $B2^U_{\min}$, (6.75)) required to complete all orders on or before their corresponding due dates, then the following constraints guarantee nondelayed completion of all orders.

Order-to period nondelayed assignment constraints

$$x_{jt} = 0;\ j \in J;\ t \in T\text{: } t > d_j,\ D^1 = \emptyset,\ B \geq B_{\min} \quad (11.18)$$

$$\sum_{t=a_j}^{d_j} x_{jt} = 1;\ j \in J\text{: } D^1 = \emptyset,\ B \geq B_{\min}, \quad (11.19)$$

where

$$B_{\min} = \min \Bigg\{ B \geq 0\text{: } (11.2), (11.7), (11.8),$$

$$\sum_{j \in J_i} n_j p_{ij} x_{jt} \leq c_{it} \sum_{g \in G_i} \mu_{gi};\ i \in I, t \in T,$$

$$\sum_{j \in J, t \in T: t > d_j} x_{jt} = 0 \Bigg\}. \quad (11.20)$$

Finally, constraint (11.21) (cf. (6.89)) presented below ensures a delayed assignment of customer orders.

Order-to-period delayed assignment constraints

$$\sum_{j \in J(d) \cap J_i, t \in D_0\text{: } d \leq \underline{d} < t,\ \psi_{i\underline{d}} > 1} x_{jt} \geq 1;\ i \in I, \quad (11.21)$$

where $\underline{d} = \min\{d\text{: } d \in D_1 \cap D^1\}$ is the earliest potentially unsatisfied due date.

In addition, the following bounds on maximum machine assignments $M_{\max}$ can be added to model *MPS*:

$$\underline{M} \leq M_{\max} \leq \overline{M}. \quad (11.22)$$

The lower bound $\underline{M}$ and upper bound $\overline{M}$ on $M_{\max}$ are defined as follows:

$$\underline{M} = \sum_{i \in I} \min\left\{\gamma(i): \sum_{g=1}^{\gamma(i)} \mu_{gi} \geq \sum_{j \in J_i} n_j p_{ij}/hH\right\} \tag{11.23}$$

$$\overline{M} = \max_{d \in D}\left\{\sum_{i \in I} \min\left\{m_i, \min\left\{\Gamma(id): \sum_{g=1}^{\Gamma(id)} \mu_{gi} \geq \sum_{j \in J_i : j \in J(d)} n_j p_{ij}/c_i\right\}\right\}\right\}. \tag{11.24}$$

The lower bound $\underline{M}$ is calculated assuming that all orders are evenly distributed over a monthly planning horizon with full processing time H available in each stage in each daily planning period, The upper bound $\overline{M}$ assumes that each order is completed exactly on its due date with the same limited processing time c_i available in each stage i in every period t.

The cutting constraints (11.14) to (11.16), (11.18), (11.19), (11.21), and (11.22) should be added to model *MPS* to reduce computational effort required to find an optimal solution.

Note that, if all machines have unit capacity, that is, $\mu_{gi} = 1$ for all $i \in I$, $g \in G_i$, then in (11.12), (11.13), and (11.17), $\sum_{g \in G_i} \mu_{gi} = m_i$, (cf. (6.73), (6.77), and (6.84)), and the cutting constraints (11.14) to (11.16), (11.18), (11.19), and (11.21) reduce to the corresponding constraints (6.80) to (6.82) and (6.87) to (6.89) from Section 6.5.1.

11.4 SHORT-TERM MACHINE ASSIGNMENT AND SCHEDULING

In this section, two-level machine assignment and scheduling is presented and the corresponding MIP formulations are provided for short-term machine scheduling.

The customer orders assigned to one planning period are split into production lots, each to be processed as a separate job. Each order $j \in J$ is divided into $\lceil n_j/b_j \rceil$ production lots: $\lfloor n_j/b_j \rfloor$ lots, each of size b_j and at most one lot of size $n_j - \lfloor n_j/b_j \rfloor b_j$ ($\lfloor \cdot \rfloor$ is the greatest integer not greater than $\cdot$).

Let $K = \{1, \ldots, n\}$, be the set of indices of all lots, and $\tilde{b}_k$, the size of lot $k \in K$, where $n = \sum_{j \in J:\, x_{jt}=1} \lceil n_j/b_j \rceil$ is the total number of production lots of all customer orders assigned to one planning period $t \in T$. Each production lot must be processed in various stages; however, some lots may bypass some stages. Let $e_k \in I$ and $f_k \in I$, $f_k \geq e_k$ be the number of the first and the last processing stage for lot k, respectively. Denote by $K_i \subset K$ the subset of lots that must visit stage i, and let $\tilde{p}_{ik} > 0$ be the processing time in stage i of each product in lot $k \in K_i$.

All products in a lot are processed consecutively on the same machine and transported together between the machines, and therefore each lot can be scheduled as a separate job. The actual processing time in stage i of lot $k \in K_i$ depends on capacity μ_{gi} of parallel machine $j \in J_i$ selected for processing and is $\tilde{p}_{ik}\lceil \tilde{b}_k/\mu_{gi} \rceil$. In

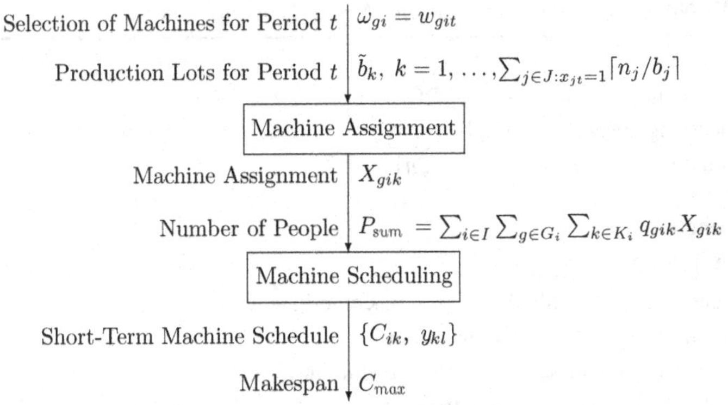

Selection of Machines for Period t | $\omega_{gi} = w_{git}$

Production Lots for Period t | $\tilde{b}_k,\ k = 1, \ldots, \sum_{j \in J: x_{jt}=1} \lceil n_j / b_j \rceil$

Machine Assignment

Machine Assignment | X_{gik}

Number of People | $P_{\text{sum}} = \sum_{i \in I} \sum_{g \in G_i} \sum_{k \in K_i} q_{gik} X_{gik}$

Machine Scheduling

Short-Term Machine Schedule | $\{C_{ik},\ y_{kl}\}$

Makespan | C_{max}

Figure 11.3 A two-level machine assignment and scheduling.

addition, processing of production lot k on machine g in stage i requires q_{gik} operators to be assigned.

For each planning period, given the set of production lots of all customer orders assigned to this period and the set of machines selected for assignment in this period, the objective of short-term machine assignment and scheduling is to determine the assignment of the production lots to machines and a detailed processing schedule to minimize the number of people as a primary optimality criterion and to minimize makespan as a secondary criterion.

The two-level machine assignment and scheduling is shown in Figure 11.3. First at the top level the machine assignment is solved to best allocate a daily subset of production lots among the selected machines and to minimize the number of people required to complete the daily production plan. Then, at the base level the shortest assembly schedule is found for prefixed assignment of the production lots. Figure 11.3 also illustrates the linkage between the medium-term aggregate scheduling and the short-term machine assignment.

11.4.1 Machine Assignment

The binary program *MA* for machine assignment is presented below (for notation used, see Table 11.2).

Model MA: *Machine Assignment*

Minimize

$$P_{\text{sum}} = \sum_{i \in I} \sum_{g \in G_i} \sum_{k \in K_i} q_{gik} X_{gik} \tag{11.25}$$

subject to

1. *Product Assignment Constraints*:
   - in every stage each lot is assigned to exactly one selected machine,
   - at least one lot is assigned to every selected machine,

Table 11.2 Notation: Short-Term Machine Assignment and Scheduling

Indices

g	= parallel machine in stage i, $g \in G_i = \{1, \ldots, m_i\}$
i	= processing stage, $i \in I = \{1, \ldots, m\}$
k	= production lot, $k \in K = \{1, \ldots, n\}$

Input parameters

$\tilde{b}_k$	= size of lot k
c_i	= processing time available on each machine in stage i
e_k, f_k	= first and last processing stage for lot k
$\tilde{p}_{ik}$	= processing time in stage i of each product in lot k
q_{gik}	= number of operators required for processing of lot k on machine g in stage i
H	= length of the daily scheduling horizon
K_i	= subset of production lots to be processed in stage i
ω_{gi}	= 1, if at the medium-term production scheduling level machine g in stage i has been selected for assignment; otherwise $\omega_{gi} = 0$

Decision variables

C_{ik}	= completion time of production lot k in stage i (timing variable)
X_{gik}	= 1, if production lot k is assigned to machine $g \in G_i$ in stage $i \in I$; otherwise $X_{gik} = 0$ (machine assignment variable)
y_{kl}	= 1, if lot k precedes lot l; otherwise $y_{kl} = 0$ (lot sequencing variable)

Objective functions

$C_{\max}$	= schedule length
P_{sum}	= total number of people required for assignment

– the total processing time of all lots assigned to each machine cannot exceed processing time available during the scheduling horizon,

$$\sum_{g \in G_i} X_{gik} = 1; \;\; i \in I, k \in K_i \tag{11.26}$$

$$X_{gik} \leq \omega_{gi}; \;\; i \in I, g \in G_i, k \in K_i \tag{11.27}$$

$$\sum_{k \in K_i} X_{gik} \geq \omega_{gi}; \;\; i \in I, g \in G_i \tag{11.28}$$

$$\sum_{k \in K_i} \tilde{p}_{ik} \lceil \tilde{b}_k / \mu_{gi} \rceil X_{gik} \leq \omega_{gi} c_i; \;\; i \in I, g \in G_i \tag{11.29}$$

2. *Variable Integrality Conditions*:

$$X_{gik} \in \{0, 1\}; \;\; i \in I, g \in G_i, k \in K_i. \tag{11.30}$$

The objective (11.25) is to minimize the number of people assignments, P_{sum}.

The duration c_i, (11.29), of processing time available on each machine in stage i can be calculated as (see also (5.1) in Section 5.2 and Section 6.4.3)

$$c_i = H - (1-\alpha)\max_{k\in K_i}\left(\sum_{i'\in I: i'<i} \tilde{p}_{i'k}\lceil \tilde{b}_k/\mu_{1i'}\rceil\right)$$
$$- (1-\alpha)\max_{k\in K_i}\left(\sum_{i'\in I: i'>i} \tilde{p}_{i'k}\lceil \tilde{b}_k/\mu_{1i'}\rceil\right)$$
$$- \alpha\min_{k\in K_i}\left(\sum_{i'\in I: i'<i} \tilde{p}_{i'k}\lceil \tilde{b}_k/\mu_{1i'}\rceil\right)$$
$$- \alpha\min_{k\in K_i}\left(\sum_{i'\in I: i'>i} \tilde{p}_{i'k}\lceil \tilde{b}_k/\mu_{1i'}\rceil\right), \qquad (11.31)$$

where $0 \le \alpha \le 1$. The coefficient α reflects the machine idle time during waiting for the first production lot from upstream stages and the machine idle time during processing of the last production lot at downstream stages.

11.4.2 Machine Scheduling

The MIP model MS/A for machine scheduling with prefixed machine assignments $X^A_{gik};\ i \in I, g \in G_i, k \in K_i$ is presented below (for notation used, see Table 11.2), where the formulation is a priori strengthened by the addition of cutting constraints.

Model MS|A: *Machine Scheduling with Prefixed Machine Assignment*

Minimize

$$C_{\max} \qquad (11.32)$$

subject to

1. *Product Noninterference Constraints*:

– no two different production lots that have been assigned to the same machine can be processed simultaneously,

$$C_{ik} + Hy_{kl} \ge C_{il} + \tilde{p}_{ik}\lceil \tilde{b}_k/\mu_{gi}\rceil;$$
$$i \in I, g \in G_i, k, l \in K_i: k < l,\ X^A_{gik}X^A_{gil} = 1 \qquad (11.33)$$

$$C_{il} + H(1-y_{kl}) \ge C_{ik} + \tilde{p}_{il}\lceil \tilde{b}_l/\mu_{gi}\rceil;$$
$$i \in I, g \in G_i, k, l \in K_i: k < l,\ X^A_{gik}X^A_{gil} = 1 \qquad (11.34)$$

2. *Product Completion Constraints*:

- each lot must be completed on the selected machines in the first stage, and in all downstream stages of its processing route,
- each production lot bypasses stages not in its processing route,
- the completion time of each lot in its first processing stage cannot be greater than the sum of the processing times of all lots assigned to the same machine,

$$C_{e_k k} \geq \tilde{p}_{e_k k} \lceil \tilde{b}_k / \mu_{g e_k} \rceil; \;\; k \in K, g \in G_{e_k}: X^A_{g e_k k} = 1 \tag{11.35}$$

$$C_{ik} - C_{i-1k} \geq \tilde{p}_{ik} \lceil \tilde{b}_k / \mu_{gi} \rceil;$$

$$i \in I, g \in G_i, k \in K_i: i > e_k, X^A_{gik} = 1 \tag{11.36}$$

$$C_{ik} = C_{i-1k}; \;\; i \in I, k \in K: i > 1, \tilde{p}_{ik} = 0 \tag{11.37}$$

$$C_{e_k k} \leq \sum_{l \in K_{e_k}: X^A_{g e_k l} = 1} \tilde{p}_{e_k l} \lceil \tilde{b}_l / \mu_{g e_k} \rceil; \;\; g \in G_{e_k}, k \in K: X^A_{g e_k k} = 1 \tag{11.38}$$

3. *Maximum Completion Time Constraints*:

- each lot must be completed sufficiently early to have all of its remaining tasks finished by the end of the schedule,
- the maximum completion time of the last task determines the minimized makespan,
- the makespan is not less than the maximum workload of a machine and must not exceed the scheduling horizon,

$$C_{ik} + \sum_{i' \in I: i' > i} \; \sum_{g \in G_{i'}: X^A_{g i' k} = 1} \tilde{p}_{i'k} \lceil \tilde{b}_k / \mu_{gi'} \rceil \leq C_{\max}; \;\;\; i \in I, k \in K_i: i < f_k \tag{11.39}$$

$$C_{f_k k} \leq C_{\max}; \;\; k \in K \tag{11.40}$$

$$\sum_{k \in K_i: X_{gik} = 1} \tilde{p}_{ik} \lceil \tilde{b}_k / \mu_{gi} \rceil + \min_{k \in K_i} \left(\sum_{i' \in I: i' < i} \tilde{p}_{i'k} \lceil \tilde{b}_k / \mu_{1i'} \rceil \right)$$

$$+ \min_{k \in K_i} \left(\sum_{i' \in I: i' > i} \tilde{p}_{i'k} \lceil \tilde{b}_k / \mu_{1i'} \rceil \right) \leq C_{\max} \leq H; \;\; i \in I, g \in G_i \tag{11.41}$$

4. *Variable Nonnegativity and Integrality Conditions*:

$$C_{ik} \geq 0; \;\; i \in I, k \in K_i \tag{11.42}$$

$$C_{\max} \geq 0; \tag{11.43}$$

$$y_{kl} \in \{0, 1\}; \;\; k, l \in K: k < l. \tag{11.44}$$

The cutting constraints (11.38), (11.39), and (11.41) may help to reduce the computational effort required to find a proven optimal solution to the MIP model MS/A.

11.5 COMPUTATIONAL EXAMPLES

In this section numerical examples and computational results are presented to illustrate possible applications of the proposed approach. The examples are modeled after a real-world distribution center for high-tech products (Fig. 11.4), where finished products are assembled for shipping to customers.

The distribution center can be modeled as a flexible flow shop consisting of six processing stages with parallel multicapacity machines and a finite capacity output buffer. In the distribution center, 10 product types of three product groups are assembled. The processing stages are the following: material preparation stage, where all materials required for assembly of each product are prepared, postponement stage, where products for some orders are customized, three flashing/flexing stages, one for each group of products, where required software is downloaded, and a packing stage, where products and required accessories are packed for shipping. Each flashing/flexing stage consists of parallel multicapacity machines. The direct shipping lines shown in Figure 11.4 are SMT lines (see Chapter 2) directly linked with flashing/flexing and packing stages to fulfill the customer orders beginning with the assembly

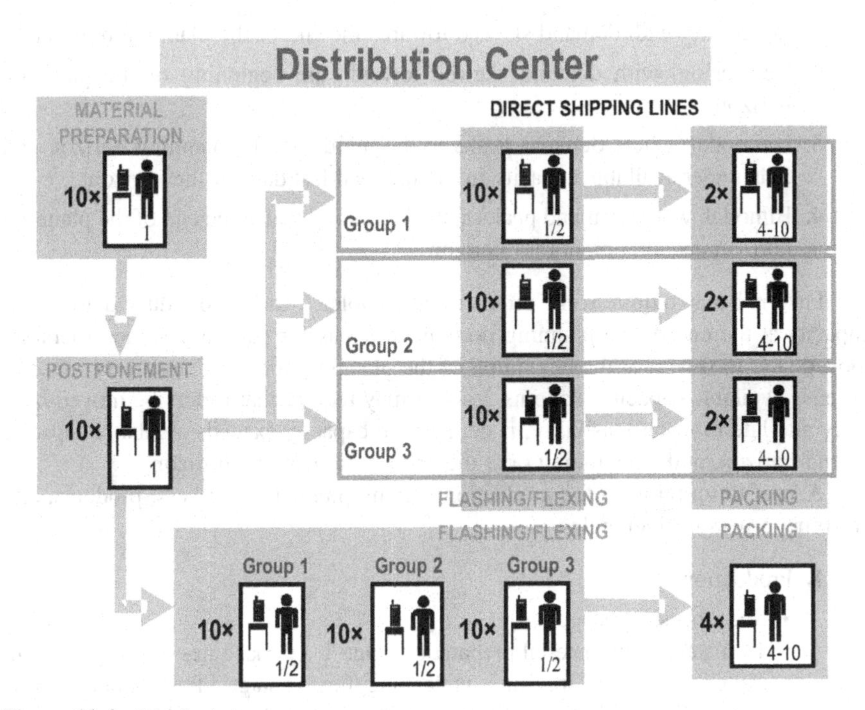

Figure 11.4 Distribution center.

of customer-required printed wiring boards. However, the SMT lines are not considered in the computational examples presented in this section, while the flashing/flexing stages and the packing stages connected with direct shipping lines are considered along with the corresponding stages of the distribution center. After all processing stages a finite output buffer is introduced to hold completed products before delivery to the customer. The finite output buffer limits the possibility of order reallocation to earlier periods with excess of capacity, which can affect tardiness. On the other hand, as the output buffer size decreases, completion dates of customer orders are approaching due dates, which results in greater fluctuations of machine and people assignments required to complete all customer orders.

Customer orders require processing in at most four stages: material preparation stage, postponement stage, one flashing/flexing stage, and packing stage (see Fig. 11.4). However, some orders do not need postponement. Each customer order must be completed in one day.

Customer orders are split into production lots of fixed sizes, each to be processed as a separate job. The processing time of each production lot in the flashing/flexing stage depends on the capacity of the machine selected for assignment. A daily production consists of various production lots from different customer orders and each lot requires a different processing route.

In the computational experiments four test problems are constructed with the following four regular patterns of demand (see Fig. 11.5), which best model the actual demand patterns that are encountered in the distribution center:

1. Increasing, with demand skewed towards the end of the planning horizon.
2. Decreasing, with demand skewed towards the beginning of the planning horizon.
3. Unimodal, where demand peaks in the middle of the planning horizon and falls under available capacity in the first and last days of the horizon.
4. Bimodal, where demand peaks at the beginning and at the end of the planning horizon and slumps in mid-horizon.

Pattern 1 may require some orders to be completed earlier to reduce demand on capacity at the end of the planning horizon, whereas for pattern 2 some orders are moved later in time if at the beginning of the planning horizon demand on capacity exceeds available capacity. Patterns 3 and 4 may require that orders be moved both early and late to reach feasibility, if demand on capacity exceeds available capacity in mid-horizon, at the beginning or at the end of the planning horizon.

A brief description of the production system, production process, products, and customer orders is given below.

1. Production system
    - Six processing stages
    - 10 machines in material preparation stage 1, 10 machines in postponement stage 2, 20 parallel machines in flashing/flexing stage 3 for product group 1, 20 parallel machines in flashing/flexing stage 4 for product group 2, 20

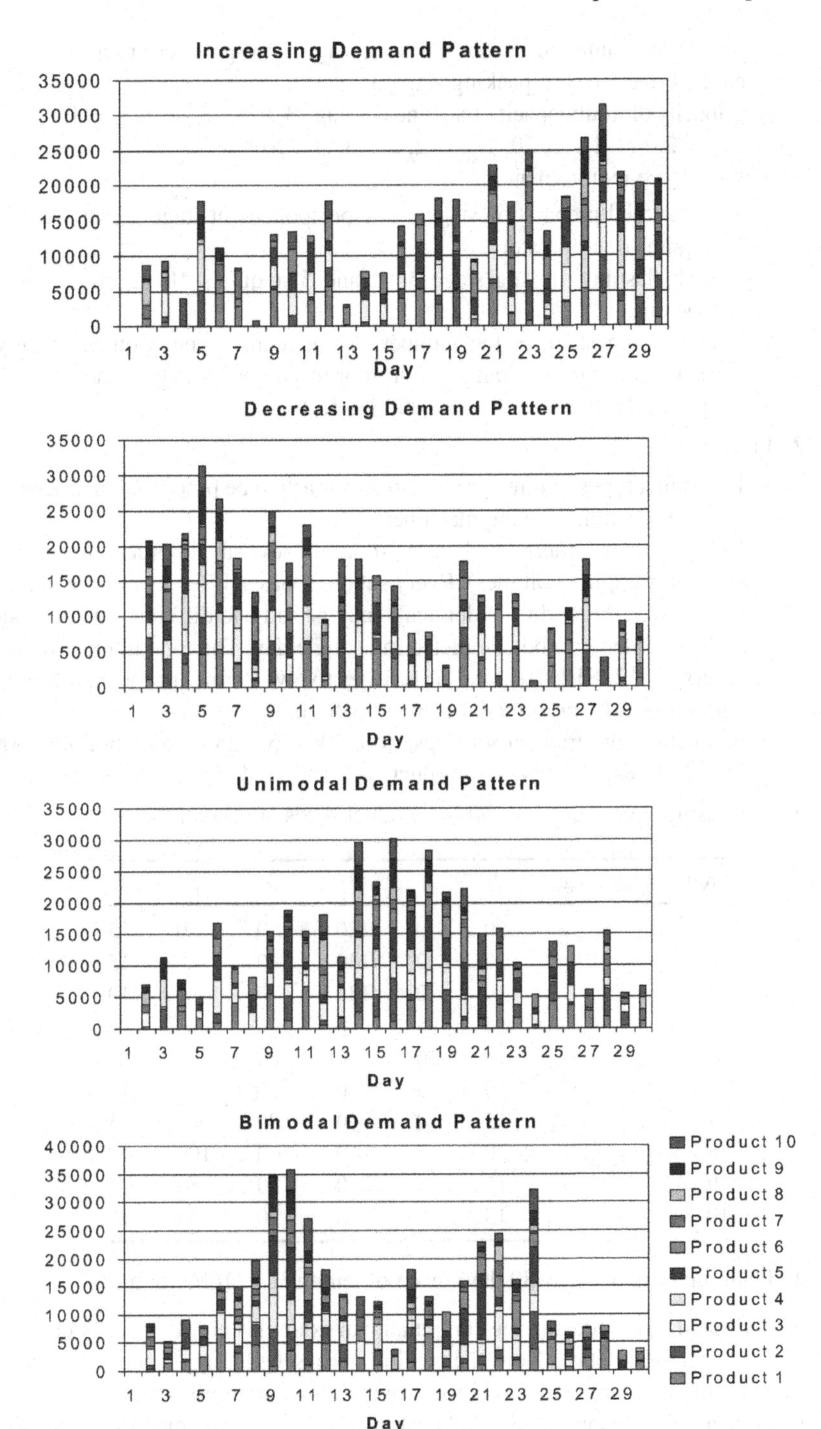

Figure 11.5 Demand patterns.

parallel machines in flashing/flexing stage 5 for product group 3, and 10 parallel machines in packing stage 6

- Capacity of multicapacity machines in stage 3, 4, 5: $\mu_{g3} = 8$, $g = 1, \ldots, 20$, $\mu_{g4} = 2$, $g = 1, \ldots, 20$, $\mu_{g5} = 2$, $g = 1, \ldots, 20$
- Workforce requirements:
  - in material preparation stage 1 and postponement stage 2 each machine requires one operator
  - each flashing/flexing stage 3, 4, and 5, requires 10 operators for all machines
  - in packing stage 6, the number of operators depends on the type of packed product and ranges from four to 10 operators per machine
- Output buffer capacity $B = 45{,}000$ products

2. Products

- 10 product types of three product groups, each to be processed on a separate group of flashing/flexing machines
- 100 customer orders, each consisting of several suborders (customer-required shipping volumes). Every suborder has a different volume ranging from 5 to 5200 products, identical ready period and different due date, and each suborder is to be completed in a single day. The total number of suborders is 812, 812, 704, and 671, respectively, for the increasing, decreasing, unimodal, and bimodal demand pattern,
- Production and transfer lot sizes: 500, 500, 750, 250, 250, 250, 500, 500, 750, 250, respectively, for product type 1, 2, 3, 4, 5, 6, 7, 8, 9, 10

3. Processing times (in seconds) for product types are shown below

Product type/stage	1	2	3	4	5	6
1	20	0	120	0	0	15
2	20	0	140	0	0	15
3	10	0	120	0	0	10
4	15	5	0	220	0	15
5	15	10	0	280	0	15
6	10	5	0	220	0	10
7	15	10	0	280	0	15
8	20	5	0	0	100	15
9	15	0	0	0	80	10
10	15	0	0	0	100	10

4. Planning horizon: $h = 30$ days, each of length $H = 1080$ minutes

Notice that the suborders in the computational examples play the role of orders in the mathematical formulation. Now, the medium-term scheduling objective is to determine an assignment of customer suborders over the planning horizon.

In the example presented below all customer orders are assumed to be ready for completion at the beginning of the planning horizon, that is, $a_j = 1, j \in J$, and the order due dates are distributed over the entire horizon according to the demand patterns shown in Figure 11.5. In the objective function (11.1) of model *MPS* the weights used for unscheduled orders, tardy orders, early order, and maximum machine assignments were $\lambda_1 = 0.898$, $\lambda_2 = 0.1$, $\lambda_3 = 0.001$, and $\lambda_4 = 0.001$, respectively.

Figure 11.6 presents the medium-term aggregate production schedules for the example with various demand patterns, where, for each planning period, detailed values of the decision variables have been replaced with aggregate values of the total production. The corresponding machine assignments are shown in Figure 11.7. The results indicate that, for all demand patterns with the exception of increasing demand, the optimal production schedules are directly driven by the demand for products at the end of the planning horizon where the demand on capacity does not exceed available capacity. This, however, results in greater fluctuations of machine assignments at the end of the planning horizon for all demand patterns with the exception of increasing demand. In contrast, the machine assignments are leveled as long as the demand on capacity exceeds available capacity, that is, at the beginning and end of the planning horizon, respectively, for the decreasing and increasing demand pattern, and in mid-horizon for the unimodal demand pattern.

As an example of short-term machine assignment and scheduling, a detailed daily machine schedule was found for the increasing demand pattern and day 15 of the medium-term schedule. The customer orders (suborders) for various product types assigned to day 15 were split into 34 production lots of different sizes to be scheduled during a daily horizon:

Order No.	20	20	21	22	23	24	25	26	27	48	48	49	50	50	50	50	51	51	51	51	51	52	52	52	53	54	54	55	55	55	55	81	81	100
Product type No.	2	2	3	3	3	3	3	3	3	5	5	5	5	5	5	5	6	6	6	6	6	6	6	6	6	6	6	6	6	6	6	9	9	10
Lot No.	1	2	3	4	5	6	7	8	9	10	11	12	13	14	15	16	17	18	19	20	21	22	23	24	25	26	27	28	29	30	31	32	33	34
Lot size	500	335	210	425	740	15	255	130	700	250	140	50	250	250	250	20	250	250	250	250	55	250	250	195	90	250	100	250	250	250	30	750	520	45

The Gantt chart in Figure 11.8 shows the detailed machine schedule for day 15. In Figure 11.8, MP, PT, F1, F2, F3, and PK indicate Material Processing, Postponement, three groups of Flashing/Flexing, and Packing, respectively. In the Gantt chart, different numbers indicate production lots of 10 different product types. The daily schedule length is $C_{\max} = 1049$ minutes.

The characteristics of the integer and mixed integer programs for the example with increasing demand pattern and the solution results are summarized in

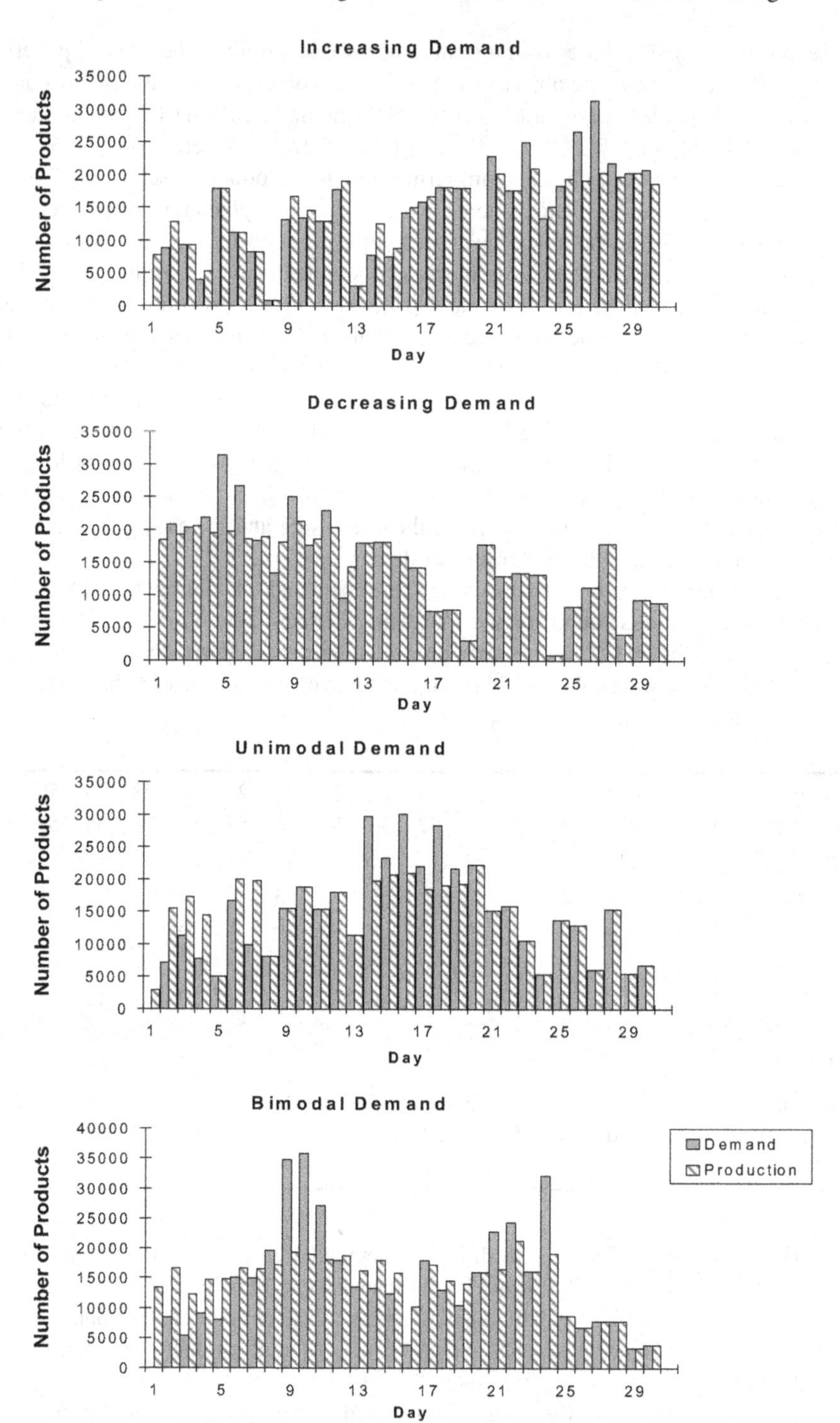

Figure 11.6 Medium-term production schedules for various demand patterns.

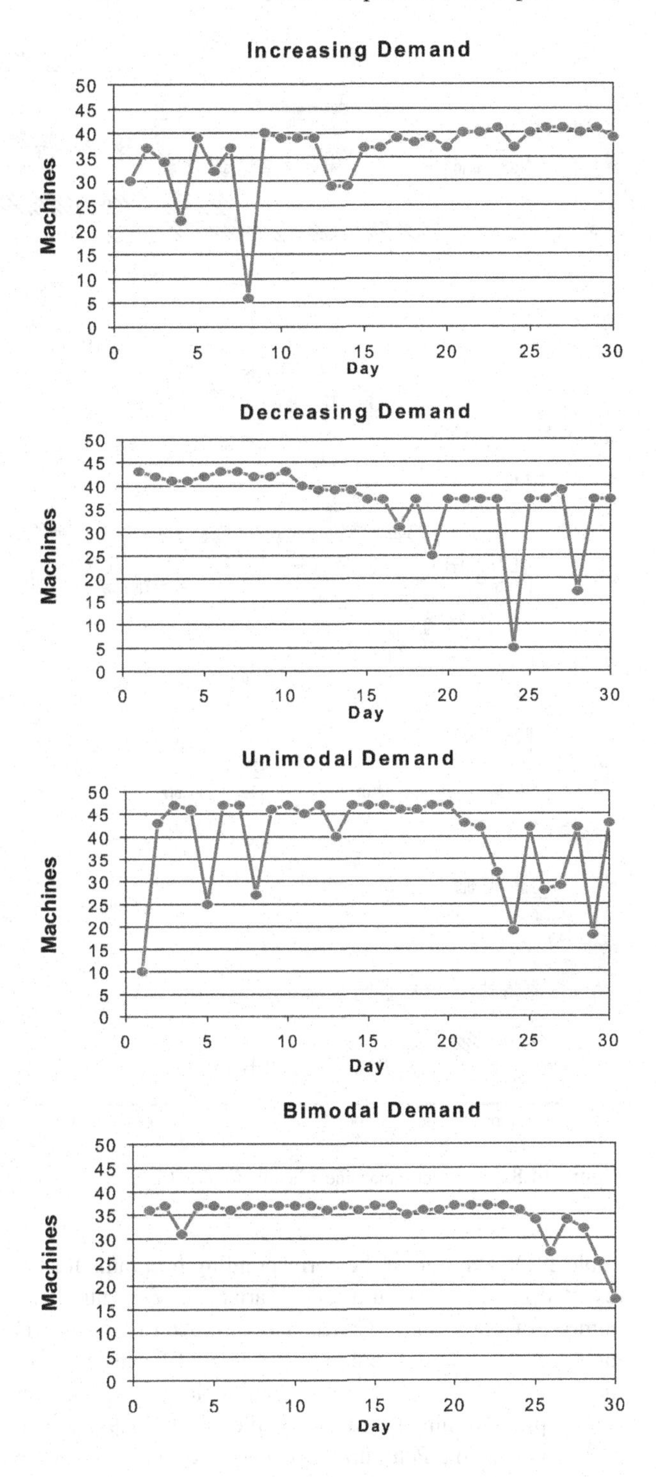

Figure 11.7 Machine assignments for various demand patterns.

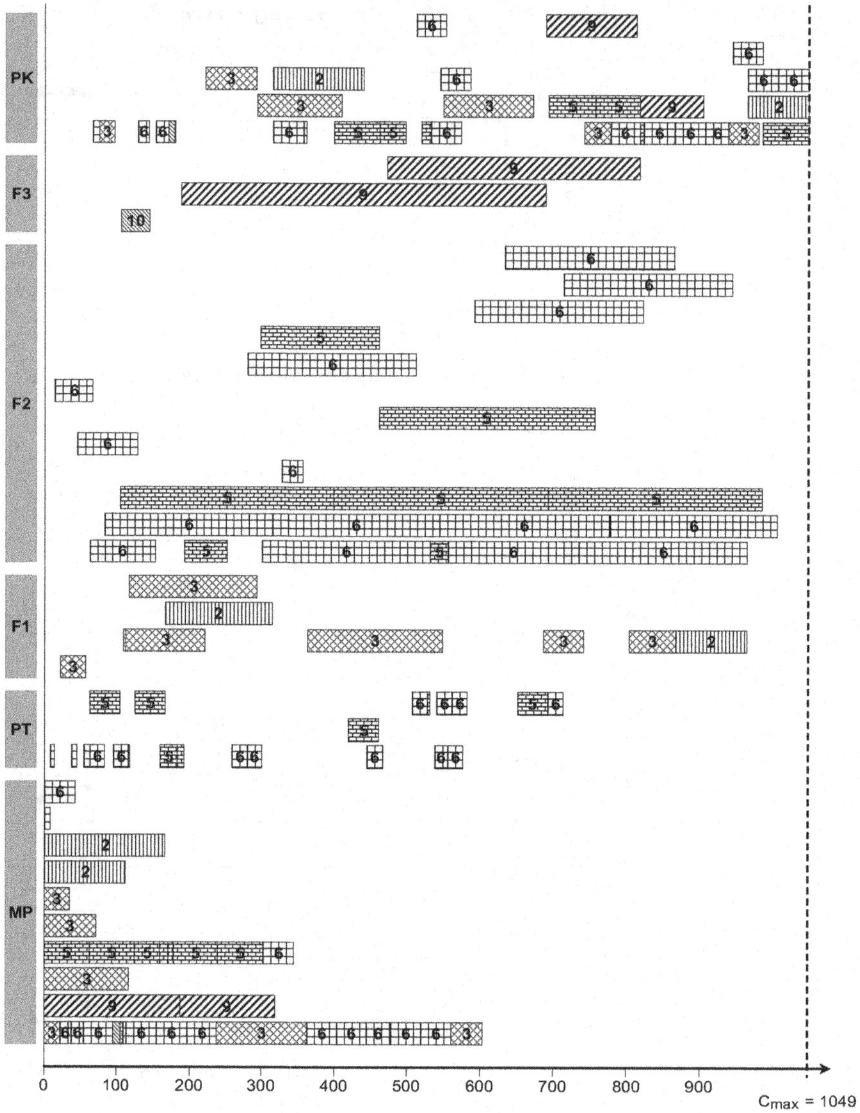

Figure 11.8 Short-term machine schedule for day 15.

Table 11.3. The size of the corresponding programs for the example problem is represented by the total number of variables, Var., number of binary variables, Bin., number of constraints, Cons., and number of nonzero elements in the constraint matrix, Nonz. The last column of Table 11.3 presents the proven optimal values of the objective functions. The computational experiments were performed using AMPL programming language and the CPLEX v.9.1 solver (with the default settings) on a laptop with a Pentium IV processor running at 1.8 GHz and with 1 GB RAM.

Table 11.3 Computational Results for Increasing Demand Pattern

Problem	Var.	Bin.	Cons.	Nonz.	Solution values
Medium-term scheduling	18,557	18,556	4554	179,840	0,16,41[a]
Machine assignment	885	885	191	2553	$P_{sum} = 53$
Machine scheduling	526	323	2300	5652	$C_{max} = 62940$s

[a] -,-,- number of tardy suborders, early suborders, maximum machine assignments.

11.5.1 Material Availability

In the previous examples the materials required for completing all customer orders were assumed to be available at the beginning of the monthly planning horizon, that is, the ready periods for all orders were identical and equal to the first day of the month. In this subsection the impact of material availability on aggregate scheduling is analyzed by reducing the length of the interval between the order due date d_j and ready period a_j (see the maximum earliness of customer order E_{max}, introduced in Section 6.4.1). In the computational experiments the ready period is $a_j = \max(1, d_j - 3)$ for each order j, that is, the length of interval between the order due date and ready period is not greater than 3 days (the maximum earliness of orders, $E_{max} = 3$). When order ready periods and due dates are closer, the limited order earliness due to the limited material availability restricts reallocation of orders to the earlier periods with surplus of capacity, which may result in a greater number of tardy orders. On the other hand, both the input inventory of raw materials waiting for processing in the system and the output inventory of the finished products completed before the due dates and waiting for delivery to the customers can be reduced when order ready periods and due dates are closer (see Section 6.4.1).

The characteristics of IP model *MPS* for the test problems and the solution results are summarized in Tables 11.4 and 11.5, respectively, for the basic and strengthened

Table 11.4 Computational Results: Basic Integer Program *MPS*

Demand	Ready period	Var.	Bin.	Cons.	Nonz.	Solution[a]	GAP[b]
Increasing	$a_j = 1$	27,873	27,872	3722	208,022	0,16,41	–
Increasing	$a_j = \max(1, d_j - 3)$	15,210	15,209	3722	106,691	0,31,42	6.80
Decreasing	$a_j = 1$	27,873	27,872	3722	208,022	1,16,42	1.81
Decreasing	$a_j = \max(1, d_j - 3)$	20,861	20,860	3722	152,654	1,22,41	1.95
Unimodal	$a_j = 1$	24,525	24,524	3614	181,754	0,17,38	–
Unimodal	$a_j = \max(1, d_j - 3)$	16,025	16,024	3614	114,246	0,43,45	6.29
Bimodal	$a_j = 1$	23,502	23,501	3581	173,651	0,23,37	–
Bimodal	$a_j = \max(1, d_j - 3)$	15,712	15,711	3581	111,709	0,40,42	4.31

[a] -,-,- number of tardy suborders, early suborders, maximum machine assignments, respectively.

[b] % GAP after 3600 seconds of CPU time.

Table 11.5 Computational Results: Strengthened Integer Program *MPS*

Demand	Ready period	Var.	Bin.	Cons.	Nonz.	Solution values[a]
Increasing	$a_j = 1$	18,557	18,556	4554	179,840	0,16,41
Increasing	$a_j = \max(1, d_j - 3)$	5894	5893	4618	52,845	0,30,42
Decreasing	$a_j = 1$	20,675	20,674	4230	181,627	1,15,42
Decreasing	$a_j = \max(1, d_j - 3)$	20,861	20,860	3880	267,447	1,21,42
Unimodal	$a_j = 1$	13,976	13,975	4338	127,318	0,17,38
Unimodal	$a_j = \max(1, d_j - 3)$	5476	5475	4395	46,824	0,39,47
Bimodal	$a_j = 1$	13,138	13,137	4377	160,004	0,23,37
Bimodal	$a_j = \max(1, d_j - 3)$	5348	5347	4333	45,684	0,40,42

[a] -,-,- number of tardy suborders, early suborders, maximum machine assignments, respectively.

(with cuts derived in Section 11.3.1) integer program *MPS*. The size of the IP model is represented by the total number of variables, Var., number of binary variables, Bin., number of constraints, Cons., and number of nonzero coefficients, Nonz., in the constraint matrix. The last two columns of Table 11.4 present the proven optimal solution values and % GAP after 3600s of CPU time, if optimality of the solution is not proven. Table 11.4 shows that for the basic integer program the CPLEX solver was not always able to prove optimality within the allowed 3600 s of CPU time; however, the best solutions were found much earlier than the time limit. On the other hand, Table 11.5 indicates that when the integer program *MPS* was strengthened by the addition of cutting constraints presented in Section 11.4, the CPLEX solver was capable of finding proven optimal solutions for all test examples.

For the example data the optimal medium-term production schedule (Fig. 11.9) is driven directly by the demand for products as long as the demand on capacity does not exceed available capacity. The results indicate that, for increasing and decreasing demand patterns, the optimal production is very close to demand at the beginning and end, respectively, of the planning horizon, whereas for unimodal demand, the production level is very close to demand at the beginning and end of the horizon.

The results also indicate that both the input inventory of raw materials (Fig. 11.10) and the output inventory of the finished products (Fig. 11.11) are smaller when materials are supplied only a few days before customer-required shipping days. The length of interval between material supply and due dates, however, must be sufficiently large to avoid additional tardy orders due to limited order earliness. In the example, the minimum length of the interval (the maximum earliness $E_{\max}$) was found to be 3 days, for which the numbers of tardy orders were identical for both $a_j = 1$ and $a_j = \max(1, d_j - 3)$, $j \in J$. The number of tardy orders increases as the interval length decreases and is less than 3 days.

In the example only the input inventory of product-specific materials was considered, with no common materials for different product types. It should be noted that Figure 11.10 compares the actual input inventory when all materials are available in period 1 with the inventory of materials supplied three days before each customer

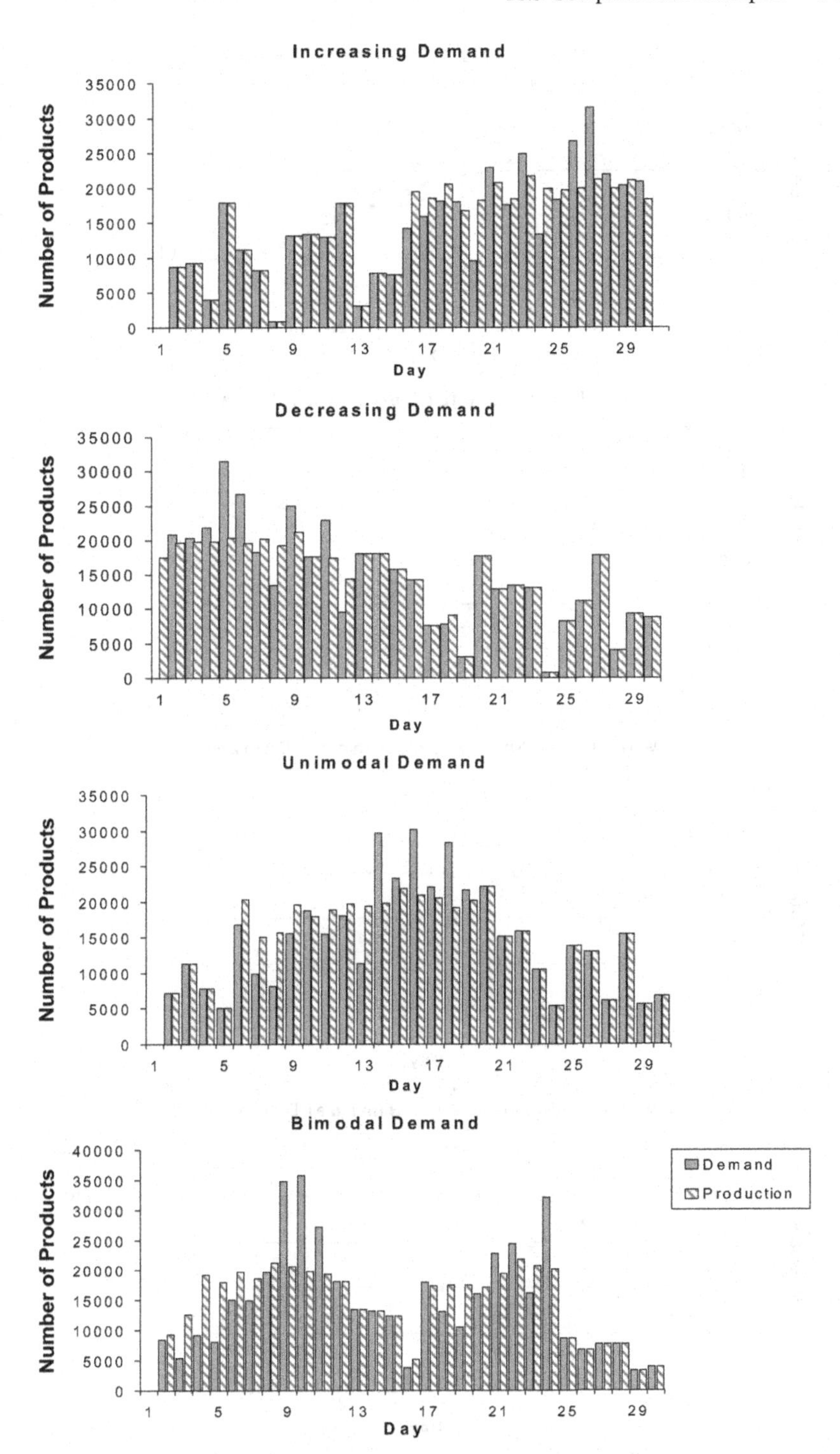

Figure 11.9 Medium-term production schedules for limited material availability.

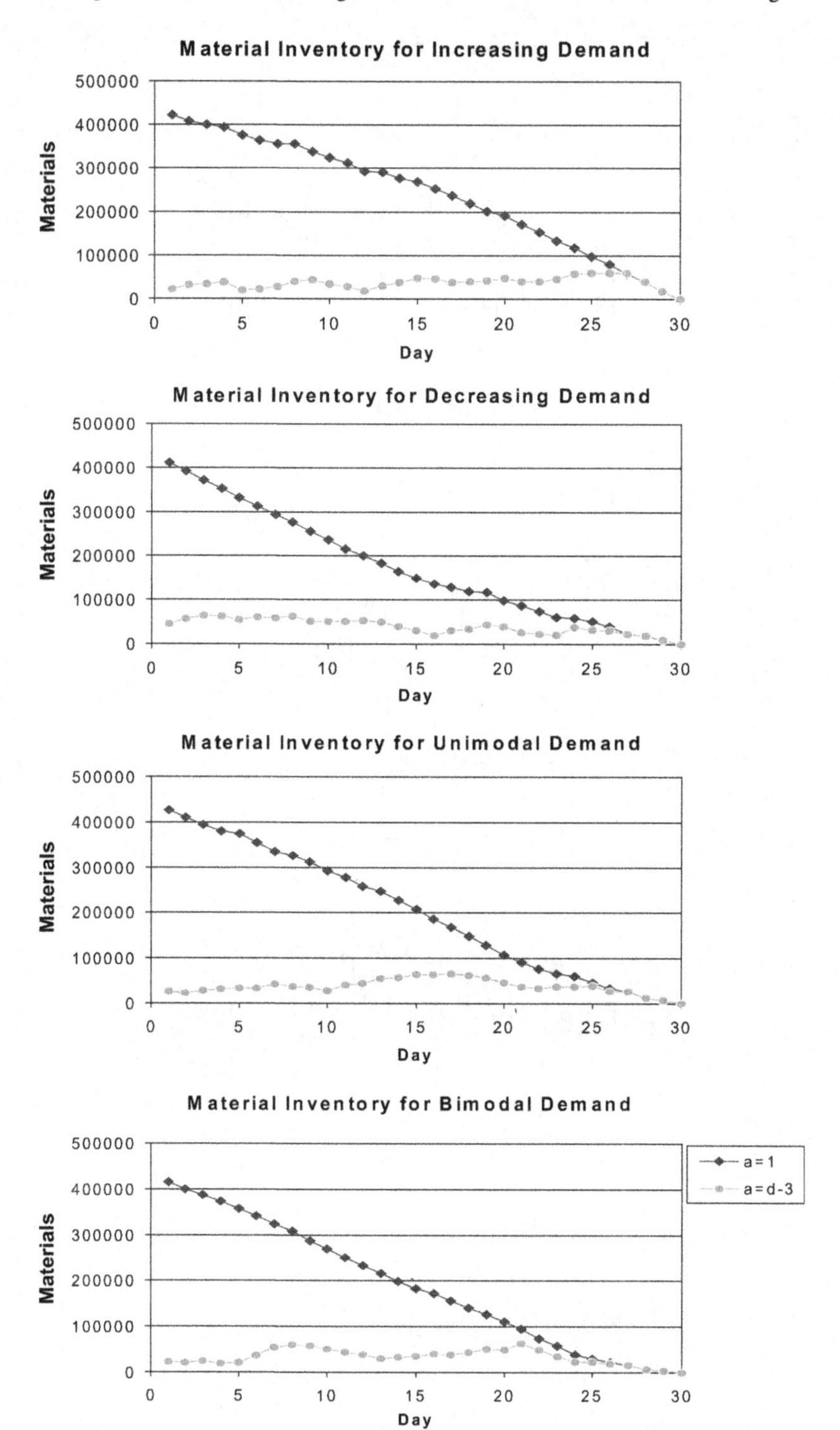

Figure 11.10 Product-specific material inventory vs. material availability: $a_j = 1$ or $a_j = \max(1, d_j - 3)$.

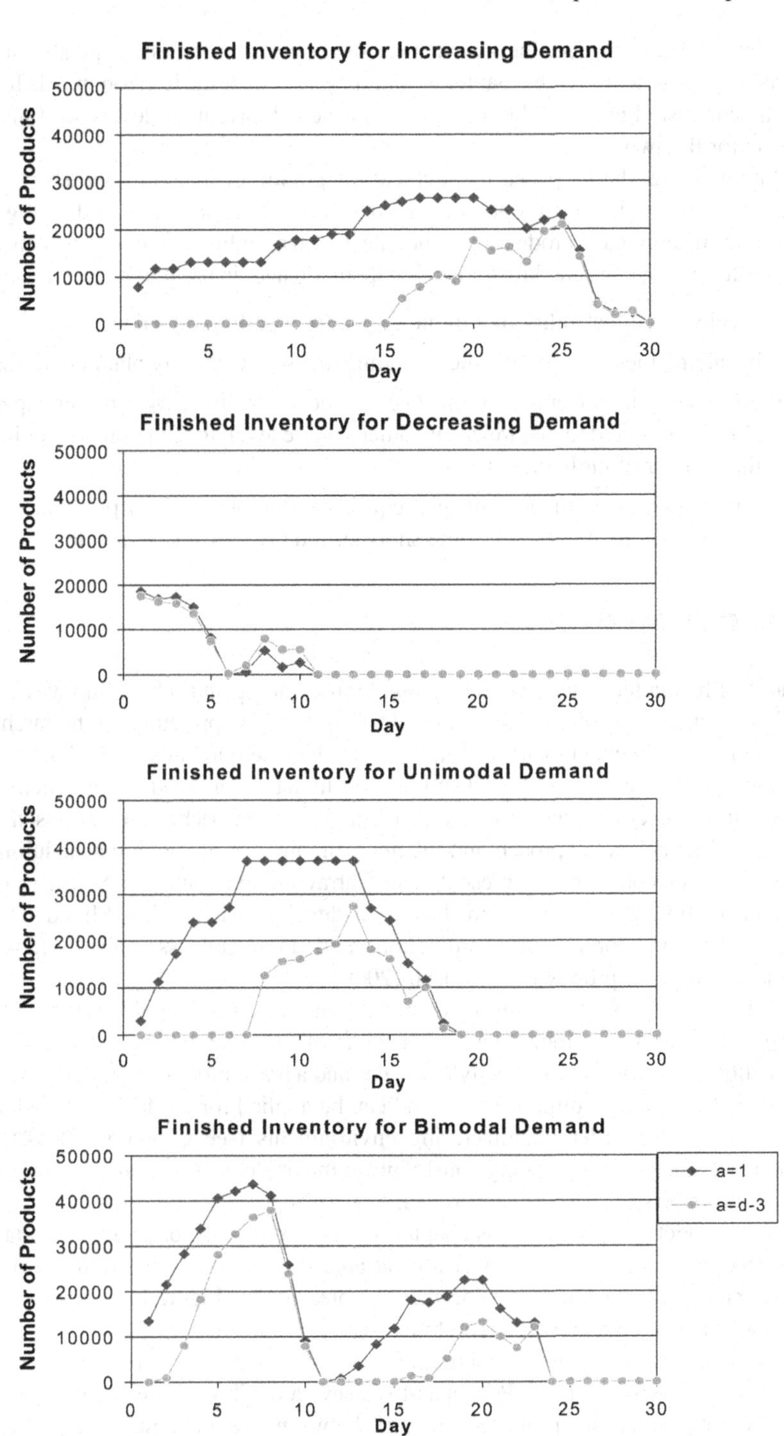

Figure 11.11 Finished inventory vs. material availability: $a_j = 1$ or $a_j = \max(1, d_j - 3)$.

order due date. In practice, the actual level of the input inventory depends on the material supply schedule (see Chapter 8) and may vary over the horizon in a different way. In contrast, Figure 11.11, compares the actual inventory levels of finished products for the two cases.

Summarizing, the proposed hierarchical integration of medium- and short-term scheduling and the integer programming approach are capable of coordinating the medium-term aggregate production schedule for a monthly planning horizon and the short-term machine schedule for a daily operation and, in particular, are capable of

- checking the feasibility of a monthly and/or a daily production,
- balancing the machine (people) assignments over a monthly planning horizon,
- minimizing the inventory of finished products (e.g., the output buffer capacity (11.20)) required to maximize customer service level, for example, to minimize the number of tardy orders,
- minimizing the number of people required to complete a daily production and minimizing the makespan of the short-term daily machine schedule.

11.6 COMMENTS

The idea of hierarchical integration of production planning and scheduling was introduced by Anthony (1965) and Hax and Meal (1975). The principles of hierarchical planning for supply chain management have also been applied in so-called advanced planning systems (e.g., Stadtler, 2005). In the literature on production scheduling in make-to-order environments different hierarchical approaches are proposed; for example, a hierarchical approach and integer programs for production scheduling in a make-to-order company are presented in Carravilla and Pinho de Sousa (1995). However, computational results are based on developed heuristics. Mixed integer programs for the problem of coordination fabrication and assembly in make-to order production are proposed in Kolisch (2000).

The material presented in this chapter is based on the results published by Sawik (2006). The proposed formulations have been applied to production scheduling in a flexible flow shop with multicapacity machines and a batch processing mode. The formulations, however, are quite versatile and can be applied for production scheduling in various make-to-order manufacturing environments (see Chapters 1 to 4), for example, by adjusting the capacity constraints or the objective functions. In particular, multiperiod orders can also be considered.

In the models proposed in this chapter, the size of production and transportation lots is a fixed parameter and not a decision variable to be optimized in a framework of a lot-sizing model. In the hierarchical framework presented in Figure 11.1, however, the batching level can be removed or combined with the short-term scheduling level in a joint lot-sizing and scheduling level (e.g., Dauzere-Peres and Lasserre, 1997; Kimms, 1997; Drexl and Kimms, 1997; Pochet and Wolsey, 2006; Tempelmeier and Derstroff, 1996). Furthermore, the proposed linkage between the medium-term aggregate production scheduling and the short-term machine scheduling can be modified and different relationships between variable values of the *MPS* model and parameters of the remaining models can be applied.

The proposed hierarchical framework can also be adapted to another ranking among the optimality criteria (see Chapter 6), for example, if a lexicographic approach is applied (e.g., Sawik, 2007a).

EXERCISES

11.1 Consider the weighted-sum integer program *MPS* for the quad-objective aggregate production scheduling. Transform the monolithic model *MPS* into a sequence of four single-objective integer programs, using the lexicographic optimization approach.

11.2 In the hierarchical approach (Exercise 11.1) consider the top-level problem of the minimizing the number of unscheduled (rejected) customer orders. Introduce new variables: $\overline{m_i},\ i \in I$, and $\overline{B}$, respectively, the number of unit-capacity machines to be added in stage i, the additional capacity of the output buffer. Reformulate the top-level problem to minimize the total number of additional machines, $\sum_{i \in I} \overline{m_i}$, and/or the additional buffer capacity, $\overline{B}$, so that all customer orders are completed during the planning horizon.

11.3 Consider model *MPS* and assume that all m_i machines in each processing stage should be assigned. However, the amount of processing time available on every machine in each stage, $c_{it},\ i \in I,\ t \in T$, is a variable that depends on the unknown length H of each planning period (e.g., depends on the number of shifts per day). In the hierarchical approach (Exercise 11.1) assume that the objective of the top-level problem is to complete all customer orders during the planning horizon in such a way that the maximum amount of processing time available on each machine over all stages and all periods, $\max_{i \in I, t \in T}(c_{it})$, is minimized. Formulate the mixed integer program for the top level problem.

(**a**) How can the required length H of each planning period be determined, using the optimal values of c_{it} obtained by solving the top-level problem?

11.4 Assuming that total processing time available on each machine in each stage is constant over the entire planning horizon, that is, $c_{it} = c_i \forall\, t \in T$, determine the minimum length of the planning horizon, h, required to complete all customer orders.

11.5 Assume that in (11.1) the fourth objective of the weighted-sum mixed integer program *MPS* is to level people assignments (workforce size) over the planning horizon, instead of machine assignments. Let $P_{\max}$ be the maximum number of people assignments in a planning period.

(**a**) How should the monolithic model *MPS* for the aggregate production scheduling be modified to account for the new objective function, in which $M_{\max}$ is replaced by $P_{\max}$?

(**b**) In the hierarchical approach (Exercise 11.1) modify the MIP formulation of the base-level problem such that instead of machine assignments, people assignments are leveled over the planning horizon.

(**c**) Let $W_{\max}$ be the maximum workload of a machine (i.e., the total processing time of all lots assigned to the machine). In the short-term machine assignment and scheduling, modify the binary program *MA* for machine assignment to balance the machine workloads in each planning period.

Chapter 12

Coordinated Scheduling in Supply Chains with a Single Supplier

12.1 INTRODUCTION

In customer-driven supply chains the procurement and production policies should be coordinated in an efficient manner to achieve a high customer service level at minimum cost. This chapter deals with coordinated medium-term scheduling of manufacturing and supply of parts and assembly of finished products in a customer-driven supply chain with a single supplier and single producer. The supply chain consists of the supplier stage made up of identical production lines in parallel, the producer stage which is a flexible assembly line (FAL) made up of several assembly stages in series, and a set of customers. The supplier manufactures and delivers product-specific materials (parts) to the producer where finished products are assembled according to customer orders, and then delivered to the customers to meet their demands. The overall problem is how to coordinate manufacturing and supply of parts and assembly of products with respect to limited capacities and required customer service level such that the total supply chain inventory holding cost (or maximum total inventory level) and the production line start-up costs at the supplier stage and parts shipment costs are minimized. Two approaches are proposed and compared: an integrated (monolithic) approach where the coordinated manufacturing, supply, and assembly schedules are determined simultaneously, and a hierarchical approach, where first an assignment of customer orders over the planning horizon and thereby the finished products assembly schedule is determined, and then the parts manufacturing and supply schedules are found. Furthermore, the two alternative measures of the total inventory, inventory holding cost and maximum inventory level, are compared.

The following time-indexed IP and MIP models are presented in this chapter:

MME for minimization of the maximum earliness of customer orders

SMSA for integrated scheduling of manufacturing, supply, and assembly

Scheduling in Supply Chains Using Mixed Integer Programming. By Tadeusz Sawik

SSA(I_{max}) for scheduling supply of parts and assembly of products to minimize maximum level of total inventory

SSA(Φ_{sum}) for scheduling supply of parts and assembly of products to minimize total inventory holding cost

SMS for scheduling of manufacturing and supply of parts

Numerical examples modeled after a real-world coordinated scheduling in a customer-driven supply chain in the electronics industry and computational results are reported in Section 12.6. The computational examples indicate that both integrated and hierarchical approaches are capable of finding good coordinated supply chain schedules for large problems in a reasonable computation time, using commercially available software for mixed integer programming.

12.2 PROBLEM DESCRIPTION

The supply chain consists of three distinct stages (Fig. 12.1): manufacturer/supplier of product-specific materials (parts), producer where finished products are assembled according to customer orders, and a set of customers who generate final demand for the products.

In the manufacturing stage product-specific parts are manufactured for all product types that are assembled at the producer stage. The manufacturing stage is made up of M identical production lines in parallel and an unlimited output buffer for storing the parts waiting for shipment to the producer. Let K be the set of product-specific part types and q_k processing time of part type $k \in K$. The planning horizon consists of h planning periods (e.g., working days) of equal length H (e.g., hours or minutes) and let $T = \{1, \dots h\}$ be the set of planning periods. In each period at most one part type can be manufactured on each production line. When a production line switches from one part type to another a start-up time should be considered at the beginning of the period. The start-up times are sequence-independent and are assumed to be equal for all part types. Let σ be the start-up time of each production line.

The manufactured parts are next transported to the producer at most once per period. Different part types can be shipped together so that a shipping cost arises only once per shipment. The size of each shipment is limited by the minimum and

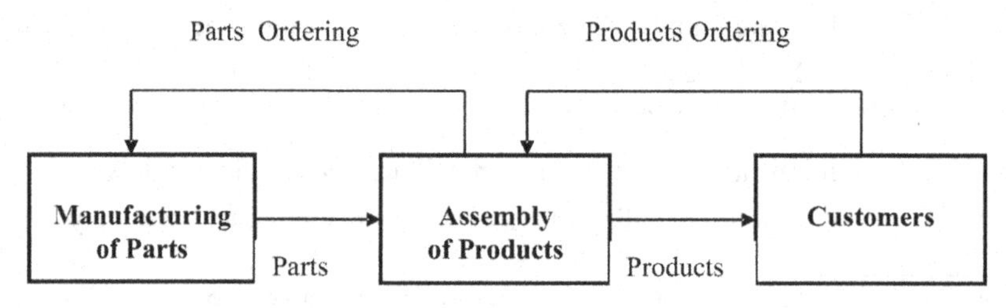

Figure 12.1 A customer-driven supply chain with a single supplier.

the maximum allowed shipping lot, $\underline{S}$ and $\overline{S}$, respectively. The parts manufactured in period t can be shipped to the producer in the same period, and the transportation time is assumed to be constant and equal to one period for every shipping lot. As a result the parts manufactured in period t can be used for product assembly in period $t+1$, at the earliest.

The producer stage is a flexible assembly line consisting of several assembly stages, an unlimited input buffer for storing the supplied parts waiting for assembly, and an unlimited output buffer for storing the finished products waiting for delivery to the customers. In this stage various types of products are assembled in a make-to-order environment responding directly to customer orders.

Let J be the set of single-product customer orders known ahead of time, K the set of product types (identical with the set of product-specific part types), and J_k the subset of orders for product type $k \in K$ (i.e., requiring part type k to be assembled). Each order $j \in J$ is described by a triple (a_j, d_j, n_j), where a_j is the order ready date (e.g., the earliest period of material availability), d_j is the customer-requested due date (e.g., customer-required shipping date), and n_j is the size of the order (quantity of ordered products of a specific type).

Each order must be completed in a single planning period (e.g., during one day) and requires processing in various assembly stages; however, some orders may bypass some stages. Let $p_{gj} \geq 0$ be the processing time in stage $g \in G$ of each product in order $j \in J$ and c_{gt} the total processing time available in stage $g \in G$ in period $t \in T$. The assumption of single-period orders, which simplifies the mathematical formulations, can be easily relaxed to allow for the inclusion of multiperiod orders. Large, multiperiod orders that require more than one planning period to complete can be split into single-period suborders to be allocated among consecutive planning periods (see Sections 5.5, 6.2, and 7.2).

It is assumed that each product requires one unit of the corresponding product-specific part (e.g., one printed wiring board of a specific design per electronic device of the corresponding type). Furthermore, the manufacturing and supplies of common materials for different product types are not explicitly taken into account. As a result, for each order j the required quantity of product-specific parts type k (such that $j \in J_k$) equals the quantity n_j of the ordered products k. The above assumptions make the inventory calculations clearer and can be relaxed by introducing for each type of product the associated unit requirements for each type of part, see Chapter 8. Accordingly, the models proposed in this chapter can be further enhanced to simultaneously consider both product-specific and common materials.

The overall problem is how to coordinate manufacturing and supply of parts and assembly of products with respect to limited capacities and required customer service level such that the total supply chain inventory holding cost (or maximum total inventory level) and the production line start-up and parts shipping costs are minimized.

Two approaches are proposed: a monolithic approach, where integrated manufacturing, supply, and assembly schedules are determined simultaneously, and a hierarchical approach, where first the product assembly schedule is found, and then the corresponding schedules for manufacturing and supply of parts are determined. Both approaches begin with determining the latest delivery dates of required parts to ensure meeting all customer due dates at a minimum total inventory of parts and finished products at the producer.

12.2.1 Conditions for Feasibility of Customer Due Dates

The following necessary condition must hold to complete all the customer orders by their due dates (see also, critical load index (5.4) in Section 5.2.1)

$$PCR(t) = \max_{g \in G, \tau \in T: \tau \le t} \left(\frac{\sum_{j \in J:\ \tau \le a_j \le d_j \le t} n_j p_{gj}}{\sum_{t'=\tau}^{t} c_{gt'}} \right) \le 1;\ t \in T \qquad (12.1)$$

where $PCR(t)$ is the production cumulative capacity ratio until period t.

If $PCR(t) \le 1$, then for any period $\tau \le t$, the cumulative demand on production capacity of all the customer orders with due dates not greater than t and ready dates not less than τ (the numerator in Equation (12.1)) does not exceed the cumulative capacity available in periods τ through t (the denominator in Equation (12.1)).

A similar feasibility condition can be derived for the manufacturing stage for the worst-case scenario, when each production line must be started up in every period.

$$MCR = \max_{t \in T} \left(\frac{\sum_{k \in K} q_k \left(\sum_{j \in J_k : d_j \le t+1} n_j - I_{k0}^1 - I_{k0}^2 \right)}{M(H - \sigma)t} \right) \le 1 \qquad (12.2)$$

where MCR is the manufacturing worst-case cumulative capacity ratio.

If $MCR \le 1$, then for any period t, the cumulative demand on capacity necessary to manufacture all parts required for the customer orders due not later than $t + 1$ (the numerator in Equation (12.2)) does not exceed the cumulative capacity available in the manufacturing stage in periods 1 through t under the worst-case scenario (the denominator in Equation (12.2)).

The feasibility conditions (12.1) and (12.2) would be sufficient for all orders to be completed by the customer-requested due dates if the orders were continuously divisible among the consecutive time periods so that periods could be filled exactly to their capacities.

If the feasibility conditions are not satisfied for customer-requested due dates, then later due dates can be committed for some orders to reach (12.1) and (12.2), when the order acceptance and due dates setting decisions are made, see Chapter 5.

12.3 SUPPLY CHAIN INVENTORY

In this section various types of supply chain inventories are described and an integer program is presented for determining the latest delivery dates of product-specific parts, while meeting all customer due dates at a minimum total input and output inventory of producer. The latest delivery dates of parts are next used as the earliest release dates of customer orders. The notation used is presented in Table 12.1.

Table 12.1 Notation: Coordinated Scheduling in a Supply Chain with a Single Supplier

Indices		
g	=	assembly stage, $g \in G$
j	=	customer order, $j \in J$
k	=	product type or product-specific part type $k \in K$
t	=	planning period, $t \in T$
Input parameters		
a_j, d_j, n_j	=	ready date, due date, size of order j
c_{gt}	=	processing time available in assembly stage g in period t
h	=	planning horizon (number of planning periods)
p_{gj}	=	processing time in stage g of each product in order j
q_k	=	processing time for one unit of part type k
H	=	length of each planning period
I_{k0}^1, I_{k0}^2	=	the beginning inventory of part type k at supplier, producer, respectively
J_k	=	subset of customer orders for product type k (i.e. requiring part type k)
M	=	number of parallel production lines in the manufacturing stage
$\underline{S}, \overline{S}$	=	the minimum, the maximum shipping lot, respectively
σ	=	the start-up time of production line in the manufacturing stage
$\varphi_k^1, \varphi_k^2, \varphi_k^3$	=	inventory holding cost per unit of part stored by the supplier, transported to the producer or stored by the producer, per unit of finished product stored by the producer, respectively
ψ_u, ψ_z	=	unit cost of part shipment, production line start-up, respectively
Decision variables		
r_{kt}	=	manufacturing lot of part type k in period t (lot sizing variable)
s_{kt}	=	transportation lot of part type k in period t (lot sizing variable)
u_t	=	1, if a shipment of parts is scheduled for period t, otherwise $u_t = 0$ (shipment period selection variable)
x_{jt}	=	1, if customer order j is assigned to planning period t; otherwise $x_{jt} = 0$ (order-to-period assignment variable)
y_{kt}	=	number of parallel production lines set up for processing part type k in period t (set-up variable)
z_{kt}	=	number of parallel production lines started up in period t to process part type k after processing another part type (start-up variable)

In the supply chain under study, the following three types of inventory can be distinguished:

- I_{kt}^1 supplier output inventory of manufactured part type k in period t
- I_{kt}^2 producer input inventory of supplied part type k in period t, including parts transported in period t from supplier to producer
- I_{kt}^3 producer output inventory of finished product type k in period t

$$I_{kt}^1 = I_{k0}^1 + \sum_{\tau=1}^{t} (r_{k\tau} - s_{k\tau});\ k \in K, t \in T \tag{12.3}$$

$$I_{kt}^2 = I_{k0}^2 + \sum_{\tau=1}^{t} \left(s_{k\tau} - \sum_{j \in J_k} n_j x_{j\tau} \right);\ k \in K, t \in T \tag{12.4}$$

$$I_{kt}^3 = \sum_{\tau=1}^{t} \sum_{j \in J_k : d_j > t} n_j x_{j\tau};\ k \in K, t \in T, \tag{12.5}$$

where I_{k0}^1, I_{k0}^2 are the beginning inventory of part type k at supplier and producer, respectively.

The total supply chain inventory $I(t)$ in every period t is the sum of the three types of parts and products inventories (12.3) to (12.5), that is,

$$\begin{aligned} I(t) &= \sum_{k \in K} (I_{kt}^1 + I_{kt}^2 + I_{kt}^3) \\ &= \sum_{k \in K} \left(I_{k0}^1 + I_{k0}^2 + \sum_{\tau=1}^{t} r_{k\tau} - \sum_{\tau=1}^{t} \sum_{j \in J_k : a_j \le \tau \le d_j \le t} n_j x_{j\tau} \right);\ t \in T. \end{aligned} \tag{12.6}$$

The total supply chain inventory can be optimized by minimizing either its maximum level $I_{\max}$ or its holding cost Φ_{sum} defined below.

$$I_{\max} = \max_{t \in T} \sum_{k \in K} \left(I_{k0}^1 + I_{k0}^2 + \sum_{\tau=1}^{t} r_{k\tau} - \sum_{\tau=1}^{t} \sum_{j \in J_k : a_j \le \tau \le d_j \le t} n_j x_{j\tau} \right) \tag{12.7}$$

$$\Phi_{\text{sum}} = \sum_{k \in K} \sum_{t \in T} (\varphi_k^1 I_{kt}^1 + \varphi_k^2 I_{kt}^2 + \varphi_k^3 I_{kt}^3). \tag{12.8}$$

In this chapter the above two alternative measures of the total inventory will be considered and compared.

The inventory holding costs φ_k^1 and φ_k^2, φ_k^3, respectively, of parts stored at supplier and of parts, finished products stored at producer, satisfy the inequalities $\varphi_k^1 < \varphi_k^2 < \varphi_k^3;\ k \in K$, due to the value-added activities down through the supply chain.

Denote by $E_{\max}$ the maximum earliness of orders and assume that for each customer order $j \in J$, product-specific parts are delivered exactly $E_{\max}$ periods ahead of the order due date d_j. Then the cumulative supply of parts to the producer in periods 1 through t (including initial input inventories) are

$$\sum_{j \in J : d_j \le t + E_{\max}} n_j$$

and the cumulative deliveries of products to the customers in periods 1 through t are

$$\sum_{\tau=1}^{t} \sum_{j \in J: d_j \le t,\ d_j - E_{\max} \le \tau \le t} n_j x_{j\tau}.$$

The total producer inventory in period t is the difference of the above two terms

$$\sum_{k \in K} (I_{kt}^2 + I_{kt}^3) = \sum_{j \in J: d_j \le t + E_{\max}} n_j - \sum_{\tau=1}^{t} \sum_{j \in J: d_j \le t,\ d_j - E_{\max} \le \tau \le t} n_j x_{j\tau}.$$

The last formula can be rewritten as

$$\sum_{k \in K} (I_{kt}^2 + I_{kt}^3) = \sum_{j \in J: d_j \le t} n_j \left(1 - \sum_{\tau = d_j - E_{\max}}^{t} x_{j\tau}\right) + \sum_{j \in J: t+1 \le d_j \le t + E_{\max}} n_j.$$

The first summation term is the inventory of product-specific parts for customer orders due by period t, and the second term is the inventory of product-specific parts and finished products of customer orders due after period t. If all customer orders are completed on or before their due dates (i.e., 100% of service level), the first term is equal to zero and the last formula reduces to (see also (6.41) in Section 6.4.1)

$$\sum_{k \in K} (I_{kt}^2 + I_{kt}^3) = \sum_{j \in J: t+1 \le d_j \le t + E_{\max}} n_j.$$

Hence, if no tardy orders are allowed, the total producer inventory decreases with $E_{\max}$, that is, both the input inventory of product-specific parts and the output inventory of finished products can be reduced when the latest delivery dates of the required parts and the due dates of customer orders are closer. As a result, the maximum level of the total producer inventory can be implicitly minimized by minimizing the maximum earliness $E_{\max}$ of early orders, subject to no tardiness constraints.

The time-indexed MIP model *MME* presented below can be applied to determine for each customer order the latest delivery date of the required parts such that the minimum total producer inventory is achieved with no tardy orders, assuming that the feasibility conditions (12.1) and (12.2) are satisfied (see also model *ILE* in Section 6.4.1).

Model MME: *Minimization of the Maximum Earliness of Customer Orders*

Minimize

$$E_{\max} \tag{12.9}$$

subject to

1. *Order-to-Period Nondelayed Assignment Constraints*:
   - each customer order is assigned to exactly one planning period not later than its due date,

$$\sum_{t \in T:\ a_j \le t \le d_j} x_{jt} = 1;\ j \in J \tag{12.10}$$

2. *Assembly Capacity Constraints*:

- in every period the demand on capacity at each assembly stage cannot be greater than the maximum available capacity in this period,
- in period $t = 1$ the demand on each part type k cannot be greater than the initial inventory of this part type,

$$\sum_{j \in J} n_j p_{gj} x_{jt} \le c_{gt}; \; g \in G, t \in T \tag{12.11}$$

$$\sum_{j \in J_k: \, a_j = 1} n_j x_{j1} \le I^2_{k0}; \; k \in K \tag{12.12}$$

3. *Maximum Earliness Constraints*:

- for each early order j assigned to period $t < d_j$, its earliness $(d_j - t)$ cannot exceed the maximum earliness $E_{\max}$ to be minimized

$$(d_j - t)x_{jt} \le E_{\max}; \; j \in J, t \in T: a_j \le t \le d_j \tag{12.13}$$

4. *Nonnegativity and Integrality Conditions*:

$$x_{jt} \in \{0, 1\}; \; j \in J, t \in T: a_j \le t \le d_j \tag{12.14}$$

$$E_{\max} \ge 0. \tag{12.15}$$

The objective (12.9) represents the minimum value of the maximum earliness, that is, of the maximum difference between order due date and the latest delivery period of the required parts, such that no tardy order occurs. If for some customer orders the required parts were delivered later than $E_{\max}$ periods ahead of the due date, the limited order earliness due to the later parts availability could restrict a reallocation of the orders to the earlier periods with surplus of capacity. Consequently, tardy orders or even infeasible schedules could occur, with some customer orders unscheduled during the planning horizon.

Let

$$\max \{a_j, d_j - E_{\max}\} \tag{12.16}$$

be the latest delivery date of the required parts for each customer order j. In the sequel the latest delivery dates (12.16) are considered to be the earliest release dates of customer orders.

Having solved the integer program *MME* to determine the earliest release dates (12.16) for customer orders, the next step is to find a nondelayed schedule for assembly of finished products and the corresponding manufacturing and supply schedules for parts.

12.4 COORDINATED SUPPLY CHAIN SCHEDULING: AN INTEGRATED APPROACH

In this section the weighted-sum, time-indexed MIP model *SMSA* is presented for the integrated scheduling of products assembly and parts manufacturing and supply in the customer-driven supply chain. The following two variables used in monolithic model *SMSA* need to be additionally explained (for definitions of all decision variables, see Table 12.1).

- The set-up variable y_{kt} that describes a state of manufacturing system in each planning period t, that is, the number of production lines set up for manufacturing part type k.
- The start-up variable z_{kt} that represents the number of production lines that are started up to manufacture part type k in period t, that is, the number of lines set up in period t for part type k, which has not been set up for this part type in period $t-1$. The start-up variable can take a positive value only if the corresponding set-up variable has in period t a higher value than in period $t-1$, that is,

$$z_{kt} = \max\{0, y_{kt} - y_{kt-1}\}.$$

The objective function (12.17) represents the weighted sum of the maximum level of total inventory, the number of start-ups of production lines in the manufacturing stage, and the number of part shipments from that stage to the producer stage. The weights λ_1, λ_2 represent the relative importance of material shipments and production line start-ups with respect to maximum level of parts and products inventory.

The objective function (12.18) represents total supply chain cost of holding the inventory of parts and products, of material shipments, and of production line start-ups. The unit shipment cost ψ_u and start-up cost ψ_z represent a trade-off between the numbers of part shipments and production line start-ups and the total inventory holding cost.

Model SMSA: *Scheduling Manufacturing, Supply, and Assembly*

Minimize

$$I_{\max} + \lambda_1 \sum_{t \in T} u_t + \lambda_2 \sum_{k \in K} \sum_{t \in T} z_{kt} \tag{12.17}$$

or

$$\Phi_{\text{sum}} + \psi_u \sum_{t \in T} u_t + \psi_z \sum_{k \in K} \sum_{t \in T} z_{kt} \tag{12.18}$$

subject to

1. *Order-to-Period Nondelayed Assignment Constraints*: (12.10)
2. *Assembly Capacity Constraints*: (12.11)

3. *Manufacturing Line Set-up and Start-up Constraints*:
   - in every period total number of production lines set up for manufacturing different part types is not greater than total number M of available lines,
   - all production lines set up for part type k in period 1 should be started up to manufacture this part type,
   - in every period $t > 1$, the number of production lines started up for part type k cannot be less than the difference between the number of lines set up for this part type in periods t and $t-1$,
   - in every period $t > 1$, the number of production lines started up for part type k cannot be greater than the number of lines set up for part type k in this period and cannot be greater than the number of lines set up for the other part types or idle in period $t-1$,

$$\sum_{k \in K} y_{kt} \leq M; \ t \in T \tag{12.19}$$

$$z_{k1} = y_{k1}; \ k \in K \tag{12.20}$$

$$z_{kt} \geq y_{kt} - y_{k,t-1}; \ k \in K, t \in T : t > 1 \tag{12.21}$$

$$z_{kt} \leq y_{kt}; \ k \in K, t \in T : t > 1 \tag{12.22}$$

$$z_{kt} \leq M - y_{k,t-1}; \ k \in K, t \in T : t > 1 \tag{12.23}$$

4. *Manufacturing Capacity Constraints*:
   - in every period t the production volume of part type k cannot be greater than the maximum volume corresponding to the capacity assigned to part type k in this period,

$$r_{kt} \leq \lfloor (H - \sigma)/q_k \rfloor z_{kt} + \lfloor H/q_k \rfloor (y_{kt} - z_{kt}); \ k \in K, t \in T \tag{12.24}$$

   where $\lfloor \cdot \rfloor$ is the greatest integer not greater than $\cdot$.
5. *Material Manufacturing and Shipment Constraints*:
   - parts can be supplied only in periods scheduled for shipment, and each shipping lot is limited by its minimum and maximum allowed size, $\underline{S}$ and $\overline{S}$, respectively.
   - for each part type k and period t the cumulative shipping lots in periods 1 through t cannot be greater than the initial stocks and the cumulative production of this part type in periods 1 through t,

$$s_{kt} \leq \overline{S} u_t; \ k \in K, t \in T \tag{12.25}$$

$$\sum_{k \in K} s_{kt} \leq \overline{S}; \ t \in T \tag{12.26}$$

$$\sum_{k \in K} s_{kt} \geq \underline{S} u_t; \ t \in T \tag{12.27}$$

$$\sum_{\tau=1}^{t} s_{k\tau} \leq I_{k0}^{1} + \sum_{\tau=1}^{t} r_{k\tau}; \ k \in K, t \in T \tag{12.28}$$

6. *Material Demand Satisfaction Constraints*:
   - for every period t the cumulative production of product type k in periods 1 through t cannot be greater than the initial stocks and the cumulative supplies of part type k in periods 1 through $t-1$,
   - total production of each part type is equal to total demand for this part type less the initial total stocks,
   - total supplies of each part type are equal to total demand for this part type less the initial producer stocks,

$$\sum_{j \in J_k} \sum_{\tau=1}^{t} n_j x_{j\tau} \leq I_{k0}^2 + \sum_{\tau=1}^{t-1} s_{k\tau};\ k \in K, t \in T{:}t < h \tag{12.29}$$

$$\sum_{t=1}^{h-1} r_{kt} = \sum_{j \in J_k} n_j - I_{k0}^1 - I_{k0}^2;\ k \in K \tag{12.30}$$

$$\sum_{t=1}^{h-1} s_{kt} = \sum_{j \in J_k} n_j - I_{k0}^2;\ k \in K \tag{12.31}$$

7. *Inventory Constraints*:
   - if objective (12.17) is selected, then

$$\sum_{k \in K} \left(I_{k0}^1 + I_{k0}^2 + \sum_{\tau=1}^{t} r_{k\tau} - \sum_{\tau=1}^{t} \sum_{j \in J_k : a_j \leq \tau \leq d_j \leq t} n_j x_{j\tau} \right) \leq I_{\max};\ t \in T \tag{12.32}$$

   - if objective (12.18) is selected, then (12.3) to (12.5) and (12.8).
8. *Coordinating Constraints*:
   - cumulative assembly of each product type k in periods 1 through t cannot be greater than the initial total stocks of part type k and the cumulative production of this part type in periods 1 through $t-1$,

$$\sum_{j \in J_k} \sum_{\tau=1}^{t} n_j x_{j\tau} \leq I_{k0}^1 + I_{k0}^2 + \lfloor H/q_k \rfloor \sum_{\tau=1}^{t-1} (y_{k\tau} - z_{k\tau})$$

$$+ \lfloor (H-\sigma)/q_k \rfloor \sum_{\tau=1}^{t-1} z_{k\tau};\ k \in K, t \in T{:}t > 1 \tag{12.33}$$

9. *Nonnegativity and Integrality Conditions*: (12.14)

$$r_{kt} \geq 0;\ k \in K, t \in T \tag{12.34}$$

$$s_{kt} \geq 0;\ k \in K, t \in T \tag{12.35}$$

$$u_t \in \{0, 1\};\ t \in T \tag{12.36}$$

$$y_{kt} \geq 0, \ \ integer; \ k \in K, t \in T \tag{12.37}$$

$$z_{kt} \geq 0, \ \ integer; \ k \in K, t \in T \tag{12.38}$$

$$I_{\max} \geq 0 \text{ (if objective (12.17) is selected).} \tag{12.39}$$

The coordinating constraint (12.33) has been introduced to strengthen the mixed integer program *SMSA* as it directly links the assignment of production lines to part types over the planning horizon (variables y_{kt} and z_{kt}) and the assignment of customer orders to planning periods (variables x_{jt}). Thus, (12.33) directly coordinates the schedules of the supplier and the producer. Constraint (12.33) is implied by constraints (12.24), (12.28), and (12.29), and any feasible solution that satisfies (12.33) also satisfies (12.24), (12.28), and (12.29).

Note that in the nondelayed assignment constraints (12.10), the ready dates a_j are replaced with the earliest release dates determined by (12.16). The resulting nondelayed assembly schedule for the finished products ensures that all committed due dates are met (cf. Section 12.2.1). Hence, the ordered products can be shipped to the customers exactly on the due dates.

12.4.1 Lower and Upper Bounds

The following lower bounds can be derived on the total number of shipments for the entire planning horizon and on the remaining number of shipments for each planning period $t \in T$.

$$\sum_{t \in T} u_t \geq \left\lceil \frac{\sum_{j \in J} n_j - \sum_{k \in K} I_{k0}^2}{\overline{S}} \right\rceil \tag{12.40}$$

$$\sum_{\tau = t}^{h} u_\tau \geq \frac{\sum_{j \in J} n_j - \sum_{k \in K} (I_{k0}^2 + \sum_{\tau=1}^{t-1} s_{k\tau})}{\overline{S}}; \ t \in T, \tag{12.41}$$

where $\lceil \cdot \rceil$ is the least integer not less than $\cdot$.

Similarly, an upper bound on the total number of start-ups can be derived

$$\sum_{k \in K} \sum_{t \in T} z_{kt} \leq H(h-1)M/\sigma - \sum_{k \in K} q_k \left(\sum_{j \in J_k} n_j - I_{k0}^1 - I_{k0}^2 \right) \Big/ \sigma. \tag{12.42}$$

The bounds (12.40) to (12.42) can be added to model *SMSA* to reduce the computational effort required to find the optimal solution.

12.5 COORDINATED SUPPLY CHAIN SCHEDULING: A HIERARCHICAL APPROACH

In this section a hierarchical approach is proposed for scheduling products assembly and parts manufacturing and supply in the customer-driven supply chain. The

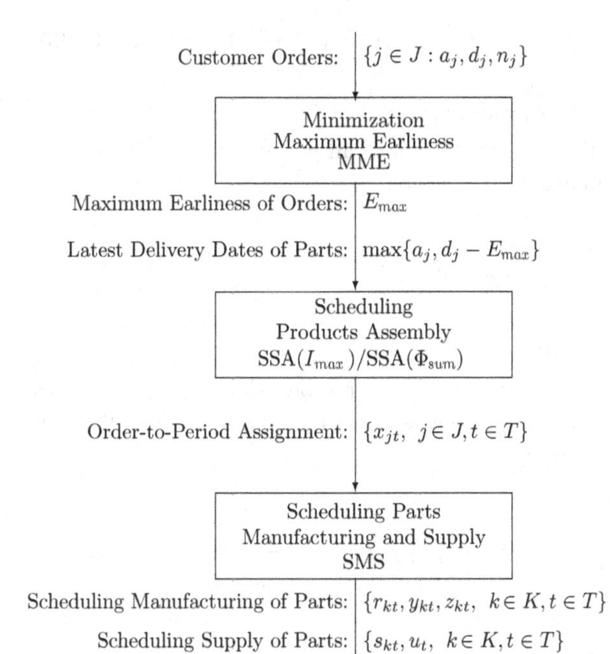

Figure 12.2 Coordinated scheduling of manufacturing, supply, and assembly: a hierarchical approach.

proposed hierarchical approach can be applied within a pull planning framework and the hierarchical decomposition scheme is based on the lexicographic approach to the multi-objective supply chain scheduling, where the primary objective is to minimize supply chain inventory. First, at the top level the earliest release dates of customer orders are found to minimize the maximum earliness of orders or equivalently the producer maximum inventory of parts and finished products. Then, at the middle level a nondelayed assignment of customer orders to planning periods is determined to minimize the selected measure of total supply chain inventory, $I_{\max}$, (12.7), or Φ_{sum}, (12.8), for the worst-case manufacturing scenario, when each production line must be started up in every period. Finally, given the assignment of customer orders and the corresponding assembly schedule of the finished products, at the base level the integrated schedule of parts manufacturing and supply is determined to minimize the selected measure ((12.17) or (12.18)) of the total supply chain inventory and the number of line start-ups in the manufacturing stage and material shipments from that stage to the producer stage (see Fig. 12.2).

The two alternative middle-level problems are defined below as time-indexed, mixed integer programs *SSA*($I_{\max}$) and *SSA*(Φ_{sum}), respectively, for objectives (12.17) and (12.18).

Model SSA($I_{\max}$): *Scheduling Supply and Assembly to Minimize Maximum Level of Total Inventory*

Minimize

$$I_{\max} \tag{12.43}$$

subject to (12.10), (12.11), (12.14), (12.25) to (12.32), (12.34) to (12.36), (12.39), and

1. *Manufacturing Capacity Constraints*:
   - in every period t the total production of all part types cannot be greater than the maximum available capacity,

$$\sum_{k\in K} r_{kt}/\rho_k \leq M;\ t \in T \tag{12.44}$$

2. *Coordinating Constraints*:
   - the cumulative demand on capacity for production of all part types required by period t cannot be greater than the total capacity of the supplier available by period $t-1$

$$\sum_{k\in K}\left(\sum_{j\in J_k}\sum_{\tau=1}^{t} n_j x_{j\tau} - I_{k0}^1 - I_{k0}^2\right)\Big/\rho_k \leq M(t-1);\ t \in T{:}t > 1, \tag{12.45}$$

where $\rho_k = \lfloor (H-\sigma)/q_k \rfloor$ is the one machine period worst-case manufacturing rate for part type k.

Model SSA(Φ_{sum}): *Scheduling Supply and Assembly to Minimize Total Inventory Holding Cost*

Minimize

$$\Phi_{\text{sum}} \tag{12.46}$$

subject to (12.3)–(12.5), (12.8), (12.10), (12.11), (12.14), (12.25)–(12.31), (12.34)–(12.36), (12.44), (12.45).

Note that constraint (12.32) in $SAS(I_{\max})$ defines the maximum inventory $I_{\max}$, whereas constraints (12.3) to (12.5) and (12.8) in $SAS(\Phi_{\text{sum}})$ define the inventory holding cost Φ_{sum}.

The solution to $SSA(I_{\max})$ or $SSA(\Phi_{\text{sum}})$ determines the optimal assignment of customer orders to planning periods $\{x_{jt},\ j \in J, t \in T\}$ and thereby the optimal assembly schedule for the finished products and the corresponding delivery schedule of products to customers. Denote by N_{kt} the optimal cumulative production of product type k in periods 1 through t

$$N_{kt} = \sum_{j\in J_k}\sum_{\tau=1}^{t} n_j x_{j\tau};\ k \in K, t \in T. \tag{12.47}$$

Furthermore, the solution to the middle-level problem determines also the optimal inventory of finished products over the planning horizon, assuming that all the products ordered by customers are shipped to the customers exactly on the requested due dates

$$I^3_{kt} = N_{kt} - \sum_{j\in J_k:\, d_j\le t} n_j;\ k \in K,\ t \in T. \tag{12.48}$$

Now, the base-level problem can be formulated as the following weighted-sum, time-indexed, mixed integer program *SMS*.

Model SMS: *Scheduling Manufacturing and Supply*

Minimize (12.17) or (12.18) subject to

1. *Manufacturing Line Set-up and Start-up Constraints*: (12.19) to (12.23)
2. *Manufacturing Capacity Constraints*: (12.24)
3. *Material Manufacturing and Shipment Constraints*: (12.25) to (12.28)
4. *Material Demand Satisfaction Constraints*: (12.30), (12.31)
   - for each period t and part type k, the cumulative supplies in periods 1 through $t-1$ cannot be less than the net cumulative demand (after removal of initial stocks) for this part type in periods 1 through t,

$$\sum_{\tau=1}^{t-1} s_{k\tau} \ge N_{kt} - I^2_{k0};\ k \in K,\ t \in T \tag{12.49}$$

5. *Inventory Constraints*:

   if objective (12.17) is selected:
   - in every period the total system inventory of parts and finished products cannot exceed its maximum level $I_{\max}$,

$$\sum_{k\in K}\left(I^1_{k0} + I^2_{k0} + \sum_{\tau=1}^{t} r_{k\tau}\right) - \sum_{j\in J:\, d_j\le t} n_j \le I_{\max};\ t \in T \tag{12.50}$$

   if objective (12.18) is selected:
   - supplier output inventory of manufactured parts I^1_{kt} - (12.3),
   - producer output inventory of finished products I^3_{kt} - (12.48),
   - producer input inventory of supplied parts, including parts transported in period t from supplier to producer

$$I^2_{kt} = I^2_{k0} + \sum_{\tau=1}^{t} s_{k\tau} - N_{kt};\ k \in K,\ t \in T \tag{12.51}$$

6. *Coordinating Constraints*:

- the cumulative production of each part type by period $t-1$ cannot be less than the net cumulative demand for this part type by period t,

$$\lfloor H/q_k \rfloor \sum_{\tau=1}^{t-1} (y_{k\tau} - z_{k\tau}) + \lfloor (H-\sigma)/q_k \rfloor \sum_{\tau=1}^{t-1} z_{k\tau}$$
$$\geq N_{kt} - I_{k0}^1 - I_{k0}^2;\ k \in K, t \in T : t > 1 \qquad (12.52)$$

7. *Nonnegativity and Integrality Conditions*: (12.34) to (12.39).

Let us note that the finished products inventory (12.48) is fixed and hence only material inventory is minimized in the objective functions (12.17) or (12.18) of SMS.

12.6 COMPUTATIONAL EXAMPLES

In this section numerical examples and computational results are presented to illustrate possible applications of the two proposed approaches (integrated and hierarchical) and MIP models for the coordinated scheduling of manufacturing, supply, and assembly in a supply chain with a single supplier. The examples are modeled after a real-world supply chain of high-tech products. The supply chain consists of a single manufacturer/supplier of product-specific materials (electronic components), distribution center (see Fig. 5.3) where finished products (electronic devices) are assembled according to customer orders, and a set of customers who generate final demand for the products.

The manufacturing stage consists of 22 parallel production lines, each capable of producing all part types and the distribution center is a flexible flow shop made up of six processing stages with parallel machines. Each type of product (each customer order) requires processing in at most four stages: 1, 2, 3 or 4 or 5, and 6.

A brief description of the manufacturing and assembly stages of the supply chain, parts, products, and the customer orders is given below.

1. Planning horizon: $h = 30$ periods (days), each of length $H = 2 \times 9$ hours.

2. Manufacturing

- $M = 22$ parallel production lines, each to be set up at most once per planning period, with identical start-up time $\sigma = 9000$ seconds.
- Parts
  - 10 product-specific part types
  - processing times q_k (in seconds) for part types $k = 1, \ldots, 10$: $q_1 = 65, q_2 = 70, q_3 = 80, q_4 = 85,$ $q_5 = 90, q_6 = 95, q_7 = 65, q_8 = 80, q_9 = 90, q_{10} = 85.$

3. Supply
- at most one shipment of various parts per period,
- the minimum and the maximum allowed shipping lot: $\underline{S} = 1000$, and $\overline{S} = 50000$ parts,
- transportation time constant and equal to one period for every shipping lot.

4. Assembly
- six assembly stages in series with parallel machines: $m_g = 10$ parallel machines in each stage $g = 1, 2$; $m_g = 20$ parallel machines in each stage $g = 3, 4, 5$; and $m_g = 10$ parallel machines in stage $g = 6$.
- for each assembly stage g, the available processing time c_{gt} is the same in every period t:

$$c_{gt} = c_g = \alpha_g H m_g;\ g = 1, \ldots, 6,\ t = 1, \ldots, 30$$

where parameter $\alpha_g \in (0, 1)$ reflects the idle time of each machine waiting for the first production lot from upstream stages $1, \ldots, g-1$, preceding stage g and the idle time during processing of the last production lot at downstream stages following stage g, assuming that all production lots of each customer order are completed in a single period.

- Products
  - 10 product types of three product groups, each to be processed on a separate group of machines (in stage 3 or 4 or 5),
  - processing times (in seconds) for product types:

Product type/stage	1	2	3	4	5	6
1	20	0	120	0	0	15
2	20	0	140	0	0	15
3	10	0	160	0	0	10
4	15	5	0	120	0	15
5	15	10	0	140	0	15
6	10	5	0	160	0	10
7	15	10	0	180	0	15
8	20	5	0	0	120	15
9	15	0	0	0	140	10
10	15	0	0	0	160	10

  - 805 single-product customer orders ranging from 5 to 9620 products, with various requested due dates, each to be completed in a single period. The total monthly demand is 535,000 products and the distribution of demand is shown in Figure 12.3.

5. Initial inventory
- The initial supplier stocks of manufactured parts are equal to an average two-day manufacturing volumes

$$I_{k0}^1 = 2\lfloor HM/10q_k \rfloor;\ k = 1, \ldots, 10$$

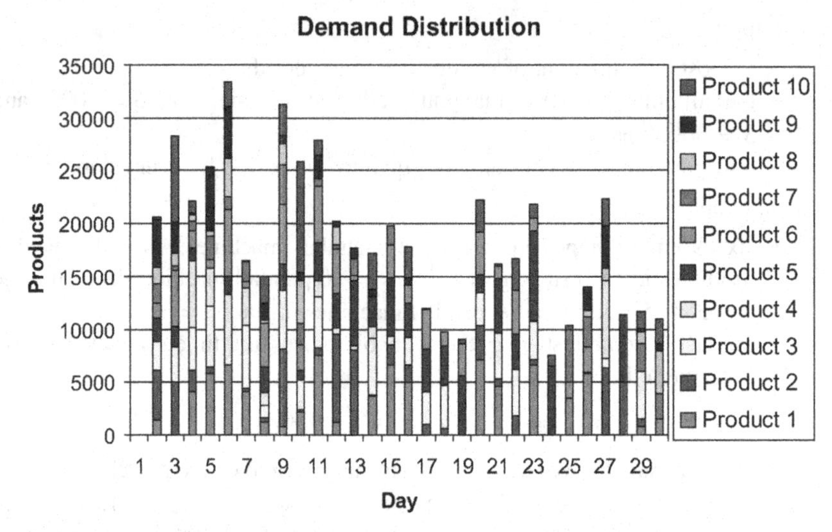

Figure 12.3 Distribution of demand.

- The initial producer stocks of supplied parts ensure an average two-day assembly of the finished products

$$I_{k0}^2 = 2 \max_{g \in G,\, j \in J_k, k \in K} \lfloor c_g / 10 p_{gj} \rfloor; \; k = 1, \ldots, 10.$$

The characteristics of the MIP models for the two approaches (integrated and hierarchical) and for various weights in the objective functions are summarized in Table 12.2 and Table 12.3. The size of mixed integer programs *SMSA*, *SSA*, and *SMS* (enhanced with bounds (12.40) to (12.42)) is represented by the total number of variables, Var., number of binary variables, Bin., number of integer variables, Int., and number of constraints, Cons. The counts presented in the tables are taken from the models after presolving. The last columns of the tables present the solution values: $I_{\max}$, maximum total inventory level; Φ_{sum}, total inventory holding cost; $U_{\text{sum}} = \sum_{t \in T} u_t$, total number of shipments; and $Z_{\text{sum}} = \sum_{k \in K} \sum_{t \in T} z_{kt}$, total number of start-ups. The solution value $I_{\max}$ or Φ_{sum} is presented along with the corresponding associated value of Φ_{sum} or $I_{\max}$ (in parentheses), respectively, for the objective function (12.17) or (12.18). The computational experiments were performed using AMPL programming language and the CPLEX v.9.1 solver (with the default settings) on a laptop with a Pentium IV processor running at 1.8 GHz and with 1 GB RAM. The computation time was limited to 1800 CPU seconds.

Similar computational experiments were also conducted using the basic mixed integer programs *SMSA*, *SAS*, and *SMS* without coordinating constraints, (12.33), (12.45), and (12.52), respectively. The results indicate that introduction of (12.33), (12.45), and (12.52) strengthened the formulations and reduced CPU time up to 20%.

While a subset of nondominated solutions was determined for the objective function (12.18) for both the weighted-sum mixed integer programs *SMSA* and *SMS*,

Table 12.2 Computational Results: Integrated Approach

Model	Var.	Bin.	Int.	Cons.	Solution values[a]
MME	9822	9821		9996	$E_{max} = 2$
Objective function (12.17): $I_{max} + \lambda_1 U_{sum} + \lambda_2 Z_{sum}$					
SMSA(0.05,0.95)[b]	3508	2307	600	3425	$I_{max} = 120348$, ($\Phi_{sum} = 3462$), $U_{sum} = 23$, $Z_{sum} = 36$
SMSA(0.25,0.75)[b]	3508	2307	600	3425	$I_{max} = 120348$, ($\Phi_{sum} = 3505$), $U_{sum} = 24$, $Z_{sum} = 37$
SMSA(0.75,0.25)[b]	3508	2307	600	3425	$I_{max} = 120348$, ($\Phi_{sum} = 3543$), $U_{sum} = 20$, $Z_{sum} = 40$
SMSA(0.95,0.05)[b]	3508	2307	600	3425	$I_{max} = 120348$, ($\Phi_{sum} = 3648$), $U_{sum} = 20$, $Z_{sum} = 52$
SMSA(0.1,0.1)[b]	3508	2307	600	3425	$I_{max} = 120348$, ($\Phi_{sum} = 3672$), $U_{sum} = 21$, $Z_{sum} = 43$
SMSA(1,1)[b]	3508	2307	600	3425	$I_{max} = 120348$, ($\Phi_{sum} = 3537$), $U_{sum} = 19$, $Z_{sum} = 31$
SMSA(10,10)[b]	3508	2307	600	3425	$I_{max} = 120348$, ($\Phi_{sum} = 3568$), $U_{sum} = 19$, $Z_{sum} = 32$
SMSA(100,100)[b]	3508	2307	600	3425	$I_{max} = 120348$, ($\Phi_{sum} = 3681$), $U_{sum} = 18$, $Z_{sum} = 31$
Objective function (12.18): $\Phi_{sum} + \psi_u U_{sum} + \psi_z Z_{sum}$					
SMSA(0.05,0.95)[c]	3507	2307	600	3395	$\Phi_{sum} = 2133$, ($I_{max} = 120348$), $U_{sum} = 29$, $Z_{sum} = 103$
SMSA(0.25,0.75)[c]	3507	2307	600	3395	$\Phi_{sum} = 2125$, ($I_{max} = 120348$), $U_{sum} = 29$, $Z_{sum} = 119$
SMSA(0.75,0.25)[c]	3507	2307	600	3395	$\Phi_{sum} = 2100$, ($I_{max} = 120348$), $U_{sum} = 28$, $Z_{sum} = 159$
SMSA(0.95,0.05)[c]	3507	2307	600	3395	$\Phi_{sum} = 2100$, ($I_{max} = 120348$), $U_{sum} = 28$, $Z_{sum} = 185$
SMSA(0.1,0.1)[c]	3507	2307	600	3395	$\Phi_{sum} = 2097$, ($I_{max} = 120348$), $U_{sum} = 29$, $Z_{sum} = 169$
SMSA(1,1)[c]	3507	2307	600	3395	$\Phi_{sum} = 2142$, ($I_{max} = 120389$), $U_{sum} = 28$, $Z_{sum} = 99$
SMSA(10,10)[c]	3507	2307	600	3395	$\Phi_{sum} = 2332$, ($I_{max} = 120348$), $U_{sum} = 28$, $Z_{sum} = 40$
SMSA(100,100)[c]	3507	2307	600	3395	$\Phi_{sum} = 3078$, ($I_{max} = 130516$), $U_{sum} = 14$, $Z_{sum} = 28$

[a] E_{max} = maximum earliness; I_{max} = maximum total inventory level; $\Phi_{sum} = \sum_{k \in K} \sum_{t \in T} (0.001 I^1_{kt} + 0.002 I^2_{kt} + 0.004 I^3_{kt})$ = total inventory holding cost; $U_{sum} = \sum_{t \in T} u_t$ = total number of shipments; $Z_{sum} = \sum_{k \in K} \sum_{t \in T} z_{kt}$ = total number of start-ups.

[b] Weights (λ_1, λ_2) in the objective function (12.17).

[c] Unit costs (ψ_u, ψ_z) in the objective function (12.18).

Table 12.3 Computational Results: Hierarchical Approach

Model	Var.	Bin.	Int.	Cons.	Solution values[a]
Objective function (2.17): $I_{max} + \lambda_1 U_{sum} + \lambda_2 Z_{sum}$					
SSA(I_{max})	2908	2307	-	1958	$I_{max} = 127652$, ($\Phi_{sum} = 2971$, $U_{sum} = 29$)
SMS(0.05,0.95)[b]	1230	29	600	2479	$I_{max} = 120348$, ($\Phi_{sum} = 3336$), $U_{sum} = 24$, $Z_{sum} = 35$
SMS(0.25,0.75)[b]	1230	29	600	2479	$I_{max} = 120348$, ($\Phi_{sum} = 3484$), $U_{sum} = 23$, $Z_{sum} = 40$
SMS(0.75,0.25)[b]	1230	29	600	2479	$I_{max} = 120348$, ($\Phi_{sum} = 3560$), $U_{sum} = 20$, $Z_{sum} = 41$
SMS(0.95,0.25)[b]	1230	29	600	2479	$I_{max} = 120348$, ($\Phi_{sum} = 3628$), $U_{sum} = 20$, $Z_{sum} = 46$
SMS(0.1,0.1)[b]	1230	29	600	2479	$I_{max} = 120348$, ($\Phi_{sum} = 3729$), $U_{sum} = 21$, $Z_{sum} = 36$
SMS(1,1)[b]	1230	29	600	2479	$I_{max} = 120348$, ($\Phi_{sum} = 3564$), $U_{sum} = 20$, $Z_{sum} = 32$
SMS(10,10)[b]	1230	29	600	2479	$I_{max} = 120348$, ($\Phi_{sum} = 3556$), $U_{sum} = 20$, $Z_{sum} = 31$
SMS(100,100)[b]	1230	29	600	2479	$I_{max} = 120348$, ($\Phi_{sum} = 3544$), $U_{sum} = 20$, $Z_{sum} = 31$
Objective function (2.18): $\Phi_{sum} + \psi_u U_{sum} + \psi_z Z_{sum}$					
SSA(Φ_{sum})	2907	2307	-	1928	$\Phi_{sum} = 2281$, ($I_{max} = 127652$, $U_{sum} = 29$)
SMS(0.05,0.95)[c]	1229	29	600	2446	$\Phi_{sum} = 2139$, ($I_{max} = 120497$), $U_{sum} = 29$, $Z_{sum} = 108$
SMS(0.25,0.75)[c]	1229	29	600	2446	$\Phi_{sum} = 2115$, ($I_{max} = 120444$), $U_{sum} = 29$, $Z_{sum} = 136$
SMS(0.75,0.25)[c]	1229	29	600	2446	$\Phi_{sum} = 2098$, ($I_{max} = 120348$), $U_{sum} = 29$, $Z_{sum} = 167$
SMS(0.95,0.05)[c]	1229	29	600	2446	$\Phi_{sum} = 2095$, ($I_{max} = 120348$), $U_{sum} = 29$, $Z_{sum} = 192$
SMS(0.1,0.1)[c]	1229	29	600	2446	$\Phi_{sum} = 2096$, ($I_{max} = 120348$), $U_{sum} = 29$, $Z_{sum} = 181$
SMS(1,1)[c]	1229	29	600	2446	$\Phi_{sum} = 2144$, ($I_{max} = 120444$), $U_{sum} = 29$, $Z_{sum} = 102$
SMS(10,10)[c]	1229	29	600	2446	$\Phi_{sum} = 2344$, ($I_{max} = 121542$), $U_{sum} = 28$, $Z_{sum} = 41$
SMS(100,100)[c]	1229	29	600	2446	$\Phi_{sum} = 3142$, ($I_{max} = 132514$), $U_{sum} = 14$, $Z_{sum} = 28$

[a] I_{max} = maximum total inventory level; $\Phi_{sum} = \sum_{k \in K} \sum_{t \in T} (0.001 I^1_{kt} + 0.002 I^2_{kt} + 0.004 I^3_{kt})$ = total inventory holding cost;
$U_{sum} = \sum_{t \in T} u_t$ = total number of shipments; $Z_{sum} = \sum_{k \in K} \sum_{t \in T} z_{kt}$ = total number of start-ups.

[b] Weights (λ_1, λ_2) in the objective function (2.17).

[c] Unit costs (ψ_u, ψ_z) in the objective function (2.18).

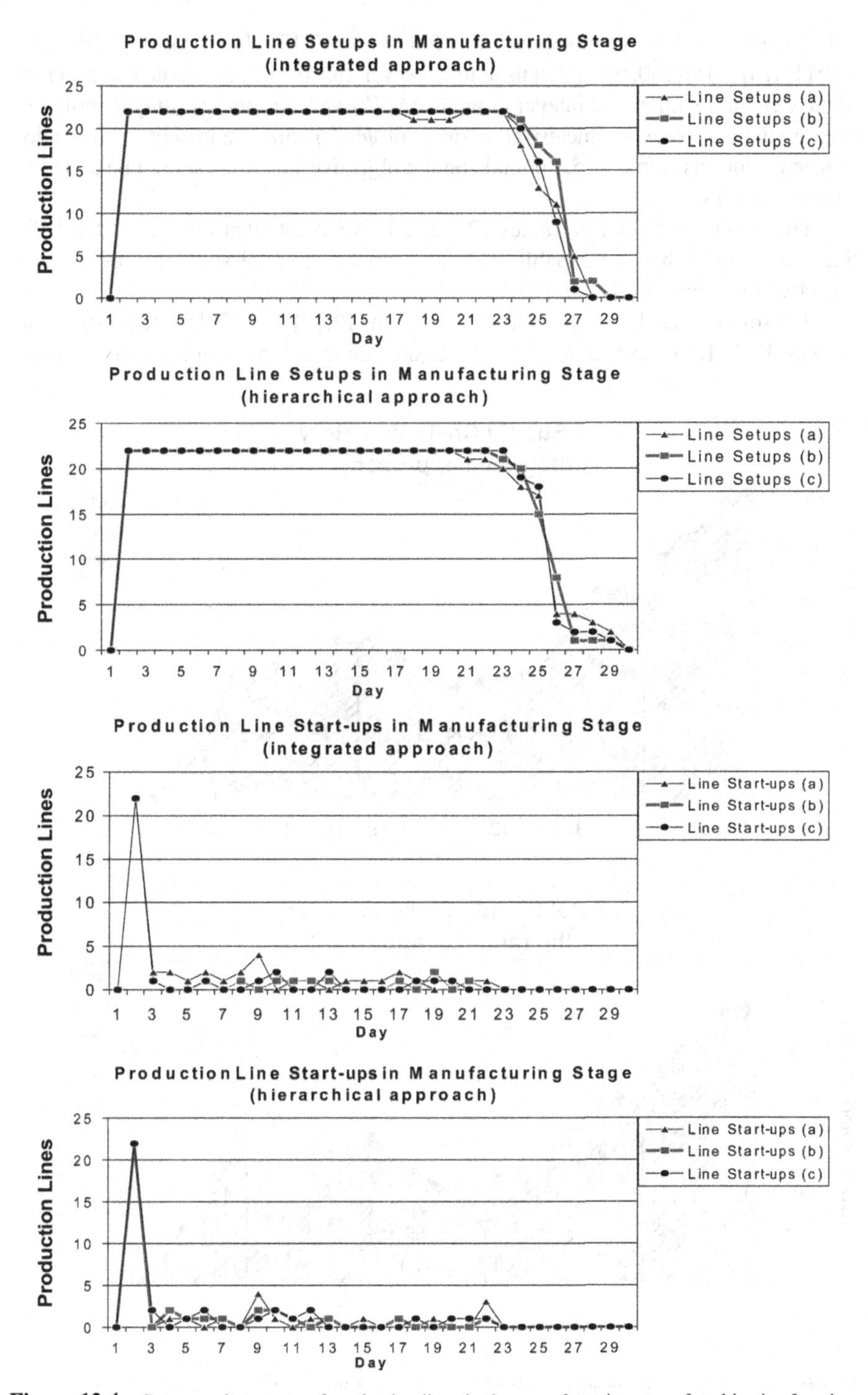

Figure 12.4 Setups and start-ups of production lines in the manufacturing stage for objective function (12.17): (a) $\lambda_1 = 1, \lambda_2 = 1$, (b) $\lambda_1 = 10, \lambda_2 = 10$, (c) $\lambda_1 = 100, \lambda_2 = 100$.

only a single nondominated solution was found for (12.17) (for *SMSA*(100,100) and *SMS*(10,10), *SMS*(100,100)). Let us note, however, that the nondominated solution set of the weighted-sum mixed integer program *SMSA* or *SMS* cannot be fully determined even by the complete parameterization on λ (or ψ). To compute unsupported nondominated solutions, some upper bounds on the objective functions should be added to *SMSA* and *SMS*.

The results presented in Tables 12.2 and 12.3 indicate that for the limited CPU time both approaches are capable of finding proven optimal solutions for the two objective functions (12.17) and (2.18).

The solution results for the objective function (12.17) and (12.18) are illustrated in Figures 12.4, 12.5, 12.6, and 12.7. The figures compare the results obtained using

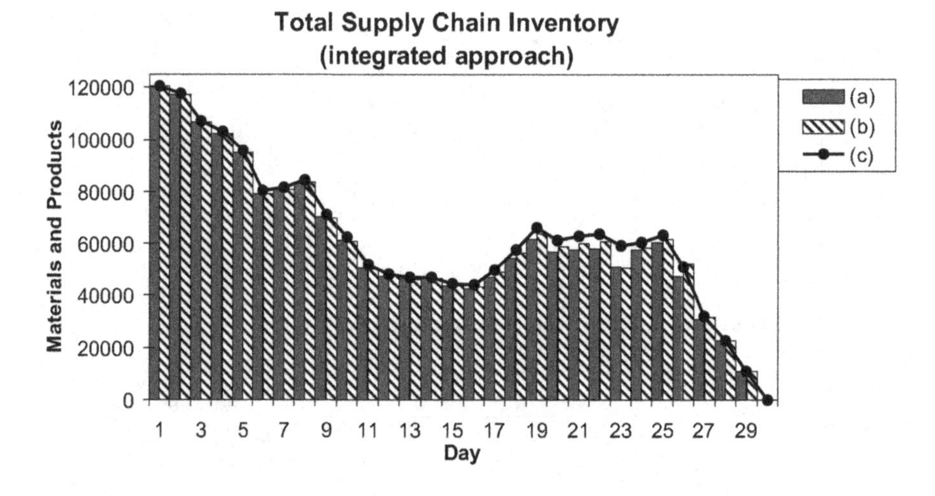

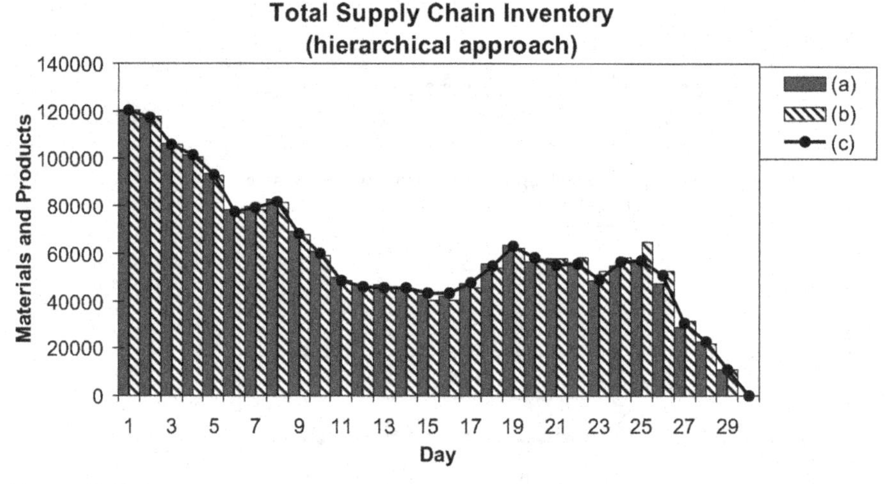

Figure 12.5 Total supply chain inventory for objective function (12.17): (a) $\lambda_1 = 1, \lambda_2 = 1$, (b) $\lambda_1 = 10, \lambda_2 = 10$, (c) $\lambda_1 = 100, \lambda_2 = 100$.

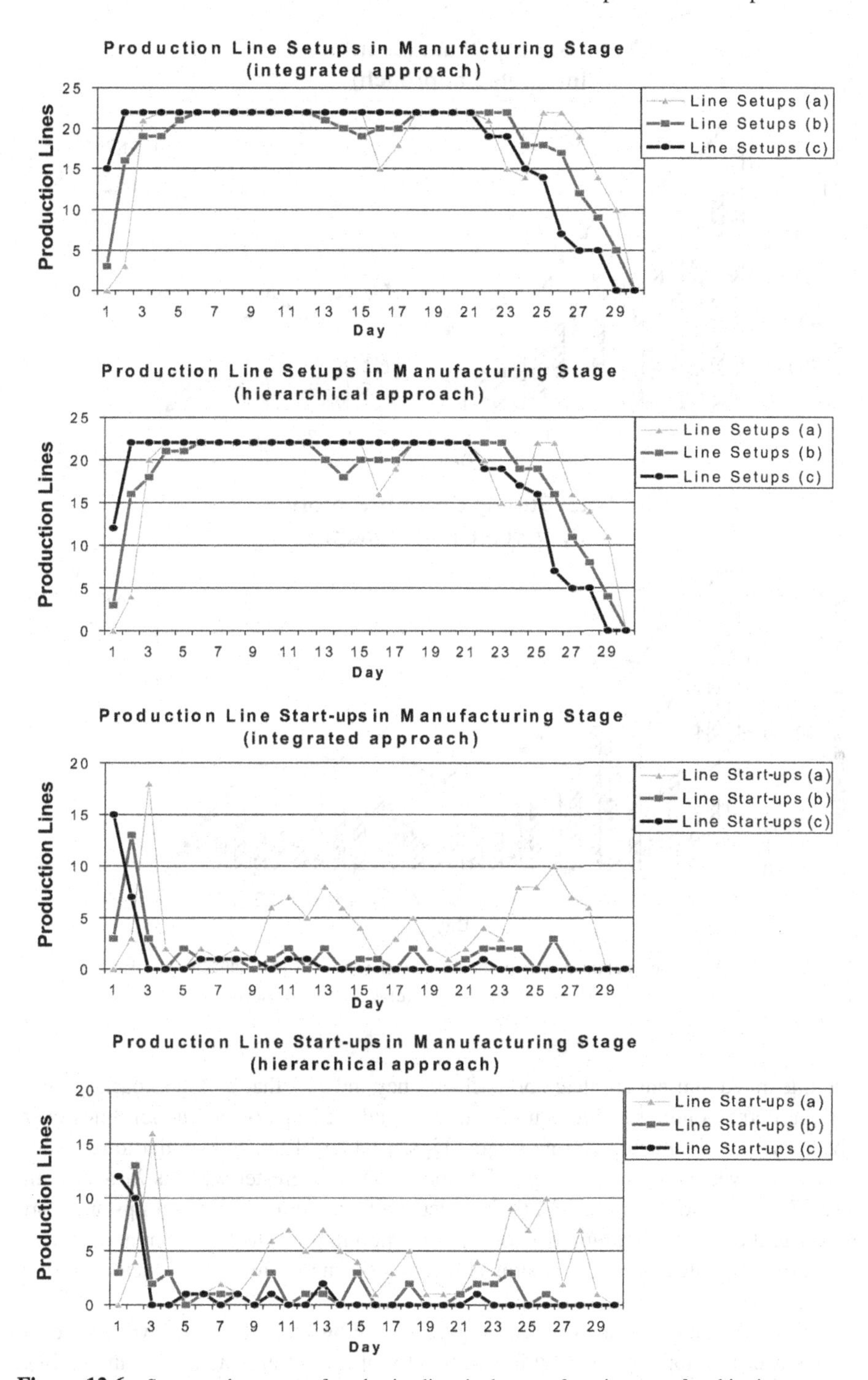

Figure 12.6 Setups and start-ups of production lines in the manufacturing stage for objective function (12.18) with $\varphi_1 = 0.001$, $\varphi_2 = 0.002$, $\varphi_3 = 0.004$: (a) $\psi_u = 1$, $\psi_z = 1$, (b) $\psi_u = 10$, $\psi_z = 10$, (c) $\psi_u = 100$, $\psi_z = 100$.

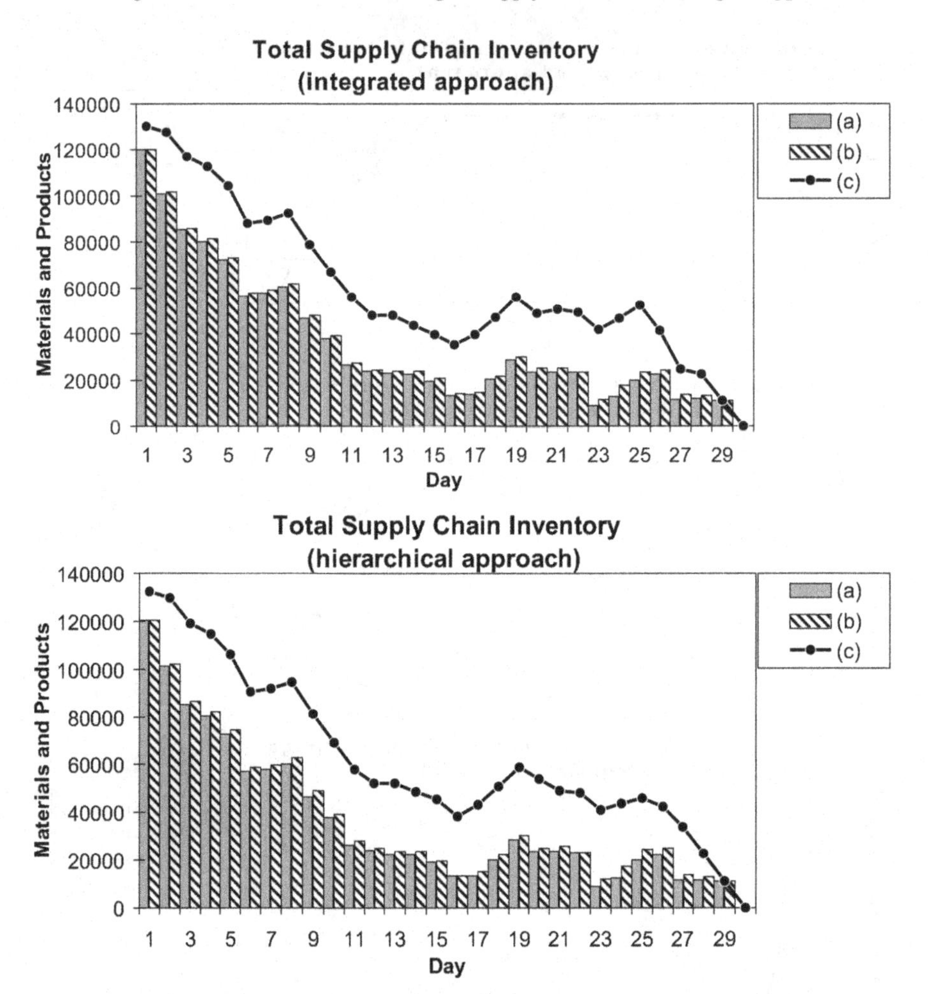

Figure 12.7 Total supply chain inventory for objective function (12.18) with $\varphi_1 = 0.001$, $\varphi_2 = 0.002$, $\varphi_3 = 0.004$: (a) $\psi_u = 1$, $\psi_z = 1$, (b) $\psi_u = 10$, $\psi_z = 10$, (c) $\psi_u = 100$, $\psi_z = 100$.

the integrated and hierarchical approaches. They indicate that both approaches yield similar corresponding results (similar set ups and start ups of production lines over the horizon in the manufacturing stage—Figs. 12.4 and 12.6, and similar total inventory level over the horizon—Figs. 12.5 and 12.7). The greater weights λ_1 and λ_2 in (12.17) or ψ_u and ψ_z in (12.18), the smaller the fluctuations of line set ups and start ups over the horizon. Comparison of the corresponding results for different objective functions indicates that (12.17) aims at leveling of line set ups and reduction of line start ups.

Comparison of Figure 12.5 and Figure 12.7 shows that the total inventory level varies over the horizon virtually independently of the weights λ_1 and λ_2 in (12.17), whereas for objective (12.18) the lower total inventory levels are achieved for smaller unit costs ψ_u and ψ_z. The latter result is due to the more frequent line start-ups and part supplies required to meet fluctuating demand for parts for the lower start-up and shipping unit costs, which leads to reducing of the total inventory.

Of the two different criteria proposed to measure the total supply chain inventory, minimization of the maximum inventory level leads to leveling over the horizon of production of parts and of line set ups and start ups in the manufacturing stage. At the same time the supplier and the producer inventories of parts are higher and the shipping lots are larger. The inventories of parts and the shipping lots are smaller for minimization of the total inventory holding cost. In general, the objective function (12.17) aims at leveling of production at the manufacturing stage at the expense of greater fluctuations and a higher inventory of parts and larger shipping lots.

12.7 COMMENTS

The coordination of raw materials manufacturing and supply, and finished goods production and distribution is one of the main issues of supply chain management, in particular, in customer-driven supply chains, for example, Thomas and Griffin (1996), Erenguc et al. (1999), Kolisch (2000), Shapiro (2001), Schneeweiss (1999), Z.-L. Chen (2004), Z.-L. Chen and Vairaktarakis (2005), Kaczmarczyk et al. (2006), Z.-L. Chen and Pundoor (2006), Z.-L. Chen and Hall (2007), Arshinder et al. (2008), van der Vaart and van Donk (2008), Kreipl and Dickersbach (2008).

The majority of research on supply chain coordination is devoted to developing joint economic lot size models, where the objective is to determine joint economic ordering and production lot sizes to simultaneously minimize total cost of material ordering and holding and manufacturing setup and finished products holding, for example, Goyal and Deshmukh (1992) and W. Lee (2005). However, the models are based on various simplifying assumptions such as a single finished product with a constant or a piece-wise linear demand pattern. Therefore, the models cannot be directly applied in a complex, multiproduct make-to-order environment, with an arbitrary demand pattern for different finished products. Coordination between the procurement and production policies in a just-in-time environment is proposed by Goyal and Deshmukh (1997). Their model, again restricted to a single product, aims at the minimization of total variable cost and determining the lot sizes for the product and raw material order sizes in a multistage batch environment.

Simplified models for integrated scheduling of production and distribution operations are studied by Z.-L. Chen and Vairaktarakis (2005) and Z.-L. Chen and Pundoor (2006), where transportation time and cost are considered. The authors have analyzed the computational complexity of various cases of the problem and have developed heuristics for NP-hard cases. Hall and Potts (2003) consider a different set of models that treats both delivery lead time and transportation cost, but assume that delivery is done instantaneously without any transportation time. Agnetis et al. (2006) study models for resequencing jobs, using limited buffers, between a supplier and several producers with different ideal sequences. The issues of conflict and cooperation between the suppliers and the producer are studied by Z.-L. Chen and

Hall (2007), where classical scheduling objectives are considered: minimization of the total completion time and of the maximum lateness. Moon et al. (2008) propose the evolutionary search approach to find feasible solutions for a simplified integrated process planning and scheduling in a supply chain, and Chauhan et al. (2007) consider real-time scheduling of flow shops with the no-wait constraint in a supply chain environment.

An operational, noncentralized coordination mechanism between a producer and a supplier within a supply chain, based on the theory of hierarchical planning and mixed integer programming, is presented in Schneeweiss and Zimmer (2004). An MIP formulation for the optimal scheduling of industrial supply chains is presented by Amaro and Barbosa-Póvoa (2008). Liao and Rittscher (2007) develop a multiobjective programming model, integrating supplier selection, procurement lot sizing, and carrier selection decisions for a single purchasing item.

In order to more precisely predict the supply chain performance, the stochastic nature of demand, processing, and transportation times (e.g. Sawik, 1977), and of failures and other important events should be explicitly treated. Some recent stochastic modeling and optimization approaches for manufacturing and supply chain management are presented in Gershwin et al. (2002) and Shanthikumar et al. (2003). On the other hand, a review and discussion of the literature on supply chain management by Thomas and Griffin (1996) is concluded with the need for research that addresses supply chain issues at an operational level and that uses deterministic rather than stochastic models. A similar conclusion can also be found in a more recent paper on supply chain scheduling by Z.-L. Chen and Hall (2007).

The material presented in this chapter is based on the results published by Sawik (2009c). The proposed approach can be modified and enhanced (see Chapters 13 and 14) to consider additional features of customer-driven supply chains, such as

- the manufacturing and supplies of both product-specific and common parts, when products contain parts of different types.
- the manufacturing and supplies by different suppliers at different locations and with different shipping capacity limits, transportation times, and costs.

In practice supply chain schedule needs to be updated whenever the customer orders are modified or new orders arrive, the supplies of parts are delayed, or the assembly of finished products is subject to losses. The proposed approaches and MIP models can also be applied for reactive scheduling on a rolling horizon basis (see Chapter 7), in response to various disruptions in a supply chain.

EXERCISES

12.1 Consider a set of customer orders such that problem *MME* has no feasible solution, that is, some orders need to be delayed to complete the set of orders during the planning horizon. In this case, to minimize the total input and output inventory level, problem *MME* of minimizing the maximum earliness can be replaced by the following problem *MMET* of minimizing the sum of maximum earliness and maximum tardiness, $E_{\max} + T_{\max}$,

Model MMET: *Minimization of the maximum earliness and maximum tardiness of customer orders*

Minimize

$$E_{\max} + T_{\max}$$

subject to (12.11)–(12.13), (12.15), and

$$\sum_{t \in T:\, t \geq a_j} x_{jt} = 1;\, j \in J$$

$$(t - d_j)x_{jt} \leq T_{\max};\, j \in J,\, t \in T: t \geq a_j$$

$$x_{jt} \in \{0, 1\};\, j \in J,\, t \in T: t \geq a_j$$

$$T_{\max} \geq 0.$$

How should the MIP model SMSA be modified to account for the delayed customer orders?

(**a**) What changes are required in models *MME* and *MMET*, when customer-requested due date d_j is replaced by customer-requested due date window, $[d_j - \delta_{j\max}, d_j + \delta_{j\max}]$, during which the customer must receive the order, where $\delta_{j\max} > 0$ is the maximum deviation from the requested due date?

12.2 Assuming that M production lines of the supplier in model *SMSA* are not identical, replace parameters q_k, σ by q_{kl}, σ_l, $l = 1, \ldots M$, and integer set-up and start-up variables y_{kt}, z_{kt} by binary variables, respectively, y_{klt}, z_{klt}, $l = 1, \ldots M$ and formulate the modified mixed integer program for the integrated scheduling of manufacturing, supply, and assembly.

12.3 Formulate coordinating constraints (12.33) for the integrated scheduling of manufacturing, supply, and assembly.

(**a**) In the case of a supply chain with two suppliers and one producer.

(**b**) In the case of a supply chain with one supplier and two producers.

12.4 In the case of nonidentical production lines at suppliers (Exercise 12.2), formulate coordinating constraints (12.33) for the two cases from Exercise 12.3.

12.5 Compare the results of computational examples illustrated in Figures 12.3 to 12.6, and explain why for the objective function (12.17) the number of production line start-ups is smaller than for the objective (12.18), in contrast to the total supply chain inventory, which is smaller for the latter objective, except for the case with largest shipment and start-up unit costs.

Chapter 13

Coordinated Scheduling in Supply Chains with Assignment of Orders to Suppliers

13.1 INTRODUCTION

This chapter deals with coordinated multi-objective scheduling of manufacturing and delivery of parts and assembly of finished products in a customer-driven supply chain. The supply chain consists of multiple suppliers (manufacturers) of parts, a single producer of finished products, and a set of customers who generate final demand for the products. Each supplier has a number of identical production lines in parallel for manufacturing of parts, and the producer has a flexible assembly line (FAL) for assembly of products. In the supply chain, the following static and deterministic coordinated scheduling problem is considered. Given a set of customer orders, the problem objective is to determine which orders are to be provided with parts by each supplier, find a schedule for manufacturing parts at each supplier and for the delivery of parts from each supplier to the producer, and find a schedule for product assembly for each order by the producer, such that a high customer service level is achieved and the total cost of supply chain inventory holding, production line start-ups, and the part shipments is minimized. Selection of the part supplier for each customer order is combined with due date setting for some orders, subject to available capacity of the suppliers and the producer, to maximize the number of orders that can be completed by customer-requested due dates.

An integrated (monolithic) approach, where manufacturing, supply, and assembly schedules are determined simultaneously is compared with a hierarchical approach. The hierarchical approach is based on the lexicographic optimization and decomposition of the complex problem of multi-objective production, manufacturing, and supply scheduling into a hierarchy of much simpler decision-making problems. In

Scheduling in Supply Chains Using Mixed Integer Programming. By Tadeusz Sawik

the hierarchical approach, first a supplier of parts is selected for each order, the maximal subset of orders that can be completed by the customer requested due dates is found, and for the remaining orders delayed due dates are determined to satisfy capacity constraints. Then the assignment of orders to planning periods over the horizon is found to minimize the holding cost of finished product inventory of the customer orders completed before their due dates. Finally, the manufacturing and delivery schedules of the required parts are determined independently for each supplier to minimize the total cost of holding the inventory of parts, of production line start-ups, and of parts shipments from the supplier to the producer. The objective function integrates both the supply chain performance and the customer service level.

The following time-indexed IP and MIP models are presented in this chapter:

INT for integrated, multi-objective scheduling of manufacturing, supply, and assembly

INT_λ for integrated, multi-objective scheduling of manufacturing, supply, and assembly using the weighted-sum approach

INT^λ for integrated, multi-objective scheduling of manufacturing, supply, and assembly using the reference point approach

INT(1), INT(2), INT(3) for integrated, multi-objective scheduling of manufacturing, supply, and assembly using the lexicograhic approach

RCA for rough-cut capacity allocation

NDS for nondelayed scheduling of customer orders

SMD(i) for scheduling of manufacturing and delivery of parts for supplier i

Numerical examples modeled after real-world coordinated scheduling in a customer-driven supply chain in the electronics industry and computational results are reported in Section 13.7.

13.2 PROBLEM DESCRIPTION

The supply chain under consideration (Fig. 13.1) consists of m manufacturers/suppliers of parts, a single producer where finished products are assembled according

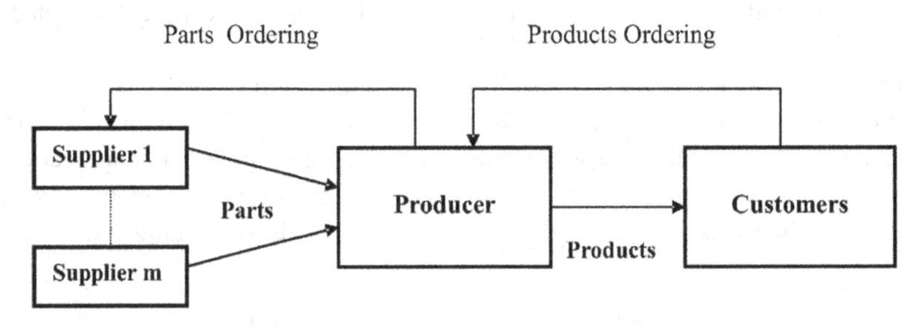

Figure 13.1 A customer-driven supply chain with multiple suppliers.

to customer orders, and a set of customers who generate final demand for the products. In the supply chain, various types of products are assembled by the producer using different part types supplied by the manufacturers (for the notation used, see Table 13.1). Let J be the set of customer orders known ahead of time, K the set of

Table 13.1 Notation: Multi-Objective Scheduling in a Supply Chain with Multiple Suppliers

Indices		
g	=	assembly stage, $g \in G$
i	=	supplier, $i \in I = \{1, \ldots, m\}$
j	=	customer order, $j \in J = \{1, \ldots, n\}$
k	=	part type $k \in K$
l	=	product type, $l \in L$
t	=	planning period, $t \in T = \{1, \ldots, h\}$
Input parameters		
a_j, d_j, D_j, n_j	=	ready date, customer-requested due date, committed due date, size of order j
b_{kl}	=	unit requirement of product type l for part type k
c_{gt}	=	capacity of each machine in assembly stage g in period t
$C_g(t, d)$	=	$m_g \sum_{\tau \in T:\, t \le \tau \le d} c_{g\tau}$, cumulative capacity available in assembly stage g in periods t through d
$c^i(t)$	=	manufacturing and delivery capacity of supplier i in period t
$C^i(d)$	=	$\sum_{t \in T:\, t \le d} c^i(t)$, cumulative manufacturing and delivery capacity of supplier i until period d
e_i	=	the earliness of manufacturing start time for supplier i
H	=	length of each planning period
J_l	=	the subset of orders for product type l
m_g	=	number of parallel machines in assembly stage g
M_i	=	number of parallel production lines at supplier i
p_{gj}	=	processing time in assembly stage g of each product in order j
q_{ik}	=	processing time for one unit of part type k at supplier i
$\underline{S}, \overline{S}$	=	the minimum, maximum shipment capacity, respectively
$T^{(i)}$	=	$\{-e_i, \ldots, -1, 1, \ldots, h\}$, enlarged, ordered set of planning periods
σ_i	=	the start-up time of each production line at supplier i
ρ_{ik}	=	$\lfloor (H - \sigma_i)/q_{ik} \rfloor$ the one machine-period manufacturing worst-case rate of part type k at supplier i
ϱ_{gj}	=	$\lfloor c_g/p_{gj} \rfloor$ the one machine-period production rate at assembly stage g for products in order j
θ_i	=	the transportation time from supplier i to producer
$\varphi_{1ik}, \varphi_{2ik}, \varphi_{3k}, \varphi_{4j}$	=	inventory holding cost per unit of part stored by the supplier, transported to the producer, stored by the producer, per unit of finished product stored by the producer, respectively
ψ_{1ik}, ψ_{2i}	=	unit cost of production line start-up, part shipment, respectively

part types, L the set of product types, and J_l the subset of orders for product type $l \in L$. Each product type $l \in L$ requires $b_{kl} \geq 0$ parts of type $k \in K$.

Each order $j \in J$ is described by a triple (a_j, d_j, n_j), where a_j is the order ready date (e.g. the earliest period of material availability), d_j is the customer-requested due date (e.g., customer-required shipping date), and n_j is the size of the order (quantity of ordered products of a specific type).

The planning horizon consists of h planning periods (e.g., working days) of equal length H (e.g., hours or minutes) and let $T = \{1, \ldots, h\}$ be the set of planning periods. It is assumed that each customer order must be fully completed in exactly one planning period (e.g., during one day); however, this assumption can be easily relaxed. Large, multiperiod orders that require more than one planning period for completion can be split into single-period suborders to be allocated among consecutive planning periods, see Section 13.4.4.

The suppliers manufacture and deliver parts to the producer. The manufacture of parts by supplier $i \in I$, at the earliest can be started e_i periods in advance of period $t = 1$, when the producer can start the assembly of products. The earliness e_i of manufacturing start time for supplier i represents the maximum manufacturing capacity of supplier i available in advance of the start of the assembly schedule. Each supplier i has M_i identical production lines in parallel, capable of manufacturing all part types and let q_{ik} be the processing time for part type $k \in K$ at supplier i. In each period at most one part type can be manufactured on each production line. When a production line switches from one part type to another, a start-up time should be considered at the beginning of the period. The start-up times are sequence-independent and are assumed to be equal for all part types. Let σ_i be the start-up time of each production line of supplier i.

The manufactured parts are next transported to the producer at most once per period. Different part types can be shipped together so that a shipping cost arises only once per shipment. Each delivery shipment is limited by the minimum and the maximum capacity, respectively, of $\underline{S}$ and $\overline{S}$ parts. The transportation time of a shipment from supplier i to the producer is assumed to be constant and equal to θ_i periods. The parts manufactured by supplier i in period t can be shipped to the producer in the same period and can be used for the assembly of products in period $t + \theta_i$, at the earliest.

The producer has a flexible assembly line that consists of several assembly stages in series and each stage $g \in G$ is made up of $m_g \geq 1$ parallel identical machines. Each customer order requires processing in various assembly stages; however, some orders may bypass some stages. Let $p_{gj} \geq 0$ be the processing time in stage g of each product in order j, and $n_j\, p_{gj}$ is the total processing time required to complete order j in stage g.

The problem objective is to determine which orders are to be provided with parts by each supplier, find a schedule for parts manufacture at each supplier and for the delivery of the parts from each supplier to the producer, and find a schedule for product assembly for each order by the producer, such that a high customer service level is achieved and the total cost of supply chain inventory holding, production line start-ups, and part shipments is minimized. The selection of part supplier for each customer order is combined with a due date setting for some orders to maximize the number of orders that can be completed by customer-requested due dates.

13.3 CONDITIONS FOR FEASIBILITY OF CUSTOMER DUE DATES

In this section some necessary conditions are derived under which all customer due dates can be met (see also Sections 5.2.1 and 12.2.1). Let c_{gt} be the total processing time available in period t on each machine in stage g and $C_g(t, d) = m_g \sum_{\tau \in T:\, t \leq \tau \leq d} c_{g\tau}$ the total cumulative capacity available in stage g in periods t through d. A necessary condition to have a feasible production schedule with all the customer orders completed by the producer during the planning horizon is that for each assembly stage $g \in G$ the total demand on capacity does not exceed total available capacity, that is,

$$\max_{g \in G} \left(\frac{\sum_{j \in J} n_j p_{gj}}{C_g(1, h)} \right) \leq 1. \tag{13.1}$$

Furthermore, all the customer orders can be completed by the producer on or before their due dates if the following necessary condition holds (cf. critical load index (5.4))

$$PCR(d) = \max_{g \in G}(PCR_g(d)) \leq 1;\ d \in T \tag{13.2}$$

where $PCR_g\,(d)$ is the producer cumulative capacity ratio for due date d with respect to processing stage g

$$PCR_g(d) = \max_{t \in T: t \leq d} \left(\frac{\sum_{j \in J: t \leq a_j \leq d_j \leq d} n_j p_{gj}}{C_g(t, d)} \right);\ d \in T,\ g \in G \tag{13.3}$$

Notice that if $PCR_g\,(d) \leq 1$, then for any period $t \leq d$, the cumulative demand on capacity in stage g of all the orders with due dates not greater than d and ready dates not less than t (the numerator in Equation (13.3)) does not exceed the cumulative capacity available in this stage in periods t through d (the denominator in Equation (13.3)).

The amount c_{gt} of processing time available in period t on each machine in stage g must take into account the flow shop configuration of the production system and the transfer lot sizes to ensure that all products are completed in the same planning period at all upstream stages $1, \ldots, g-1$ and at all downstream stages $g+1, \ldots$ (e.g. (5.1)). As a result the available capacity c_{gt} is smaller than simply the available machine hours in period t.

Assuming that the producer capacity is time-invariant, that is, $c_{gt} = c_g \forall\, t \in T$, the cumulative capacity $C_g(t, d)$ in (13.3) can be calculated as

$$C_g(t, d) = c_g m_g (d - t + 1);\ g \in G,\ d, t \in T: t \leq d \tag{13.4}$$

A similar feasibility condition can be derived for the suppliers under the worst case scenario, when each production line starts up in every period.

Let $c^i(t)$ be the manufacturing and delivery capacity of supplier i in period t and let $C^i(d) = \sum_{t \in T: t \leq d} c^i(t)$ be the cumulative manufacturing and delivery capacity of supplier i until period d. In addition, denote by $J(i) \subset J$ the subset of customer orders for which the required parts are provided by supplier i, where $\bigcup_{i \in I} J(i) = J$ and $\bigcap_{i \in I} J(i) = \emptyset$.

The required parts for all customer orders can be manufactured and delivered by their due dates if the following necessary condition holds:

$$SCR = \max_{i \in I} SCR(i) \leq 1 \tag{13.5}$$

where $SCR(i)$ is the cumulative capacity ratio for supplier i:

$$SCR(i) = \max_{d \in T} \left(\frac{\sum_{k \in K} q_{ik} \sum_{l \in L} \sum_{j \in J_l \cap J(i): d_j \leq d} b_{kl} n_j}{C^i(d)} \right) \tag{13.6}$$

If $SCR \leq 1$, then for all due dates d, the cumulative demand on capacity necessary to manufacture and deliver all parts required for the customer orders due not later than d (the numerator in Equation (13.6)) does not exceed the suppliers' cumulative capacity available by period d (the denominator in Equation (13.6)).

For the worst-case scenario, where each production line starts up in every period, $c^i(t)$ can be calculated as

$$c^i(t) = \begin{cases} 0 & \textit{if } t \leq -e_i + \theta_i \\ M_i(H - \sigma_i) & \textit{if } t > -e_i + \theta_i, \end{cases} \tag{13.7}$$

that is, no parts from supplier i can be delivered earlier than θ_i periods after the earliest period $-e_i$ of manufacturing.

If the allocation of customer orders among the suppliers for parts provision is not known beforehand, an approximate cumulative capacity ratio can be defined as

$$\tilde{SCR} = \max_{d \in T} \left(\frac{\sum_{k \in K} \tilde{q}_k \sum_{l \in L} \sum_{j \in J_l: d_j \leq d} b_{kl} n_j}{\sum_{i \in I} C^i(d)} \right) \tag{13.8}$$

where $\tilde{q}_k$ is the average processing time required to manufacture one unit of part type k.

If all customer orders were continuously allocated among consecutive time periods so that all periods could be filled exactly to their capacities, the necessary conditions (13.2) and (13.5) could become sufficient for all orders to be completed by their due dates. However, owing to the discrete nature of indivisible, in particular single-period, customer orders that cannot be continuously allocated among consecutive periods, it is possible that some planning periods will not be filled exactly to their capacities. As a result the necessary conditions (13.2) and (13.5) are not sufficient for all orders to be scheduled by their due dates.

The necessary conditions for feasibility of the customer-requested due dates are checked at the top level of the proposed hierarchical approach (see Section 13.6), where the primary objective is to maximize customer service level. If the feasibility conditions are satisfied for all requested due dates, then no tardy order occurs. Otherwise, to reach feasibility, later due dates must be committed for some customer orders to minimize the number of tardy orders, that is, the orders with committed due dates later than the requested due dates.

13.4 COORDINATED SUPPLY CHAIN SCHEDULING: AN INTEGRATED APPROACH

In this section an integrated approach and multi-objective MIP monolithic model *INT* is presented for the coordinated scheduling of manufacturing and supply of parts and assembly of products in a customer-driven supply chain.

13.4.1 Decision Variables

Let $T^{(i)}$ be the enlarged, ordered set of planning periods with e_i periods added before period $t = 1$, that is,

$$T^{(i)} = \{-e_i, \ldots, -1\} \cup T = \{-e_i, \ldots, -1, 1, \ldots, h\}.$$

The various decision variables (Table 13.2) used in model *INT* to coordinate different types of schedules in the supply chain are additionally explained below.

- Manufacturing schedule for each supplier $i \in I$:
  - the set-up variable y_{ikt} that describes a state of manufacturing system of each supplier in each planning period, that is, the number of production lines of supplier i set up to manufacture part type k in period $t \in T^{(i)}$,
  - the start-up variable z_{ikt} that represents the number of production lines of supplier i which are started up to manufacture part type k in period $t \in T^{(i)}$, that is, the number of lines of supplier i, set up in period $t \in T^{(i)}$ for part type k, which has not been set up for this part type in the previous period $prev(t, T^{(i)})$. The start-up variable can take a positive value only if the corresponding set-up variable has in period t higher value than in period $prev(t, T^{(i)})$, that is,

$$z_{ikt} = \max\{0, y_{ikt} - y_{ik,prev(t,T^{(i)})}\}.$$

Table 13.2 Decision Variables: Multi-Objective Scheduling in a Supply Chain with Multiple Suppliers

Model INT

r_{ikt} = manufacturing lot at suplier i of part type k in period t (manufacturing lot sizing variable)

s_{ikt} = delivery lot from supplier i of part type k in period t (delivery lot sizing variable)

u_{it} = 1, if delivery of parts from supplier i is scheduled for period t, otherwise $u_{it} = 0$ (delivery period selection variable)

y_{ikt} = number of parallel production lines of supplier i set up for manufacturing part type k in period t (production line set-up variable)

z_{ikt} = number of parallel production lines of supplier i started up in period t to manufacture part type k after processing another part type (production line start-up variable)

X_{ijt} = 1, if customer order j is supplied with parts by supplier i and assigned for processing to planning period t; otherwise $X_{ijt} = 0$ (order-to-period and to supplier assignment variable)

Model RCA

v_{ij} = 1, if order j is accepted with its customer-requested due date and selected for supplier i; $v_{ij} = 0$ if order j needs delaying to be accepted (order acceptance and supplier selection variable)

w_{ijt} = 1, if delayed order j is selected for supplier i and is assigned the adjusted due date t, $(t > d_j)$; otherwise $w_{ijt} = 0$ (due date assignment and supplier selection variable)

Model NDS

x_{jt} = 1, if customer order j is assigned to planning period t; otherwise $x_{jt} = 0$ (order-to-period assignment variable)

- the manufacturing lot sizing variable r_{ikt} that represents the lot size of each part type manufactured by each supplier in each planning period, that is, the number of parts type k manufactured in period $t \in T^{(i)}$ by supplier i.

• Delivery schedule for each supplier $i \in I$:

  - the delivery lot sizing variable s_{ikt} that represents the lot size of each part type delivered from each supplier to the producer in each planning period, that is, the number of parts type k delivered in period $t \in T^{(i)}$ from supplier i,
  - the delivery period selection variable $u_{it} = 1$, if delivery of parts from supplier i is scheduled for period $t \in T^{(i)}$, otherwise $u_{it} = 0$.

• Assembly schedule for customer orders:

  - the order assignment and supplier selection variable $X_{ijt} = 1$, if customer order j is supplied with parts by supplier i and assigned for processing to planning period $t \in T$; otherwise $X_{ijt} = 0$.

13.4.2 Objective Functions: Customer Service Level vs. Total Cost

The objective of the coordinated scheduling is to allocate customer orders among the suppliers of parts, to determine for each supplier a schedule for parts manufacture and a schedule of supplying the parts to the producer, and to find for the producer an assembly schedule for the products ordered, so as to maximize customer service level or equivalently to minimize the number of tardy orders, as a primary optimality criterion. Simultaneously, to achieve a low unit cost, the total supply chain inventory holding cost and cost of production line start-ups and part shipments are minimized. Thus, the secondary objective functions are minimization of the total inventory holding cost and minimization of production line start-ups and part shipment costs. The three objective functions, numbered according to their decreasing importance, are defined below.

The Number of tardy orders

$$f_1 = \sum_{i \in I} \sum_{j \in J} \sum_{t \in T: t > d_j} X_{ijt}. \tag{13.9}$$

The cost of inventory holding

$$\begin{aligned} f_2 = & \sum_{i \in I} \sum_{k \in K} \sum_{t \in T} \sum_{\tau \in T^{(i)}: \tau \leq t} \varphi_{1ik}(r_{ik\tau} - s_{ik,next(\tau,T^{(i)},\theta_i)}) \\ & + \sum_{i \in I} \sum_{k \in K} \sum_{t \in T: t \geq next(-e_i, T^{(i)}, \theta_i)} \varphi_{2ik} \theta_i s_{ikt} \\ & + \sum_{i \in I} \sum_{k \in K} \sum_{t \in T} \varphi_{3k} \left(\sum_{\tau \in T^{(i)}: \tau \leq t} s_{ik\tau} - \sum_{l \in L} \sum_{j \in J_l} \sum_{\tau=1}^{t} b_{kl} n_j X_{ij\tau} \right) \\ & + \sum_{i \in I} \sum_{t \in T} \sum_{\tau=1}^{t} \sum_{j \in J: d_j > t} \varphi_{4j} n_j X_{ij\tau} \end{aligned} \tag{13.10}$$

where φ_{1ik}, φ_{2ik}, φ_{3k}, and φ_{4j} are the unit holding costs of parts stored at supplier i, of parts in transit from supplier i to the producer, of parts stored at producer, and of finished products stored at the producer, respectively.

The cost of production line start-ups and part shipments

$$f_3 = \sum_{i \in I} \sum_{k \in K} \sum_{t \in T^{(i)}} \psi_{1ik}(\sigma_i / q_{ik}) z_{ikt} + \sum_{i \in I} \sum_{t \in T^{(i)}} \psi_{2i} \theta_i u_{it}, \tag{13.11}$$

where ψ_{1ik}, ψ_{2i} are the unit start-up and shipment costs, respectively.

For each supplier i the start-up cost is measured by the total production of parts lost due to line start-ups, whereas the fixed shipment cost is proportional to the total transportation time.

Notice that the total cost of transporting parts from suppliers to the producer consists of variable in-transit holding cost and a fixed shipment cost.

13.4.3 A Multi-Objective Monolithic Model

The time-indexed MIP monolithic model *INT* for the integrated multi-objective scheduling of manufacturing, supply, and assembly is presented below.

Model INT: *Integrated, Multi-Objective Scheduling of Manufacturing, Supply, and Assembly*

$$\min\ f_1, f_2, f_3 \tag{13.12}$$

subject to

1. *Order-to-Period and Supplier Assignment Constraints*:
   - each customer order is assigned to exactly one supplier and to exactly one planning period,

$$\sum_{i \in I} \sum_{t \in T: a_j \le t} X_{ijt} = 1;\ j \in J \tag{13.13}$$

2. *Producer Capacity Constraints*:
   - in every period the demand on capacity at each assembly stage cannot be greater than the maximum available capacity in this period,

$$\sum_{i \in I} \sum_{j \in J} n_j p_{gj} X_{ijt} \le m_g c_{gt};\ g \in G,\ t \in T \tag{13.14}$$

3. *Manufacturing Line Set-up and Start-up Constraints for Each Supplier*:
   - in every period total number of production lines set up for manufacturing different part types is not greater than total number M_i of available lines,
   - all production lines set up for part type k in the beginning period should be started up to manufacture this part type,
   - in every period t, the number of production lines started up for part type k cannot be less than the difference between the number of lines set up for this part type in period t and in the previous period $prev(t, T^{(i)})$,
   - in every period t, the number of production lines started up for part type k cannot be greater than the number of lines set up for part type k in this period and cannot be greater than the number of lines set up for the other part types or idle in the previous period $prev(t, T^{(i)})$,

$$\sum_{k \in K} y_{ikt} \le M_i;\ i \in I,\ t \in T^{(i)} \tag{13.15}$$

$$z_{i,k,-e_i} = y_{i,k,-e_i};\ i \in I,\ k \in K \tag{13.16}$$

$$z_{ikt} \ge y_{ikt} - y_{ik,prev(t,T^{(i)})};\ i \in I,\ k \in K,\ t \in T^{(i)}: t > -e_i \tag{13.17}$$

$$z_{ikt} \le y_{ikt};\ i \in I,\ k \in K,\ t \in T^{(i)} \tag{13.18}$$

$$z_{ikt} \le M_i - y_{ik,prev(t,T^{(i)})};\ i \in I,\ k \in K,\ t \in T^{(i)}: t > -e_i \tag{13.19}$$

4. *Manufacturing Capacity Constraints for Each Supplier*:
   - in every period t the production volume of part type k cannot be greater than the maximum volume corresponding to the capacity assigned to part type k in this period,

$$r_{ikt} \leq \lfloor (H - \sigma_i)/q_{ik} \rfloor z_{ikt} + \lfloor H/q_{ik} \rfloor (y_{ikt} - z_{ikt});$$
$$i \in I, k \in K, t \in T^{(i)} \qquad (13.20)$$

   where $\lfloor \cdot \rfloor$ is the greatest integer not greater than $\cdot$.

5. *Part Manufacturing and Delivery Constraints for Each Supplier*:
   - for each part type the cumulative delivery by period t cannot be greater than the cumulative manufacturing of this part type by period $prev(t, T^{(i)}, \theta_i)$, that is, manufactured not later than θ_i periods before period t,
   - the total delivery of each part type k from each supplier i is equal to its total production over the entire plannning horizon,
   - parts can be delivered only in periods scheduled for delivery and each shipment is limited by its minimum and maximum capacity, $\underline{S}$ and $\overline{S}$, respectively,
   - no delivery from supplier i can be scheduled before period $next(-e_i, T^{(i)}, \theta_i)$, that is, no delivery can be scheduled earlier than θ_i periods after the earliest period $-e_i$ of manufacturing,

$$\sum_{\tau \in T^{(i)}: \tau \leq t} s_{ik\tau} \leq \sum_{\tau \in T^{(i)}: \tau \leq prev(t, T^{(i)}, \theta_i)} r_{ik\tau};\ i \in I, k \in K, t \in T^{(i)}:$$
$$next(-e_i, T^{(i)}, \theta_i) \leq t < h \qquad (13.21)$$

$$\sum_{t \in T^{(i)}} s_{ikt} = \sum_{t \in T^{(i)}: t \leq prev(h, T^{(i)}, \theta_i)} r_{ikt};\ i \in I, k \in K \qquad (13.22)$$

$$s_{ikt} \leq \overline{S} u_{it};\ i \in I, k \in K, t \in T^{(i)} \qquad (13.23)$$

$$\sum_{k \in K} s_{ikt} \leq \overline{S};\ i \in I, t \in T^{(i)} \qquad (13.24)$$

$$\sum_{k \in K} s_{ikt} \geq \underline{S} u_{it};\ i \in I, t \in T^{(i)} \qquad (13.25)$$

$$s_{ikt} = 0;\ i \in I, k \in K, t \in T^{(i)}: t < next(-e_i, T^{(i)}, \theta_i) \qquad (13.26)$$

$$u_{it} = 0;\ i \in I, t \in T^{(i)}: t < next(-e_i, T^{(i)}, \theta_i) \qquad (13.27)$$

6. *Part Demand Satisfaction Constraints*:
   - for each part type k the cumulative deliveries from supplier i by period t cannot be less than the cumulative demand for this part type by this period,

– total delivery of each part type k from all suppliers is equal to the total demand for this part type,

$$\sum_{\tau \in T^{(i)}:\tau \le t} s_{ik\tau} \ge \sum_{l \in L} \sum_{j \in J_l} \sum_{\tau=1}^{t} b_{kl} n_j X_{ij\tau};\; i \in I,\, k \in K,\, t \in T^{(i)} : t < h \tag{13.28}$$

$$\sum_{i \in I} \sum_{t \in T^{(i)}} s_{ikt} = \sum_{l \in L} \sum_{j \in J_l} b_{kl} n_j;\; k \in K \tag{13.29}$$

7. *Coordinating Constraints*:

– the cumulative requirement for each part type k from each supplier i of the orders assigned to periods 1 through t and to supplier i (the right-hand side of inequality (13.30)) cannot be greater than the total production at supplier i of this part type by period $prev(t, T^{(i)}, \theta_i)$, that is, manufactured at i not later than θ_i periods before period t (the left-hand side of inequality (13.30))

$$\lfloor H/q_{ik} \rfloor \sum_{\tau \in T^{(i)}:\tau \le prev(t, T^{(i)}, \theta_i)} (y_{ik\tau} - z_{ik\tau})$$

$$+ \lfloor (H - \sigma_i)/q_{ik} \rfloor \sum_{\tau \in T^{(i)}:\tau \le prev(t, T^{(i)}, \theta_i)} z_{ik\tau} \ge \sum_{l \in L} \sum_{j \in J_l} \sum_{\tau=1}^{t} b_{kl} n_j X_{ij\tau};$$

$$i \in I,\, k \in K,\, t \in T^{(i)} : t \ge next(-e_i, T^{(i)}, \theta_i) \tag{13.30}$$

8. *Nonnegativity and Integrality Conditions*:

$$r_{ikt} \ge 0;\; i \in I,\, k \in K,\, t \in T^{(i)} \tag{13.31}$$

$$s_{ikt} \ge 0;\; i \in I,\, k \in K,\, t \in T^{(i)} \tag{13.32}$$

$$u_{it} \in \{0, 1\};\; i \in I,\, t \in T^{(i)} \tag{13.33}$$

$$y_{ikt} \ge 0,\ integer;\; i \in I,\, k \in K,\, t \in T^{(i)} \tag{13.34}$$

$$z_{ikt} \ge 0,\ integer;\; i \in I,\, k \in K,\, t \in T^{(i)} \tag{13.35}$$

$$X_{ijt} \in \{0, 1\};\; i \in I, j \in J,\, t \in T : a_j \le t. \tag{13.36}$$

The coordinating constraint (13.30) has been introduced to strengthen the mixed integer program *INT* as it directly links the assignment of production lines to part types over the planning horizon (variables y_{ikt} and z_{ikt}) and the assignment of customer orders to planning periods and to suppliers (variable X_{ijt}). Thus, coordinating constraint (13.30) directly coordinates the suppliers' schedules and the producer's schedule. Coordinating constraint (13.30) is implied by constraints (13.20), (13.21), and (13.28), and any feasible solution that satisfies (13.30) also satisfies (13.20), (13.21), and (13.28).

Note that in model *INT* all parts required for each customer order are assumed to be provided by exactly one supplier (0-1 order assignment variable X_{ijt}). In practice, parts required for a customer order may be partially provided by different suppliers. Then the corresponding binary assignment variable X_{ijt} should be redefined as a non-negative order allocation variable denoting the fraction of all required parts purchased from supplier i to fulfill customer order j assigned to planning period t.

13.4.4 Multiperiod Orders

Model *INT* is capable of scheduling single-period orders only, where each order can be fully processed in a single time period. In some cases, in addition to single-period orders, also large-sized, divisible orders should be considered, where each order cannot be completed in one period and must be split into single-period portions to be processed in a subset of consecutive time periods. The large orders are referred to as multiperiod orders (e.g., Sections 5.5, 6.2, 6.3).

In some practical cases, the two types of customer orders must be scheduled simultaneously. Denote by $J1 \subseteq J$, and $J2 \subseteq J$ the subset of indivisible, single-period orders and divisible, multiperiod orders, respectively, where $J1 \cup J2 = J$, and $J1 \cap J2 = \emptyset$.

The MIP model *INT* can be enhanced to simultaneously schedule both the single and the multiperiod orders. In addition to order assignment and supplier selection binary variable X_{ijt}, a new order allocation continuous variable ζ_{ijt} must be introduced, where $\zeta_{ijt} \in [0, 1]$ denotes a fraction of a multiperiod order j assigned to period t and provided with parts by supplier i.

Model *INT* can be enhanced as follows. In definition (13.10) of f_2 and in constraints (13.13), (13.14), (13.28), and (13.30), the binary assignment variable X_{ijt} should be replaced with continuous allocation variable ζ_{ijt}. Then the modified (13.13) will ensure that each order is assigned to exactly one supplier and is completed during the planning horizon. In addition, the following new constraints should be added to model *INT*:

Order-to-Period and Supplier Assignment Constraints:

- each single-period order is assigned to exactly one planning period and to exactly one supplier,

$$\sum_{i \in I} \sum_{t \in T: a_j \leq t} X_{ijt} = 1; \; j \in J1,$$

- each multiperiod order is assigned to a subset of consecutive planning periods and a single supplier,

$$X_{ij[(\tau_1+\tau_2)/2]} \geq X_{ij\tau_1} + X_{ij\tau_2} - 1; \; i \in I, j \in J2, \tau_1, \tau_2 \in T: a_j \leq \tau_1 < \tau_2$$

Order Allocation Constraints:

- each single-period order must be completed in a single period,

$$X_{ijt} = \zeta_{ijt}; \; i \in I, j \in J1, t \in T: a_j \leq t,$$

- each multiperiod order is allocated among all the periods that are selected for its assignment,

$$\zeta_{ijt} \le X_{ijt};\ i \in I, j \in J2, t \in T : a_j \le t,$$

Nonnegativity Conditions:

$$\zeta_{ijt} \in [0, 1];\ i \in I, j \in J, t \in T : a_j \le t.$$

While the focus in this chapter is on single-period customer orders, the ideas presented in the sequel may as well be applied for scheduling multiperiod orders.

13.5 SELECTED MULTI-OBJECTIVE SOLUTION APPROACHES

This section provides a brief description of selected approaches that can be used to determine the nondominated solution set of the multi-objective mixed integer program *INT*.

13.5.1 Weighted-Sum Program

The nondominated solution set of the multi-objective mixed integer program *INT* can be partially determined by the parameterization on λ of the following weighted-sum program.

Model INT_λ

$$\min \sum_{\iota=1}^{3} \lambda_\iota f_\iota \qquad (13.37)$$

subject to (13.13)–(13.36), where $\lambda_1 > \lambda_2 > \lambda_3 > 0$, $\lambda_1 + \lambda_2 + \lambda_3 = 1$.

It is well known, however, that the nondominated solution set of a multi-objective integer program such as INT_λ cannot be fully determined even if the complete parameterization on λ is attempted (e.g., Steuer, 1986). To compute unsupported nondominated solutions, some upper bounds on the objective functions should be added to INT_λ (e.g., Alves and Climaco, 2007).

13.5.2 Reference Point-Based Scalarizing Program

Let $\underline{f} = (\underline{f}_1, \underline{f}_2, \underline{f}_3)$ be a reference point in the criteria space such that $\underline{f}_\iota < f_\iota, \forall \iota$ for all feasible solutions satisfying (13.13) to (13.36), and denote by ε a small positive value. The nondominated solution set of the multi-objective program *INT* can be found by the parameterization on λ of the following mixed integer program INT^λ.

Model INT^{λ}

$$\min\left\{\delta+\varepsilon\sum_{\iota=1}^{3}f_{\iota}\right\} \tag{13.38}$$

subject to (13.13)–(13.36) and

$$\lambda_{\iota}(f_{\iota}-\underline{f}_{\iota})\leq\delta;\quad \iota=1,2,3 \tag{13.39}$$

$$\delta\geq 0. \tag{13.40}$$

The MIP model INT^{λ} is based on the augmented $\lambda-$weighted Chebyshev metric $\min_{\iota=1,2,3}\{\lambda_{\iota}|f_{\iota}-\underline{f}_{\iota}|+\varepsilon(f_1+f_2+f_3)\}$, where $\lambda_{\iota}\geq 0, \forall\iota$ and $\sum_{\iota=1}^{3}\lambda_{\iota}=1$.

13.5.3 Lexicographic Approach

Considering the relative importance of the three objective functions (13.9) to (13.11), the multi-objective mixed integer program *INT* can replaced with a sequence *INT(1)*, *INT(2)*, *INT(3)* of the following three single-objective mixed integer programs to be solved subsequently.

Model *INT* (ι), ι = 1, 2, 3

$$\min\ f_{\iota} \tag{13.41}$$

subject to (13.13)–(13.36) and

$$f_l=f_{\iota}^{*};\ l<\iota:\iota>1, \tag{13.42}$$

where f_{ι}^{*} is the optimal solution value to the mixed integer program $INT(\iota)$, $\iota=1,2$.

The hierarchical decomposition scheme for the multi-objective coordinated supply chain scheduling proposed in the next section is based on the lexicographic approach.

13.6 COORDINATED SUPPLY CHAIN SCHEDULING: A HIERARCHICAL APPROACH

In this section the MIP formulations are presented for hierarchical scheduling of customer orders, manufacturing, and supplies of parts. The proposed hierarchical approach can be applied within a pull planning framework in customer-driven supply chains. The hierarchical decomposition scheme is based on the lexicographic approach (see Section 13.5.3). However, in addition to the vertical decomposition driven by relative importance of the objective functions, it also utilizes the horizontal decomposition enabled by the allocation of customer orders among suppliers. The

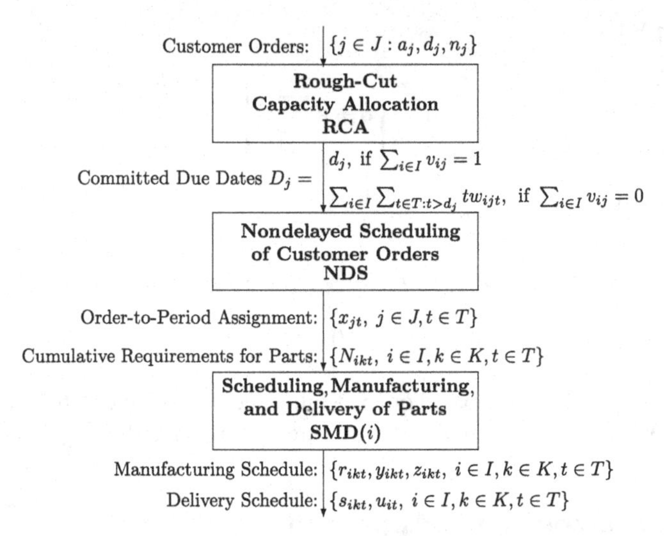

Figure 13.2 Coordinated scheduling of customer orders and manufacturing and delivery of parts: a hierarchical approach.

hierarchical framework consists of the following three decision-making problems to be solved sequentially (Fig. 13.2):

1. Rough-cut capacity allocation, *RCA*
2. Nondelayed scheduling of customer orders, *NDS*
3. Scheduling manufacturing and delivery of parts from each supplier, *SMD*(i).

Unlike the detailed formulations *INT(1)* and *INT(2)* of the lexicographic approach, the top-level problem *RCA* accounts for the cumulative capacity constraints of the producer, and both the top- and the middle-level problems *RCA* and *NDS* contain the worst-case capacity constraints of the suppliers, where all production lines are assumed to be started up in each planning period. However, a feasible solution to *RCA* that satisfies the cumulative capacity constraints of the producer would not necessary lead to a feasible, nondelayed schedule of customer orders at the middle-level *NDS*, if greater capacity of the suppliers were allowed at the top level. This is due to the fact that some periods may not be filled exactly to their capacities since the discrete and, in particular, single-period orders cannot be continuously allocated among the consecutive periods.

Furthermore, to decrease the computational effort required for scheduling customer orders for the producer, the top-level problem *RCA* sets delayed due dates for some orders to meet conditions (13.2) and (13.5), and by this the middle-level problem is reduced to a nondelayed scheduling problem. This coincides with typical industrial practice, where the order acceptance and due date setting decisions made by the sales departments precede any scheduling decisions. However, in contrast to the common practice, where the sales and the production departments often work independently, the committed due dates determined at the top level must satisfy the available capacity of both the suppliers and the producer.

The objective of the top-level problem *RCA* is to select a supplier of parts for each customer order, to determine the maximal subset of orders that can be completed by the customer-requested due dates, and to update (delay) due dates for the remaining orders to meet the capacity constraints.

Given the supplier of parts and the due date for each customer order, the objective of the middle-level problem *NDS* is to determine a feasible, nondelayed assignment of orders to planning periods over the planning horizon to minimize the finished product inventory holding cost of customer orders completed before their due dates.

Finally, given the cumulative requirements for all part types from each supplier, the objective of the base-level problem *SMD*(i) is to determine the integrated manufacturing and delivery schedule of parts independently for each supplier such that the weighted cost of manufacturing, parts supply, and holding the inventory of parts in the supply chain is minimized.

The time-indexed IP and MIP models *RCA*, *NDS*, and *SMD*(i) are presented below (for notation used and definitions of the decision variables, see Tables 13.1 and 13.2).

13.6.1 Rough-Cut Capacity Allocation

The time-indexed binary program *RCA* contains the following two decision variables (see also, model *DDS* in Section 5.3).

- Order acceptance and supplier selection variable: $v_{ij} = 1$, if order j is accepted with its customer-requested due date and assigned to supplier i; $v_{ij} = 0$ if order j needs to be delayed to be accepted,
- Due date assignment and supplier selection variable: $w_{ijt} = 1$, if order j unacceptable with requested due date (with $\sum_{i \in I} v_{ij} = 0$) is assigned a later due date t, $(t > d_j)$ and is selected for supplier i; otherwise $w_{ijt} = 0$.

Model RCA: *Rough-Cut Capacity Allocation*

Minimize the number of tardy orders

$$f_1 = \sum_{j \in J} \left(1 - \sum_{i \in I} v_{ij} \right) \qquad (13.43)$$

subject to

1. *Supplier Selection, Order Acceptance, and Due Date Setting Constraints*:
   - each customer order is assigned to exactly one supplier and is either accepted with its requested due date or is assigned a later due date to reach production and delivery schedule feasibility,

- no customer order with a due date earlier than the first delivery of parts from a supplier can be assigned to that supplier,

$$\sum_{i \in I} v_{ij} + \sum_{i \in I} \sum_{t \in T: t > d_j} w_{ijt} = 1; \quad j \in J \tag{13.44}$$

$$v_{ij} = 0; \quad i \in I,\ j \in J: d_j < next(-e_i, T^{(i)}, \theta_i) \tag{13.45}$$

$$w_{ijt} = 0; \quad i \in I,\ j \in J, t \in T: d_j < t < next(-e_i, T^{(i)}, \theta_i) \tag{13.46}$$

2. *Producer Cumulative Capacity Constraints*:
   - for any period $t \le d$, the cumulative demand on capacity in stage g of all the customer orders accepted with requested (or adjusted) due dates not greater than d and ready dates (or requested due dates, respectively) not less than t must not exceed the cumulative capacity available in this stage in periods t through d

$$\sum_{i \in I} \sum_{j \in J: t \le a_j \le d_j \le d} n_j p_{gj} v_{ij} + \sum_{i \in I} \sum_{j \in J} \sum_{\tau \in T: t \le d_j < \tau \le d} n_j p_{gj} w_{ij\tau}$$

$$\le C_g(t, d); \quad d, t \in T,\ g \in G: t \le d \tag{13.47}$$

3. *Supplier Cumulative Capacity Constraints*:
   - the cumulative demand on manufacturing capacity for parts required in periods 1 through t for the customer orders assigned to supplier i cannot be greater than the available cumulative capacity of this supplier by period t,

$$\sum_{k \in K} \sum_{l \in L} b_{kl} q_{ik} \left(\sum_{j \in J_l: d_j \le t} n_j v_{ij} + \sum_{j \in J_l} \sum_{\tau \in T: d_j < \tau \le t} n_j w_{ij\tau} \right)$$

$$\le C^i(t); \quad i \in I,\ t \in T \tag{13.48}$$

4. *Integrality Conditions*:

$$v_{ij} \in \{0, 1\}; \quad i \in I,\ j \in J \tag{13.49}$$

$$w_{ijt} \in \{0, 1\}; \quad i \in I,\ j \in J,\ t \in T: t > d_j. \tag{13.50}$$

The assumption of the time-invariant capacity of the producer (13.4) and the worst-case capacity of the suppliers (13.7) allows the producer cumulative capacity constraints (13.47) and the supplier cumulative capacity constraints (13.48) to be replaced with the following constraints (13.51) and (13.52), (13.53), (13.54), respectively,

$$\sum_{i\in I}\left(\sum_{j\in J:t\le a_j\le d_j\le d} n_j v_{ij}/\varrho_{gj} + \sum_{j\in J}\sum_{\tau\in T:t\le d_j<\tau\le d} n_j w_{ij\tau}/\varrho_{gj}\right)$$

$$\le m_g(d-t+1);\quad d,t\in T,\ g\in G: t\le d, \tag{13.51}$$

$$\sum_{l\in L}\sum_{j\in J_l:d_j\le t} b_{kl} n_j v_{ij}/\rho_{ik} + \sum_{l\in L}\sum_{j\in J_l}\sum_{\tau\in T:d_j<\tau\le t} b_{kl} n_j w_{ij\tau}/\rho_{ik}$$

$$\le \sum_{\tau=-e_i}^{prev(t,T^{(i)},\theta_i)} y_{ik\tau};\quad i\in I, k\in K, t\in T^{(i)}: t\ge next(-e_i, T^{(i)}, \theta_i) \tag{13.52}$$

$$\sum_{k\in K} y_{ikt}\le M_i;\quad i\in I,\ t\in T^{(i)} \tag{13.53}$$

$$y_{ikt}\ge 0,\ integer;\quad i\in I, k\in K, t\in T^{(i)}, \tag{13.54}$$

where

$\varrho_{gj} = \lfloor c_g/p_{gj}\rfloor$ is the one machine-period production rate at assembly stage g for products in order j, and

$\rho_{ik} = \lfloor (H-\sigma_i)/q_{ik}\rfloor$ is the one machine-period worst-case manufacturing rate for part type k at supplier i.

The capacity constraints (13.47) and (13.48) ensure that each order $j\in J$ can be completed on or before its requested due date d_j (if $\sum_{i\in I} v_{ij}=1$) or on its delayed due date $t>d_j$ (if $\sum_{i\in I} v_{ij}=0$ and $\sum_{i\in I} w_{ijt}=1$), see (13.2), (13.3) and (13.5), (13.6) in Section 13.3. If conditions (13.2) and (13.5) hold for all customer-requested due dates, then problem *RCA* reduces to supplier selection for each order and the objective function f_1 (13.43) takes on a value of zero, since $\sum_{i\in I} v_{ij}=1\ \forall j\in J$ and $\sum_{i\in I} w_{ijt}=0\ \forall j\in J, t\in T: t>d_j$.

The solution to the binary program *RCA* determines the maximal subset $\{j\in J: \sum_{i\in I} v_{ij}=1\}$ of customer orders accepted with the customer-requested due dates d_j and the subset of remaining orders $\{j\in J: \sum_{i\in I} v_{ij}=0\}$ with updated (delayed) due dates. Denote by D_j, the requested or updated due date for each order $j\in J$, that is,

$$D_j=\begin{cases} d_j & \text{if } \sum_{i\in I} v_{ij}=1\\ \sum_{i\in I}\sum_{t\in T:t>d_j} t w_{ijt} & \text{if } \sum_{i\in I} v_{ij}=0. \end{cases}$$

Furthermore, a supplier has been selected for each order to provide the required parts and we let $J(i)=\{j\in J: v_{ij}=1 \text{ or } \sum_{t\in T:t>d_j} w_{ijt}=1\}\subset J$ be the subset of orders assigned to supplier $i\in I$.

13.6.2 Scheduling of Customer Orders

Given the updated due dates D_j, $j \in J$ and the allocation $J(i)$, $i \in I$ of the customer orders among the suppliers, the next decision step is to determine a nondelayed assembly schedule for the producer, that is, the assignment of customer orders to planning periods by the committed due dates (deadlines) over the horizon to minimize the finished products inventory holding cost.

The basic order assignment and supplier selection variable X_{ijt} used in the monolithic model *INT* is replaced below by the order assignment variable $x_{jt} = 1$ if customer order j is assigned to planning period t; otherwise $x_{jt} = 0$.

The time-indexed IP model *NDS* for a nondelayed scheduling of customer orders is presented below.

Model NDS: *Nondelayed Scheduling of Customer Orders*

Minimize the holding cost of finished product inventory

$$f_{20} = \sum_{t \in T} \sum_{\tau=1}^{t} \sum_{j \in J: d_j > t} \varphi_{4j} n_j x_{j\tau} \tag{13.55}$$

subject to

1. *Order-to-Period Nondelayed Assignment Constraints*:
   - each customer order is assigned to exactly one planning period not later than its due date,

$$\sum_{t \in T: a_j \le t \le D_j} x_{jt} = 1; \quad j \in J \tag{13.56}$$

2. *Producer Capacity Constraints*:
   - in every period the demand on capacity at each assembly stage cannot be greater than the maximum available capacity in this period,

$$\sum_{j \in J} n_j x_{jt} / \varrho_{gj} \le m_g; \quad g \in G, \ t \in T \tag{13.57}$$

3. *Supplier Cumulative Capacity Constraints*:
   - the cumulative demand on manufacturing capacity for parts required in periods 1 through t for the customer orders assigned to supplier i cannot be greater than the available cumulative capacity of this supplier by period t,

$$\sum_{l \in L} \sum_{j \in J_l \cap J(i)} \sum_{\tau \in T: a_j \le \tau \le D_j, \tau \le t} b_{kl} n_j x_{j\tau} / p_{ik} \le \sum_{\tau=-e_i}^{prev(t, T^{(i)}, \theta_i)} y_{ik\tau};$$

$$i \in I, k \in K, t \in T: t \ge next(-e_i, T^{(i)}, \theta_i) \tag{13.58}$$

$$\sum_{k \in K} y_{ikt} \le M_i; \quad i \in I, \ t \in T^{(i)} \tag{13.59}$$

4. *Integrality Conditions*:

$$x_{jt} \in \{0, 1\}; \quad j \in J, t \in T : a_j \leq t \leq D_j \tag{13.60}$$

$$y_{ikt} \geq 0, \textit{ integer}; \, i \in I, k \in K, t \in T^{(i)}. \tag{13.61}$$

In order to better utilize the manufacturing capacity of suppliers, the worst-case manufacturing rates ρ_{ik} can be replaced with the average manufacturing rates $\overline{\rho}_{ik}$, where $\overline{\rho}_{ik} = \lfloor (H - \sigma_i \eta_i / h)/q_{ik} \rfloor$ is the one machine-period average manufacturing rate for part type k at supplier i, and $\eta_i \leq h$ is the average number of production line start-ups at supplier i over the planning horizon.

13.6.3 Scheduling Manufacturing and Delivery of Parts

The solution to the *NDS* model determines the optimal assignment of customer orders to planning periods $\{x_{jt}, \; j \in J, t \in T\}$ and thereby the optimal assembly schedule for the finished products. Denote by N_{ikt} the optimal cumulative usage in periods 1 through t of part type k delivered by supplier i, that is, the optimal cumulative requirement for part type k provided by supplier i

$$N_{ikt} = \sum_{l \in L} \sum_{j \in J_l \cap J(i)} \sum_{\tau=1}^{t} b_{kl} n_j x_{j\tau}; \quad i \in I, k \in K, t \in T. \tag{13.62}$$

The base-level problem can now be formulated. Given the cumulative requirements $N_{ikt}, \; t \in T$ for all part types $k \in K$ from each supplier $i \in I$ by each period t, determine the manufacturing and the delivery schedule of parts, independently for each supplier. In the time-indexed MIP model *SMD(i)* presented below, the supplier subscript i is a fixed parameter that can be suppressed.

Model SMD(*i*): *Scheduling Manufacturing and Delivery of Parts for Supplier i*

Minimize the weighted cost of parts manufacturing, supplies, and inventory holding

$$\begin{aligned}
\gamma f_{2i} + (1-\gamma) f_{3i} = \gamma \Bigg(& \sum_{k \in K} \sum_{t \in T} \sum_{\tau \in T^{(i)} : \tau \leq t} \varphi_{1ik} (r_{ik\tau} - s_{ik,next(\tau, T^{(i)}, \theta_i)}) \\
& + \sum_{k \in K} \sum_{t \in T : t \geq next(-e_i, T^{(i)}, \theta_i)} \varphi_{2ik} \theta_i s_{ikt} \\
& + \sum_{k \in K} \sum_{t \in T} \varphi_{3k} \Big(\sum_{\tau \in T^{(i)} : \tau \leq t} s_{ik\tau} - N_{ikt} \Big) \Bigg) \\
& + (1-\gamma) \sum_{t \in T^{(i)}} \Bigg(\sum_{k \in K} \psi_{1ik} (\sigma_i / q_{ik}) z_{ikt} + \psi_{2i} \theta_i u_{it} \Bigg),
\end{aligned} \tag{13.63}$$

where $\quad 0 < \gamma < 1$

subject to

1. *Manufacturing Line Set-up and Start-up Constraints*: (13.15) to (13.19) for supplier i
2. *Manufacturing Capacity Constraints*: (13.20) for supplier i
3. *Part Manufacturing and Delivery Constraints*: (13.21) to (13.27) for supplier i
4. *Part Demand Satisfaction Constraints*:
   - for each part type k the cumulative deliveries from supplier i by period t cannot be less than the cumulative demand for this part type by this period,
   - total delivery of each part type k from supplier i is equal to the total demand of this part type from this supplier,

$$\sum_{\tau \in T^{(i)}:\, \tau \le t} s_{ik\tau} \ge N_{ikt};\ \ k \in K,\, t \in T: t < h \tag{13.64}$$

$$\sum_{t \in T^{(i)}} s_{ikt} = N_{ikh};\ \ k \in K \tag{13.65}$$

5. *Coordinating Constraints*:
   - the total production of each part type at supplier i by period $prev(t, T^{(i)}, \theta_i)$, that is, manufactured at i not later than θ_i periods before period t, must not be less than the cumulative requirement for this part type by period t of the orders assigned to supplier i

$$\lfloor H/q_{ik} \rfloor \sum_{\tau \in T^{(i)}:\, \tau \le prev(t, T^{(i)}, \theta_i)} (y_{ik\tau} - z_{ik\tau}) + \lfloor (H - \sigma_i)/q_{ik} \rfloor \sum_{\tau \in T^{(i)}:\, \tau \le prev(t, T^{(i)}, \theta_i)} z_{ik\tau} \ge N_{ikt};\ \ k \in K,\, t \in T^{(i)}: t \ge next(-e_i, T^{(i)}, \theta_i) \tag{13.66}$$

6. *Nonnegativity and Integrality Conditions*: (13.31) to (13.35) for supplier i.

The coordinating constraint (13.66) is implied by (13.20), (13.21), and (13.64), and it directly links the assignment of production lines of supplier i to part-types over the planning horizon with the cumulative requirement for the part types from supplier i, (cf. (13.30)).

In the objective function (13.63), f_{2i} and f_{3i} are the inventory holding cost of parts manufactured by supplier i and stored by the supplier, being in transit to the producer or stored by the producer and the cost to supplier i of production line start-ups and part shipments, respectively.

Recall that in the hierarchical approach the total supply chain inventory holding cost is split among two decision levels. Thus, the final values of the objective functions

f_2 (13.10) and f_3 (13.11) can be calculated as the sums of their partial solution values, determined for the middle-level problem *NDS* and the base-level problems *SMD*(i), $i \in I$

$$f_2 = f_{20} + \sum_{i \in I} f_{2i} \tag{13.67}$$

$$f_3 = \sum_{i \in I} f_{3i}. \tag{13.68}$$

13.7 COMPUTATIONAL EXAMPLES

In this section numerical examples and computational results are presented to illustrate possible applications of the two proposed approaches (integrated and hierarchical) and IP and MIP models for the coordinated scheduling of manufacturing, supply, and assembly in a supply chain with multiple suppliers with selection of a single supplier of parts for each customer order. The examples are modeled after a real-world customer-driven supply chain for high-tech products, see Figures 13.3 and 13.4 (see also, Fig. 6.4). The supply chain consists of three manufacturers/suppliers of parts (electronic components), a distribution center (see Fig. 5.3) where finished products (electronic devices) are assembled according to customer orders, and a set of customers who generate final demand for the products.

A brief description of the planning horizon, manufacturers/suppliers, producer, part types, product types, and customer orders is given below.

1. Planning horizon: $h = 14$ periods (days), each of length $H = 2 \times 9$ hours

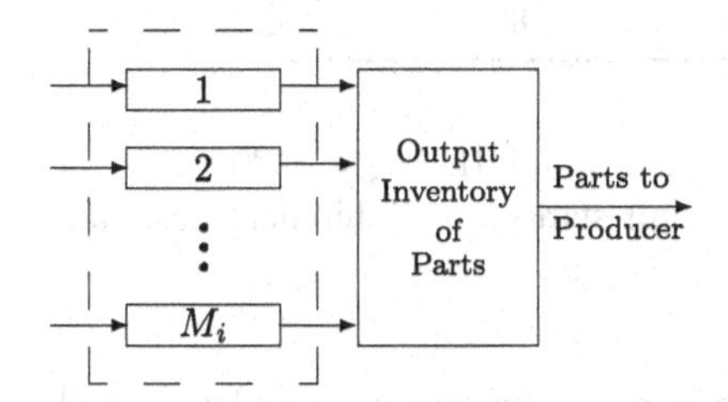

Figure 13.3 Manufacturer/supplier i.

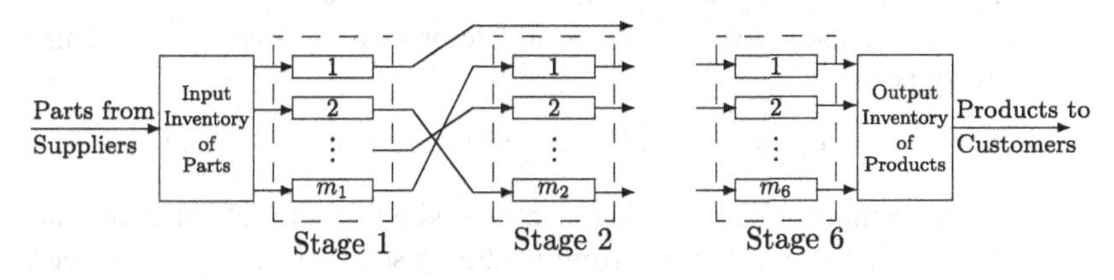

Figure 13.4 Producer.

2. Manufacturers/suppliers

- $m = 3$ suppliers:
  - supplier 1 with transportation time $\theta_1 = 1$ period, $M_1 = 12$ production lines and identical start-up time $\sigma_1 = 10{,}000$ seconds for each line
  - supplier 2 with transportation time $\theta_2 = 2$ periods, $M_2 = 11$ production lines and identical start-up time $\sigma_2 = 8000$ seconds for each line
  - supplier 3 with transportation time $\theta_3 = 3$ periods, $M_3 = 10$ production lines and identical start-up time $\sigma_3 = 9000$ seconds for each line
- the earliness of the manufacturing start in advance of the assembly start is $e_i = 2$ days for each supplier $i = 1, 2, 3$
- the minimum and the maximum allowed shipment from each supplier: $\underline{S} = 5000$, and $\overline{S} = 50{,}000$ parts
- Parts
  - 10 part types
  - manufacturing times (in seconds) for part types:

Part type/supplier	1	2	3
1	65	60	70
2	70	75	65
3	80	75	85
4	85	80	90
5	90	95	85
6	95	90	100
7	65	60	70
8	75	70	65
9	90	80	100
10	85	80	90

- the unit inventory holding costs are $\varphi_{1ik} = 0.001$, $\varphi_{2ik} = 0.0015$, $\varphi_{3k} = 0.002$, $\varphi_{4j} = 0.004 \ \forall\, i, k, j$, and the unit start-up and shipment costs are $\psi_{1ik} = 0.004$, $\psi_{2i} = 1 \ \forall\, i, k$

3. Producer

- six assembly stages in series with parallel machines: $m_g = 10$ parallel machines in each stage $g = 1, 2$; $m_g = 20$ parallel machines in each stage $g = 3, 4, 5$; and $m_g = 10$ parallel machines in stage $g = 6$
- for each assembly stage g the available processing time c_{gt} is the same in every period t:

$$c_{gt} = c_g = \alpha_g H m_g; \ \ g = 1, \ldots, 6, \ \ t = 1, \ldots, 14$$

where parameter $\alpha_g \in (0, 1)$ reflects the idle time of each machine waiting for the first production lot from upstream stages $1, \ldots, g-1$, preceding stage g and the idle time during processing of the last production lot at downstream stages following stage g, assuming that all production lots of each customer order are completed in a single period

- Products
  - 10 product types of three product groups, each to be processed on a separate group of machines (in the assembly stage 3 or 4 or 5),
  - assembly times (in seconds) for product types:

Product type/stage	1	2	3	4	5	6
1	20	0	120	0	0	15
2	20	0	140	0	0	15
3	10	0	120	0	0	10
4	15	5	0	120	0	15
5	15	10	0	180	0	15
6	10	5	0	120	0	10
7	15	10	0	180	0	15
8	20	5	0	0	100	15
9	15	0	0	0	80	10
10	15	0	0	0	100	10

4. Customers: $n = 521$ customer orders ranging from 5 to 8309 products, with various requested due dates, each to be completed in a single period; the total demand is 258,414 products

In the above example the set L of product types is identical with the set K of product-specific part types (i.e., $K = L = \{1, \ldots, 10\}$) and each product type l requires one unit of the corresponding product-specific part type k such that $k = l$, (i.e., $b_{kl} = 1,\ k \in K, l \in L\colon k = l,\ b_{kl} = 0,\ k \in K, l \in L\colon k \neq l$), for example, a single printed wiring board of a specific design is required for each electronic device of the corresponding type. As a result, for each order $j \in J_l$ the required quantity of product-specific part type $k = l$ equals the quantity n_j of the ordered products l.

The characteristics of the IP and MIP models for integrated and hierarchical approaches are summarized in Tables 13.3–13.5. The size of each program is

Table 13.3 Computational Results for the Monolithic Model *INT*: Weighted-Sum Program INT_λ

$\lambda_1, \lambda_2, \lambda_3$	Var.	Bin.	Int.	Cons.	Nonz.	f_1, f_2, f_3	GAP%[a]
0.97,0.02,0.01	23,262	21,402	960	4384	500,446	2,793,119	11.08
0.75,0.15,0.1	23,262	21,402	960	4384	500,446	7,774,101	12.09
0.7,0.2,0.1	23,262	21,402	960	4384	500,446	10,765,114	12.37
0.6,0.3,0.1	23,262	21,402	960	4384	500,446	13,738,104	9.22
0.5,0.3,0.2	23,262	21,402	960	4384	500,446	17,741,92	11.87
0.4,0.35,0.25	23,262	21,402	960	4384	500,446	24,735,108	13.47

[a]GAP% for 7200 CPU seconds on a PC Pentium IV, 1.8 GHz, 1 GB RAM, CPLEX 11.

Table 13.4 Computational Results for the Monolithic Model *INT*: Lexicographic Approach

Model	Var.	Bin.	Int.	Cons.	Nonz.	Solution value	CPU (GAP%)[a]
INT(1)	23,262	21,402	960	4384	500,446	$f_1 = 1, (f_2 = 1739, f_3 = 231)$	933
INT(2)	23,262	21,402	960	4385	510,658	$f_2 = 836, (f_1 = 1, f_3 = 160)$	(7.16%)
INT(3)	23,262	21,402	960	4386	532,656	$f_3 = 151$	(10.04%)

[a]CPU seconds for proving optimality on a PC Pentium IV, 1.8 GHz, 1 GB RAM/CPLEX 11 or GAP% if optimality not proven in 7200 CPU seconds.

Table 13.5 Computational Results for the Hierarchical Approach

Model	Var.	Bin.	Int.	Cons.	Nonz.	Solution values	CPU[a]
RCA	12,255	11,775	480	1837	1,332,549	$f_1 = 1$	30
NDS	3939	3539	400	1004	49,884	$f_{20} = 209$	2320
$\gamma = 0.05$ in (13.63)							
SMD(1)	645	15	320	1291	9944	$f_{21} = 283, f_{31} = 24$	1210
SMD(2)	592	11	299	1141	7902	$f_{22} = 321, f_{32} = 33$	96
SMD(3)	621	11	320	1175	7959	$f_{23} = 432, f_{33} = 38$	57
$f_2 = f_{20} + \sum_{i \in I} f_{2i} = 1245, f_3 = \sum_{i \in I} f_{3i} = 95$							
$\gamma = 0.25$ in (13.63)							
SMD(1)	645	15	320	1291	9944	$f_{21} = 204, f_{31} = 33$	548
SMD(2)	592	11	299	1141	7902	$f_{22} = 261, f_{32} = 41$	11
SMD(3)	621	11	320	1175	7959	$f_{23} = 353, f_{33} = 50$	45
$f_2 = f_{20} + \sum_{i \in I} f_{2i} = 1027, f_3 = \sum_{i \in I} f_{3i} = 124$							
$\gamma = 0.5$ in (13.63)							
SMD(1)	645	15	320	1291	9944	$f_{21} = 179, f_{31} = 47$	562
SMD(2)	592	11	299	1141	7902	$f_{22} = 255, f_{32} = 45$	8
SMD(3)	621	11	320	1175	7959	$f_{23} = 343, f_{33} = 56$	6
$f_2 = f_{20} + \sum_{i \in I} f_{2i} = 986, f_3 = \sum_{i \in I} f_{3i} = 148$							
$\gamma = 0.75$ in (13.63)							
SMD(1)	645	15	320	1291	9944	$f_{21} = 175, f_{31} = 52$	73
SMD(2)	592	11	299	1141	7902	$f_{22} = 252, f_{32} = 49$	11
SMD(3)	621	11	320	1175	7959	$f_{23} = 341, f_{33} = 58$	3
$f_2 = f_{20} + \sum_{i \in I} f_{2i} = 977, f_3 = \sum_{i \in I} f_{3i} = 159$							
$\gamma = 0.95$ in (13.63)							
SMD(1)	645	15	320	1291	9944	$f_{21} = 175, f_{31} = 60$	487
SMD(2)	592	11	299	1141	7902	$f_{22} = 250, f_{32} = 56$	21
SMD(3)	621	11	320	1175	7959	$f_{23} = 341, f_{33} = 64$	56
$f_2 = f_{20} + \sum_{i \in I} f_{2i} = 975, f_3 = \sum_{i \in I} f_{3i} = 180$							

[a]CPU seconds for proving optimality on a PC Pentium IV, 1.8 GHz, 1 GB RAM/CPLEX 11.

represented by the total number of variables, Var., number of binary variables, Bin., number of integer variables, Int., number of constraints, Cons., and number of nonzero elements in the constraint matrix, Nonz. The counts presented in the tables are taken from the models after presolving. The last two columns of the tables present the solution values and CPU time in seconds required to find optimal solution and to prove its optimality or GAP% if optimality is not proven within a specified CPU time limit. The computational experiments have been performed on a PC Pentium IV, 1.8 GHz, 1 GB RAM using AMPL/CPLEX 11 with a traditional branch and bound search and strong branching option.

The solution results for the integrated approach and the characteristics of weighted-sum MIP model INT_λ for selected λ and of the lexicographic MIP models *INT(1)*, *INT(2)*, and *INT(3)* are summarized, in Tables 13.3 and 13.4. For the integrated approach no proven optimal solution was determined in two hours of CPU time, and GAP was over 10% for the best solutions found. On the other hand, Table 13.4 indicates that proven optimal solution to *INT(1)* was found in less than 16 minutes of CPU time, in two hours a feasible solution with GAP over 7% and over 10% was determined for *INT(2)* and *INT(3)*, respectively. In Table 13.4 the solution value of each objective function for the lexicographic approach is presented along with the corresponding associated values of the remaining objective functions (in parentheses).

The solution results for the hierarchical approach and the characteristics of integer programs *RCA* and *NDS* and of mixed integer programs $SMD(i)$ for $i \in I$ are summarized in Table 13.5 with different weight γ applied in the objective function (13.63) of $SMD(i)$.

While the integrated approach is capable of finding a nondominated solution set of the coordinated scheduling problem, the hierarchical approach was able to find the best solution with respect to the primary objective function f_1, the number of tardy orders. On the other hand, the hierarchical approach leads to a greater total supply chain inventory holding cost f_2, because parts and products inventories are considered separately at two decision levels and, in addition, the cost of holding the inventory of parts are split among all subproblems $SMD(i)$, $i \in I$, of the base level. Furthermore, f_{2i} that includes the in-transit holding cost increases with the transportation time θ_i.

Unlike for the monolithic MIP model *INT*, the computational effort required for the hierarchical approach is relatively small and proven optimal solutions have been reached in a short CPU time for each level problem. The coordinating constraints (13.30) and (13.66) have reduced CPU time up to 20%.

Examples of the aggregate production schedule ($\sum_{i \in I} \sum_{j \in J} n_j X_{ijt}$, $t \in T$ and $\sum_{j \in J} n_j x_{jt}$, $t \in T$) for the producer and the aggregate manufacturing schedule ($\sum_{k \in K} r_{ikt}$, $t \in T^{(i)}$) and the delivery schedule ($\sum_{k \in K} s_{ikt}$, $t \in T^{(i)}$) for each supplier $i \in I$ are shown in Figure 13.5 for the integrated approach with the weighted-sum MIP model INT_λ and in Figure 13.6 for the hierarchical approach.

The computational results have led to some observations on properties of the optimal coordinated schedules in the customer-driven supply chain. For example:

- The shorter the transportation time from a supplier, the greater the total demand for parts of customer orders assigned to that supplier, and also the greater is

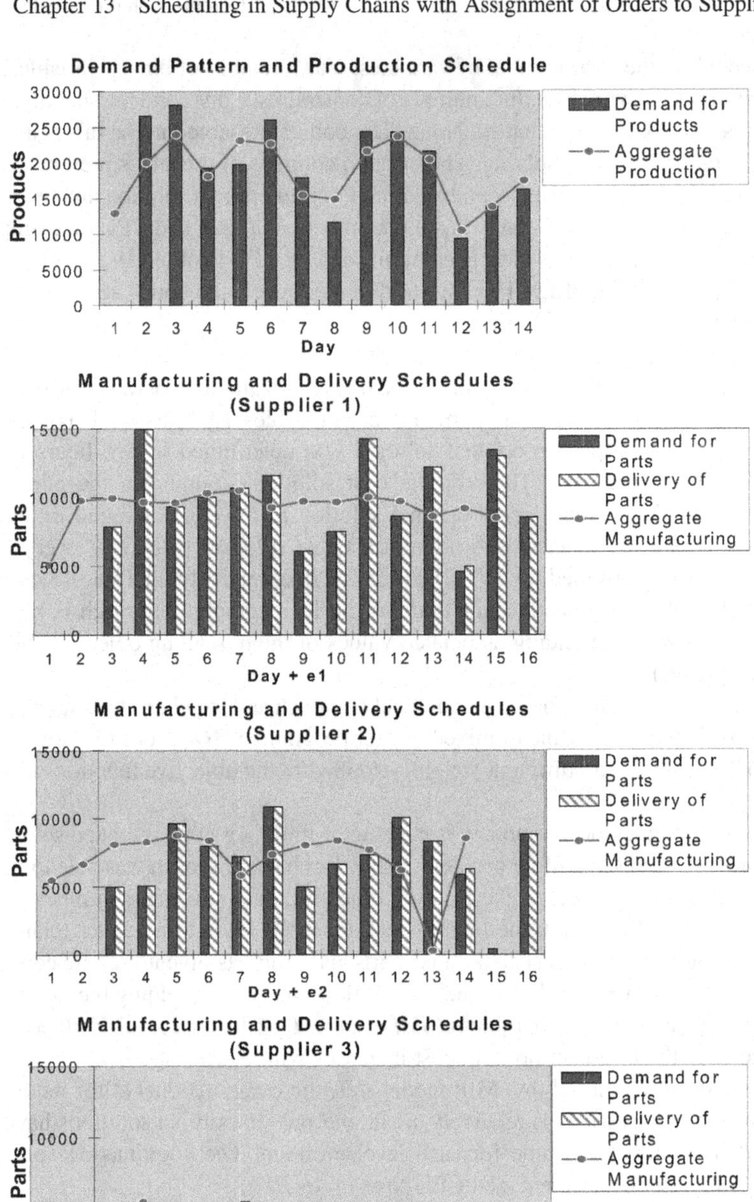

Figure 13.5 Manufacturing, delivery, and production schedules for the monolithic approach and the weighted-sum program INT_λ with $\lambda_1 = 0.97$, $\lambda_2 = 0.02$, $\lambda_3 = 0.01$.

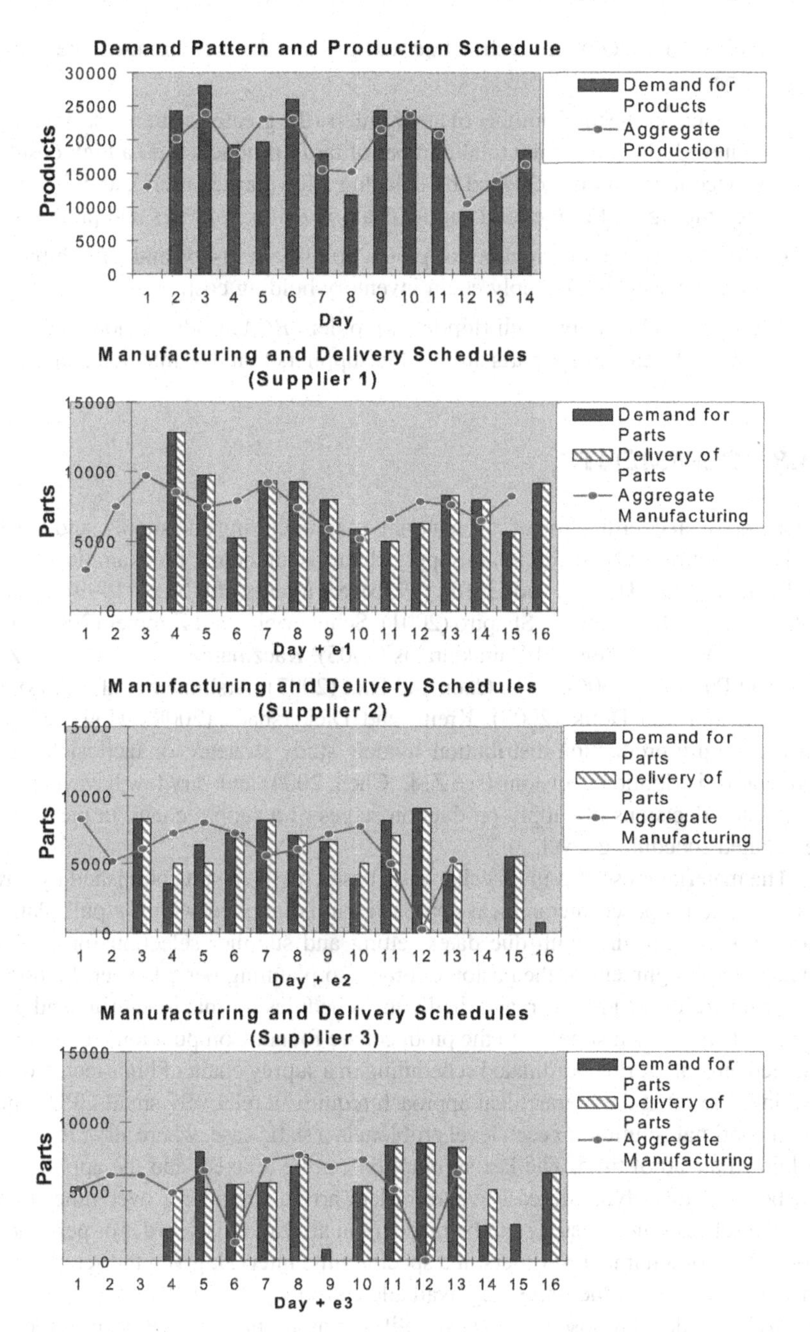

Figure 13.6 Manufacturing, delivery, and production schedules for the hierarchical approach with $\gamma = 0.5$ in (13.63).

the total number of production line start-ups and the more frequent the delivery of parts.

- The smaller the total number of tardy orders (the greater customer service level) achieved, the greater the total number of tardy products (i.e., a high customer service level can be achieved by delaying a few, large orders), and the greater the holding cost of the total supply chain inventory of parts and products.
- The smaller the total number of production line start-ups and part shipments, the greater the total supply chain inventory holding cost.
- The hierarchical approach (top-level problem *RCA*) leads to more equal allocation of demand for parts among the suppliers than the integrated approach.

13.8 COMMENTS

The literature on coordination of raw materials manufacturing and supply, and finished goods production and distribution in supply chains is abundant, for example, Chandra and Fisher (1994), Thomas and Griffin (1996), Sarmiento and Nagi (1999), Erengüc et al. (1999), Kolisch (2000), Shapiro (2001), Schneeweiss and Zimmer (2004), Z.-L. Chen (2004), Z.-L. Chen and Vairaktarakis (2005), Kaczmarczyk et al. (2006), Z.-L. Chen and Pundoor (2006), Z.-L. Chen and Hall (2007), Arshinder et al. (2008), van der Vaart and van Donk (2008), Kreipl and Dickersbach (2008). However, most existing supply-production-distribution models study strategic or tactical levels of production-distribution decisions (see Z.-L. Chen, 2004) and very few have addressed coordinated decisions of supply-production stages of a supply chain, in particular at the detailed scheduling level.

The material presented in this chapter is based on the results published by Sawik (2009b). The proposed hierarchical approach can be applied within a pull planning framework as it begins with due dates setting and supplier selection for customer orders, then assignment of the customer orders to planning periods over the horizon is determined for the producer, and finally the schedules for manufacturing and delivery of parts from each supplier to the producer are found. Computational experiments modeled on real-world coordinated scheduling in a supply chain of high-tech products have indicated that the hierarchical approach requires a relatively small CPU time to find the optimal solution for each level problem in a static case, where all order arrivals are known ahead of time. The last assumption can be relaxed, and the approach can also be used in a dynamic case where orders arrive irregularly over time. In this case, the scheduling decisions can be made upon arrival of each order or periodically upon arrival of a number of orders in a specific time interval, given the set of already scheduled orders and the remaining available capacity.

In the models proposed various simplifying assumptions have been introduced. For example, it has been assumed that each supplier is capable of manufacturing all the part types. In a more general setting, each supplier may only be prepared to manufacture a subset of part types and provide with the parts the corresponding subset of orders. In this case, supplier selection and supplier capacity constraints of *INT* and

RCA should be modified appropriately. Furthermore, the proposed approaches can be enhanced to simultaneously consider the single- and multiperiod orders (see Section 13.4.4). Finally, the two approaches can also be applied for different machine configurations of both parts manufacturers and the finished products producer. In particular, different models for machine setups and start-ups can be applied, with manufacturing of more than one part type in a single planning period (e.g., Drexl and Kimms, 1997; Pochet and Wolsey, 2006).

EXERCISES

13.1 Assuming that each supplier i is capable of manufacturing only a subset $K_i \subset K$ of all part types, introduce appropriate changes in model *INT* for the integrated scheduling of manufacturing, supply, and assembly.

13.2 Assuming that production lines of each supplier i in model *INT* are not identical, replace parameters q_{ik}, σ_i by $q_{ikl}, \sigma_{il}, \ i \in I, l = 1, \ldots, M_i$, and integer set-up and start-up variables y_{ikt}, z_{ikt} by binary variables, respectively $y_{iklt}, z_{iklt}, \ i \in I, l = 1, \ldots, M_i$ and formulate the modified mixed integer program for the integrated scheduling of manufacturing, supply, and assembly.

- **(a)** Compare the size of the two mixed integer programming formulations: with integer set-up and start-up variables and with binary set-up and start-up variables, respectively, for identical and nonidentical production lines.

13.3 Formulate coordinating constraints (13.30) for the integrated scheduling of manufacturing, supply, and assembly.

- **(a)** In the case of a supply chain with multiple producers.
- **(b)** In the case of a supply chain with one supplier and multiple producers.

13.4 Assuming that multiperiod customer orders can be arbitrarily allocated among different (adjacent and not adjacent) planning periods, how should the integer program *NDS* for nondelayed scheduling of customer orders be modified?

- **(a)** What are the implications of noncontiguous allocation of multiperiod orders among planning periods on the other formulations presented in this chapter?

13.5 Comparing the results of computational examples illustrated in Figures 13.5 and 13.6, answer the following questions.

- **(a)** Why are the schedules of customer orders (i.e., the aggregate production schedules) obtained by both the integrated approach and the hierarchical approach very close to one another?
- **(b)** Why does the integrated approach place fewer orders for parts to suppliers with a longer transportation time?

Chapter 14

Coordinated Scheduling in Supply Chains without Assignment of Orders to Suppliers

14.1 INTRODUCTION

This chapter deals with the coordinated bi-objective scheduling of manufacturing and delivery of parts and assembly of finished products in a customer-driven supply chain. The supply chain consists of multiple suppliers (manufacturers) of parts, a single producer of finished products, and a set of customers who generate final demand for the products. Each supplier has a number of parallel identical machines for manufacturing parts, and the producer has a flexible assembly line (FAL) for assembly of products.

In the supply chain, the following static and deterministic coordinated scheduling problem is considered. Given a set of customer orders, the problem objective is to determine a coordinated medium-term schedule for the manufacture of parts by each supplier, for the delivery of parts from each supplier to the producer, and for the assembly of products for each order by the producer, such that a high revenue or maximum service level is achieved at minimum cost of holding the total supply chain inventory of parts and products.

An integrated (monolithic) approach, where the coordinated manufacturing, supply, and assembly schedules are determined simultaneously, is compared with a hierarchical approach. In the hierarchical approach, first the subset of orders that can be completed by the customer-requested due dates is found and for the remaining orders delayed due dates are determined to minimize the lost revenue due to order tardiness or to minimize the total number of tardy products. The due date setting decisions are subject to the due date feasibility constraints ensuring the existence of a feasible deadline schedule for customer orders and a feasible manufacturing and

---

Scheduling in Supply Chains Using Mixed Integer Programming. By Tadeusz Sawik

supply schedule for parts. Then the deadline assignment of customer orders to planning periods over the horizon is found to minimize the holding cost of finished product inventory of the customer orders completed before their due dates. Finally, the manufacturing and delivery schedules of the required parts are determined for all suppliers to minimize the total cost of holding the inventory of parts.

The following time-indexed IP and MIP models are presented in this chapter:

INTb for integrated, bi-objective scheduling of manufacturing, supply, and assembly

$INTb_\lambda$ for integrated, bi-objective scheduling of manufacturing, supply, and assembly using the weighted-sum approach

$INTb^\lambda$ for integrated, bi-objective scheduling of manufacturing, supply, and assembly using the reference point approach

DDA for due date assignment

ODS for order deadline scheduling

MDS for part manufacturing and delivery scheduling

Numerical examples modeled after a real-world coordinated scheduling in a customer-driven supply chain in the electronics industry and computational results are reported in Section 14.6.

The coordinated supply chain scheduling problem considered in this chapter is closely related to that presented in Chapter 13. However, the selection of a single supplier of parts required for each customer order is not considered. In contrast to the hierarchical approach presented in Chapter 13 which begins with the combined due date setting and selection of a single parts supplier for each customer order, in the approach proposed in this chapter the allocation of demand for parts among suppliers is postponed until the final scheduling of manufacturing and delivery of parts. Each customer order can be partially provided with the required parts by different suppliers. Furthermore, unlike the production lines for parts manufacturing with nonnegligible start-up times, now the parallel machines at suppliers operate with no start-ups. In addition to maximum service level, a maximum revenue is used as an alternative primary objective function. Finally, to determine the subsets of nondominated solutions for the integrated approach the weighted-sum program is compared with the reference point-based Chebyshev program, in which the relative importance of the objective functions is weighted in the constraint set.

14.2 PROBLEM DESCRIPTION

The problem description is similar to that presented in Section 13.2. Briefly, the supply chain consists of m manufacturers/suppliers of parts, a single producer where finished products are assembled according to customer orders, and a set of customers who generate final demand for the products (see Fig. 13.1). In the supply chain various types of products are assembled by the producer using different part types supplied by the manufacturers (for the basic notation used see Table 13.1).

Let J be the set of customer orders known ahead of time, K the set of part types, L the set of product types, J_l the subset of orders for product type $l \in L$, and I_k the subset of suppliers capable of producing part type k. Each product type $l \in L$ requires $b_{kl} \geq 0$ parts of type $k \in K$.

Each supplier i has M_i parallel identical machines with negligible start-up times (i.e., $\sigma_i = 0 \; \forall i \in I$) and let $\rho_{ik} = \lfloor H/q_{ik} \rfloor$ be the one machine-period manufacturing rate for part type $k \in K$ at supplier $i \in I_k$. The producer has a flexible assembly line that consists of several assembly stages in series with parallel identical machines and let $\varrho_{gj} = \lfloor c_g/p_{gj} \rfloor$ be the one machine-period production rate at assembly stage g for products in order j.

The problem objective is to determine a coordinated medium-term schedule for the manufacture of parts by each supplier, for the delivery of parts from each supplier to the producer, and for the assembly of products for each order by the producer, such that a high revenue or a maximum service level is achieved at minimum cost of holding the total supply chain inventory of parts and products.

An integrated (monolithic) approach, where the coordinated manufacturing, supply, and assembly schedules are determined simultaneously is compared with a hierarchical approach based on lexicographic optimization of the bi-objective scheduling problem and its decomposition.

14.3 COORDINATED SUPPLY CHAIN SCHEDULING: AN INTEGRATED APPROACH

In this section an integrated approach and bi-objective, time-indexed MIP monolithic model *INTb* is presented for the coordinated scheduling of manufacturing and supply of parts and assembly of products in a customer-driven supply chain with multiple suppliers.

14.3.1 Decision Variables

Let $T^{(i)}$ be the enlarged, ordered set of planning periods with e_i periods added before period $t = 1$, that is,

$$T^{(i)} = \{-e_i, \ldots, -1\} \cup T = \{-e_i, \ldots, -1, 1, \ldots, h\}.$$

The various decision variables (Table 14.1) used in model *INTb* to coordinate different types of schedules in the supply chain are additionally explained below.

- Manufacturing schedule for each supplier $i \in I$:
  - the manufacturing lot sizing variable r_{ikt} that represents the lot size for each part type manufactured by each supplier in each planning period, that is, the number of parts type k manufactured in period $t \in T^{(i)}$ by supplier i.

Table 14.1 Decision Variables: Bi-Objective Scheduling in a Supply Chain with Multiple Suppliers

Model *INTb*

r_{ikt} = manufacturing lot at supplier i of part type k in period t (manufacturing lot sizing variable)

s_{ikt} = delivery lot from supplier i of part type k in period t (delivery lot sizing variable)

u_{it} = 1, if delivery of parts from supplier i is scheduled for period t, otherwise $u_{it} = 0$ (delivery period selection variable)

x_{jt} = 1, if customer order j is assigned to planning period t; otherwise $x_{jt} = 0$ (order-to-period assignment variable)

Model *DDA*

v_j = 1, if order j is accepted with its customer-requested due date; $v_j = 0$ if order j needs delaying to be accepted (order acceptance variable)

w_{jt} = 1, if delayed order j is assigned the adjusted due date t, $(t > d_j)$; otherwise $w_{jt} = 0$ (due date assignment variable)

- Delivery schedule for each supplier $i \in I$:
  - the delivery lot sizing variable s_{ikt} that represents the lot size of each part type delivered from each supplier to the producer in each planning period, that is, the number of parts type k delivered from the supplier i to the producer in period $t \in T^{(i)}$,
  - the delivery period selection variable $u_{it} = 1$, if delivery of parts from the supplier i to the producer is scheduled for period $t \in T^{(i)}$, otherwise $u_{it} = 0$.
- Customer order schedule:
  - the order-to-period assignment variable $x_{jt} = 1$, if customer order j is assigned for processing to planning period $t \in T$; otherwise $x_{jt} = 0$.

14.3.2 Objective Functions: Lost Revenue vs. Inventory Holding Cost

Sales departments often apply revenue management principles for order selection and due date setting. The objective is to maximize a revenue function (see Section 5.4.1), or alternatively, to minimize the total loss revenue due to tardiness of orders.

Most often customers value short lead times (due dates) over long lead times. Setting delayed due dates results in reduction of revenue. Let ν_{jt} be the loss of revenue per unit of each product in the delayed order j completed in period t, later than the customer-requested due date d_j. The lost revenue increases with an increase in the delay of completion dates with respect to requested due dates, i.e.,

$$\nu_{jt+1} > \nu_{j,t};\ j \in J, t \in T : t > d_j.$$

We assume that setting delayed due date results in reduction of revenue proportional to the delay. Per unit loss of revenue ν_{jt} increases by some percent for each day of delay

$(t - d_j)$ of completion of order j with respect to the customer-requested date d_j, for example,

$$v_{jt} = v_j \alpha_j (t - d_j);\ j \in J, t \in T: t \geq d_j,$$

where v_j is per unit revenue for a nondelayed order j and $0 < \alpha_j < 1$ is the rate of per unit daily loss of revenue for order j.

In addition, a fixed loss $\beta_j (0 < \beta_j < 1)$ of unit revenue may be applied for each delayed product in order j, that is,

$$v_{jt} = v_j (\beta_j + \alpha_j (t - d_j));\ j \in J, t \in T: t > d_j.$$

The objective of the integrated scheduling is to determine for each supplier a schedule of manufacturing parts and a schedule of supply of parts to the producer, and to find for the producer a schedule of assembly of the ordered products, so as to maximize revenue or equivalently to minimize the lost revenue due to tardiness of orders, as a primary optimality criterion. Simultaneously, to achieve a low unit cost, the total supply chain inventory holding cost is minimized. Thus, the secondary objective function is minimization of the total inventory holding cost. The two objective functions, f_1, f_2, numbered according to their decreasing importance, are defined below.

The lost revenue

$$f_1 = \sum_{j \in J} \sum_{t \in T: t > d_j} v_{jt} n_j x_{jt}. \tag{14.1}$$

The cost of inventory holding

$$\begin{aligned} f_2 = & \sum_{k \in K} \sum_{i \in I_k} \sum_{t \in T} \sum_{\tau \in T^{(i)}:\ \tau \leq t} \varphi_{1ik} (r_{ik\tau} - s_{ik,\mathrm{next}(\tau, T^{(i)}, \theta_i)}) \\ & + \sum_{k \in K} \sum_{i \in I_k} \sum_{t \in T: t \geq \mathrm{next}(-e_i, T^{(i)}, \theta_i)} \varphi_{2ik} \theta_i s_{ikt} \\ & + \sum_{k \in K} \sum_{i \in I_k} \sum_{t \in T} \varphi_{3k} \left(\sum_{\tau \in T^{(i)}: \tau \leq t} s_{ik\tau} - \sum_{l \in L} \sum_{j \in J_l} \sum_{\tau = 1}^{t} b_{kl} n_j x_{j\tau} \right) \\ & + \sum_{t \in T} \sum_{\tau = 1}^{t} \sum_{j \in J: d_j > t} \varphi_{4j} n_j x_{j\tau}, \end{aligned} \tag{14.2}$$

where φ_{1ik}, φ_{2ik}, φ_{3k}, and φ_{4j} are the unit holding costs of parts stored at supplier i, of parts in transit from supplier i to the producer, of parts stored at the producer, and of finished products stored at the producer, respectively. The in-transit holding cost represents the cost of transportation of parts from suppliers to the producer.

14.3.3 A Bi-Objective Monolithic Model

This subsection presents the bi-objective MIP monolithic model *INTb* for the integrated supply chain scheduling.

Model *INTb*: *Integrated, Bi-Objective Scheduling of Manufacturing, Supply, and Assembly*

$$\min f_1, f_2 \tag{14.3}$$

subject to

1. *Order-to-Period Assignment Constraints*:
   - each customer order is assigned to exactly one planning period,

$$\sum_{t \in T: a_j \le t} x_{jt} = 1; \; j \in J \tag{14.4}$$

2. *Producer Capacity Constraints*:
   - in every period t the total number of machine-periods of production at each assembly stage g cannot be greater than the number of machines available at this stage,

$$\sum_{j \in J} n_j x_{jt} / \varrho_{gj} \le m_g; \; g \in G, t \in T \tag{14.5}$$

3. *Manufacturing Capacity Constraints for Each Supplier*:
   - in every period t the total number of machine-periods of production of all part types at each supplier i cannot be greater than the total number of machines available at this supplier,

$$\sum_{k \in K: i \in I_k} r_{ikt} / \rho_{ik} \le M_i; \; i \in I, t \in T^{(i)} \tag{14.6}$$

4. *Part Manufacturing and Delivery Constraints for Each Supplier*:
   - the cumulative delivery of each part type by period t cannot be greater than the cumulative manufacturing of this part type by period prev(t, $T^{(i)}$, θ_i), that is, manufactured not later than θ_i periods before period t,
   - the total delivery of each part type is equal to the total manufacturing of this part type,
   - parts can be delivered only in periods scheduled for delivery and each shipment is limited by its minimum and maximum capacity, $\underline{S}$ and $\overline{S}$, respectively,

– no delivery from supplier i can be scheduled before period next$(-e_i, T^{(i)}, \theta_i)$, that is, no delivery can be scheduled earlier than θ_i periods after the earliest period $-e_i$ of manufacturing,

$$\sum_{\tau \in T^{(i)}: \tau \le t} s_{ik\tau} \le \sum_{\tau=-e_i}^{\text{prev}(t, T^{(i)}, \theta_i)} r_{ik\tau};\ k \in K, i \in I_k, t \in T^{(i)}:$$

$$\text{next}(-e_i, T^{(i)}, \theta_i) \le t < h \quad (14.7)$$

$$\sum_{t \in T^{(i)}} s_{ikt} = \sum_{\tau=-e_i}^{\text{prev}(h, T^{(i)}, \theta_i)} r_{ikt};\ k \in K, i \in I_k \quad (14.8)$$

$$s_{ikt} \le \overline{S} u_{it};\ k \in K, i \in I_k, t \in T^{(i)} \quad (14.9)$$

$$\sum_{k \in K} s_{ikt} \le \overline{S};\ i \in I, t \in T^{(i)} \quad (14.10)$$

$$\sum_{k \in K} s_{ikt} \ge \underline{S} u_{it};\ i \in I, t \in T^{(i)} \quad (14.11)$$

$$s_{ikt} = 0;\ k \in K, i \in I_k, t \in T^{(i)}: t < \text{next}(-e_i, T^{(i)}, \theta_i) \quad (14.12)$$

$$u_{it} = 0;\ i \in I, t \in T^{(i)}: t < \text{next}(-e_i, T^{(i)}, \theta_i) \quad (14.13)$$

5. *Part Demand Satisfaction Constraints*:
   – the cumulative deliveries of each part type k by period t from all suppliers $i \in I_k$ cannot be less than the cumulative usage of this part type in periods 1 through t,
   – total delivery of each part type k from all suppliers $i \in I_k$ is equal to the total demand for this part type,

$$\sum_{i \in I_k} \sum_{\tau \in T^{(i)}: \tau \le t} s_{ik\tau} \ge \sum_{l \in L} \sum_{j \in J_l} \sum_{\tau=1}^{t} b_{kl} n_j x_{j\tau};\ k \in K, t \in T: t < h \quad (14.14)$$

$$\sum_{i \in I_k} \sum_{t \in T^{(i)}} s_{ikt} = \sum_{l \in L} \sum_{j \in J_l} b_{kl} n_j;\ k \in K \quad (14.15)$$

6. *Coordinating Constraints*:
   – the cumulative requirement for each part type k of the orders assigned to periods 1 through t (right-hand side of (14.16)) cannot be greater than the total production of this part type at each supplier $i \in I_k$ by period prev$(t, T^{(i)}, \theta_i)$, that is, manufactured at i not later than θ_i periods before period t (left-hand side of (14.16))

$$\sum_{i \in I_k: \text{next}(-e_i, T^{(i)}, \theta_i) \le t} \ \sum_{\tau=-e_i}^{\text{prev}(t, T^{(i)}, \theta_i)} r_{ik\tau} \ge \sum_{l \in L} \sum_{j \in J_l} \sum_{\tau=1}^{t} b_{kl} n_j x_{j\tau};\ k \in K, t \in T \quad (14.16)$$

7. *Nonnegativity and Integrality Conditions*:

$$r_{ikt} \geq 0; \; k \in K, i \in I_k, t \in T^{(i)} \tag{14.17}$$

$$s_{ikt} \geq 0; \; k \in K, i \in I_k, t \in T^{(i)} \tag{14.18}$$

$$u_{it} \in \{0, 1\}; \; i \in I, t \in T^{(i)} \tag{14.19}$$

$$x_{jt} \in \{0, 1\}; \; j \in J, t \in T: a_j \leq t. \tag{14.20}$$

The coordinating constraint (14.16) has been introduced to strengthen the mixed integer program *INTb* as it directly links the manufacturing of parts over the planning horizon at each supplier (variable r_{ikt}) and the assignment of customer orders to planning periods (variable x_{jt}). Thus, coordinating constraint (14.16) directly coordinates the suppliers' schedules and the producer's schedule (cf. (13.30)). Coordinating constraint (14.16) is implied by constraints (14.7) and (14.14), and any feasible solution that satisfies (14.16) also satisfies (14.7) and (14.14).

The mathematical model presented in this section can be modified or enhanced to consider additional features of the bi-objective supply chain scheduling that can be met in practice. In particular, the primary objective function can be modified. In order to achieve a high customer service level the number of orders completed later than the customer-requested dates should be minimized, that is, the maximum service level can be achieved by minimizing the number of tardy orders

$$f_1^O = \sum_{j \in J} \sum_{t \in T: t > d_j} x_{jt}. \tag{14.21}$$

A surrogate objective function that also considers the total lost revenue due to tardiness of orders is minimization of the total number of tardy products

$$f_1^P = \sum_{j \in J} \sum_{t \in T: t > d_j} n_j x_{jt}. \tag{14.22}$$

14.4 SELECTED BI-OBJECTIVE SOLUTION APPROACHES

This section provides a brief description of selected approaches that can be used to determine the nondominated solution set of the bi-objective mixed integer program *INTb*.

14.4.1 Weighted-Sum Program

The nondominated solution set of the bi-objective program *INTb* can be partially determined by the parameterization on λ of the following weighted-sum program.

Model $INTb_\lambda$

$$\min \lambda f_1 + (1 - \lambda) f_2 \tag{14.23}$$

subject to (14.4)–(14.20),
where $0.5 \leq \lambda \leq 1$.

14.4.2 Reference Point-Based Scalarizing Program

Let $\underline{f} = (\underline{f}_1, \underline{f}_2)$ be a reference point in the criteria space such that $\underline{f}_\iota < f_\iota$, $\iota = 1, 2$ for all feasible solutions satisfying (14.4) to (14.20), and denote by ε a small positive value. The nondominated solution set of the bi-objective program *INTb* can be found by the parameterization on λ of the following mixed integer program $INTb^\lambda$.

Model $INTb^\lambda$

$$\min\{\delta + \varepsilon(f_1 + f_2)\} \tag{14.24}$$

subject to (14.4)–(14.20) and

$$\lambda(f_1 - \underline{f}_1) \leq \delta \tag{14.25}$$

$$(1 - \lambda)(f_2 - \underline{f}_2) \leq \delta \tag{14.26}$$

$$\delta \geq 0, \tag{14.27}$$

where $0.5 \leq \lambda \leq 1$.

MIP model $INTb^\lambda$ is based on the augmented λ-weighted Chebyshev metric $\min\{\lambda|f_1 - \underline{f}_1| + \varepsilon(f_1 + f_2), (1 - \lambda)|f_2 - \underline{f}_2| + \varepsilon(f_1 + f_2)\}$.

14.5 COORDINATED SUPPLY CHAIN SCHEDULING: A HIERARCHICAL APPROACH

In this section MIP formulations are presented for a hierarchical scheduling of customer orders and manufacturing and supplies of parts. The proposed hierarchical approach can be applied within a pull planning framework in customer-driven supply chains. The hierarchical decomposition scheme is based on the lexicographic approach, i.e., it is driven by the relative importance of the objective functions. The hierarchical framework consists of the following three decision-making problems to be solved sequentially (Fig. 14.1):

1. Due date assignment, *DDA*
2. Order deadline scheduling, *ODS*
3. Part manufacturing and delivery scheduling, *MDS*

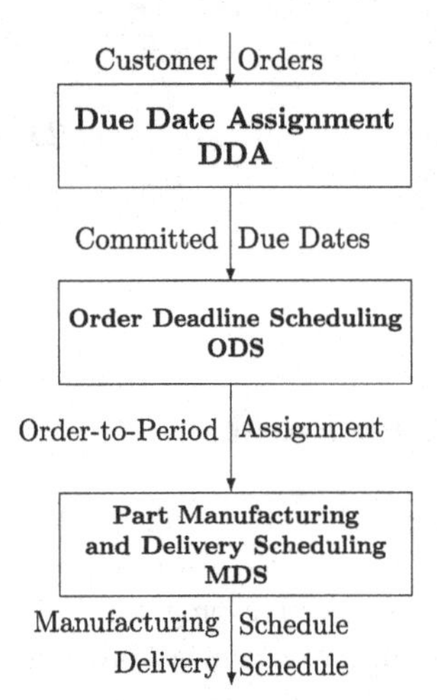

Figure 14.1 Coordinated supply chain scheduling: a hierarchical approach.

The top-level problem *DDA* accounts for the cumulative capacity of the producer, and both the top- and the middle-level problems *DDA* and *ODS* contain the capacity constraints of the suppliers.

The objective of the top-level problem *DDA* is to determine a subset of orders that can be completed by the customer-requested due dates and to update (delay) due dates for the remaining orders such that the capacity constraints are met and the lost revenue due to the orders delay is minimized. At this level the necessary conditions (see Section 13.3) for feasibility of the customer-requested due dates are checked. If the feasibility conditions (constraints (14.30) and (14.31) of *DDA*) are satisfied for all requested due dates, then no tardy order occurs and the total revenue is maximal. Otherwise, to reach feasibility later due dates must be committed for some customer orders to minimize the lost revenue for the orders with committed due dates later than the requested due dates. All committed due dates are next interpreted as deadlines that cannot be violated at the order scheduling level.

Given the committed due date for each customer order, the objective of the middle-level problem *ODS* is to determine a feasible, nondelayed assignment of orders to planning periods over the planning horizon to minimize the finished product inventory holding cost of customer orders completed before their due dates.

Finally, given the cumulative requirements for all part types over the planning horizon, the objective of the base-level problem *MDS* is to determine the integrated manufacturing and delivery schedule of parts for each supplier such that the total cost of holding the inventory of parts in the supply chain is minimized.

The time-indexed MIP models *DDA*, *ODS*, and *MDS* are presented below.

14.5.1 Due Date Assignment

In addition to the manufacturing lot sizing variable r_{ikt}, the time-indexed MIP model *DDA* contains the following two binary decision variables.

- Order acceptance variable: $v_j = 1$, if order j is accepted with its customer-requested due date; $v_j = 0$ if order j needs delaying to be accepted.
- Due date setting variable: $w_{jt} = 1$, if order j unacceptable with requested due date (with $v_j = 0$) is assigned a later due date t, $(t > d_j)$; otherwise $w_{jt} = 0$.

Model *DDA*: *Due Date Assignment*

Minimize the lost revenue

$$f_1 = \sum_{j \in J} \sum_{t \in T: t > d_j} \nu_{jt} n_j w_{jt}, \tag{14.28}$$

subject to

1. *Order Acceptance and Due Date Setting Constraints*:
   - each customer order is either accepted with its requested due date or is assigned a later due date to reach production and delivery schedule feasibility,

$$v_j + \sum_{t \in T: t > d_j} w_{jt} = 1; \ j \in J \tag{14.29}$$

2. *Producer Cumulative Capacity Constraints*:
   - for any period $t \leq d$, the cumulative demand on capacity in stage g of all the customer orders accepted with requested due dates d_j (or adjusted due dates τ) not greater than d and ready dates a_j (or requested due dates d_j, respectively) not less than t must not exceed the cumulative capacity available in this stage in periods t through d

$$\sum_{j \in J: t \leq a_j \leq d_j \leq d} n_j v_j / \varrho_{gj} + \sum_{j \in J} \sum_{\tau \in T: t \leq d_j < \tau \leq d} n_j w_{j\tau} / \varrho_{gj}$$
$$\leq m_g(d - t + 1); \ d, t \in T, \ g \in G: t \leq d \tag{14.30}$$

3. *Supplier Cumulative Capacity Constraints*: (14.6) and
   - the cumulative demand on part type k required in periods 1 through t cannot be greater than the total cumulative production of this part type at each supplier $i \in I_k$ by period prev$(t, T^{(i)}, \theta_i)$, that is, manufactured at i not later than θ_i periods before period t,

$$\sum_{l \in L} \sum_{j \in J_l: d_j \leq t} b_{kl} n_j v_j + \sum_{l \in L} \sum_{j \in J_l} \sum_{\tau \in T: d_j < \tau \leq t} b_{kl} n_j w_{j\tau}$$
$$\leq \sum_{i \in I_k: \text{next}(-e_i, T^{(i)}, \theta_i) \leq t} \ \sum_{\tau = -e_i}^{\text{prev}(t, T^{(i)}, \theta_i)} r_{ik\tau}; \ k \in K, t \in T \tag{14.31}$$

4. *Nonnegativity and Integrality Conditions*: (14.17) and

$$v_j \in \{0, 1\};\ j \in J \tag{14.32}$$

$$w_{jt} \in \{0, 1\};\ j \in J, t \in T: t > d_j. \tag{14.33}$$

The objective function (14.28) represents the lost revenue of orders for which the committed due dates are later than due dates requested by the customers. Alternatively, the objective function f_1 (14.28) can be replaced with f_1^O (14.21) or f_1^P (14.22), defined below

$$f_1^O = \sum_{j \in J} (1 - v_j) \tag{14.34}$$

$$f_1^P = \sum_{j \in J} n_j(1 - v_j). \tag{14.35}$$

The capacity constraints (14.30) and (14.31) ensure that each order $j \in J$ can be completed on or before its requested due date d_j (if $v_j = 1$) or on its delayed due date $t > d_j$ (if $v_j = 0$ and $w_{jt} = 1$). If the capacity constraints hold for all customer-requested due dates, then the objective function f_1 (14.28) takes on the value of zero, since $w_{jt} = 0\ \forall j \in J, t \in T: t > d_j$.

The solution of the MIP model *DDA* determines the subset $\{j \in J: v_j = 1\}$ of customer orders accepted with the customer-requested due dates d_j and the subset of the remaining orders $\{j \in J: v_j = 0\}$ with the postponed due dates such that the lost revenue due to the delay of orders is minimized. Denote by D_j, the requested or updated due date for each order $j \in J$, that is,

$$D_j = \begin{cases} d_j & \text{if } v_j = 1 \\ \sum_{t \in T: t > d_j} t w_{jt} & \text{if } v_j = 0. \end{cases}$$

The updated due dates D_j are interpreted as the deadlines at the order scheduling level.

14.5.2 Scheduling of Customer Orders

Given the updated due dates $D_j, j \in J$, the next decision step is to determine a nondelayed assembly schedule for the producer, that is, the assignment of customer orders to planning periods by the committed due dates (deadlines) over the horizon to minimize the finished products inventory holding cost.

The order assignment variables x_{jt} introduced in the monolithic bi-objective MIP model *INTb* are the basic variables used in the time-indexed MIP model *ODS* presented below.

Model *ODS*: *Deadline Scheduling of Customer Orders*

Minimize the holding cost of finished product inventory

$$f_2' = \sum_{t \in T} \sum_{\tau=1}^{t} \sum_{j \in J: d_j > t} \varphi_{4j} n_j x_{j\tau} \tag{14.36}$$

subject to

1. *Order-to-Period Nondelayed Assignment Constraints*:
   - each customer order is assigned to exactly one planning period not later than its due date,

$$\sum_{t \in T: a_j \le t \le D_j} x_{jt} = 1; \; j \in J \tag{14.37}$$

2. *Producer Capacity Constraints*:
   - in every period the demand on capacity at each assembly stage cannot be greater than the maximum available capacity in this period,

$$\sum_{j \in J} n_j x_{jt} / \varrho_{gj} \le m_g; \; g \in G, t \in T \tag{14.38}$$

3. *Supplier Cumulative Capacity Constraints*: (14.6) and
   - the cumulative demand on part type k required for customer orders assigned to periods 1 through t cannot be greater than the total cumulative production of this part type at all suppliers $i \in I_k$ by the corresponding periods $prev(t, T^{(i)}, \theta_i)$, that is, manufactured at i not later than θ_i periods before period t,

$$\sum_{l \in L} \sum_{j \in J_l} \sum_{\tau \in T: a_j \le \tau \le D_j, \tau \le t} b_{kl} n_j x_{j\tau} \le \sum_{i \in I_k: \text{next}(-e_i, T^{(i)}, \theta_i) \le t} \; \sum_{\tau=-e_i}^{\text{prev}(t, T^{(i)}, \theta_i)} r_{ik\tau};$$
$$k \in K, t \in T \tag{14.39}$$

4. *Nonnegativity and Integrality Conditions*: (14.17)

$$x_{jt} \in \{0, 1\}; \; j \in J, t \in T: a_j \le t \le D_j. \tag{14.40}$$

14.5.3 Scheduling Manufacturing and Delivery of Parts

The solution to *ODS* determines the optimal assignment of customer orders to planning periods $\{x_{jt}, j \in J, t \in T\}$ and thereby the optimal aggregate schedule for the finished products. Denote by N_{kt} the optimal cumulative usage in periods 1 through t of each part type k, that is, the optimal cumulative requirement for each part type k

$$N_{kt} = \sum_{l \in L} \sum_{j \in J_l} \sum_{\tau=1}^{t} b_{kl} n_j x_{j\tau}; \; k \in K, t \in T. \tag{14.41}$$

Now the base-level problem can be formulated as a time-indexed mixed integer program *MDS*. Given the cumulative requirements $N_{kt}, t \in T$ for all part types $k \in K$ by each period t, determine the schedules for manufacturing and delivery of parts for each supplier.

Model *MDS*: *Manufacturing and Delivery Scheduling*

Minimize the cost of holding the supply chain inventory of parts

$$\begin{aligned} f_2'' = & \sum_{k \in K} \sum_{i \in I_k} \sum_{t \in T} \sum_{\tau \in T^{(i)}: \tau \le t} \varphi_{1ik} (r_{ik\tau} - s_{ik,\text{next}(\tau, T^{(i)}, \theta_i)}) \\ & + \sum_{k \in K} \sum_{i \in I_k} \sum_{t \in T: t \ge \text{next}(-e_i, T^{(i)}, \theta_i)} \varphi_{2ik} \theta_i s_{ikt} \\ & + \sum_{k \in K} \sum_{t \in T} \varphi_{3k} \left(\sum_{i \in I_k} \sum_{\tau \in T^{(i)}: \tau \le t} s_{ik\tau} - N_{kt} \right) \end{aligned} \tag{14.42}$$

subject to

1. *Manufacturing Capacity Constraints*: (14.6)
2. *Part Manufacturing and Delivery Constraints*: (14.7) to (14.13)
3. *Part Demand Satisfaction Constraints*:
   - the cumulative deliveries of each part type k by period t cannot be less than the cumulative demand for this part type by this period,
   - the total delivery of each part type k is equal to the total demand of this part type,

$$\sum_{i \in I_k} \sum_{\tau \in T^{(i)}: \tau \le t} s_{ik\tau} \ge N_{kt}; \; k \in K, t \in T: t < h \tag{14.43}$$

$$\sum_{i \in I_k} \sum_{t \in T^{(i)}} s_{ikt} = N_{kh}; \; k \in K \tag{14.44}$$

4. *Coordinating Constraints*:
   - for each part type k and each period t the total production at each supplier $i \in I_k$ manufactured by period prev(t, $T^{(i)}$, θ_i), that is, not later than θ_i periods before t cannot be less than the cumulative requirement for this

part type by period t,

$$\sum_{i \in I_k : \text{next}(-e_i, T^{(i)}, \theta_i) \le t} \sum_{\tau = -e_i}^{\text{prev}(t, T^{(i)}, \theta_i)} r_{ik\tau} \ge N_{kt}; \; k \in K, t \in T \qquad (14.45)$$

5. *Nonnegativity and Integrality Conditions*: (14.17) to (14.19).

Coordinating constraint (14.45) is implied by (14.7) and (14.43), and it directly links the manufacturing of parts over the planning horizon with the producer cumulative requirement for parts, (cf. (14.16)).

In the objective function (14.42), f_2'' is the inventory holding cost of manufactured parts stored by the suppliers, in transit to the producer, or stored by the producer.

Recall that in the hierarchical approach the total supply chain inventory holding cost is split among two decision levels. Thus, the final value of the objective function f_2 (14.2) can be calculated as the sum of its partial solution values, determined for the middle-level problem *ODS* and the base-level problem *MDS*,

$$f_2 = f_2' + f_2''. \qquad (14.46)$$

Note that, unlike models *SMD(i)* (Section 13.6.3) for scheduling manufacturing and delivery of parts independently for each supplier i, in model *MDS* all suppliers are considered together. In the hierarchical approach presented in Chapter 13, the top-level due date setting problem is combined with the selection of a single supplier of parts for each customer order (see model *RCA* in Section 13.6.1), whereas in this chapter, the allocation of demand for parts among the suppliers is performed at the base level. As a result, the decomposition of the base-level problem into m independent scheduling problems is no longer possible.

14.6 COMPUTATIONAL EXAMPLES

In this section numerical examples and computational results are presented to illustrate possible applications of the two proposed approaches (integrated and hierarchical) and IP and MIP models for the coordinated scheduling of manufacturing, supply, and assembly in a supply chain with multiple suppliers without selection of a single part supplier for each customer order. The examples are modeled after a real-world customer-driven supply chain for high-tech products described in Section 13.7. The only difference is that the machine start-up times at suppliers are now negligible. (For a brief description of the planning horizon, manufacturers/suppliers, producer, part types, product types, and customer orders, see Section 13.7.)

Similar to the example data in Section 13.7, the set of product types is identical with the set of product-specific part types and for each order the required quantity of product-specific part types equals the quantity of the ordered products.

In the computational experiments the revenue parameters are identical for all orders and take on the following values:

$$\nu_j = 1,\ \beta_j = 0.2,\ \alpha_j = 0.02;\ j \in J,$$

that is, for each product in a nondelayed order the revenue ν_j is 100%, and for each product in a delayed order the fixed loss of revenue β_j is 20%, and the daily loss of revenue α_j is 2%.

The characteristics of the weighted-sum program $INTb_\lambda$ and Chebyshev program $INTb^\lambda$ for the integrated approach with the primary objective of maximizing the service level or the revenue are summarized in Tables 14.2 to 14.5. The results for the hierarchical approach are shown in Table 14.6. The size of each program is represented by the total number of variables, Var., number of binary variables, Bin., number of constraints, Cons., and number of nonzero elements in the constraint matrix, Nonz. The counts presented in the tables are taken from the models after presolving. The last two columns of the tables present the solution values and CPU time in seconds required to find the optimal solution and to prove its optimality or GAP% if optimality is not proven within a specified CPU time limit. The computational experiments have been performed on a PC Pentium IV, 1.8 GHz, 1 GB RAM using AMPL/CPLEX 11 with a traditional branch and bound search and strong branching option.

The solution results for the integrated approach and the characteristics of the mixed integer programs $INTb_\lambda$ and $INTb^\lambda$ for selected λ and for different primary objective functions are summarized, respectively in Tables 14.2, 14.4 and 14.3, 14.5. For the integrated approach no proven optimal solution was determined in one hour limit of CPU time, and GAP for the best solution found ranged from 0.01% for the weighted-sum program $INTb_\lambda$ and the maximum service level as a primary objective to 9.26% for the Chebyshev program $INTb^\lambda$ and the maximum revenue. In the tables the solution value of each objective function is presented along with the corresponding associated values of the complementary objective functions (in parentheses). In total, six and 11 different nondominated solutions were found,

Table 14.2 Computational Results for the Weighted-Sum Program $INTb_\lambda$: Maximum Revenue

λ	Var.	Bin.	Cons.	Nonz.	$f_1(f_1^O, f_1^P), f_2$[a]	GAP%[b]
0.99	8236	7236	1944	180,960	199(3,825), 747	2.38%
0.9	8236	7236	1944	180,960	199(3,825), 747	2.13%
0.8	8236	7236	1944	180,960	199(3,825), 747	1.93%
0.7	8236	7236	1944	180,960	200(2,830), 746	2.30%
0.6	8236	7236	1944	180,960	201(2,835), 745	2.00%
0.5	8236	7236	1944	180,960	202(2,840), 744	1.36%

[a] $f_1(f_1^O, f_1^P), f_2$ = lost revenue (number of tardy orders, number of tardy products), inventory holding cost, respectively.

[b] GAP% for 3600 CPU seconds on a PC Pentium IV, 1.8 GHz, 1 GB RAM, CPLEX 11.

Table 14.3 Computational Results for the Chebyshev Program $INTb^{\lambda}$: Maximum Revenue

λ	Var.	Bin.	Cons.	Nonz.	$f_1(f_1^O, f_1^P), f_2$[a]	GAP%[b]
0.99	8237	7336	1916	175,637	199(2,825), 747	5.35%
0.9	8237	7336	1916	175,637	200(2,830), 746	9.26%
0.8	8237	7336	1916	175,637	201(2,835), 745	8.72%
0.7	8237	7336	1916	175,637	202(2,840), 744	1.19%
0.6	8237	7336	1916	175,637	233(5,950), 743	1.16%
0.5	8237	7336	1916	175,637	252(4,1040), 742	1.40%

[a]$f_1(f_1^O, f_1^P), f_2$ = lost revenue (number of tardy orders, number of tardy products), inventory holding cost, respectively. $(\underline{f}_1, \underline{f}_2) = (190, 640)$,

[b]GAP% for 3600 CPU seconds on a PC Pentium IV, 1.8 GHz, 1 GB RAM, CPLEX 11.

respectively, for the maximum revenue and the maximum service level as a primary objective function. The trade-off between the inventory holding cost and the number of tardy orders or the lost revenue is shown in Figure 14.2, where the two convex efficient frontiers of the bi-objective models are presented. The results emphasize the effect of varying service level/cost preference of the decisionmaker.

The solution results for the hierarchical approach and the characteristics of the mixed integer programs *DDA*, *ODS*, and *MDS* are summarized in Table 14.6. While the integrated approach is capable of finding a subset of nondominated solutions for the coordinated scheduling problem, the hierarchical approach was able to find the best solution with respect to the primary objective function. On the other hand, the hierarchical approach may lead to a greater total supply chain inventory holding cost f_2, because parts and products inventories are considered separately at two decision levels. Unlike for the integrated model, the computational effort required for the hierarchical approach is very small and proven optimal solutions have been reached in a short CPU time for each level problem. For both approaches, the coordinating constraints ((14.16) and (14.45)) have reduced CPU time up to 15%.

Table 14.4 Computational Results for the Weighted-Sum Program $INTb_{\lambda}$: Maximum Service Level

λ	Var.	Bin.	Cons.	Nonz.	$f_1^O(f_1^P, f_1), f_2$[a]	GAP%[b]
0.99	8236	7336	1944	180,960	1(1325,530), 762	0.14%
0.9	8236	7336	1944	180,960	3(11304,4178), 708	0.01%
0.8	8236	7336	1944	180,960	9(17194,5743), 679	0.30%
0.7	8236	7336	1944	180,960	14(22954,7045), 658	0.23%
0.6	8236	7336	1944	180,960	17(24659,7468), 651	0.05%
0.5	8236	7336	1944	180,960	19(28009,8474), 649	0.17%

[a]$f_1^O(f_1^P, f_1), f_2$ = number of tardy orders (number of tardy products, lost revenue), inventory holding cost, respectively.

[b]GAP% for 3600 CPU seconds on a PC Pentium IV, 1.8 GHz, 1 GB RAM, CPLEX 11.

Table 14.5 Computational Results for the Chebyshev Program $INTb^{\lambda}$: Maximum Service Level

λ	Var.	Bin.	Cons.	Nonz.	$f_1^O(f_1^P, f_1), f_2$ [a]	GAP% [b]
0.99	8237	7336	1916	175,637	2(2990,882), 728	6.98%
0.9	8237	7336	1916	175,637	3(11304,4178), 708	0.44%
0.8	8237	7336	1916	175,637	4(11914,4325), 702	0.50%
0.7	8237	7336	1916	175,637	5(19199,6215), 697	8.42%
0.6	8237	7336	1916	175,637	6(13229,6391), 692	6.25%
0.5	8237	7336	1916	175,637	7(20579,6512), 688	9.10%

[a] $f_1^O(f_1^P, f_1), f_2$ = number of tardy orders (number of tardy products, lost revenue), inventory holding cost, respectively. $(\underline{f}_1^O, \underline{f}_2) = (0, 640)$.

[b] GAP% for 3600 CPU seconds on a PC Pentium IV, 1.8 GHz, 1 GB RAM, CPLEX 11.

Comparison of the two solution approaches, the weighted-sum approach and the reference point approach, to bi-objective mixed integer programming indicates that both approaches are capable of producing similar subsets of nondominated schedules. However, the weighted-sum program computationally outperforms the Chebyshev program (in which the relative importance of the objective functions is weighted in the constraint set) and generates solutions with a smaller GAP.

Examples of the aggregate production schedule ($\sum_{j\in J} n_j x_{jt}$, $t \in T$) for the producer and the aggregate manufacturing schedule ($\sum_{k\in K} r_{ikt}$, $t \in T^{(i)}$) and the delivery schedule ($\sum_{k\in K} s_{ikt}$, $t \in T^{(i)}$) for each supplier $i \in I$ are shown in Figures 14.3 and 14.4 for the integrated approach and different primary objective functions.

Table 14.6 Computational Results for the Hierarchical Approach

Model	Var.	Bin.	Cons.	Nonz.	Solution values [a]	CPU [b]
Maximum revenue						
DDA	4363	3883	1119	397,563	$f_1 = 193(f_1^O = 4, f_1^P = 800)$	2.14
ODS	4351	3871	777	58,776	$f_2' = 150$	12.05
MDS	942	42	1278	16,022	$f_2'' = 594$	1.53
					$f_2 = f_2' + f_2'' = 744$	
Maximum service level						
DDA	4363	3883	1119	397,563	$f_1^O = 1(f_1 = 964, f_1^P = 2190)$	1.51
ODS	4382	3902	782	59,329	$f_2' = 230$	9.80
MDS	942	42	1278	16,022	$f_2'' = 593$	1.58
					$f_2 = f_2' + f_2'' = 823$	

[a] f_1, f_1^O, f_1^P, f_2 = lost revenue, number of tardy orders, number of tardy products, inventory holding cost, respectively.

[b] CPU seconds for proving optimality on a PC Pentium IV, 1.8 GHz, 1 GB RAM /CPLEX 11.

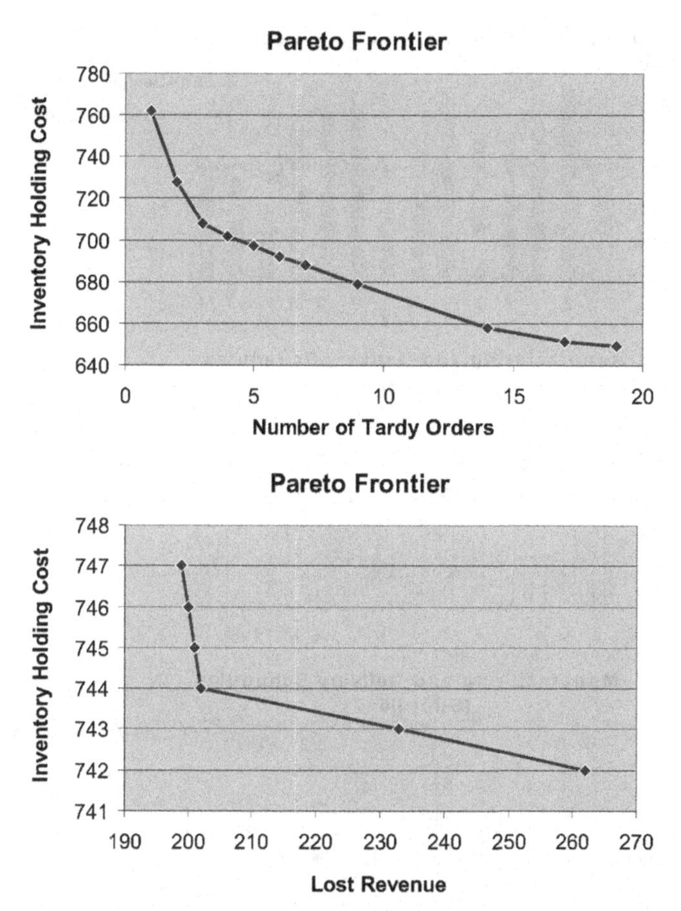

Figure 14.2 Efficient frontiers: number of tardy orders against inventory holding cost and lost revenue against inventory holding cost.

The results indicate that the shorter the transportation time from a supplier, the greater the total demand for parts of customer orders assigned to that supplier, and the delivery of parts are more frequent and better resemble the just-in-time supplies. The selection of the maximum service level as a primary objective function leads to a better leveled aggregate production of finished products and a better leveled aggregate manufacturing and delivery of parts. However, the total supply chain inventory is smaller for the maximum revenue as a primary objective function.

Figure 14.5 compares the total supply chain inventory of parts and products for the hierarchical approach and the integrated approach with different primary objective functions: maximum revenue (minimum lost revenue, f_1) and maximum service level (minimum number of tardy orders, f_1^O). The figure compares results obtained for the integrated approach with various weights λ in the weighted-sum program.

The results obtained for the hierarchical approach, which is based on the lexicographic optimization, are similar to those found for the integrated approach with

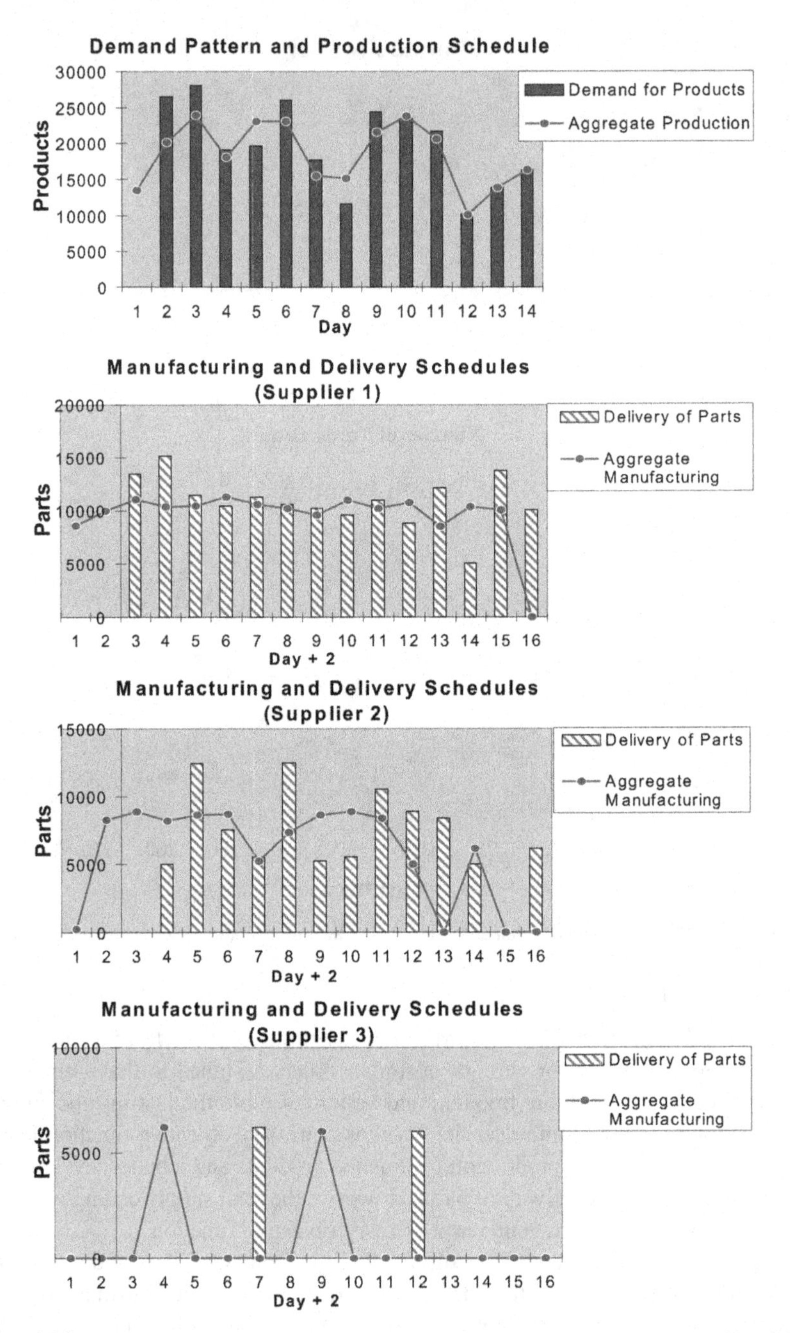

Figure 14.3 Manufacturing, delivery, and production schedules for the integrated approach (weighted-sum program $INTb_{\lambda=0.5}$, primary objective is maximum revenue).

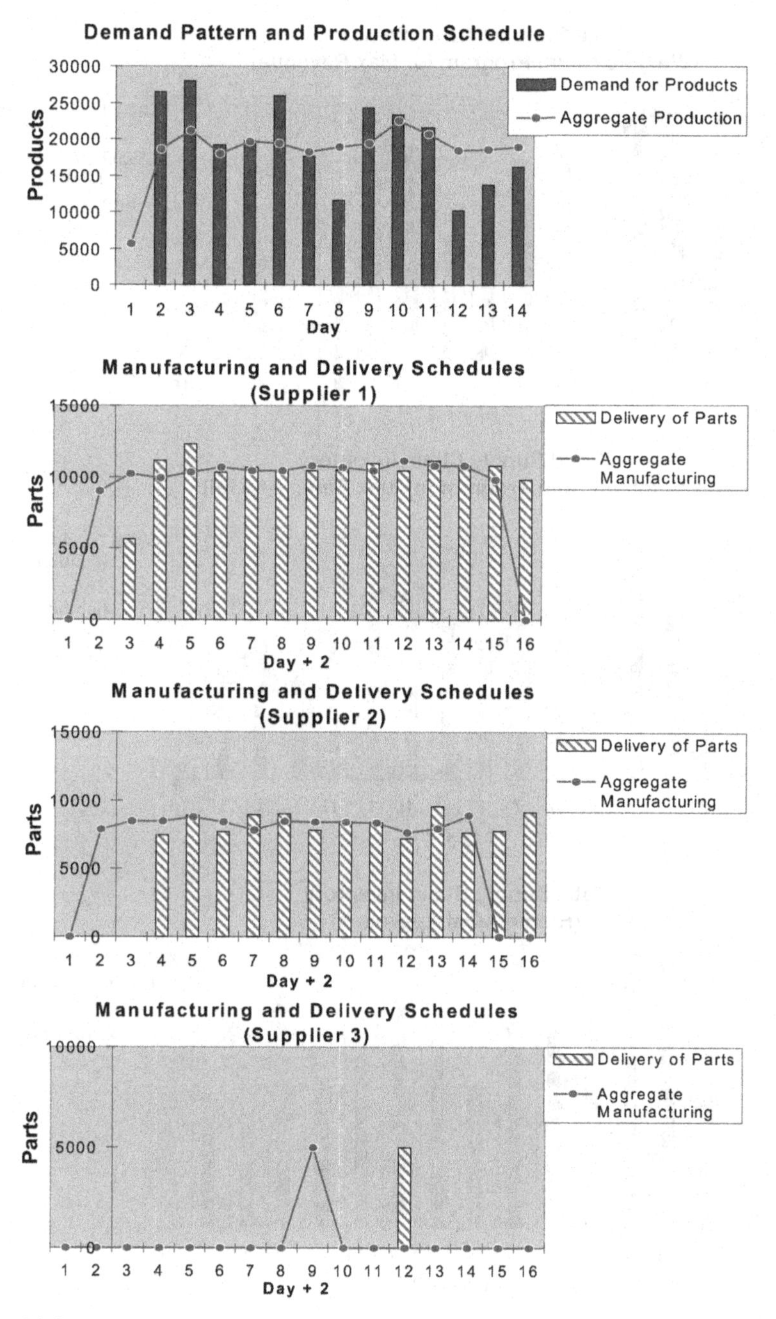

Figure 14.4 Manufacturing, delivery, and production schedules for the integrated approach (weighted-sum program $INTb_{\lambda=0.5}$, primary objective is maximum service level).

Total Supply Chain Inventory
(Weighted-Sum Program for Max Revenue)

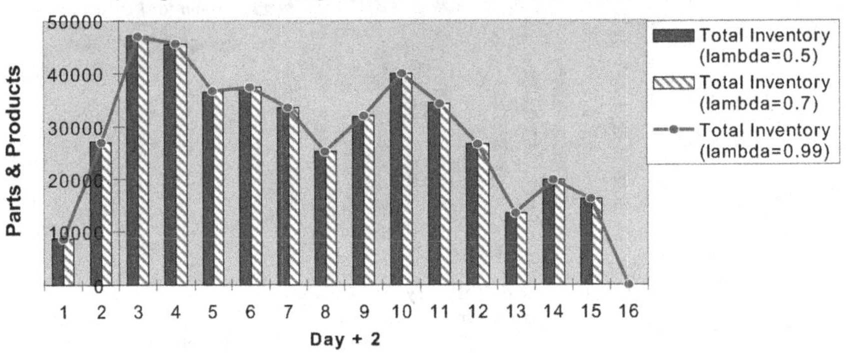

Total Supply Chain Inventory
(Weighted-Sum Program for Max Service Level)

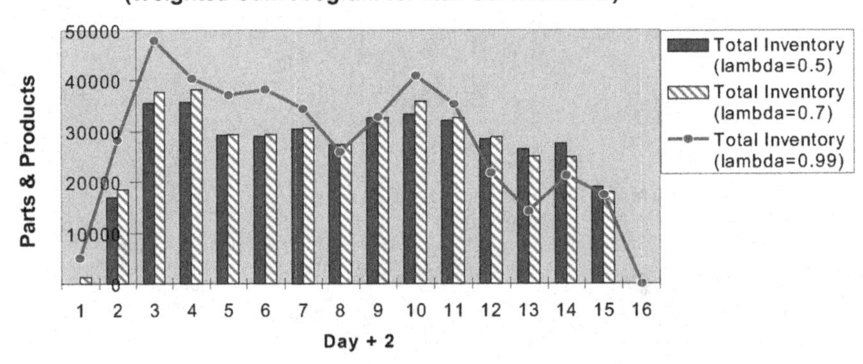

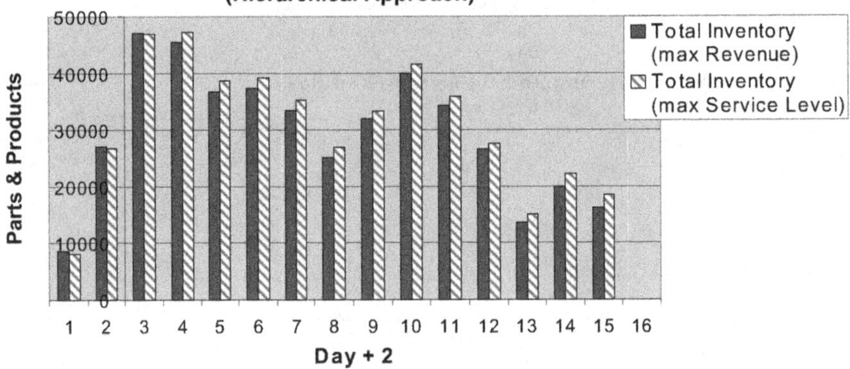

Figure 14.5 Total supply chain inventory: integrated vs. hierarchical approach.

a high weight λ given to the primary objective function. Figure 14.5 shows that the total supply chain inventory of parts and products varies similarly over the planning horizon for both the integrated and the hierarchical approach.

14.7 COMMENTS

There is a growing interest in applying the revenue management principles to coordinated supply chain scheduling, in particular to order acceptance (see Chapter 5) and scheduling decisions in a make-to-order manufacturing. In a make-to-order manufacturing environment, the sales department has to price a continuous stream of bids and requests for quotations. A large manufacturer may have to respond to over 250,000 requests for quotations every year (e.g., Talluri and van Ryzin, 2004). However, the literature on revenue management in make-to-order manufacturing is limited (e.g., Harris and Pinder, 1995; Akkan, 1997; Bertrand and van Ooijen, 2000; Barut and Sridharan, 2005; Geunes et al., 2006) and mostly restricted to the single machine case (e.g., Charnsirisaksul et al., 2004; Oğuz et al., 2010). On the other hand, the literature on deterministic scheduling with due dates objective is abundant, though the research on deadline scheduling is again mostly restricted to the single machine case, for example, Blażewicz et al. (2007).

The material presented in this chapter is based on results presented by Sawik (2010b), where the integrated and hierarchical approaches to bi-objective coordinated supply chain scheduling are compared and the two alternative primary objective functions, the maximum revenue and the maximum service level, are combined with the secondary objective of minimizing total supply chain inventory holding cost.

The proposed hierarchical approach coincides with a typical pull planning framework in a make-to-order environment and the corresponding sequential decision making. First, the due dates (deadlines for completing customer orders) are determined to ensure scheduling feasibility and a high revenue or a high service level, then the orders are assigned by the deadlines to planning periods over the horizon to minimize product inventory holding cost, and finally the schedules for manufacturing and delivery of parts from each supplier to the producer are found to minimize part inventory holding cost.

While the hierarchical approach is capable of finding a single proven optimal solution in a very short CPU time, the integrated approach, where all decision problems are solved simultaneously, provides the decisionmaker with a subset of nondominated schedules.

It is worth comparing the hierarchical approaches presented in this chapters and in Chapter 13 with respect to decisions on the allocation of demand for parts among suppliers. In Chapter 13, the allocation decision belongs to the initial decisions; at the top-level problem (*RCA*), the selection of a single part supplier for each customer order is combined with due date setting for this order. In contrast, in this chapter the allocation decision is postponed until the final scheduling of manufacturing and delivery of parts, and each customer order can be partially provided with the required parts by different suppliers (problem *MDS*). While the former approach requires a close cooperation between suppliers and the producer and may lead to a better global optimum, the latter approach is more often encountered in practice as it allows the producer to mitigate the impact of various disturbances that may occur in the supply chain.

The various simplifying assumptions introduced in the models proposed in this chapter can be relaxed in a more general setting of the coordinated supply chain scheduling with multiple suppliers (see Comments 13.8).

EXERCISES

14.1 Suppose that each supplier i has an output buffer of capacity B_i for holding completed parts waiting for shipment to the producer, and the producer has an input buffer of capacity $B1$ and an output buffer of capacity $B2$ for holding, respectively, purchased parts waiting for assembly and finished products waiting for delivery to customers. In order to account for the buffer's limited capacity, the following constraints should be added to model *INTb* for the integrated scheduling of manufacturing, supply and assembly:

$$\sum_{k \in K} \left(\sum_{\tau=-e_i}^{t} (r_{ik\tau} - s_{ik,\text{next}(\tau,T^{(i)},\theta_i)}) \right) \leq B_i; \ i \in I, t \in T^{(i)},$$

$$\sum_{i \in I} \sum_{k \in K} \left(\sum_{\tau=-e_i}^{t} s_{ik\tau} - \sum_{l \in L} \sum_{j \in J_l} \sum_{\tau=a_j}^{t} b_{kl} n_j x_{j\tau} \right) \leq B1; \ t \in T,$$

$$\sum_{j \in J} \sum_{\tau \in T: a_j \leq \tau \leq t < d_j} n_j x_{j\tau} \leq B2; \ t \in T.$$

(**a**) Formulate a mixed integer program for finding the minimum total capacity of all supply chain buffers, $B1 + B2 + \sum_{i \in I} B_i$, required to complete all customer orders during the planning horizon.

14.2 In model *INTb* replace the secondary objective function f_2, (14.2), of minimizing the supply chain inventory holding cost by the objective of minimizing the maximum total supply chain inventory level, $I_{\max}$. Formulate the secondary objective function and explain the relation between the optimal value of $I_{\max}$ and the optimal value of the total capacity of all supply chain buffers, $B1 + B2 + \sum_{i \in I} B_i$ (Exercise 14.1a.)

14.3 Apply lexicographic optimization to formulate a sequence of two mixed integer programs for the hierarchical approach to the bi-objective coordinated scheduling with the secondary objective function $f_2 = I_{\max}$ (Exercise 14.2.)

14.4 Formulate coordinating constraints (14.16) for the integrated scheduling of manufacturing, supply, and assembly.

(**a**) In the case of a supply chain with multiple producers.

(**b**) In the case of a supply chain with one supplier and multiple producers.

14.5 Comparing the results of computational examples illustrated in Figures 14.2 to 14.4, answer the following questions.

(**a**) Why does the schedule of customer orders (i.e., the aggregate production schedule) obtained by the integrated approach with the primary objective of maximizing revenue follow the profile of demand pattern, whereas for the primary objective of maximizing service level, the schedule is leveled over the planning horizon?

(b) Why, for the integrated approach with the primary objective function of maximizing service level, is the total supply chain inventory smaller than for the primary objective of maximizing revenue, except for a very small weight $1 - \lambda$ of the supply chain inventory holding cost in the weighted-sum program?

(c) Why, for the hierarchical approach and both primary objective functions, does the total supply chain inventory vary similarly over the planning horizon?

References

Abadi, I.N.K., N.G. Hall and C. Sriskandarajah (2000). Minimizing cycle time in a blocking flowshop, *Operations Research* **48**, 177–180.

Agnetis, A., D. Pacciarelli and F. Rossi (1997). Batch scheduling in a two-machine flow shop with limited buffer, *Discrete Applied Mathematics* **72**(3), 243–260.

Agnetis, A., F. Rossi and G. Gristina (1998). An exact algorithm for the batch sequencing problem in a two-machine flow shop with limited buffer, *Naval Research Logistics* **45**(2), 141–164.

Agnetis, A., N.G. Hall and D. Pacciarelli (2006). Supply chain scheduling: Sequence coordination, *Discrete Applied Mathematics* **154**, 2044–2063.

Ahmadi, J., R. Ahmadi, H. Matsuo and D. Tirupati (1995). Component fixture positioning/sequencing for printed circuit board assembly with concurrent operations, *Operations Research* **43**, 444–457.

Aissaoui, N., M. Haouari and E. Hassini (2007). Supplier selection and order lot sizing modeling: A review, *Computers & Operations Research* **34**, 3516–3540.

Akkan, C. (1997). Finite-capacity scheduling-based planning for revenue-based capacity managament, *European Journal of Operational Research* **100**, 170–179.

Alfieri, A. (2009). Workload simulation and optimisation in multi-criteria hybrid flowshop scheduling: A case study, *International Journal of Production Research* **47**(18), 5129–5145.

Alves, M.J. and J. Climaco (2007). A review of interactive methods for multiobjective integer and mixed-integer programming, *European Journal of Operational Research* **180**, 99–115.

Amaro, A.C.S. and A.P.F.D. Barbosa-Póvoa (2008). Supply chain management with optimal scheduling, *Industrial Engineering and Chemical Research* **47**(1), 116–132.

Anthony, R.N. (1965). *Planning and Control Systems. A Framework for Analysis*, Division of Research, Harvard Business School, Boston.

Arshinder, A. Kanda and S.G. Deshmukh (2008). Supply chain coordination: Perspectives, empirical studies and research directions, *International Journal of Production Economics* **115**, 316–335.

Baker, K.R. and D. Trietsch (2009). *Principles of Sequencing and Scheduling*, Wiley, Hoboken, NJ.

Balakrishnan, A. and F. Vanderbeck (1999). A tactical planning model for mixed-model electronics assembly operations, *Operations Research* **47**, 395–409.

Bard, J.F., R.W. Clayton and T.A. Feo (1994). Machine setup and component placement in printed circuit board assembly, *International Journal of Flexible Manufacturing Systems* **6**, 5–31.

Barut, S.P. and V. Sridharan (2005). Revenue management in order-driven production systems, *Decision Sciences* **36**(2), 287–316.

Basnet, C. and J.M.Y. Leung (2005). Inventory lot-sizing with supplier selection, *Computers and Operations Research* **32**, 1–14.

Berger, P.D., A. Gerstenfeld and A.Z. Zeng (2004). How many suppliers are best? A

Scheduling in Supply Chains Using Mixed Integer Programming. By Tadeusz Sawik

decision-analysis approach, *Omega: The International Journal of Management Science* **32**(1), 9–15.

Berger, P.D. and A.Z. Zeng (2006). Single versus multiple sourcing in the presence of risks, *Journal of the Operational Research Society* **57**(3), 250–261.

Bertazzi, L. (2003). Rounding off the optimal solution of the economic lot size problem, *International Journal of Production Economics* **81–82**, 385–392.

Bertrand, J.W.M., and H.P.G. van Ooijen (2000). Customer order lead times for production based on lead times and tardiness costs, *International Journal of Production Economics* **64**, 257–265.

Bilge, U. and G. Ulusoy (1995). A time window approach to simultaneous scheduling of machines and material handling system in an FMS, *Operations Research* **43**, 1058–1070.

Błażewicz, J., K. Ecker, G. Schmidt and J. Weglarz (1994). *Scheduling in Computer and Manufacturing Systems*, Springer, Berlin.

Błażewicz, J., K. Ecker, E. Pesch, G. Schmidt and J. Weglarz (2007). *Handbook on Scheduling. From Theory to Applications*, Springer, Berlin.

Brah, S.A. and J.L. Hunsucker (1991). Branch and bound algorithm for the flow shop with multiple processors, *European Journal of Operational Research* **51**, 88–99.

Campbell, H.G., R.A. Dudek and M.L. Smith (1970). A heuristic algorithm for the n job, m machine sequencing problem, *Management Science* **16**, B630–B637.

Carravilla, M.A. and J. Pinho de Sousa (1995). Hierarchical production planning in a make-to-order company: A case study, *European Journal of Operational Research* **86**, 43–56.

Chahar, K. and K. Taaffe (2009). Risk averse demand selection with all-or-nothing orders, *Omega: The International Journal of Management Science* **37**(5), 996–1006.

Chandra, P. and M.L. Fisher (1994). Coordination of production and distribution planning, *European Journal of Operational Research* **72**(3), 503–517.

Charnsirisaksul, K., P. Griffin and P. Keskinocak (2004). Order selection and scheduling with lead time flexibility, *IIE Transactions* **36**, 697–707.

Chaudhry, S.S., F.G. Forst and J.L. Zydiak (1993). Vendor selection with price breaks, *European Journal of Operational Research* **70**, 52–66.

Chauhan, S.S., V. Gordon and J.-M. Proth (2007). Scheduling in a supply chain environment, *European Journal of Operational Research* **183**(3), 961–970.

Che, Z.H. and H.S. Wang (2008). Supplier selection and supply quantity allocation of common and non-common parts with multiple criteria under multiple products, *Computers and Industrial Engineering* **55**, 110–133.

Chen, C.-Y., Z.-Y. Zhao and M.O. Ball (2001). Quantity and due date quoting available to promise, *Information Systems Frontiers* **3**(4), 477–488.

Chen, D.-S., R.G. Batson and Y. Dang (2010). *Applied Integer Programming: Modeling and Solution*, Wiley, Hoboken, NJ.

Chen, K. and P. Ji (2007). A mixed integer programming model for advanced planning and scheduling (APS), *European Journal of Operational Research* **181**, 515–522.

Chen, Z.-L. (2004). Integrated production and distribution operations: Taxonomy, models and review. In D. Simchi-Levi, S.D. Wu and Z.-J. Shen, Eds., *Handbook of Quantitative Supply Chain Analysis: Modeling in the E-Business Era*. Kluwer Academic Publishers, Boston, MA.

Chen, Z.-L. and G.L. Vairaktarakis (2005). Integrated scheduling of production and distribution operations, *Management Science* **51**(4), 614–628.

Chen, Z.-L. and G. Pundoor (2006). Order assignment and scheduling in a supply chain, *Operations Research* **54**(3), 555–572.

Chen, Z.-L. and N.G. Hall (2007). Supply chain scheduling: Conflict and cooperation in assembly system, *Operations Research* **55**(6), 1072–1089.

Cohen, M.A. and H. Kunreuther (2007). Operations risk management: Overview of Paul Kleindorfer's contributions, *Production and Operations Management* **16**(5), 525–541.

Corti, D., A. Pozzetti and M. Zorzini (2006). A capacity-driven approach to establish reliable due dates in MTO environment, *International Journal of Production Economics* **104**, 536–554.

Dauzere-Peres, S. and J.-B. Lasserre (1997). Lot streaming in job-shop scheduling, *Operations Research* **45**(4), 584–595.

Deane, R.H. and S.H. Moon (1992). Work flow control in the flexible flow line, *International Journal of Flexible Manufacturing Systems*, **8**(3–4), 217–235.

Demirtas, E.A. and O. Ustun (2008). An integrated multiobjective decision making process for supplier selection and order allocation, *Omega: The International Journal of Management Science* **36**, 76–90.

Dolgui, A., B. Finel, N. Guschinsky, G. Levin and F. Vernadat (2006). MIP approach to balancing transfer lines with blocks of parallel operations, *IIE Transactions* **38**, 869–882.

Drexl, A. and A. Kimms (1997). Lotsizing and scheduling—survey and extensions, *European Journal of Operational Research* **99**, 221–235.

Ebben, M.J.R., E.W. Hans and F.M. Olde Weghuis (2005). Workload based acceptance in job shop environments, *OR Spectrum* **27**, 107–122.

Ellis, K.P., F.J. Vittes and J.E. Kobza (2001). Optimizing the performance of a surface mount placement machine, *IEEE Transactions on Electronics Packaging Manufacturing* **24**(3), 160–170.

Erengüc, S.S., N.C. Simpson and A.J. Vakharia (1999). Integrated production/distribution planning in supply chains: An invited review, *European Journal of Operational Research*, **115**, 219–236.

Feo, T.A., J.F. Bard and S.D. Holland (1995). Facility-wide planning and scheduling of printed wiring board assembly, *Operations Research* **43**, 219–230.

Federgruen, A. and Y.-S. Zheng (1995). Efficient algorithms for finding optimal power-of-two policies for production/distribution systems with general setup costs, *Operations Research* **43**, 458–470.

Fourer, R., D.M. Gay and B.W. Kernigham (2003). *AMPL, A Modeling Language for Mathematical Programming*. Duxbury Press, Pacific Grove, CA.

Gallego, G., M. Queyranne and D. Simchi-Levi (1996). Single resource multi-item inventory systems, *Operations Research* **44**, 580–595.

Garey, M.R. and D.S. Johnson (1979). *Computers and Intractability: A Guide to the Theory of NP-Completeness*. W.H. Freeman, San Francisco.

Gershwin, S.B., Y. Dallery, Ch.T. Papadopoulos and J. MacGregor Smith, Eds. (2002). *Analysis and Modeling of Manufacturing Systems*, Springer, New York.

Geunes, J., H.E. Romeijn and K. Taaffe (2006). Requirements planning with pricing and order selection flexibility, *Operations Research* **54**(2), 394–401.

Ghosh, S. and R.J. Gagnon (1989). A comprehensive literature review and analysis of the design, balancing and scheduling of assembly systems, *International Journal of Production Research* **27**, 637–670.

Gotoh, J. and Y. Takano (2007). Newsvendor solutions via conditional value-at-risk minimization, *European Journal of Operational Research* **179**(1), 80–96.

Goyal, S.K. and S.G. Deshmukh (1992). Integrated procurement—production systems: A review, *European Journal of Operational Research* **62**, 1–10.

Goyal, S.K. and S.G. Deshmukh (1997). Integrated procurement-production system in a just-in-time environment—modeling and analysis, *Production Planning and Control* **8**(1), 31–36.

Graves, S.C., H.C. Meal, D. Stefek and A.H. Zeghmi (1983). Scheduling of reentrant flow shops, *Journal of Operations Management* **3**, 197–207.

Greene, J.T. and P.R. Sadowski (1986). A mixed integer program for loading and scheduling multiple flexible manufacturing cells, *European Journal of Operational Research* **24**, 379–386.

Guinet, A.G.P. and A.A. Solomon (1996). Scheduling hybrid flowshops to minimize maximum tardiness or maximum completion time, *International Journal of Production Research* **34**, 1643–1654.

Guschinskaya, O. and A. Dolgui (2010). Comparison of exact and heuristic methods for a transfer line balancing problem, *International Journal of Production Economics*. **120**(2), 276–286.

Güder, F., J.L. Zydiak and S.S. Chaudhry (1995). Non-stationary ordering policies for multi-item inventory systems subject to a single resource constraint, *Journal of Operational Research Society* **46**, 1145–1152.

Güder, F. and J.L. Zydiak (2000). Fixed cycle ordering policies for capacitated multiple item inventory systems with quantity discounts, *Computers & Industrial Engineering* **38**, 67–77.

Hall, N.G., M. Lesaoana and C.N. Potts (2001). Scheduling with fixed delivery dates, *Operations Research* **49**(1), 134–144.

Hall, N.G., M.E. Posner and C.N. Potts (1998a). Scheduling with finite capacity output buffers, *Operations Research* **46**, Suppl. No. 3, S84–S89.

Hall, N.G., M.E. Posner and C.N. Potts (1998b). Scheduling with finite capacity input buffers, *Operations Research* **46**, Suppl. No. 3, S154–S159.

Hall, N.G. and C.N. Potts (2003). Supply chain scheduling: Batching and delivery, *Operations Research* **51**(4), 566–584.

Harris, F.H. and J.P. Pinder (1995). Revenue management approach to demand management and order booking in assemble-to-order manufacturing, *Journal of Operations Management* **13**(4), 299–309.

Hax, A.C. and H.C. Meal (1975). Hierarchical integration of production planning and scheduling. In M.A. Geisler, Ed., *Studies in Management Science, vol. I, Logistics*, North Holland, Amsterdam, 53–69.

Hegedus, M.G. and W.J. Hopp (2001). Due date setting with supply constraints in systems using MRP. *Computers & Industrial Engineering*, **39**, 293–305.

Ho, W., X. Xu and K.D. Prasanta (2010). Multi-criteria decision making approaches for supplier evaluation and selection: A literature review, *European Journal of Operational Research* **202**, 16–24.

Hopp, W. and M. Spearman (1996). *Factory Physics: Foundations of Manufacturing Management*. McGraw-Hill, New York.

Hunsucker, J.L. and J.R. Shah (1994). Comparative performance analysis of priority rules in a constrained flow shop with multiple processors environment, *European Journal of Operational Research* **72**, 102–114.

Jackson, P.L., W.L. Maxwell and J.A. Muckstadt (1988). Determining optimal reorder intervals in capacitated production/distribution systems, *Management Science* **35**, 938–958.

Jain, A., M.E. Johnson and E. Safai (1996). Implementation of setup optimization on the shop floor, *Operations Research* **44**, 843–851.

Janak, S.L., C.A. Floudas, J. Kallrath and N. Vormbrock (2006). Production scheduling of a large-scale industrial batch plant. II. Reactive scheduling, *Industrial and Engineering Chemistry Research* **45**(25), 8253–8269.

Jayaraman, V., R. Srivastava and W.C. Benton (2006). Supplier selection and order quantity allocation: A comprehensive model, *Journal of Supply Chain Management* **35**(2), 50–58.

Jiang, J. and W. Hsiao (1994). Mathematical programming for the scheduling problem with alternate process plans in FMS, *Computers and Industrial Engineering* **27**(10), 15–18.

Jin, Z.H., K. Ohno, T. Ito and S.E. Elmaghraby (2002). Scheduling hybrid flowshops in printed circuit board assembly lines, *Production and Operations Management* **11**(2), 216–230.

Johnson, S.M. (1954). Optimal two- and three-stage production schedules with setup times included, *Naval Research Logistics Quarterly* **1**, 61–68.

Kaczmarczyk, W., T. Sawik, A. Schaller and T.M. Tirpak (2004). Optimal versus heuristic scheduling of surface mount technology lines, *International Journal of Production Research* **42**(10), 2083–2110.

Kaczmarczyk, W., T. Sawik, A. Schaller and T. Tirpak (2006). Production planning and coordination in customer driven supply chains, *Wybrane Zagadnienia Logistyki Stosowanej* **3**, 81–89.

Kasilingam, R.G. and C.P. Lee (1996). Selection of vendors: A mixed-integer programming approach, *Computers and Industrial Engineering* **31**, 347–350.

Kim, Y.-D., H.-G. Lim and M.-W. Park (1996). Search heuristics for a flowshop scheduling problem in a printed circuit board assembly process, *European Journal of Operational Research* **91**, 124–143.

Kimms, A. (1997). *Multi-Level Lot Sizing and Scheduling: Methods for Capacitated Dynamic and Deterministic Models*. Physica, Heidelberg.

Kis, T. and E. Pesch (2005). A review of exact solution methods for the non-preemptive multiprocessor flowshop problem, *European Journal of Operational Research* **164**(3), 592–608.

Kleindorfer, P.R. and G.H. Saad (2005). Managing disruption risks in supply chains, *Production and Operations Management* **14**(1), 53–68.

Kochhar, S. and R.J.T. Morris (1987). Heuristic methods for flexible flow line scheduling, *Journal of Manufacturing Systems* **6**(4), 299–314.

Kolisch, R. (2000). Integration of assembly and fabrication for make-to-order production, *International Journal of Production Economics* **68**, 287–306.

Kouvelis, P. and S. Karabati (1999). Cyclic scheduling in synchronous production lines, *IIE Transactions* **31**(8), 709–719.

Kreipl, S. and J.T. Dickersbach (2008). Scheduling coordination problems in supply chain planning, *Annals of Operations Research* **106**, 103–122.

Kumar, R. and H. Li (1995). Integer programming approach to printed circuit board assembly, *IEEE Transactions on Components Packaging and Manufacturing Technology* **B 18**, 720–727.

Kurz, M.E. and R.G. Askin (2004). Scheduling flexible flow lines with sequence-dependent setup times, *European Journal of Operational Research* **159**(1), 66–82.

Kyparisis, G.J. and C. Koulamas (2006). Flexible flowshop scheduling with uniform parallel machines, *European Journal of Operational Research* **168**(3), 985–997.

Lee, C.-Y. and G.L. Vairaktarakis (1994). Minimizing makespan in hybrid flowshops, *Operations Research Letters* **16**, 149–158.

Lee, C.-Y. and Z.-L. Chen (2001), Machine scheduling with transportation considerations, *Journal of Scheduling* **4**, 3–24.

Lee, G.-C. and Y.-D. Kim (2004). A branch-and-bound algorithm for a two-stage hybrid flowshop scheduling problem minimizing total tardiness, *International Journal of Production Research* **42**(22), 4731–4743.

Lee, W. (2005). A joint economic lot size model for raw material ordering, manufacturing setup, and finished goods delivering, *Omega: The International Journal of Management Science* **33**, 163–174.

Lewis, H.F. and S.A. Slotnick (2002). Multi-period job selection: Planning work loads to maximize profit, *Computers & Operations Research* **29**, 1081–1098.

Li, L. and Z.B. Zabinsky (2009). Incorporating uncertainty into a supplier selection problem, *International Journal of Production Economics* doi:10.1016/j.ijpe.2009.11.007.

Liao, D.-Y., S.-C. Chang, K.W. Pei and C.-M. Chang (1996). Daily scheduling for R&D semiconductor fabrication, *IEEE Transactions on Semiconductor Manufacturing* **9**, 550–560.

Liao, Z. and J. Rittscher (2007). Integration of supplier selection, procurement lot sizing and carrier selection under dynamic demand conditions, *International Journal of Production Economics* **107**, 502–510.

Linn, R. and W. Zhang (1999). Hybrid flowshop scheduling: A survey, *Computers & Industrial Engineering* **37**(1–2), 57–61.

Liu, C.-Y. and S.-C. Chang (2000). Scheduling flexible flow shops with sequence dependent setup effects, *IEEE Transactions on Robotics and Automation* **16**(4), 408–419.

Liu, J. and B.L. MacCarthy (1997). A global MILP model for FMS scheduling, *European Journal of Operational Research* **100**(3), 441–453.

Manne, A.S. (1960). On the job-shop scheduling problem, *Operations Research* **8**, 219–223.

Markland, R.E., K.H. Darby-Dowman and E.D. Minor (1990). Coordinated production scheduling for make-to-order manufacturing, *European Journal of Operational Research* **45**(2–3), 155–176.

Maxwell, W.L. and J.A. Muckstadt (1985). Establishing consistent and realistic reorder intervals in production-distribution systems, *Operations Research* **33**, 1316–1341.

McCormick, S.T., M.L. Pinedo, S. Shenker and B. Wolf (1989). Sequencing in an assembly line with blocking to minimize cycle time, *Operations Research* **37**, 925–936.

Miller, T.C. (2002). *Hierarchical Operations in Supply Chain Planning*, Springer, London.

Moon, C., Y.H. Lee, C.S. Jeong and J.S. Yun (2008). Integrated process planning and scheduling in a supply chain, *Computers and Industrial Engineering* **54**(4), 1048–1061.

Muckstadt, J.A. (1985). Planning component delivery intervals in constrained assembly systems. In S. Axsäter, Ch. Schneeweiss and E. Silver, Eds., *Multi-Stage Production Planning and Inventory Control*, Springer, Berlin, 132–149.

Muckstadt, J.A. and R.O. Roundy (1993). Analysis of multistage production systems. In S.C. Graves, A.H.G. Rinnooy Kan and P.H. Zipkin, Eds., *Handbooks in Operations Research and Management Science*, vol. 4, North-Holland, Amsterdam, 59–131.

Nawaz, M., E.E. Enscore, Jr. and I. Ham (1983). A heuristic algorithm for the m–machine, n–job flow shop sequencing problem, *Omega: The International Journal of Management Science* **11**(1), 91–95.

Nemhauser, G.L. and L.A. Wolsey (1999). *Integer and Combinatorial Optimization*, John Wiley & Sons, New York.

Norrman, A. and U. Jansson (2004). Ericsson's proactive risk management approach after a serious sub-supplier accident, *International Journal of Physical*

Distribution and Logistics Management **34**(5), 434–456.

Nowicki, E. and C. Smutnicki (1996). A fast tabu search algorithm for the permutation flow-shop problem, *European Journal of Operational Research* **91**(1), 160–175.

Nowicki, E. and C. Smutnicki (1998). The flow shop with parallel machines: A tabu search approach, *European Journal of Operational Research* **106**, 226–253.

Oğuz, C., F.S. Salman and Z.B. Yalçin (2010). Order acceptance and scheduling decisions in make-to-order systems, *International Journal of Production Economics* **125**, 200–211.

Örnek, A., S. Özpeynirci and C. Öztürk (2010). A note on "A mixed integer programming model for advanced planning and scheduling (APS)," *European Journal of Operational Research* **203**, 784–785.

Parlar, M. and D. Perry (1996). Inventory models of future supply uncertainty with single and multiple suppliers, *Naval Research Logistics* **43**, 191–210.

Pinedo, M. (2005). *Planning and Scheduling in Manufacturing and Services*. Springer, New York.

Pochet, Y. and L.A. Wolsey (2006). *Production Planning by Mixed Integer Programming*. Springer, Berlin.

Quadt, D. and H. Kuhn (2007). A taxonomy of flexible flow line scheduling procedures, *European Journal of Operational Research* **178**(3), 686–698.

Rajendran, C. and D. Chaudhuri (1992). A multistage parallel-processor flowshop problem with minimum flowtime, *European Journal of Operational Research* **57**(1), 111–122.

Rangsaritratsamee, R., W.G. Ferrel and M.B. Kurz (2004). Dynamic rescheduling that simultaneously considers efficiency and stability, *Computers and Industrial Engineering* **46**, 1–15.

Ribas, I., R. Leisten and J.M. Framinan (2010). Review and classification of hybrid flowshop scheduling problems from a production system and a solutions procedure perspective, *Computers & Operations Research* **37**, 1439–1454.

Rockafellar, R.T. and S. Uryasev (2000). Optimization of conditional value-at-risk, *Journal of Risk* **2**(3), 21–41.

Rockafellar, R.T. and S. Uryasev (2002). Conditional value-at-risk for general loss distributions, *Journal of Banking and Finance* **26**(7), 1443–1471.

Rom, W. and S.A. Slotnick (2009). Order acceptance using genetic algorithms, *Computers and Operations Research* **36**, 1758–1767.

Rosenblatt, M.J. and U.G. Rothblum (1990). On the single resource capacity problem for multi-item inventory system, *Operations Research* **38**, 686–693.

Ruiz, R. and J.A. Vazquez-Rodriguez (2010). The hybrid flow shop scheduling problem, *European Journal of Operational Research* **205**, 1–18.

Ruiz-Torres, A.J. and M. Farzad (2007). The optimal number of suppliers considering the costs of individual supplier failures, *Omega: The International Journal of Management Science* **35**(1), 104–115.

Röslof, J., I. Harjunkoski, J. Björkqvist, S. Karlsson and T. Westerlund (2001). An MILP-based reordering algorithm for complex industrial scheduling and rescheduling, *Computers and Chemical Engineering* **25**, 821–828.

Röslof, J., I. Harjunkoski, T. Westerlund and J. Isaksson (2002). Solving a large scale industrial scheduling problem using MILP combined with a heuristic procedure, *European Journal of Operational Research* **138**, 29–42.

Salvador, M.S. (1973). A solution to a special class of flowshop scheduling problems. In S.E. Elmaghraby, Ed., *Symposium on the Theory of Scheduling and Its Applications*. Berlin: Springer, 83–91.

Sanayei, A., S.F. Mousavi, M.R. Abdi and A. Mohaghar (2008). An integrated group decision-making process for supplier selection and order allocation using multi-attribute utility theory and linear programming, *Journal of the Franklin Institute* **345**(7), 731–747.

Sarmiento, A.M. and R. Nagi (1999). A review of integrated analysis of production distribution systems, *IIE Transactions* **31**, 1061–1074.

Sarykalin, S., G. Serraino and S. Uryasev (2008). Value-at-risk vs. conditional value-at-risk in risk management and optimization, *Tutorials in Operations Research*, INFORMS 2008, 270–294.

Sawik, T. (1977). Stochastic optimal control of a multi-facility, multi-product production scheduling with random times of supplies, *Control and Cybernetics* **6**(3–4), 21–35.

Sawik, T. (1990). Modelling and scheduling of a flexible manufacturing system, *European Journal of Operational Research* **45**, 177–190.

Sawik, T. (1993). A scheduling algorithm for flexible flow lines with limited intermediate buffers, *Applied Stochastic Models and Data Analysis, Special issue on Manufacturing Systems* **9**, 127–138.

Sawik, T. (1994). New algorithms for scheduling flexible flow lines. In *Proceedings of Japan-U.S.A. Symposium on Flexible Automation*. Kobe, July 11–18, **3**, 1091–1096.

Sawik, T. (1995a). Integer programming models for the design and balancing of flexible assembly systems, *Mathematical and Computer Modelling* **21**(4), 1–12.

Sawik, T. (1995b). Scheduling flexible flow lines with no in-process buffers, *International Journal of Production Research* **33**, 1359–1370.

Sawik, T. (1996). A multilevel machine and vehicle scheduling in a flexible manufacturing system, *Mathematical & Computer Modelling* **23**(7), 45–57.

Sawik, T. (1998a). A lexicographic approach to bi-objective loading of a flexible assembly system, *European Journal of Operational Research* **107**(3), 658–668.

Sawik, T. (1998b). Simultaneous loading, routing and assembly plan selection in a flexible assembly system, *Mathematical and Computer Modelling* **28**(9), 19–29.

Sawik, T. (1999). *Production Planning and Scheduling in Flexible Assembly Systems*, Springer-Verlag, Berlin.

Sawik, T. (2000a). Mixed integer programming for scheduling flexible flow lines with limited intermediate buffers, *Mathematical and Computer Modelling* **31**, 39–52.

Sawik, T. (2000b). Simultaneous versus sequential loading and scheduling of flexible assembly systems, *International Journal of Production Research* **38**, 3267–3282.

Sawik, T. (2000c). An LP-based approach for loading and routing in a flexible assembly line, *International Journal of Production Economics* **64**(1–3), 49–58.

Sawik, T. (2001). Mixed integer programming for scheduling surface mount technology lines, *International Journal of Production Research* **39**, 3219–3235.

Sawik, T. (2002a). Balancing and scheduling of surface mount technology lines, *International Journal of Production Research* **40**(9), 1973–1991.

Sawik, T. (2002b). Monolithic vs. hierarchical balancing and scheduling of a flexible assembly line, *European Journal of Operational Research* **143**(1), 115–124.

Sawik, T. (2002c). An exact approach for batch scheduling in flexible flow lines with limited intermediate buffers, *Mathematical and Computer Modelling* **36**, 461–471.

Sawik, T. (2004). Loading and scheduling of a flexible assembly system by mixed integer programming, *European Journal of Operational Research* **154**(1), 1–19.

Sawik, T. (2005a). Integer programming approach to production scheduling for make-to-order manufacturing, *Mathematical and Computer Modelling* **41**(1), 99–118.

Sawik, T. (2005b). A cyclic versus flexible approach to materials ordering in make-to-order assembly, *Mathematical and Computer Modelling* **42**(3–4), 279–290.

Sawik, T. (2006). Hierarchical approach to production scheduling in make-to-order assembly, *International Journal of Production Research* **44**(4), 801–830.

Sawik, T. (2007a). A multi-objective customer orders assignment and resource leveling in make-to-order manufacturing, *International Transactions in Operational Research* **14**(6), 491–508.

Sawik, T. (2007b). Multi-objective master production scheduling in make-to-order manufacturing, *International Journal of Production Research* **45**(12), 2629–2653.

Sawik, T. (2007c). A lexicographic approach to bi-objective scheduling of single-period orders in make-to-order manufacturing, *European Journal of Operational Research* **180**(3), 1060–1075.

Sawik, T. (2007d). Integer programming approach to reactive scheduling in make-to-order manufacturing, *Mathematical and Computer Modelling* **46**(11–12), 1373–1387.

Sawik, T. (2009a). Multi-objective due-date setting in a make-to-order environment, *International Journal of Production Research* **47**(22), 6205–6231.

Sawik, T. (2009b). Monolithic versus hierarchical approach to integrated scheduling in a supply chain, *International Journal of Production Research* **47**(21), 5881–5910.

Sawik, T. (2009c). Coordinated supply chain scheduling, *International Journal of Production Economics* **120**(2), 437–451.

Sawik, T. (2010a). Single vs. multiple objective supplier selection in a make to order environment, *Omega: The International Journal of Management Science* **38**(3–4), 203–212.

Sawik, T. (2010b). A bi-objective supply chain scheduling. In K.D. Lawrence, R.K. Klimberg and V.M. Miori, Eds., *The Supply Chain in Manufacturing, Distribution, and Transportation: Modeling, Optimization, and Applications*. CRC Press, Boca Raton, FL.

Sawik, T. (2011a). Selection of supply portfolio under disruption risks, *Omega: The International Journal of Management Science* **39**(2), 194–208.

Sawik, T. (2011b). Selection of a dynamic supply portfolio in make-to-order environment with risks, *Computers & Operations Research* **38**(4), 782–796.

Sawik, T., A. Schaller and T.M. Tirpak (2000). Issues in loading and scheduling of SMT lines, *Zeszyty Naukowe Politechniki Ślaskiej, Automatyka*, no. 129, 331–341.

Sawik, T., A. Schaller and T.M. Tirpak (2002). Scheduling of printed wiring board assembly in surface mount technology lines, *Journal of Electronics Manufacturing, special issue on Production Planning and Scheduling in Electronics Manufacturing* **11**(1), 1–17.

Schmidt, G. (2000). Scheduling with limited machine availability, *European Journal of Operational Research* **121**, 1–15.

Schneeweiss, Ch. (1999). *Hierarchies in Distributed Decision Making*, Springer-Verlag, Berlin.

Schneeweiss, Ch. and K. Zimmer (2004). Hierarchical coordination mechanism within the supply chain, *European Journal of Operational Research* **154**, 687–703.

Scholl, A. (1999). *Balancing and Sequencing of Assembly Lines*. Physica-Verlag, Heidelberg.

Shanthikumar, J.G., D.D. Yao, W. Henk and M. Zijm, Eds. (2003). *Stochastic Modeling and Optimization of Manufacturing Systems and Supply Chains*. Springer, New York.

Shapiro, J.F. (1993). Mathematical programming models and methods for production planning and scheduling. In S.C. Graves, A.H.G. Rinnooy Kan and P.H. Zipkin, Eds., *Handbook in Operations Research and Management Science: Logistics of Production and Inventory*. North-Holland, Amsterdam.

Shapiro, J.F. (2001). *Modeling the Supply Chain*. Duxbury Press, Pacific Grove, CA.

Sheffi, Y. (2005). *The Resilient Enterprise*. MIT Press, Cambridge, MA.

Silver, E.A., D.F. Pyke and R. Petersen (1998). *Inventory Management and Production Planning and Scheduling*. John Wiley & Sons, New York.

Slotnick, S.A. and T.E. Morton (1996). Selecting jobs for a heavily loaded shop with lateness penalties, *Computers & Operations Research* **23**(2), 131–140.

Slotnick, S.A. and T.E. Morton (2007). Order acceptance with weighted tardiness, *Computers and Operations Research* **34**(10), 3029–3042.

Smith, S.F. (1995). Reactive scheduling systems, In D.E. Brown and W.T. Scherer, Eds., *Intelligent Scheduling Systems*. Kluwer Academic Publishers, Boston, 155–192.

Stadtler, H. (2005). Supply chain management and advanced planning—basics, overview and challenges, *European Journal of Operational Research* **163**, 575–588.

Steuer, R.E. (1986). *Multiple Criteria Optimization: Theory, Computation, and Applications*. John Wiley & Sons, New York.

Sun, D. and D. Atkins (1997). 98%—effective lot-sizing for assembly inventory systems with backlogging, *Operations Research* **45**, 940–951.

Sun, J. and D. Xue (2001). A dynamic reactive scheduling mechanism for responding to changes of production orders and manufacturing resources, *Computers in Industry* **46**, 189–207.

Talluri, K.T. and G.J. van Ryzin (2004). *The Theory and Practice of Revenue Management*. Kluwer Academic Publishers, Boston.

Tang, C.S. (2006). Perspectives in supply chain risk management, *International Journal of Production Economics* **103**, 451–488.

Tempelmeier, H. and M.C. Derstroff (1996). A Lagrangian-based heuristic for dynamic multi-level, multi-item constrained lot sizing with setup times, *Management Science* **42**(5), 738–747.

Thomas, D.J. and P.M. Griffin (1996). Coordinated supply chain management, *European Journal of Operational Research* **94**(1), 1–15.

Tirpak, T.M. (2000). Design-to-manufacturing information management for electronics assembly, *International Journal of Flexible Manufacturing Systems* **12**(2), 189–205.

Tseng, F.T., E.F. Stafford, Jr. and J.N.D. Gupta (2004). An empirical analysis of integer programming formulations for the permutation flowshop. *Omega: The International Journal of Management Science* **32**, 285–293.

Ulusoy, G. and U. Bilge (1993). Simultaneous scheduling of machines and automated guided vehicles, *International Journal of Production Research* **31**, 2857–2873.

Uryasev, S. (2000). Conditional value-at-risk: Optimization algorithms and applications, *Financial Engineering News* **14**, February, 1–5.

Ustun, O. and E.A. Demirtas (2008). An integrated multi-objective decision making process for multi-period lot sizing with supplier selection, *Omega: The International Journal of Management Science* **36**, 509–521.

Van der Vaart, T. and D.P. Van Donk (2008). A critical review of survey-based research in supply chain integration, *International Journal of Production Economics* **111**, 42–55.

Vieira, G.E., J.W. Herrman and E. Lin (2003). Rescheduling manufacturing systems: A framework of strategies, policies and methods, *Journal of Scheduling* **6**(1), 39–62.

Vignier, A., J.C. Billaut and C. Proust (1999). Hybrid flowshop scheduling problems: State of the art, *Rairo-Recherche Operationnelle—Operations Research* **33**(2), 117–183.

Voss, S. and D.L. Woodruff (2003). *Introduction to Computational Optimization Models for Production Planning in a Supply Chain*, Springer-Verlag, Berlin.

Wagner, H.M. (1959). An integer linear-programming model for machine scheduling, *Naval Research Logistics Quarterly* **6**, 131–140.

Wang, H. (2005). Flexible flowshop scheduling: Optimum, heuristics and artificial intelligence solutions, *Expert Systems* **22**(2), 78–85.

Wang, W., P.C. Nelson and T.M. Tirpak (2000). Optimization of high-speed multistation SMT placement machines using evolutionary algorithms, *IEEE Transactions on Electronics Packaging Manufacturing* **22**(2), 137–146.

Weber, C.A. and J.R. Current (1993). A multiobjective approach to vendor selection, *European Journal of Operational Research* **68**, 173–184.

Wester, F.A.W., J. Wijngaard and W.H.M. Zijm (1992). Order acceptance strategies in a production-to-order environment with setup times and due dates, *International Journal of Production Research* **30**(4), 1313–1326.

Wilson, J.M. (1989). Alternative formulations of a flow-shop scheduling problem, *Journal of the Operational Research Society* **40**, 395–399.

Wittrock, R.J. (1985). Scheduling algorithms for flexible flow lines, *IBM Journal of Research and Development* **29**, 401–412.

Wittrock, R.J. (1988). An adaptable scheduling algorithm for flexible flow lines, *Operations Research* **36**, 445–453.

Wolsey, L.A. (1998). *Integer Programming*, John Wiley & Sons, New York.

Wu, D. and D.L. Olson (2008). Supply chain risk, simulation, and vendor selection, *International Journal of Production Economics* **114**, 646–655.

Xia, W. and Z. Wu (2007). Supplier selection with multiple criteria in volume discount environments, *Omega: The International Journal of Management Science* **35**, 494–504.

Yu, H., A.Z. Zeng and L. Zhao (2009). Single or dual sourcing: Decision-making in the presence of supply chain disruption risks, *Omega: The International Journal of Management Science* **37**, 788–800.

Yue, J., Y. Xia and T. Tran (2010). Selecting sourcing partners for a make-to-order supply chain, *Omega: The International Journal of Management Science* **38**(3–4), 136–144.

Zipkin, P.H. (2000). *Foundations of Inventory Management*, McGraw-Hill, New York.

Zorzini, M., D. Corti and A. Pozzetti (2008). Due date (DD) quotation and capacity planning in make-to-order companies: Results from an empirical analysis, *International Journal of Production Economics* **112**, 919–933.

Index

Scheduling in Supply Chains Using Mixed Integer Programming. By Tadeusz Sawik

Printed in the United States
By Bookmasters